Advances in Experimental Medicine and Biology

Proteomics, Metabolomics, Interactomics and Systems Biology

Volume 1321

This series of volumes focuses on concepts, techniques and recent advances in the field of proteomics, interactomics, metabolomics and systems biology. Recent advances in various 'omics' technologies enable quantitative monitoring of myriad various biological molecules in a high-throughput manner, and allow determination of their variation between different biological states on a genomic scale. Now that the sequencing of various genomes, from prokaryotes to humans, has provided the list and linear sequence of proteins and RNA that build living organisms, defining the complete set of interactions that sustain life constitutes one of the key challenges of the postgenomic era. This series is intended to cover experimental approaches for defining protein-protein, protein-RNA, protein-DNA and protein-lipid interactions; as well as theoretical approaches dealing with data analysis, integration and modeling and ethical issues.

More information about this series at http://www.springer.com/series/15040

Paul C. Guest
Editor

Clinical, Biological and Molecular Aspects of COVID-19

Springer

Editor
Paul C. Guest
University of Campinas
São Paulo, Brazil

ISSN 0065-2598 ISSN 2214-8019 (electronic)
Advances in Experimental Medicine and Biology
ISSN 2730-6216 ISSN 2730-6224 (electronic)
Proteomics, Metabolomics, Interactomics and Systems Biology
ISBN 978-3-030-59260-8 ISBN 978-3-030-59261-5 (eBook)
https://doi.org/10.1007/978-3-030-59261-5

This Springer imprint is published by the registered company Springer Nature Switzerland AG
The registered company address is: Gewerbestrasse 11, 6330 Cham, Switzerland

Preface

The novel coronavirus that erupted in Wuhan China at the end of 2019 (COVID-19) has led to a serious global pandemic, which is still ongoing. The virulence and infection rate of COVID-19 are profound and has required extreme social distancing measures across the globe in attempts to prevent this virus from overwhelming healthcare services and hospitals. COVID-19 appears to have the greatest risk of a more serious disease course in elderly individuals and those with co-morbid diseases, such as heart disease, asthma, and diabetes. All of this has led to an unprecedented rapid worldwide mobilization effort to identify effective treatments and develop vaccines. It is hoped that such efforts will help to control this devastating disease and in the control of future pandemics. The authors in this volume come from five of the six habitable continents from countries such as Afghanistan, Austria, Brazil, Iran, Italy, Poland, South Africa, South Korea, the United Kingdom, the United States of America, and Vietnam.

This new volume will increase the reader's understanding of the ongoing COVID-19 pandemic through a series of chapters that focus on the first wave of the pandemic. Leading experts discuss the effects of the virus in cases of co-morbidities, new treatment approaches, and mental health aspects of the pandemic and convey the results of survey studies. The book will be an excellent resource for researchers studying virology, metabolic diseases, and respiratory disorders, clinical scientists, physicians, drug companies, and healthcare services and workers. The book will also be of interest to the general population as virtually everyone has been affected by this deadly pandemic in some way.

São Paulo, Brazil — Paul C. Guest

Contents

Contributors

Mohsen Abbasi-Farajzadeh Marine Medicine Research Center, Baqiyatallah University of Medical Sciences, Tehran, Iran

Elham Abdollahi Halal Research Center of IRI, FDA, Tehran, Iran

Department of Medical Immunology and Allergy, Student Research Committee, School of Medicine, Mashhad University of Medical Sciences, Mashhad, Iran

Behnaz Abiri Department of Nutrition, Faculty of Paramedicine, Ahvaz Jundishapur University of Medical Sciences, Ahvaz, Iran

Hossein Aghamollaei Chemical Injuries Research Center, Systems biology and Poisonings Institute, Baqiyatallah University of Medical Sciences, Tehran, Iran

Hamid Asadzadeh Aghdaei Basic and Molecular Epidemiology of Gastrointestinal Disorders Research Center, Research Institute for Gastroenterology and Liver Diseases, ShahidBeheshti University of Medical Sciences, Tehran, Iran

Gholamhosein Alishiri Chemical Injuries Research Center, Systems Biology and Poisonings Institute, Baqiyatallah University of Medical sciences, Tehran, Iran

Seong Soo A. An Department of Bionanotechnology, Gachon University, Seongnam-si, Gyeonggi-do, South Korea

Sara Ashtari Basic and Molecular Epidemiology of Gastrointestinal Disorders Research Center, Research Institute for Gastroenterology and Liver Diseases, ShahidBeheshti University of Medical Sciences, Tehran, Iran

Gastroenterology and Liver Diseases Research Center, Research Institute for Gastroenterology and Liver Diseases, Shahid Beheshti University of Medical Sciences, Tehran, Iran

Mansour Babaei Health Management Research Center, Baqiyatallah University of Medical Sciences, Tehran, Iran

Eva Bagyinszky Department of Industrial and Environmental Engineering, Graduate School of Environment, Gachon University, Seongnam-si, Gyeonggi-do, South Korea

Department of Bionanotechnology, Gachon University, Seongnam-si, Gyeonggi-do, South Korea

Hamid Bakherad Department of Pharmaceutical Biotechnology and Isfahan Pharmaceutical Sciences Research Center, School of Pharmacy and Pharmaceutical Sciences, Isfahan University of Medical Sciences, Isfahan, Iran

Isabelle Pastor Bandeira Department of Medicine, Universidade da Região de Joinville (UNIVILLE), Joinville, Brazil

Behzad Behnam Pharmaceutics Research Center, Institute of Neuropharmacology, Kerman University of Medical Sciences, Kerman, Iran

Herbal and Traditional Medicines Research Center, Kerman University of Medical Sciences, Kerman, Iran

Leticia Caroline Breis Department of Medicine, Universidade da Região de Joinville (UNIVILLE), Joinville, Brazil

Giuseppe Castellano Nephrology, Dialysis and Transplantation Unit, Department of Medical and Surgical Science, University of Foggia, Foggia, Italy

Fatemeh Cheraghalipoor Department of Radiology, Health Research Center, Life Style Institute, Baqiyatallah University of Medical Sciences, Tehran, Iran

Abbas Ebadi Behavioral Sciences Research Center, Life style institute, Nursing Faculty, Baqiyatallah University of Medical Sciences, Tehran, Iran

Fakhrudin Faizi Atherosclerosis Research Center, Nursing Faculty, Baqiyatallah University of Medical Sciences, Tehran, Iran

Mojtaba Ferdosi Students' Research Committee, Nursing Faculty, Baqiyatallah University of Medical Sciences, Tehran, Iran

Marco Fiorentino Department of Emergency and Organ Transplantation, Nephrology Dialysis and Transplantation Unit, "Aldo Moro" University, Bari, Italy

Hadi Esmaeili Gouvarchin Galeh Applied Virology Research Center, Baqiyatallah University of Medical Science, Tehran, Iran

J. A. George Department of Chemical Pathology, National Health Laboratory Services, University of the Witwatersrand, Johannesburg, South Africa

Ali Ghazvini Chemical Injuries Research Center, Systems Biology and Poisonings Institute, Baqiyatallah University of Medical Sciences, Tehran, Iran

Keivan Goharimoghadam Internal Medicine, Shariati Hospital, Tehran University of Medical Sciences, Tehran, Iran

Marcus Vinícius Magno Gonçalves Professor of Neurology, Department of Medicine, Universidade da Região de Joinville (UNIVILLE), Joinville, Brazil

Armita Mahdavi Gorabi Research Center for Advanced Technologies in Cardiovascular Medicine, Tehran Heart Center, Tehran University of Medical Sciences, Tehran, Iran

Paul C. Guest Laboratory of Neuroproteomics, Department of Biochemistry and Tissue Biology, Institute of Biology, University of Campinas (UNICAMP), Campinas, Brazil

Gholam Reza Hatam Basic Sciences in Infectious Diseases Research Center, School of Medicine, Shiraz University of Medical Sciences, Shiraz, Iran

Farshad Heidari-Beni Student of Nursing, Department of Medical Surgical Nursing, School of Nursing and Midwifery, Iran University of Medical Sciences, Tehran, Iran

Mostafa Hekmat Health Management Research Center, Baqiyatallah University of Medical Sciences, Tehran, Iran

R. Hesse Department of Chemical Pathology, University of the Witwatersrand; National Health Laboratory Service, Johannesburg, South Africa

John Hulme Department of Bionanotechnology, Gachon University, Seongnam-si, Gyeonggi-do, South Korea

Morteza Izadi Health Research Center, Baqiyatallah University of Medical Sciences, Tehran, Iran

B. F. Jacobson Department of Molecular Medicine and Haematology, National Health Laboratory Services, Johannesburg, South Africa

University of the Witwatersrand, Johannesburg, South Africa

Iman Jafari Trauma Research Center, Nursing Faculty, Baqiyatallah University of Medical Sciences, Tehran, Iran

Ramezan Jafari Department of Radiology, Health Research Center, Life Style Institute, Baqiyatallah University of Medical Sciences, Tehran, Iran

Nematollah Jonaidi Jafari Health Research Center, Life Style Institute, Baqiyatallah University of Medical Sciences, Tehran, Iran

Mehdi Jafari-Oori Trauma Research Center, Nursing Faculty, Baqiyatallah University of Medical Sciences, Tehran, Iran

Tannaz Jamialahmadi Biotechnology Research Center, Pharmaceutical Technology Institute, Mashhad University of Medical Sciences, Mashhad, Iran

Department of Food Science and Technology, Quchan Branch, Islamic Azad University, Quchan, Iran

Department of Nutrition, Faculty of Medicine, Mashhad University of Medical Sciences, Mashhad, Iran

Thomas P. Johnston Division of Pharmacology and Pharmaceutical Sciences, School of Pharmacy, University of Missouri-Kansas City, Kansas City, MO, USA

Nematollah Jonaidi-Jafari Health Research Center, Baqiyatallah University of Medical Sciences, Tehran, Iran

Leila Karimi Behavioral Sciences Research Center, Life Style Institute, Nursing Faculty, Baqiyatallah University of Medical Sciences, Tehran, Iran

Goli Kazemi-Nia Sina Hospital, Ahvaz Jundishapur University of Medical Sciences, Ahvaz, Iran

Azam Khatibzadeh Health Management Research Center, Baqiyatallah University of Medical Sciences, Tehran, Iran

Jason Kidde Department of Physical Therapy, University of Utah, Salt Lake City, UT, USA

S. Louw Department of Molecular Medicine and Haematology, National Health Laboratory Services, Johannesburg, South Africa

University of the Witwatersrand, Johannesburg, South Africa

Houshyar Maghsoudi Department of Radiology, Health Research Center, Life Style Institute, Baqiyatallah University of Medical Sciences, Tehran, Iran

Mitra Mahdavian Department of Midwifery, School of Nursing and Midwifery, Islamic Azad University, Bojnourd, Iran

Somayeh Makvandi Department of Midwifery, School of Nursing and Midwifery, Islamic Azad University, Ahvaz, Iran

M. R. Maphayi Department of Chemical Pathology, National Health Laboratory Services, University of the Witwatersrand, Johannesburg, South Africa

E. S. Mayne Faculty of Health Sciences, University of the Witwatersrand, Johannesburg, South Africa

Department of Immunology, National Health Laboratory Services, Johannesburg, South Africa

Golshan Mirmomeni Hearing Research Center, Ahvaz Jundishapur University of Medical Sciences, Ahvaz, Iran

Malihe Sadat Moayed Trauma Research Center, Nursing Faculty, Baqiyatallah University of Medical Sciences, Tehran, Iran

Sayed Hussain Mosawi Medical Sciences Research Center, Ghalib University, Kabul, Afghanistan

Batool Nehrir Health Management Research Center, Nursing Faculty, Baqiyatallah University of Medical Sciences, Tehran, Iran

Mohamad Nikpouraghdam Nephrology and Urology Research Center, Baqiyatallah University of Medical sciences, Tehran, Iran

Yunes Panahi Faculty of Pharmacy, Baqiyatallah University of Medical Sciences, Tehran, Iran

Yoon Soo Park Department of Internal Medicine, Yongin Severance Hospital, Yonsei University College of Medicine, Yongin-si, Gyeonggi-do, South Korea

Jean Pierre Schatzmann Peron Department of Immunology, Institute of Biomedical Sciences, University of São Paulo (ICB-USP), São Paulo, Brazil

Scientific Platform Pasteur-USP, University of São Paulo (USP), São Paulo, SP, Brazil

Immunopathology and Allergy Post Graduate Program, School of Medicine, University of São Paulo (USP), São Paulo, SP, Brazil

T. Pillay Department of Chemical Pathology, National Health Laboratory Services, University of the Witwatersrand, Johannesburg, South Africa

Mohamad Amin Pourhoseingholi Gastroenterology and Liver Diseases Research Center, Research Institute for Gastroenterology and Liver Diseases, ShahidBeheshti University of Medical Sciences, Tehran, Iran

Farshid Rahimi-Bashar Anesthesia and Critical Care Department, Hamadan University of Medical Sciences, Hamadan, Iran

Reza Ranjbar Molecular Biology Research Center, Systems Biology and Poisonings Institute, Baqiyatallah University of Medical Sciences, Tehran, Iran

Hassan Saadat Behavioral Sciences Research Center, Life Style Institute, Baqiyatallah University of Medical Sciences, Tehran, Iran

Mahdiyar Iravani Saadi Hematology Research Center, Department of Hematology, Medical Oncology and Stem Cell Transplantation, Shiraz University of Medical Sciences, Shiraz, Iran

Mohsen Saberi Medicine, Quran and Hadith Research Center & Department of Community Medicine, Faculty of Medicine, Baqiyatallah University of Medical Sciences, Tehran, Iran

Amirhossein Sahebkar Halal Research Center of IRI, FDA, Tehran, Iran

Biotechnology Research Center, Pharmaceutical Technology Institute, Mashhad University of Medical Sciences, Mashhad, Iran

Neurogenic Inflammation Research Center, Mashhad University of Medical Sciences, Mashhad, Iran

School of Pharmacy, Mashhad University of Medical Sciences, Mashhad, Iran

Polish Mother's Memorial Hospital Research Institute (PMMHRI), Lodz, Poland

Hossein Samadinia Nanobiotechnology Research Center, Baqiyatallah University of Medical Sciences, Tehran, Iran

Rahim Sarvestani Research and Development Department, PersisGen Par Biopharma Accelerator, Tehran, Iran

Thozhukat Sathyapalan Academic Diabetes, Endocrinology and Metabolism, Hull York Medical School, University of Hull, Hull, UK

Marco Antônio Machado Schlindwein Department of Medicine, Universidade da Região de Joinville (UNIVILLE), Joinville, Brazil

Alireza Shahriary Chemical Injuries Research Center, Systems Biology and Poisonings Institute, Baqiyatallah University of Medical Sciences, Tehran, Iran

Asma Shojaee Student Research Committee, Nursing Faculty, Baqiyatallah University of Medical Sciences, Tehran, Iran

Sajad Shojaee Basic and Molecular Epidemiology of Gastrointestinal Disorders Research Center, Research Institute for Gastroenterology and Liver Diseases, Shahid Beheshti University of Medical Sciences, Tehran, Iran

Gastroenterology and Liver Diseases Research Center, Research Institute for Gastroenterology and Liver Diseases, Shahid Beheshti University of Medical Sciences, Tehran, Iran

Hussein Solomon Department of Political Studies & Governance, University of Free State, Bloemfontein, South Africa

Arno Tausch University of Innsbruck, Innsbruck, Tirol, Austria

Fataneh Tavasolian Department of Immunology, Faculty of Medical Sciences, Tarbiat Modares University, Tehran, Iran

Mohammadreza Vafa Department of Nutrition, School of Public Health, Iran University of Medical Sciences, Tehran, Iran

Ensieh Vahedi Chemical Injuries Research Center, Systems Biology and Poisonings Institute, Baqiyatallah University of Medical Sciences, Tehran, Iran

Amir Vahedian-Azimi Trauma Research Center, Nursing Faculty, Baqiyatallah University of Medical Sciences, Tehran, Iran

J. L. Vaughan Department of Molecular Medicine and Haematology, National Health Laboratory Service and Faculty of Health Sciences, University of the Witwatersrand, Johannesburg, South Africa

Giau Van Vo Department of Industrial and Environmental Engineering, Graduate School of Environment, Gachon University, Seongnam-si, Gyeonggi-do, South Korea

Department of Bionanotechnology, Gachon University, Seongnam-si, Gyeonggi-do, South Korea

School of Medicine, Vietnam National University Ho Chi Minh City, Ho Chi Minh City, Vietnam

D. J. van der Westhuizen Department of Chemical Pathology, University of Cape Town, Cape Town, South Africa

National Health Laboratory Service, Cape Town, South Africa

T. M. Wiggill Department of Molecular Medicine and Haematology. National Health Laboratory Service and Faculty of Health Sciences, University of the Witwatersrand, Johannesburg, South Africa

Hamed Zare Cellular and Molecular Research Center, Birjand University of Medical Sciences, Birjand, Iran

Part I

Reviews

SARS-CoV-2 (COVID-19): Beginning to Understand a New Virus

1

Giau Van Vo, Eva Bagyinszky, Yoon Soo Park, John Hulme, and Seong Soo A. An

Abstract

Within the last two decades, several members of the Coronaviridae family demonstrated epidemic potential. In late 2019, an unnamed genetic relative, later named SARS-CoV-2 (COVID-19), erupted in the highly populous neighborhoods of Wuhan, China. Unchecked, COVID-19 spread rapidly among interconnected communities and related households before containment measures could be enacted. At present, the mortality rate of COVID-19 infection worldwide is 6.6%. In order to mitigate the number of infections, restrictions or recommendations on the number of people that can gather in a given area have been employed by governments worldwide. For governments to confidently lift these restrictions as well as counter a potential secondary wave of infections, alternative medications and diagnostic strategies against COVID-19 are urgently required. This review has focused on these issues.

Keywords

COVID-19 · SARS-CoV-2 · Semen · Wuhan · Coronavirus

1.1 Introduction

The Novel SARS-CoV-2 (COVID-19) is a zoonotic coronavirus belonging to Coronaviridae family of viruses, including those which caused severe respiratory syndrome (SARS-CoV) and Middle East respiratory syndrome (MERS-CoV).

G. Van Vo
Department of Industrial and Environmental Engineering, Graduate School of Environment, Gachon University, Seongnam-si, Gyeonggi-do, South Korea

Department of Bionanotechnology, Gachon University, Seongnam-si, Gyeonggi-do, South Korea

School of Medicine, Vietnam National University Ho Chi Minh City, Ho Chi Minh City, Vietnam

E. Bagyinszky
Department of Industrial and Environmental Engineering, Graduate School of Environment, Gachon University, Seongnam-si, Gyeonggi-do, South Korea

Department of Bionanotechnology, Gachon University, Seongnam-si, Gyeonggi-do, South Korea

Y. S. Park
Department of Internal Medicine, Yongin Severance Hospital, Yonsei University College of Medicine, Yongin-si, Gyeonggi-do, South Korea

J. Hulme (✉) · S. S. A. An (✉)
Department of Bionanotechnology, Gachon University, Seongnam-si, Gyeonggi-do, South Korea
e-mail: seongaan@gachon.ac.kr

P. C. Guest (ed.), *Clinical, Biological and Molecular Aspects of COVID-19*, Advances in Experimental Medicine and Biology 1321, https://doi.org/10.1007/978-3-030-59261-5_1

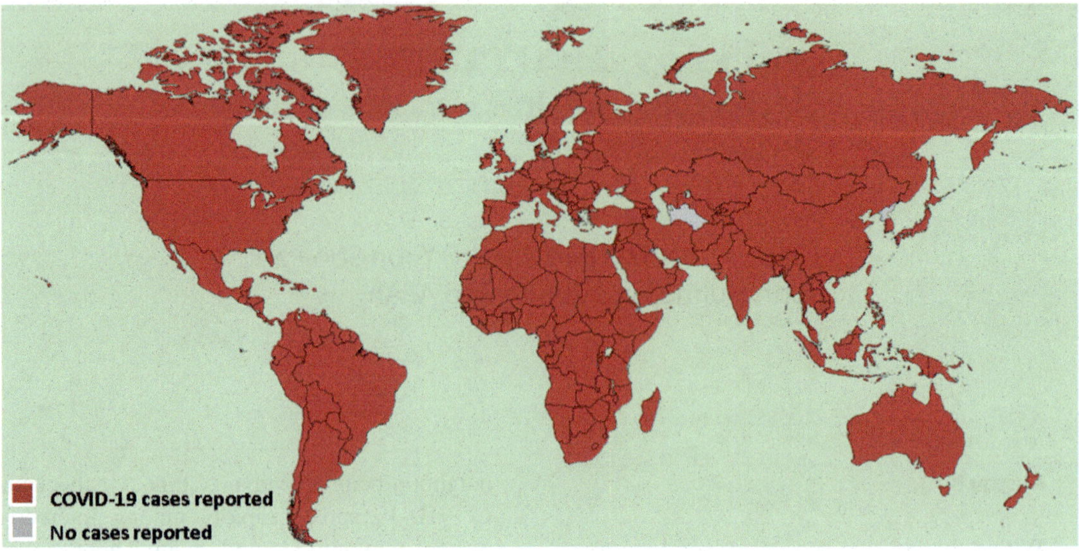

Fig. 1.1 Centers for Diseases Control-confirmed COVID-19 cases globally. (Adapted from https://www.cdc.gov/coronavirus/2019-ncov/global-covid-19/world-map.html)

In late 2019, SARS-CoV-2 emerged in Wuhan province, China, infecting significant portions of the population. The highly infectious characteristics of SARS-CoV-2 permitted its rapid spread in Hubei province and other parts of the country [1, 2]. In the first weeks of January 2020, containment of this novel virus by local authorities was naturally fragmented due to the novelty and evolving diagnostic procedures, resulting in nondisclosure of the level of infection to the public and to the broader scientific community. Since this time, the virus has emerged on every habitable continent, infecting more than 4.9 million people and killing more than 320,000 individuals in 218 countries and territories (as of May19, 2020, Johns Hopkins University website; https://coronavirus.jhu.edu/map.html) (Fig. 1.1).

COVID-19 remains a highly infectious disease even after incurring significant mutations in its genome during the last 5 months [3, 4], placing it front and center in the sights of the World Health Organization (WHO) as one of the most dangerous pandemics faced in the last 100 years. Hence, this perspective work aims to provide a timely overview of the COVID-19 outbreak and explore alternative avenues of testing and potential treatment approaches.

1.2 SARS-CoV-2 Phylogenetic Analysis

COVID-19 is caused by the SARS-CoV-2 RNA virus belonging to the Coronaviridae family. It contains a single-stranded positive-sense RNA, ranging from 26 to 32 kilobases (kb) in length. Coronaviruses are considered zoonotic, residing in many different animal species, such as camels, bats, mice, and dogs [5]. The SARS-CoV-2 viral genome includes 6–11 open reading frames (ORFs). The first ORF constitutes the largest part of the genome (67%) and encodes nonstructural proteins. The rest of the ORFs encode accessory and structural proteins. Four main structural proteins have been distinguished known as the spike (S), envelope (E), matrix (M), and nucleocapsid (N) proteins. The spike surface proteins are responsible for binding of the virus to host cell receptors and determine the host tropism and capacity of transmission [6].

Initial next-generation sequencing of the virus was performed on throat swabs and bronchoalveolar fluid collected from nine patients by Lu et al. [5]. An additional sequencing study was also performed by Zhou et al. on samples isolated from five patients [7]. Reverse transcription and full annotation of the patient sample revealed that the viral genomes were 99.9% similar [5, 7]. Analysis of the SARS-CoV-2 genome revealed a 97–99% similarity to a SARS-like betacoronavirus (SL-CoVZC45) and another bat coronavirus (CoVZXC21). Genetically, SARS-CoV-2 demonstrated less genetic similarity with those causing SARS (~79%) and MERS (~50%) [5]. Phylogenetic analysis showed that SARS-CoV-2 belongs to subgenus Sarbecovirus of the betacoronavirus genus. This study also examined the spike protein sequence of SARS-CoV-2, which could be divided into S1 and S2 domains. These domains are thought to play a role in receptor-binding and membrane fusion. The SARS-CoV-2 S2 domain shared relatively high sequence homology with that of bat coronaviruses (93%), whereas the S1 domain showed only 68% homology. However, SARS-CoV-2 and SARS-CoV shared 50 conserved amino acids inside the S1 domain (located in C-terminal area), which were different in bat coronaviruses. Structural predictions revealed that the SARS-CoV-2 receptor-binding domain was similar to that of SARS-CoV, and previous studies suggested that angiotensin-converting enzyme 2 (ACE2) was a potential receptor for this virus [5]. Infectivity assays confirmed ACE2 as a receptor for SARS-CoV-2 [7]. In these studies, HeLA cells were infected with the virus, and cells which expressed the ACE2 receptor were more susceptible to infection. However, additional studies may be needed to determine or confirm the mechanism of transmission [8].

1.3 Possible Origin of SARS-CoV2

Whether it is a product of natural evolution, a laboratory experiment, or otherwise engineered, the possible origins of SARS-CoV-2 have been debated since the beginning of the outbreak. A potential way to end the debate is to identify an intermediate host or hosts that can successfully transmit the virus to humans and vice versa [7]. Since the first cases of COVID-19 infections were thought to stem from the conjoined seafood and live animal market in Wuhan [1], the COVID-19 disease is generally considered a zoonotic disease (like SARS). Through the International Food Safety Authorities Network (INFOSAN), the national food safety authorities are seeking more information on the potential for persistence of COVID-19 in foods traded internationally as well as on the potential role of foods in the transmission of the virus. Currently, investigations are underway to evaluate the viability and survival time of SARS-CoV-2 [9–11]. As a general rule, the consumption of raw or undercooked animal products should be avoided. Raw meat, milk, or animal organs should be handled with care to avoid cross contamination with uncooked foods. However, human-to-human transmission has also been confirmed.

This novel virus responsible for the ongoing COVID-19 epidemic was initially thought to be a product of laboratory recombination as strongly claimed on the 3rd of February 2020 (https://jameslyonsweiler.com/2020/02/02/moderately-strong-confirmation-of-a-laboratory-origin-of-2019-ncov/) online by the chief executive officer of the Institute for Pure and Applied Knowledge. The claim stated that SARS-CoV-2 had a unique inserted sequence (1378 bp) located in the middle of its spike protein gene that did not match other coronaviruses but was similar to a sequence in the pShuttle-SN expression vector, commonly used in research laboratories. However, this hypothesis was rejected by Hao et al. [12]. These researchers aligned several coronavirus sequences and found that the "unique" sequence from SARS-CoV-2 was also found in these other viruses with a high sequence identity. This suggested that SARS-CoV-2 arose via a natural mechanism.

On the other hand, similar to the case for SARS-CoV and MERS-CoV [13], the bat is the most likely species of origin for COVID-19 due to the high degree of whole-genome identity with a bat coronavirus (BatCoV RaTG13) [7]. Notably, a SARS-like coronavirus was initially described close to when the COVID-19 outbreak occurred in lung samples from two dead Malayan pangolins [14], and these findings were confirmed by Zhang and colleagues [15]. This study suggests that the pangolin species may be a natural reservoir of coronaviruses. Other possible reservoirs include camels and civets (Fig. 1.2).

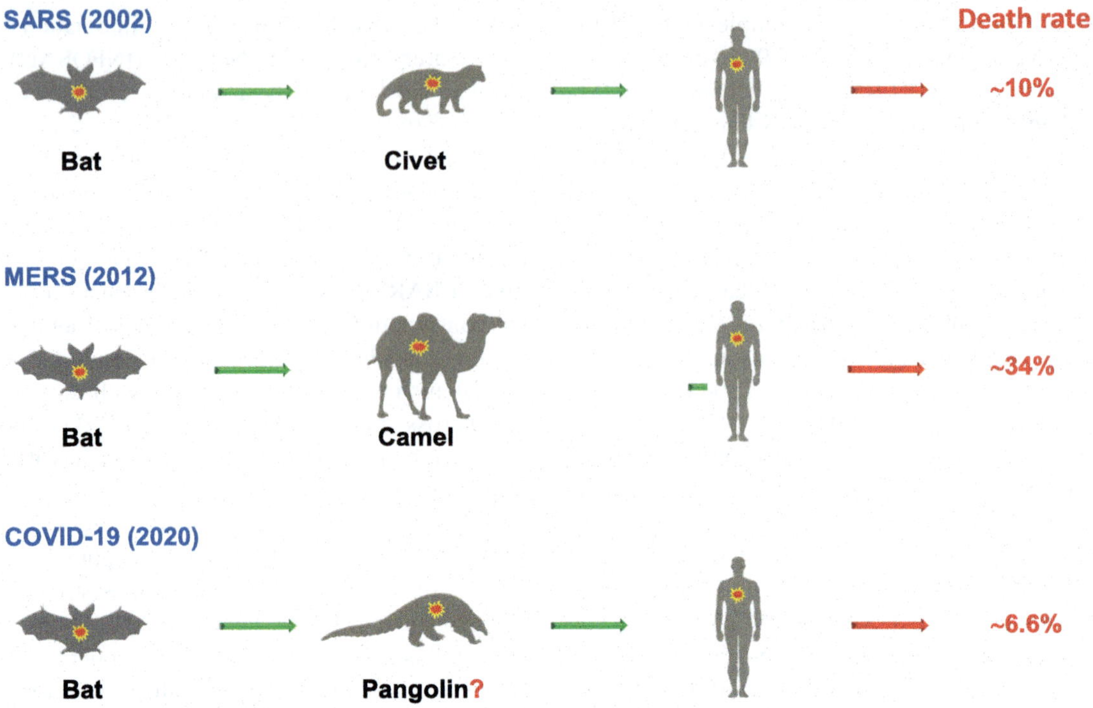

Fig. 1.2 Possible origin of the coronaviruses including SARS, MERS, and COVID-19

1.4 Main Properties of COVID-19 Infection

SARS-CoV-2 belongs to the betacoronaviridae family which usually displays a round/elliptic shape with a diameter of 60–140 nm. Similar to most coronaviruses, SARS-CoV-2 could be sensitive to heat, ultraviolet rays, and lipid solvents. Coronaviruses could infect several animal species (such as camels, cats, cattle, pangolins, and bats), and infection may be transmitted to humans [16]. The stability of SARS-CoV and SARS-CoV-2 may be similar, with capabilities of surviving for up to 3 h in aerosols, 4 h on copper, and up to 1 day on cardboard [17]. The virus has a longer viability on steel and plastic surfaces and could survive for approximately 2–3 days. Higher temperature and humidity may result in reduction in viral transmission not only in influenza and SARS but also in the case of COVID-19 [18].

The structure of SARS-CoV-2 is similar to that of other coronaviruses, which contain the proteins mentioned above (spike, envelope, membrane, and nucleocapsid proteins), along with different enzymes, such as RNA polymerase, helicase, or papain [19]. The spike proteins on the viral surface play a significant role in target recognition and membrane fusion. These proteins also play a critical role in human-to-human transmission of the virus. The spike protein has two subunits. One of these contains the receptor-binding domain (RBD), which enhances the attachment between the virus and host cell, while the second subunit appears to be involved in membrane fusion [20]. Similar to SARS-CoV, the RBD is conserved, but there may be important differences between the two viral strains. In the RBD, there are five critical amino acid residues at positions 442, 472, 479, 487, and 491 which could be involved in the human-to-human transmission. These critical residues (apart from tyrosine 491) may not be conserved between SARS-CoV and SARS-CoV-2. Three-dimensional modeling has revealed that the spike protein of SARS-CoV-2 could interact with the ACE2 receptor [21], resulting in the virus-host interaction. The entry of the virus into the cell could then enhance the immune and inflammation response of the host [22]. The mechanism of entry involves fusion of the viral envelope with the host cell, allowing it to enter the endosomal system. Next, the virus is uncoated and releases its RNA into the host, and this is translated to the viral replicase polyproteins pp1a and pp1ab, which are then cleaved by viral proteinases. The polymerase then carries out discontinuous transcription to produce subgenomic mRNAs and these are translated into the viral proteins. Finally, the viral proteins and RNA genome are assembled into new virus particles in the endoplasmic reticulum, allowing these to be released from the cell via the secretory pathway [19].

1.5 Typical Clinical Features of COVID-19 Cases

Several studies on the pneumonia outbreak caused by COVID-19 infection have been reported in the general population. In the first report, Huang et al. indicated that the infected patients had a history of exposure to the Huanan Seafood Wholesale Market [2]. Various symptoms such as fever, nonproductive cough, dyspnea, myalgia, fatigue, normal or decreased leukocyte counts, and radiographic evidence of pneumonia could be seen in the infected cases, with organ dysfunction and death occurring in severe cases [2]. In the Wuhan region, 99 cases were confirmed to be infected with COVID-19 among groups of humans in close contact, with more severe effects seen in older people with comorbidities [23]. Case reports confirmed human-to-human transmission causing the novel coronavirus–infected pneumonia (NCIP) [11]. The clinical characteristics of COVID-19 in pregnancy and intrauterine vertical transmission potential have also been reported [24]. This study revealed that the symptoms of pregnant women with COVID-19 pneumonia were diverse, with the main ones being fever and cough but with no evidence for vertical transmission in late pregnancy [24]. Table 1.1 presents the typical clinical and imaging findings of COVID-19 as well as those of common cold, influenza, and other coronavirus diseases including SARS and MERS [11, 25–29].

Coronaviridae represent an important family of animal and human viruses that are in permanent circulation [32, 33]. The first SARS coronavirus had a low impact on global morbidity and mortality, with more than 8000 recognized cases and 774 deaths [34, 35]. In contrast, the MERS coronavirus remained localized in Saudi Arabia, with smaller numbers in other countries of the Middle East and South Korea [36]. Studies on both of these viral outbreaks highlighted the potential danger of nosocomial transmission to health-care personnel, which could be critical in such epidemics by incapacitating the normal caregivers [37]. A common feature of SARS, MERS, and COVID-19 infection is the presence of severe acute respiratory syndrome, although the current estimated fatality rate of COVID-19 (6.6%, as of May 19, 2020) is lower than that of SARS (~10%) and MERS (~36%) [33, 35]. However, COVID-19 has led to emergence of much larger numbers of infected patients (approximately five million, compared with approximately 11,000 for SARS and MERS combined, as of May 19, 2020). Similar to SARS-CoV, the entry of SARS-CoV-2 into host cells depends on the recognition and binding of the spike protein to the ACE2 receptor on host cells [39–45]. The high affinity of this protein to the ACE2 receptor likely contributes to the quick spreading of virus. This receptor is expressed at high levels in human organs, such as lung alveolar epithelial cells and enterocytes of the small intestine, which are potentially the targets of COVID-19 infection [46]. In addition, COVID-19 transmission is currently known to be similar to that of other coronaviruses (such as SARS and MERS) in which human-to-human transmission transfer occurs through droplets, contact, and fomites. Based on the transmission modes of SARS and MERS, avoiding close contact with people suffering from acute respiratory infections and frequent handwashing are recommended. In addition, SARS-CoV-2 particles have also been detected in stool samples of some patients, suggesting that a more possible fecal-oral transmission occurs [47]. This suggests that multiple shedding routes of SARS-CoV-2 might exist.

Table 1.1 Comparison of typical clinical and imaging findings between common cold, influenza, SARS, MERS, and COVID-19

Disease	Respiratory symptom	Constitutional syndromes	CT imaging findings	Diagnostic methods	Refs.
Common cold	Stuffy nose, runny nose, and sneezing	No clear	X-ray	PCR and pan-viral DNA microarrays	[30, 31]
Influenza	Stuffy nose, runny nose, sore throat, and dry cough	High fever, muscle ache, and malaise	Small patch ground-glass opacity and consolidation with subpleural and/or peribronchial distribution	PCR, RT-PCR, immunofluorescence, direct (DFA) or indirect (IFA) florescent antibody staining, and viral tissue cell culture	[25]
SARS	Cough and dyspnea	Fever, chill, malaise, headache, and diarrhea	Subpleural ground-glass opacity and consolidation, prominent lower lobe involved, and interlobular septal and intralobular septa thickening	RT-PCR, rRT-PCR, RT-LAMP, rRT-LAMP, and coronavirus detection kit	[26, 27]
MERS	Sore throat, dry cough, and dyspnea	Fever, chill, and rigor	Bilateral, basilar, and subpleural airspace, extensive ground-glass opacity, and occasional septal thickening and pleural effusions	RT-PCR, rRT-PCR, RT-LAMP, rRT-LAMP, and coronavirus detection kit	[28, 29]
Mild SARS-COVID-19	Possible cough and sore throat	Fever	Multifocal patchy ground-glass opacity with subpleural distribution	RT-PCR, rRT-PCR, RT-LAMP, rRT-LAMP, and coronavirus detection kit	[11]
Severe SARS-COVID-19	Breathless and respiratory failure	Fever, muscle ache, confusion, and headache	Diffuse heterogeneous consolidation with ground-glass opacity		[11]

1.6 Immune Response to COVID-19 Infection

The innate immune response against viral infection typically involves production of the interferons, IFNα and IFNβ. In the case of MERS-CoV, the timing of interferon production appears to be critical in determining whether it protects against infection [48]. Interferon administration 1 day after infection was protective in a mouse model, whereas delayed interferon treatment did not block viral replication and actually increased lung inflammation, resulting in fatal pneumonia. Although the mechanisms of the immune response triggered by MERS infection and immune evasion strategies have not yet been fully studied, these findings indicate that the use of interferon-based therapies should be considered with caution [49]. A compromised immune system may have been the cause of the high case-fatality rates in MERS-infected patients [50]. This is consistent with the findings of previous studies which indicate that older individuals with compromised immune systems are more susceptible to the effects of SARS infection [51, 52].

Coronaviruses have evolved multiple immune evasion mechanisms to limit the early induction of interferon. SARS-CoV encodes an enzyme that adds a 2′ O-methyl group to the viral RNA, thereby evading detection by the viral sensor molecule, MDA5 [53, 54]. Similarly, coronaviruses encode proteases that antagonize the stimulator of interferon genes (STING) [55], which is thought to be the case for SARS-CoV-2. However, neutralizing antibodies that target the receptor-binding domain of the spike protein can competitively inhibit viral entry into host cells, as shown in the case of SARS-CoV [56]. Most initial efforts to generate vaccines against SARS-CoV-2 have focused on the corresponding spike protein and its receptor-binding domain. Interestingly, the affinity of the SARS-CoV-2 spike protein for its receptor, ACE2, is approximately 15 nM, which is over 20 times greater compared to the SARS-CoV spike protein [41, 57]. Based on this, the development of neutralizing antibodies may prove to be more difficult for COVID-19 infection. However, some neutralizing antibodies developed for SARS-CoV do cross-react with SARS-CoV-2, and these bind to the protein core rather than the receptor-binding domain [58–60]. It is possible that some of these may be useful for vaccine development considering the sequence similarity of SARS-CoV-2 and SARS-CoV and the facts that they use the same ACE2 receptor for cellular entry and cause similar acute respiratory syndromes [61].

1.7 Diagnostic Strategies

The genome sequence of SARS-CoV-2 was released online in early January 2020 (GenBank accession number MN908947), which facilitated the development of nucleotide-based diagnostic methods. Real-time polymerase chain reaction (PCR) was verified as a routine method for detection [62]. Corman et al. developed a PCR method based on the similarity between SARS-CoV and SARS-CoV-2. The probes and primers targeted the receptor-binding gene (spike protein gene), as well as the E and N genes, which allowed successfully discrimination between the SARS-CoV and SARS-CoV-2 viruses [63]. Chu et al. designed a quantitative real-time PCR test for the highly conserved ORF1b and N genes of the SARS-CoV-2 genome [38]. For this, SARS-CoV PCR products were also cloned into plasmids as references, and additional viral genomes were tested, such as those from MERS-CoV, avian influenza, human influenza, and parainfluenza. Patients with COVID-19 were positive for both SARS-CoV-2 genes and not those from the other viruses. The detection of the N gene appeared to be more sensitive, compared to ORF1b. Thus, N gene detection may be more effective in the case of SARS-CoV-2, and testing for the Orf1b gene could be used for confirmation.

Next-generation sequencing (NGS) has been regularly used for viral disease diagnosis and could detect all kinds of DNA/RNA fragments from viral infections. In addition, NGS approaches could play a role in the discovery of novel viruses. RNA-based NGS has also been suggested for rapid diagnosis of respiratory viruses. Compared to PCR methods, NGS may have higher specificity and could detect viruses even in the latency phase of infection [64]. Screening for ACE2 receptor expression in patients may also be helpful in disease risk assessment as this may help to identify the possible and main routes of infection. For this, bulk RNA sequencing and single-cell transcriptomic analysis was performed on patient samples taken from mucosa or the oral cavity [40]. This revealed that the ACE2 receptor was expressed in oral tissues, especially in the tongue. The finding that the expression of the ACE2 receptor is also elevated in the lungs lends further support that blocking its interaction with the virus may be a possible therapy against COVID-19 infection (see below) [39].

Currently, no proteomic biomarkers of COVID-19 infection have been validated, but several candidates are possible. Routine blood tests have revealed that procalcitonin and C-reactive protein (CRP) may be elevated in several disease cases but not in all patients [23, 65]. Cytokines may also be potential marker candidate of COVID-19 infections. Huang et al. analyzed the cytokine and chemokine levels in plasma of patients using a Human Cytokine Standard 27-Plex Assays panel and the Bio-Plex 200 device [2]. Several inflammatory markers showed elevated levels in affected patients, such as interleukin (IL-)1β, IL-1RA, IL-7, IL-8, IL-9, IL-10, fibroblast growth factor (FGF), tumor necrosis factor alpha (TNFα), vascular endothelial growth factor VEGF, IFNγ, and IFNγ-inducible protein 10 (IP10). The same study also showed that patients who needed intensive care presented higher levels of plasma markers such as granulocyte colony-stimulating factor (GCSF), IP10, monocyte chemoattractant protein-1 (MCP1), macrophage inflammatory protein 1A (MIP1A), and TNFα, compared to patients who did not need critical care. These data revealed that elevated concentrations of cytokines may correlate with disease severity.

1.8 Potential of SARS-COV-2 Diagnostic Testing in Male Semen

Coronaviruses are a large family of viruses that cause illness ranging from the common cold to more severe diseases such as MERS and SARS [66–68]. The SARS-CoV-2 virus was originally discovered when viral metagenomic analysis was carried out on three bronchoalveolar lavage specimens from Chinese adult patients with unexplained severe pneumonia [1]. Currently, SARS-CoV-2 can be detected in human clinical specimens including saliva, nasal fluid, blood, feces, and urine by NGS, real-time PCR, cell culture, and electron microscopy [1]. The Centers for Disease Control (CDC) has recommended that clinical virology laboratories should not attempt viral isolation from specimens collected from COVID-19 patients under investigation (https://www.cdc.gov/coronavirus/nCoV-2019/guidelines-clinical-specimens.html). Even though several countries are currently producing the test kit for the coronavirus after the availability of the SARS-CoV-2 genome, diagnostic testing for this virus should be carried out according to CDC guidelines.

Currently, to determine if someone has been infected with SARS-CoV-2, samples from the throat, coughed-up sputum, or lung fluid are used. Importantly, recent studies have indicated that semen from males infected with viruses could be considered as a primary route of human transmission [69–71]. In addition, the hypothesis that the ACE2 receptor could serve as a mediator for endocytosis of the SARS-CoV coronavirus [72] has led to the suggestion that SARS-CoV-2 may enter host cells and tissues through a similar mechanism (Fig. 1.3). From the protein expression atlas (www.proteinatlas.org), high levels of the ACE2 precursor are found in kidney, intestines, and testis (Fig. 1.3) and not in female reproductive system. This latter finding could be the possible explanation for the higher rate of death in males than females [23]. Hence, SARS-CoV-2 isolation from the semen or reproductive tracts of infected men may be considered as the only direct and definitive approach for proving infectivity or their potential function as reservoir. A study which carried out semen testing found that 6 patients (15.8%) had positive results for COVID-19, including 4 of 15 patients (26.7%) who were at the acute stage of infection and 2 of 23 patients (8.7%) who were recovering [73]. This latter finding is particularly noteworthy as it suggests that sexual transmission is possible. Therefore, it is important to determine if the viral infection is sexually transmitted to determine the risk, especially in the case of males who have recovered and may still harbor the virus.

1.9 ACE-2: The SARS-CoV-2 Receptor Identified

As stated above, the coronavirus spike protein receptor-binding domain appears to be responsible for the interaction between the virus and the host cells, and one receptor for this is ACE2 [44, 74]. ACE2 is a transmembrane protein enzyme with its active site on the extracellular surface of cells. ACE2 plays a role in conversion of angiotensin II to angiotensin 1–7, which may play a role in regulation of cardiac function and blood pressure. In addition, overexpression of ACE2 may be involved in diseases such as cardiac dysfunction or diabetes [75]. Hence, many studies have been carried out by targeting ACE2 for treatment of circulatory diseases. In animal models of hypertension, reduced ACE2 expression or activity has been reported. Angiotensin 1–7 production via ACE2 cleavage could have several protective roles, including vasodilatory, antiproliferative, or anti-oxidative stress effects. Since hypertension is a major health issue in diseases such as diabetes, heart failure, liver failure, and pulmonary injury [76, 77], ACE2 and angiotensin 1–7 have been suggested as potential therapeutic targets. In addition, higher expressions of ACE2 have been found in renal and liver cancer and could therefore have prognostic value. On the

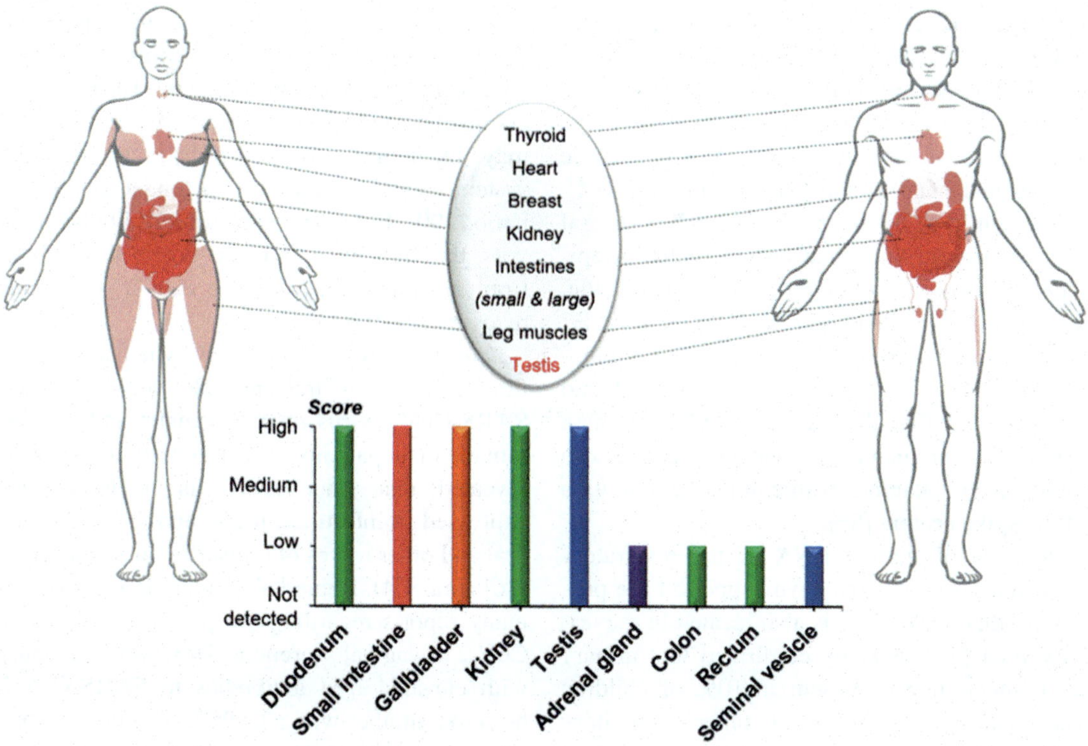

Fig. 1.3 Selective ACE2 expression in thyroid, heart, breast, kidney, intestines, leg muscles, and testis and schematic overview of the ACE2 expression are shown for each of the nine tissues. Color-coding is based on tissue groups, and no protein expression data was observed in cerebral cortex, cerebellum, hippocampus, caudate, thyroid gland, parathyroid gland, nasopharynx, bronchus, lung, oral mucosa, salivary gland, esophagus, stomach, liver, pancreas, urinary bladder, epididymis, prostate, vagina, ovary, fallopian tube, endometrium, cervix, uterine, placenta, breast heart muscle, smooth muscle, skeletal muscle, soft tissue, adipose tissue, skin, appendix, spleen, lymph node, tonsil, and bone marrow tissues. The proteomic analysis is combined with RNA-Seq on the organ, tissue, and cellular level, and all data are freely accessible on the Human Protein Atlas web portal, www.proteinatlas.org

other hand, overexpression of ACE2 may be protective against oxidative stress [78].

The possibility that ACE2 expression may increase susceptibility to SARS-CoV-2 infection may be linked to the finding that older individuals show greater effects of viral infection [79]. This is due to the fact that ACE2 expression may be increased in adults in their 40s–50s since it may be associated with several protective mechanisms against age-related diseases. In addition, ACE2 is an important regulator of Ras signaling, which has been suggested as a modulator of the aging process. These findings suggest that children are less susceptible to the effects of SARS-CoV-2 infection due to lower expression of ACE2 [80, 81]. In addition, elderly individuals are known to have reduced innate immune functions, resulting in a higher mortality rate [82, 83].

A study of approximately 40 patients with SARS-CoV-2 infections in early January 2020 found that the majority of these were male with underlying diseases, such as diabetes, hypertension, or cardiovascular disorders [2]. This was supported by a follow-up study which showed that over two-thirds of the patients were male with an average age of 55.5 years [23]. Together, these findings suggest that males with underlying comorbidities have a greater risk of infection and death from COVID-19.

The initial infection site for COVID-19 infection is the upper respiratory system where ACE2 sites are abundant, followed by the lower respiratory tract where numbers of these receptors are lower, and this can lead to bronchiolitis, secondary pneumonia, or both [43]. In addition, the reported symptoms of COVID-19 infection in the early phase of disease progression can include indigestion, nausea, diarrhea, or vomiting [84]. Coincidently, ACE2 expression is high in several cells of the gastrointestinal system, including epithelial cells in the esophagus, enterocytes, colon, and ileum [85]. In addition, high levels of ACE2 are also found in the kidneys [42]. Receptor binding of SARS-CoV-2 was recently modelled, and this revealed that the receptor-binding domain has a stronger interaction with ACE2 due to a unique phenylalanine residue in the flexible loop of the spike protein [86].

Reports of finding RNA of the respiratory syncytial virus in urine have suggested the possibility that COVID-19 is also located in the urinary tract [87]. A study confirmed that urinary tract infection was present in 10% of children with acute bronchiolitis or other respiratory viruses, such as influenza virus or enterovirus, and coronavirus was detected in 7% of the cases [87]. Since high levels of ACE2 expressions have been reported in kidney, ACE2 could be involved in controlling the different renal functions, suggesting a potential route of SARS-CoV-2 infection or clearance [42].

ACE2 has been reported to have high expression levels in testis and seminal vesicles, especially in the Leydig and Sertoli cells [88]. Rat experiments showed that ACE2 may be a constitutive product of mature Leydig cells and may play a role in testicular functions, such as spermatogenesis [88]. Even though no evidence has been found regarding the potential involvement of COVID-19 in the testicular system, high ACE2 levels in testis and seminal vesicles may provide an additional route of infection or reservoir. Several viruses have been detected in the seminal fluid with long survival times, which has suggested that they may be sexually transmitted [69–71]. As an example, the Ebola virus was found to be sexually transmittable since a Liberian male survivor of Ebola infected his female partner with the virus. This study also revealed that infectious Ebola virus may be present at least for 179 days in survivors after disease onset [89]. Additional male survivors were investigated, confirming the presence of virus in their semen [90, 91]. The researchers involved in this study suggested that the virus may hide in the testes and thereby escape immune system detection, leading to a longer incubation period [90]. In a larger study of 220 Ebola survivors, the viral RNA could be detected in semen from several months to more than 2 years after recovery [92, 93].

Currently, we do not know whether seminal fluid can act as transmissible vector for the SARS-CoV-2 virus in symptomatic and asymptomatic male patients [2, 23]. Investigations have revealed that since ACE2 is predominantly expressed in intestines, testis, and kidney, fecal-oral and other routes of transmission are possible [86] (Fig. 1.4). Remarkably, there have not been many reports regarding the presence of SARS-CoV-2 in vaginal secretions. However, a neonate with elevated IgM antibodies to SARS-CoV-2 born to a mother with a COVID-19 infection has been reported [10]. Therefore, there is still a controversy regarding whether or not COVID-19 can be transmitted in utero from an infected mother to her infant before birth.

1.10 Current Treatment Strategies

Currently, there are no useful therapies available against COVID-19 infection although intensive efforts are underway to develop drugs and vaccines. Studies on the therapeutics used against SARS and MERS could also provide useful leads in developing new treatments against COVID-19. In addition to targeting the interaction between the ACE2 receptor and the viral spike protein, other possible therapeutic targets could be the two proteases involved in proteolysis and virion packaging: the coronavirus main proteinase (3CLpro) and the papain-like protease (PLpro) [93]. Traditional Chinese medicines may also be helpful by maintaining body health, and research is ongoing regarding the efficacy of this approach in treating COVID-19. Some Chinese medicines have been suggested to have some impact on disease by either preventing mild disease symptoms

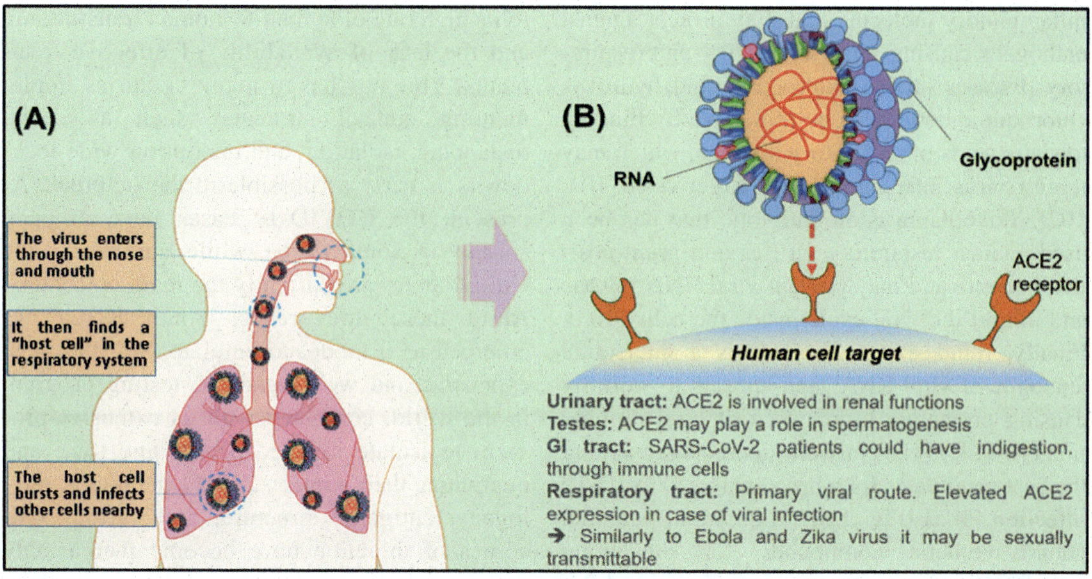

Fig. 1.4 (**a**) Transmission of the coronavirus disease and (**b**) potential routes of SARS-COV-2 infection in terms of ACE2 expression

progressing to more severe ones or speeding up recovery from infection. However, such potential effects have been questioned in the case of COVID-19, and more scientific studies are needed to prove or disprove efficacy against the virus [94].

Currently, antibodies and vaccines are under development, which target the spike protein. Tian et al. discovered a monoclonal antibody, CR3022, which could bind to the receptor-binding domain of SARS-CoV-2 although it was initially used as SARS-CoV antibody [59]. In the case of SARS-CoV-2, CR0322 may recognize an epitope that does not overlap with the ACE2 binding site. Instead, CR0322 may be a neutralizing antibody and could therefore offer some protection against the disease via another mechanism.

Given the likely role of ACE2 in SARS-CoV-2 entry into cells, inhibitors of this molecule could play a significant role in the prevention or therapy against COVID-19 infection. There are several active compounds such as scutellarin, hesperetin, nicotianamine, and glycyrrhizin that may reduce ACE2 activity/expression or inhibit ACE2 and spike protein binding [95]. Chloroquine phosphate is an old drug used for the treatment of malaria or rheumatoid arthritis and may also be useful against COVID-19 infection. This drug could protect against the virus-cell fusion, enhance endosomal functions, or interfere with glycosylation of coronavirus receptors. In addition, it may also have anti-inflammatory effects. Another antiparasitic drug called diminazene aceturate (DIZE) is known to exert vasodilatory effects in experimental models of hypertension, myocardial infarction, diabetes, and atherosclerosis, and it has mild antiviral properties [96]. Remdesivir is an adenosine analogue that can be used to inhibit viral gene expression. It was successfully used against different viral infections, including SARS and MERS [97]. Additional treatment strategies may inhibit the coronavirus proteases, prevent viral replication, and enhance the immune protection of the host [98]. The 3C protease and pyridoxal 5′-phosphate (PLP) inhibitors have also been used with some success against SARS and MERS. Ivermectin is a Food and Drug Administration (FDA)-approved antiparasitic drug, which could be used against different kinds of viruses such as HIV and Zika. Studies on Vero/hSLAM cells infected with virus suggested that ivermectin may significantly reduce viral RNA levels in less than 48 h [99]. In addition, melatonin is an anti-oxidative and anti-inflammatory molecule that may protect against pathogens causing acute lung injury and respiratory diseases [100]. Chloroquine and hydroxychloroquine could also be protective by binding to sialic acids of respiratory systems, which may inhibit virus interactions with target cells [101, 102]. Teicoplanin is an antibiotic that has been used against respiratory viruses and gram-positive bacteria and may prevent viral RNA release and inhibit the viral cycle inside the cells [103]. Finally, remdesivir (GS-5734) is a nucleotide analogue of viral RNAs and appears to work by causing premature termination of viral replication [104]. It is currently considered to be among the best candidates for a drug against COVID-19 infection [105] (Fig. 1.5). In addition, potential natural/synthetic compounds for preventing COVID-19 infection are presented in Table 1.2.

1.11 Future Perspectives

Compared with the SARS and MERS viral outbreaks, the COVID-19 pandemic has thus far shown a lower mortality rate but with a drastically higher infection capability. In just a few months, this pandemic has resulted in a global challenge for disease prevention and control due to its high rate of human-to-human transmission and the lack of availability of effective treatments. This has led to many countries implementing control strategies such as social distancing, isolation, and institution-wide lockdowns as early as possible in the outbreak. At present, the COVID-19 cases have dropped sharply in South Korea, while Europe and the United States are currently the main epicenters. Amid these dire trends, South Korea has emerged as a model to emulate with the most expansive and well-organized testing program in the world, combined with its extensive program to isolate infected people and trace and quarantine their contacts. As the genomics, phylogeny, antigenic structure, and various outcomes of infection have become increasingly understood, the development of diagnostic strategies, new therapeutic approaches, and vaccines against COVID-19 has become a worldwide effort. Perhaps one positive outcome of these efforts is that this collective mobilization has led to the scientific, technical, and financial support to combat this current pandemic at a speed that has never been seen before. It is hoped that this will help to lay the foundations to control this current outbreak and provide a strong base to tackle future pandemics.

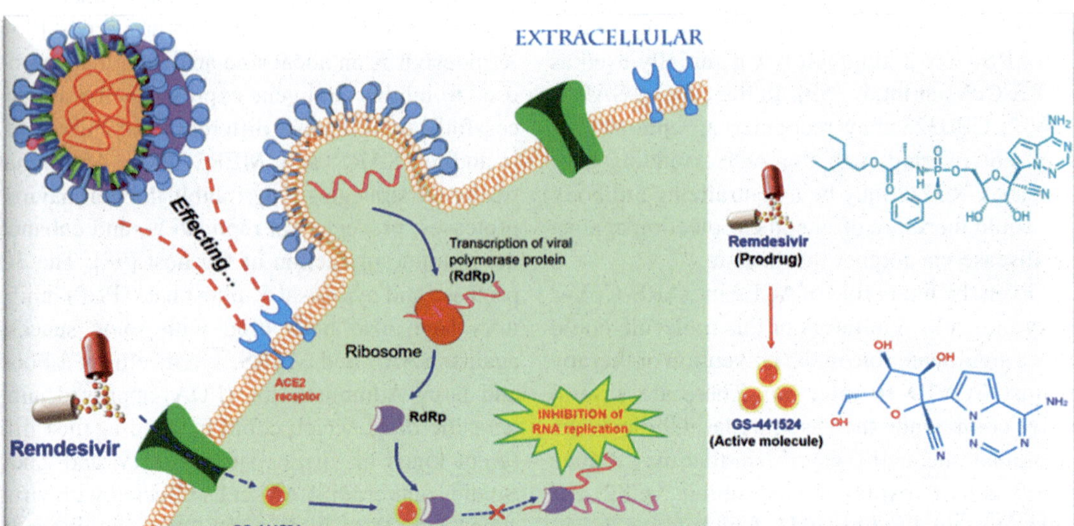

Fig. 1.5 Potential repurposed remdesivir drug candidate for treatment of COVID-19

Table 1.2 Suggested potential drug candidates against SARS-COV-2

Drug candidate	Description	Possible role in COVID-19	Refs.
CR3022	Monoclonal antibody	May bind spike protein epitope	[59]
Chloroquine phosphate	Old drug against malaria and rheumatoid arthritis	May prevent virus-cell fusion and anti-inflammatory effects	[96]
Remdesivir	Adenosine analogue; results in premature termination of viral transcription	Useful against RNA viruses (SARS, MERS); may inhibit COVID-19 infection	[98]
Baicalin	Natural compound from Scutellaria baicalensis Georgi; could have anti-inflammatory effect	Binds to ACE2; antiviral effect against SARS-COV. Possible effects against COVID-19	[59, 86]
Scutellarin	Natural compounds from Erigeron breviscapus; could have anti-inflammatory effect	Could bind to ACE2; may prevent COVID-19 binding	
Hesperetin	Bioflavonoid from citruses; inhibits 3C-like protease	Possible ACE2 inhibitor	
Nicotianamine	Can be isolated from soybean	May block COVID-19 through inhibiting ACE2	
Glycyrrhizin	Isolated from licorice root; protective against SARS virus	Interacts with ACE2; further studies are needed on COVID-19	
Ivermectin	Popular antiviral drug, HIV inhibitor, and promising candidate against SARS-COV-2	May inhibit transport of viral proteins through IMPα/β1 pathway	[100]
Melatonin	Safe drug against several respiratory diseases	May reduce vessel permeability and improve life quality of patients with SARS-COV-2	[101]
Chloroquine (CQ) and hydroxychloroquine (HCQ)	Effective against several RNA viruses	Prevention of ACE2-virus interactions	[102]
Remdesivir (GS-5734)	Nucleotide analogue of viral RNA	May result in premature viral transcription	[105]
Teicoplanin	Glycopeptide antibiotic	Prevention of the viral RNA release and viral cycle inside cells	[104]

Conflicts of Interest The author declares no competing financial interests.

Funding This research was supported by a National Research Foundation of Korea (NRF) Grants awarded by the Korean government (MEST, No. NRF-2020R1A 2B5B01002463 & NRF-2019R1G1A109740012).

References

1. Zhu N, Zhang D, Wang W, Li X, Yang B, Song J et al (2020) A novel coronavirus from patients with pneumonia in China, 2019. N Engl J Med 382:727–733
2. Huang C, Wang Y, Li X, Ren L, Zhao J, Hu Y et al (2020) Clinical features of patients infected with 2019 novel coronavirus in Wuhan, China. Lancet 395(10223):497–506
3. Pachetti M, Marini B, Benedetti F, Giudici F, Mauro E, Storici P et al (2020) Emerging SARS-CoV-2 mutation hot spots include a novel RNA-dependent-RNA polymerase variant. J Transl Med 18(1):179. https://doi.org/10.1186/s12967-020-02344-6
4. Yao H, Lu X, Chen Q, Xu K, Chen Y, Cheng L et al (2020) Patient-derived mutations impact pathogenicity of SARS-CoV-2. medRxiv. https://doi.org/10.1101/2020.04.14.20060160
5. Lu R, Zhao X, Li J, Niu P, Yang B, Wu H et al (2020) Genomic characterisation and epidemiology of 2019 novel coronavirus: implications for virus origins and receptor binding. Lancet 395(10224):565–574
6. Wu A, Peng Y, Huang B, Ding X, Wang X, Niu P et al (2020) Genome composition and divergence of the novel coronavirus (2019-nCoV) originating in China. Cell Host Microbe 27(3):325–328
7. Zhou P, Yang XL, Wang XG, Hu B, Zhang L, Zhang W et al (2020) A pneumonia outbreak associated with a new coronavirus of probable bat origin. Nature 579(7798):270–273
8. York A (2020) Host–microbiome interactions lost during flight. Nat Rev Microbiol 18(3):122. https://doi.org/10.1038/s41579-020-0328-9
9. Yang Y, Lu Q, Liu M, Wang Y, Zhang A, Jalali N et al (2020) Epidemiological and clinical features of

the 2019 novel coronavirus outbreak in China. medRxiv. https://doi.org/10.1101/2020.02.10.20021675
10. Dong L, Tian J, He S, Zhu C, Wang J, Liu C et al (2020) Possible vertical transmission of SARS-CoV-2 from an infected mother to her newborn. JAMA. https://doi.org/10.1001/jama.2020.4621. [Epub ahead of print]
11. Chan JF, Yuan S, Kok KH, To KK, Chu H, Yang J et al (2020) A familial cluster of pneumonia associated with the 2019 novel coronavirus indicating person-to-person transmission: a study of a family cluster. Lancet 395(10223):514–523
12. Hao P, Zhong W, Song S, Fan S, Li X (2020) Is SARS-CoV-2 originated from laboratory? A rebuttal to the claim of formation via laboratory recombination. Emerg Microbe Infect 9(1):545–547
13. Li W, Shi Z, Yu M, Ren W, Smith C, Epstein JH et al (2005) Bats are natural reservoirs of SARS-like coronaviruses. Science 310(5748):676–679
14. Liu P, Chen W, Chen JP (2019) Viral metagenomics revealed sendai virus and coronavirus infection of Malayan pangolins (Manis javanica). Viruses 11(11):pii: E979. https://doi.org/10.3390/v11110979
15. Zhang T, Wu Q, Zhang Z (2020) Probable pangolin origin of SARS-CoV-2 associated with the COVID-19 outbreak. Curr Biol 30(7):1346–1351.e2
16. Chen Y, Liu Q, Guo D (2020) Emerging coronaviruses: genome structure, replication, and pathogenesis. J Med Virol 92(4):418–423
17. an Doremalen N, Bushmaker T, Morris DH, Holbrook MG, Gamble A, Williamson BN et al (2020) Aerosol and surface stability of SARS-CoV-2 as compared with SARS-CoV-1. N Engl J Med 382(16):1564–1567
18. Xie J, Zhu Y (2020) Association between ambient temperature and COVID-19 infection in 122 cities from China. Sci Total Environ 724:138201. https://doi.org/10.1016/j.scitotenv.2020.138201
19. Shereen MA, Khan S, Kazmi A, Bashir N, Siddique R (2020) COVID-19 infection: origin, transmission, and characteristics of human coronaviruses. J Adv Res 24:91–98
20. Dong N, Yang X, Ye L, Chen K, Chan EWC, Yang M et al (2020) Genomic and protein structure modelling analysis depicts the origin and infectivity of 2019-nCoV, a new coronavirus which caused a pneumonia outbreak in Wuhan, China. bioRxiv. https://doi.org/10.1101/2020.01.20.913368
21. Xu X, Chen P, Wang J, Feng J, Zhou H, Li X et al (2020) Evolution of the novel coronavirus from the ongoing Wuhan outbreak and modeling of its spike protein for risk of human transmission. Sci China Life Sci 63(3):457–460
22. Rabi FA, Al Zoubi MS, Kasasbeh GA, Salameh DM, Al-Nasser AD (2020) SARS-CoV-2 and coronavirus disease 2019: what we know so far. Pathogens 9(3):pii: E231. https://doi.org/10.3390/pathogens9030231
23. Chen N, Zhou M, Dong X, Qu J, Gong F, Han Y et al (2020) Epidemiological and clinical characteristics of 99 cases of 2019 novel coronavirus pneumonia in Wuhan, China: a descriptive study. Lancet 395(10223):507–513
24. Chen H, Guo J, Wang C, Luo F, Yu X, Zhang W et al (2020) Clinical characteristics and intrauterine vertical transmission potential of COVID-19 infection in nine pregnant women: a retrospective review of medical records. Lancet 395(10226):809–815
25. Duan YN, Qin J (2020) Pre- and posttreatment chest CT findings: 2019 novel coronavirus (2019-nCoV) pneumonia. Radiology 295(1):21. https://doi.org/10.1148/radiol.2020200323
26. Lee KS (2020) Pneumonia associated with 2019 novel coronavirus: can computed tomographic findings help predict the prognosis of the disease? Korean J Radiol 21(3):257–258
27. Franquet T (2011) Imaging of pulmonary viral pneumonia. Radiology 260(1):18–39
28. Hansell DM, Bankier AA, MacMahon H, McLoud TC, Müller NL, Remy J (2008) Fleischner society: glossary of terms for thoracic imaging. Radiology 246(3):697–722
29. Malainou C, Herold S (2019) Influenza. Internist (Berl) 60(11):1127–1135
30. Schlaberg R, Queen K, Simmon K, Tardif K, Stockmann C, Flygare S et al (2017) Viral pathogen detection by metagenomics and pan-viral group polymerase chain reaction in children with pneumonia lacking identifiable etiology. J Infect Dis 215(9):1407–1415
31. Kistler A, Avila PC, Rouskin S, Wang D, Ward T, Yagi S et al (2007) Pan-viral screening of respiratory tract infections in adults with and without asthma reveals unexpected human coronavirus and human rhinovirus diversity. J Infect Dis 196(6):817–825
32. To KK, Hung IF, Chan JF, Yuen KY (2013) From SARS coronavirus to novel animal and human coronaviruses. J Thorac Dis 5(Suppl 2):S103–S108
33. Wu Z, McGoogan JM (2020) Characteristics of and important lessons from the coronavirus disease 2019 (COVID-19) outbreak in China: summary of a report of 72314 cases from the Chinese Center for Disease Control and Prevention. JAMA. https://doi.org/10.1001/jama.2020.2648. [Epub ahead of print]
34. Peiris JS, Yuen KY, Osterhaus AD, Stöhr K (2003) The severe acute respiratory syndrome. N Engl J Med 349(25):2431–2441
35. de Wit E, van Doremalen N, Falzarano D, Munster VJ (2016) SARS and MERS: recent insights into emerging coronaviruses. Nat Rev Microbiol 14(8):523–534
36. Zaki AM, van Boheemen S, Bestebroer TM, Osterhaus AD, Fouchier RA (2012) Isolation of a novel coronavirus from a man with pneumonia in Saudi Arabia. N Engl J Med 367(19):1814–1820
37. Hui DS, Azhar EI, Kim YJ, Memish ZA, Oh MD, Zumla A (2018) Middle East respiratory syndrome coronavirus: risk factors and determinants of primary, household, and nosocomial transmission. Lancet Infect Dis 18(8):e217–e227

38. Chu DKW, Pan Y, Cheng SMS, Hui KPY, Krishnan P, Liu Y et al (2020) Molecular diagnosis of a novel coronavirus (2019-nCoV) causing an outbreak of pneumonia. Clin Chem 66(4):549–555
39. Zhao Y, Zhao Z, Wang Y, Zhou Y, Ma Y, Zuo W (2020) Single-cell RNA expression profiling of ACE2, the putative receptor of Wuhan 2019-nCov. bioRxiv. https://doi.org/10.1101/2020.01.26.919985
40. Xu H, Zhong L, Deng J, Peng J, Dan H, Zeng X et al (2020) High expression of ACE2 receptor of 2019-nCoV on the epithelial cells of oral mucosa. Int J Oral Sci 12(1):8
41. Wrapp D, Wang N, Corbett KS, Goldsmith JA, Hsieh CL, Abiona O et al (2020) Cryo-EM structure of the 2019-nCoV spike in the prefusion conformation. Science 367(6483):1260–1263
42. Park SE, Kim WJ, Park SW, Park JW, Lee N, Park CY et al (2013) High urinary ACE2 concentrations are associated with severity of glucose intolerance and microalbuminuria. Eur J Encrinol 168(2):203–210
43. Jia HP, Look DC, Shi L, Hickey M, Pewe L, Netland J et al (2005) ACE2 receptor expression and severe acute respiratory syndrome coronavirus infection depend on differentiation of human airway epithelia. J Virol 79(23):14614–14621
44. Hoffmann M, Kleine-Weber H, Krüger N, Müller M, Drosten C, Pöhlmann S (2020) The novel coronavirus 2019 (2019-nCoV) uses the SARS-coronavirus receptor ACE2 and the cellular protease TMPRSS2 for entry into target cells. bioRxiv. https://doi.org/10.1101/2020.01.31.929042
45. Hamming I, Timens W, Bulthuis ML, Lely AT, Navis G, van Goor H (2004) Tissue distribution of ACE2 protein, the functional receptor for SARS coronavirus. A first step in understanding SARS pathogenesis. J Pathol 203(2):631–637
46. Zou X, Chen K, Zou J, Han P, Hao J, Han Z (2020) Single-cell RNA-seq data analysis on the receptor ACE2 expression reveals the potential risk of different human organs vulnerable to 2019-nCoV infection. Front Med 14(2):185–192
47. Yeo C, Kaushal S, Yeo D (2020) Enteric involvement of coronaviruses: is faecal-oral transmission of SARS-CoV-2 possible? Lancet Gastroenterol Hepatol 5(4):335–337
48. Channappanavar R, Fehr AR, Zheng J, Wohlford-Lenane C, Abrahante JE, Mack M et al (2019) IFN-I response timing relative to virus replication determines MERS coronavirus infection outcomes. J Clin Invest 130:3625–3639
49. Balachandran S, Roberts PC, Brown LE, Truong H, Pattnaik AK, Archer DR et al (2000) Essential role for the dsRNA-dependent protein kinase PKR in innate immunity to viral infection. Immunity 13(1):129–141
50. Breban R, Riou J, Fontanet A (2013) Interhuman transmissibility of Middle East respiratory syndrome coronavirus: estimation of pandemic risk. Lancet 382(9893):694–699
51. Rockx B, Baas T, Zornetzer GA, Haagmans B, Sheahan T, Frieman M et al (2009) Early upregulation of acute respiratory distress syndrome-associated cytokines promotes lethal disease in an aged-mouse model of severe acute respiratory syndrome coronavirus infection. J Virol 83(14):7062–7074
52. Huang KJ, Su IJ, Theron M, Wu YC, Lai SK, Liu CC et al (2005) An interferon-gamma-related cytokine storm in SARS patients. J Med Virol 75(2):185–194
53. Decroly E, Imbert I, Coutard B, Bouvet M, Selisko B, Alvarez K et al (2008) Coronavirus nonstructural protein 16 is a cap-0 binding enzyme possessing (nucleoside-2′O)-methyltransferase activity. J Virol 82(16):8071–8084
54. Züst R, Cervantes-Barragan L, Habjan M, Maier R, Neuman BW, Ziebuhr J et al (2011) Ribose 2′-O-methylation provides a molecular signature for the distinction of self and non-self mRNA dependent on the RNA sensor Mda5. Nat Immunol 12(2):137–143
55. Sun L, Xing Y, Chen X, Zheng Y, Yang Y, Nichols DB et al (2012) Coronavirus papain-like proteases negatively regulate antiviral innate immune response through disruption of STING-mediated signaling. PLoS One 7(2):e30802. https://doi.org/10.1371/journal.pone.0030802
56. van den Brink EN, Ter Meulen J, Cox F, Jongeneelen MA, Thijsse A, Throsby M et al (2005) Molecular and biological characterization of human monoclonal antibodies binding to the spike and nucleocapsid proteins of severe acute respiratory syndrome coronavirus. J Virol 79(3):1635–1644
57. Yan R, Zhang Y, Li Y, Xia L, Guo Y, Zhou Q (2020) Structural basis for the recognition of SARS-CoV-2 by full-length human ACE2. Science 367(6485):1444–1448
58. ter Meulen J, van den Brink EN, Poon LL, Marissen WE, Leung CS, Cox F et al (2006) Human monoclonal antibody combination against SARS coronavirus: synergy and coverage of escape mutants. PLoS Med 3(7):e237. https://doi.org/10.1371/journal.pmed.0030237
59. Tian X, Li C, Huang A, Xia S, Lu S, Shi Z et al (2020) Potent binding of 2019 novel coronavirus spike protein by a SARS coronavirus-specific human monoclonal antibody. Emerg Microbe Infect 9(1):382–385
60. Wang C, Li W, Drabek D, Okba NMA, van Haperen R, Osterhaus ADME et al (2020) A human monoclonal antibody blocking SARS-CoV-2 infection. bioRxiv. https://doi.org/10.1101/2020.03.11.987958
61. Iwasaki A, Yang Y (2020) The potential danger of suboptimal antibody responses in COVID-19. Nat Rev Immunol. https://doi.org/10.1038/s41577-020-0321-6. [Epub ahead of print]
62. Chen L, Liu W, Zhang Q, Xu K, Ye G, Wu W et al (2020) RNA based mNGS approach identifies a novel human coronavirus from two individual pneumonia cases in 2019 Wuhan outbreak. Emerge Microbes Infect 9(1):313–319

63. Corman VM, Landt O, Kaiser M, Molenkamp R, Meijer A, Chu DK et al (2020) Detection of 2019 novel coronavirus (2019-nCoV) by real-time RT-PCR. Euro Surveill 25(3). https://doi.org/10.2807/1560-7917.ES.2020.25.3.2000045
64. Brussel A, Brack K, Muth E, Zirwes R, Cheval J, Hebert C et al (2019) Use of a new RNA next generation sequencing approach for the specific detection of virus infection in cells. Biologicals 59:29–36
65. Habibzadeh P, Stoneman EK (2020) The novel coronavirus: a bird's eye view. Int J Occup Environ Med 11(2):65–71
66. Luk HKH, Li X, Fung J, Lau SKP, Woo PCY (2019) Molecular epidemiology, evolution and phylogeny of SARS coronavirus. Infect Genet Evol 71:21–30
67. Ramadan N, Shaib H (2019) Middle East respiratory syndrome coronavirus (MERS-CoV): a review. Germs 9(1):35–42
68. Menachery VD, Yount BL Jr, Debbink K, Agnihothram S, Gralinski LE, Plante JA et al (2015) A SARS-like cluster of circulating bat coronaviruses shows potential for human emergence. Nat Med 21(12):1508–1513
69. Feldmann H (2018) Virus in semen and the risk of sexual transmission. N Engl J Med 378(15):1440–1441
70. Mead PS, Hills SL, Brooks JT (2018) Zika virus as a sexually transmitted pathogen. Curr Opin Infect Dis 31(1):39–44
71. Fischer WA 2nd, Wohl DA (2016) Confronting Ebola as a sexually transmitted infection. Clin Infect Dis 62(10):1272–1276
72. Chu KH, Tsang WK, Tang CS, Lam MF, Lai FM, To KF et al (2005) Acute renal impairment in coronavirus-associated severe acute respiratory syndrome. Kidney Int 67(2):698–705
73. Li D, Jin M, Bao P, Zhao W, Zhang S (2020) Clinical characteristics and results of semen tests among men with coronavirus disease 2019. JAMA Netw Open 3(5):e208292–e208292
74. Letko M, Munster V (2020) Functional assessment of cell entry and receptor usage for lineage B β-coronaviruses, including 2019-nCoV. medRxiv. https://doi.org/10.1101/2020.01.22.915660
75. Tikellis C, Thomas MC (2012) Angiotensin-converting enzyme 2 (ACE2) is a key modulator of the renin angiotensin system in health and disease. Int J Pept 2012:256294. https://doi.org/10.1155/2012/256294
76. Chappell MC (2010) Does ACE2 contribute to the development of hypertension? Hypertens Res 33(2):107–109
77. Chamsi-Pasha MA, Shao Z, Tang WH (2014) Angiotensin-converting enzyme 2 as a therapeutic target for heart failure. Curr Heart Fail Rep 11(1):58–63
78. Huentelman MJ, Grobe JL, Vazquez J, Stewart JM, Mecca AP, Katovich MJ et al (2005) Protection from angiotensin II-induced cardiac hypertrophy and fibrosis by systemic lentiviral delivery of ACE2 in rats. Exp Physiol 90(5):783–790
79. del Rio C, Malani PN (2020) 2019 novel coronavirus—important information for clinicians. JAMA. https://doi.org/10.1001/jama.2020.1490. [Epub ahead of print]
80. Oudit GY, Kassiri Z, Patel MP, Chappell M, Butany J, Backx PH et al (2007) Angiotensin II-mediated oxidative stress and inflammation mediate the age-dependent cardiomyopathy in ACE2 null mice. Cardiovasc Res 75(1):29–39
81. Chappell MC, Marshall AC, Alzayadneh EM, Shaltout HA, Diz DI (2014) Update on the angiotensin converting enzyme 2-angiotensin (1-7)-MAS receptor axis: fetal programing, sex differences, and intracellular pathways. Front Endocrinol (Lausanne) 4:201. https://doi.org/10.3389/fendo.2013.00201
82. Bahadoran A, Lee SH, Wang SM, Manikam R, Rajarajeswaran J, Raju CS et al (2016) Immune responses to influenza virus and its correlation to age and inherited factors. Front Microbiol 22(7):1841. https://doi.org/10.3389/fmicb.2016.01841
83. Coates BM, Staricha KL, Koch CM, Cheng Y, Shumaker DK, Budinger GRS et al (2018) Inflammatory monocytes drive influenza a virus-mediated lung injury in juvenile mice. J Immunol 200(7):2391–2404
84. Wang D, Hu B, Hu C, Zhu F, Liu X, Zhang J et al (2020) Clinical characteristics of 138 hospitalized patients with 2019 novel coronavirus–infected pneumonia in Wuhan, China. JAMA. https://doi.org/10.1001/jama.2020.1585. [Epub ahead of print]
85. Zhang H, Kang Z, Gong H, Xu D, Wang J, Li Z et al (2020) The digestive system is a potential route of 2019-nCov infection: a bioinformatics analysis based on single-cell transcriptomes. bioRxiv. https://doi.org/10.1101/2020.01.30.927806
86. Chen Y, Guo Y, Pan Y, Zhao ZJ (2020) Structure analysis of the receptor binding of 2019-nCoV. Biochem Biophys Res Commun. pii: S0006-291X(20)30339-9. https://doi.org/10.1016/j.bbrc.2020.02.071. [Epub ahead of print]
87. Hendaus MA, Alhammadi AH, Khalifa MS, Muneer E, Chandra P (2015) Risk of urinary tract infection in infants and children with acute bronchiolitis. Paediatr Child Health 20(5):e25–e29
88. Leal MC, Pinheiro SV, Ferreira AJ, Santos RA, Bordoni LS, Alenina N et al (2009) The role of angiotensin-(1-7) receptor Mas in spermatogenesis in mice and rats. J Anat 214(5):736–743
89. Mate SE, Kugelman JR, Nyenswah TG, Ladner JT, Wiley MR, Cordier-Lassalle T et al (2015) Molecular evidence of sexual transmission of Ebola virus. N Engl J Med 373(25):2448–2454
90. Uyeki TM, Erickson BR, Brown S, McElroy AK, Cannon D, Gibbons A et al (2016) Ebola virus persistence in semen of male survivors. Clin Infect Dis 62(12):1552–1555
91. Deen GF, Broutet N, Xu W, Knust B, Sesay FR, McDonald SLR et al (2017) Ebola RNA persistence in semen of Ebola virus disease survivors – final report. N Engl J Med 377(15):1428–1437

92. Fischer WA, Brown J, Wohl DA, Loftis AJ, Tozay S, Reeves E et al (2017) Ebola virus ribonucleic acid detection in semen more than two years after resolution of acute Ebola virus infection. Open Forum Infect Dis 4(3):ofx155. https://doi.org/10.1093/ofid/ofx155
93. Morse JS, Lalonde T, Xu S, Liu WR (2020) Learning from the past: possible urgent prevention and treatment options for severe acute respiratory infections caused by 2019-nCoV. ChemBioChem 21(5):730–738
94. Cyranoski D (2020) China is promoting coronavirus treatments based on unproven traditional medicines. Nature. https://doi.org/10.1038/d41586-020-01284-x. [Epub ahead of print]
95. Gao J, Tian Z, Yang X (2020) Breakthrough: Chloroquine phosphate has shown apparent efficacy in treatment of COVID-19 associated pneumonia in clinical studies. Biosci Trends 14(1):72–73
96. Qaradakhi T, Gadanec LK, McSweeney KR, Tacey A, Apostolopoulos V, Levinger I et al (2020) The potential actions of angiotensin-converting enzyme II (ACE2) activator diminazene aceturate (DIZE) in various diseases. Clin Exp Pharmacol Physiol 47(5):751–758
97. Wang M, Cao R, Zhang L, Yang X, Liu J, Xu M et al (2020) Remdesivir and chloroquine effectively inhibit the recently emerged novel coronavirus (2019-nCoV) in vitro. Cell Res 30(3):269–271
98. Zhang L, Liu Y (2020) Potential interventions for novel coronavirus in China: a systematic review. J Med Virol 92(5):479–490
99. Caly L, Druce JD, Catton MG, Jans DA, Wagstaff KM (2020) The FDA-approved drug ivermectin inhibits the replication of SARS-CoV-2 in vitro. Antivir Res 178:104787. https://doi.org/10.1016/j.antiviral.2020.104787
100. Zhang R, Wang X, Ni L, Di X, Ma B, Niu S et al (2020) COVID-19: melatonin as a potential adjuvant treatment. Life Sci 250:117583. https://doi.org/10.1016/j.lfs.2020.117583
101. Fantini J, Di Scala C, Chahinian H, Yahi N (2020) Structural and molecular modelling studies reveal a new mechanism of action of chloroquine and hydroxychloroquine against SARS-CoV-2 infection. Int J Antimicrob Agents:105960. https://doi.org/10.1016/j.ijantimicag.2020.105960. [Epub ahead of print]
102. Meo SA, Klonoff DC, Akram J (2020) Efficacy of chloroquine and hydroxychloroquine in the treatment of COVID-19. Eur Rev Med Pharmacol Sci 24(8):4539–4547
103. Baron SA, Devaux C, Colson P, Raoult D, Rolain JM (2020) Teicoplanin: an alternative drug for the treatment of COVID-19? Int J Antimicrob Agents 55(4):105944. https://doi.org/10.1016/j.ijantimicag.2020.105944
104. Sheahan TP, Sims AC, Leist SR, Schäfer A, Won J, Brown AJ et al (2020) Comparative therapeutic efficacy of remdesivir and combination lopinavir, ritonavir, and interferon beta against MERS-CoV. Nat Commun 11(1):222. https://doi.org/10.1038/s41467-019-13940-6
105. Grein J, Ohmagari N, Shin D, Diaz G, Asperges E, Castagna A et al (2020) Compassionate use of remdesivir for patients with severe Covid-19. N Engl J Med. https://doi.org/10.1056/NEJMoa2007016. [Epub ahead of print]

2 Neurological Complications of the COVID-19 Pandemic: What Have We Got So Far?

Isabelle Pastor Bandeira,
Marco Antônio Machado Schlindwein,
Leticia Caroline Breis,
Jean Pierre Schatzmann Peron,
and Marcus Vinícius Magno Gonçalves

Abstract

The recently emerged coronavirus named severe acute respiratory syndrome coronavirus 2 (SARS-CoV-2), the causal agent of COVID-19, is the newest threat to human health. It has already infected more than 54.5 million people worldwide, currently leading to more than 1.3 million deaths. Although it causes a mild flu-like disease in most patients, lethality may increase to more than 20% in elderly subjects, especially in those with comorbidities, like hypertension, diabetes, or lung and cardiac disease, and the mechanisms are still elusive. Common symptoms at the onset of illness are fever, cough, myalgia or fatigue, headache, and diarrhea or constipation. Interestingly, respiratory viruses have also placed themselves as relevant agents for central nervous system (CNS) pathologies. Conversely, SARS-CoV-2 has already been detected in the cerebrospinal fluid. Here, we discuss several clinical features related to CNS infection during COVID-19. Patients may progress from headaches and migraines to encephalitis, stroke, and seizures with leptomeningitis. However, the pathway used by the virus to reach the brain is still unknown. It may infect the olfactory bulb by retrograde neuronal transportation from olfactory epithelium, or it could be transported by the blood. Either way, neurological complications of COVID-19 add greatly to the complex pathophysiology of the disease. Neurological signs and symptoms must alert physicians not only to worst outcomes but also to future possible degenerative diseases.

Keywords

COVID-19 · SARS-CoV-2 · Neurology · Coronavirus · Anosmia

I. P. Bandeira · M. A. M. Schlindwein · L. C. Breis
Department of Medicine, Universidade da Região de Joinville (UNIVILLE), Joinville, Brazil

J. P. S. Peron (✉)
Department of Immunology, Institute of Biomedical Sciences, University of São Paulo (ICB-USP), São Paulo, Brazil

Scientific Platform Pasteur-USP, University of São Paulo (USP), São Paulo, SP, Brazil

Immunopathology and Allergy Post Graduate Program, School of Medicine, University of São Paulo (USP), São Paulo, SP, Brazil
e-mail: jeanpierre@usp.br

M. V. M. Gonçalves (✉)
Professor of Neurology, Department of Medicine, Universidade da Região de Joinville (UNIVILLE), Joinville, Brazil

P. C. Guest (ed.), *Clinical, Biological and Molecular Aspects of COVID-19*, Advances in Experimental Medicine and Biology 1321, https://doi.org/10.1007/978-3-030-59261-5_2

Abbreviations

ACE2	angiotensin-converting enzyme 2
ADEM	acute disseminated encephalomyelitis
CD147	CD147-spike protein
CNS	central nervous system
COVID-19	coronavirus disease 2019
CoVs	coronaviruses
CSF	cerebrospinal fluid
CTSB	cathepsin B
CTSL	cathepsin L
DPP4	dipeptidyl peptidase 4
ECG	electrocardiogram
GBC	globose basal cells
GBS	Guillain-Barré syndrome
HBC	horizontal basal cells
HCoV-OC43	human coronavirus OC43
hMPV	human metapneumovirus
hRSV	human respiratory syncytial virus
MERS	Middle East respiratory syndrome
MHV	murine hepatitis virus
NIHSS	National Institutes of Health Stroke Scale
OSN	olfactory sensory neurons
PNS	peripheral nervous system
RBD	receptor-binding domain
SARS	severe acute respiratory syndrome
SARS-CoV	severe acute respiratory syndrome coronavirus
SARS-CoV-2	severe acute respiratory syndrome coronavirus 2
SP or S protein	spike proteins
STD	smell and/or taste disorders

2.1 Introduction

Viral respiratory diseases are among the most critical problems in public health as every year they are responsible for high rates of mortality [1]. The recently emerged coronavirus named severe acute respiratory syndrome coronavirus 2 (SARS-CoV-2) is the newest threat to human health. SARS-CoV-2 is an enveloped non-segmented positive-sense RNA virus that belongs to the Coronaviridae family [3]. It is closely related to previous coronaviruses of medical relevance, known as SARS-CoV and MERS-CoV. Since December of 2019, the first cases of pneumonia started to be documented in Wuhan, China [1, 2]. It has already infected more than 54.5 million people worldwide, leading to around 1.3 million deaths. The World Health Organization (WHO) officially declared a state of public health emergency of international concern in February 2020 due to the fast spread and lethality of coronavirus disease 2019 (COVID-19) [4, 5].

Although it causes mild flu-like disease in most patients, lethality may increase to 20% in elderly subjects, especially those with comorbidities, like hypertension, diabetes, or lung and cardiac disease [6], and the mechanisms are still elusive [7]. Viral replication in lung tissue leads to direct and indirect pathologies, mainly due to an exacerbated immune response and the cytokine storm produced [5]. Common symptoms at the onset of illness are fever, cough, myalgia or fatigue, headache, and diarrhea or constipation [3, 6]. Severe cases rapidly evolve to pneumonia with "ground-glass opacities" observed by computerized tomography (CT) imaging, evidencing lung infiltration and edema.

Interestingly, respiratory viruses are also capable of causing central nervous system (CNS) pathologies as seen for human respiratory syncytial virus (hRSV) [8] or human metapneumovirus (hMPV) [9]. In fact, several studies have described the association between respiratory viral infections with neurological symptoms as febrile or afebrile seizures, status epilepticus, encephalopathies, and encephalitis [1]. With regard to the recent COVID-19 epidemic, several patients have referred to the loss of the sense of smell and taste during hospitalization. This may be an important feature of COVID-19, but it is still poorly understood.

2.2 Coronaviruses and the Nervous System

Coronaviruses invade host cells through the interaction of spike proteins (SPs) with membrane receptors, such as angiotensin-converting enzyme 2 (ACE2) [10], dipeptidyl peptidase 4 (DPP4) [11], and, most recently, CD147 [12]. The ACE2 receptor was shown to have an interferon-driven expression that can be used for SARS-CoV-2 to gain access into human cells [13]. A study by Li et al. supported these findings by demonstrating ACE2 expression in many human tissues, including the brain, and this was positively correlated with interferon levels [14]. After attachment, virus particles are internalized and fused with the cell membrane, and the RNA genome is released within the cytoplasm for protein translation and replication. In this context, viral tropism and pathology intimately correlate with the expression of the aforementioned receptors throughout the body [4].

Despite their well-known respiratory effects, coronaviruses are not always confined to the respiratory tract as they may also invade intestine [15], heart tissue [16–18], and the CNS [19–21]. For example, it is already known that the human coronavirus OC43 (HCoV-OC43) gains access to the CNS through axonal transport and neuron-to-neuron propagation in experimental models [22]. Interestingly, HCoV-43 viral loads in the brain of C57Bl/6 mice reached the same levels when intra-cranioventricular and intranasal delivery were compared. The inoculation of 10^4 $TCID_{50}$ led to a time-dependent increase in brain viral load. Moreover, viral N proteins were detected by immunofluorescence, evidencing viral migration through the neurons. Noteworthy, viral proteins were detected as early as 5 days after infection [22].

Interestingly, coronaviruses may reach the CNS through the blood, either by crossing the blood-brain barrier or via the olfactory bulb and retrograde transportation, as previously demonstrated in mice [4, 23, 24]. Additionally, it has been shown that murine hepatitis virus (MHV), another type of coronavirus, may reach the CNS after intranasal delivery. Corroborating this, ablation of the olfactory nerve cells abrogated CNS infection after nasal inoculation with MHV [25]. It is noteworthy that endothelial damage can also facilitate virus access [4].

SPs are the components of SARS-CoV-2 that interact with high affinity to human ACE2 on target cells through its receptor-binding domain (RBD) [4, 26, 27]. SPs consist of an S1 subunit, which is involved in receptor recognition, and an S2 region involved in membrane fusion [26, 27]. This latter subunit must be cleaved to properly interact with ACE2, which can be mediated by transmembrane protease, serine 2 (TMPRRS2) protease, cathepsin B (CTSB), or cathepsin L (CTSL) [28]. The interaction of SARS-CoV-2 with ACE2 receptors in neurons leads to neuronal damage without substantial inflammation [4].

The olfactory epithelium is mainly composed of olfactory sensory neurons (OSNs) responsible for odor detection and transmission to the brain [28]; globose basal cells (GBCs) responsible for neurogenesis, renewing olfactory epithelium and neurons [29, 30]; horizontal basal cells (HBCs), which are quiescent cells and a stem cell reservoir [30]; and sustentacular cells, which act as structural support for OSNs [28]. OSNs in the olfactory bulb form synapses through the cribriform plate [29].

As demonstrated by Brann et al., ACE2 and TMPRSS2 are not expressed in mature OSNs but in sustentacular cells and HBCs instead, which are believed to be the target cells of SARS-CoV-2 infection [28]. Thus, infection of sustentacular cells and HBCs may damage OSNs resulting in anosmia. Another possible pathway for viruses to infect the CNS is through the olfactory bulb [31–33]. OSNs connect the nasal cavity to the CNS through the axons, which terminate in the olfactory bulb, transposing the cribriform plate [31]. On the other hand, the olfactory bulb receives dense innervation from higher brain areas to process odor information [33] and is possibly infected by coronaviruses [28]. Therefore, the olfactory deficits may occur due to other mechanisms than olfactory epithelium damage such as higher-order olfactory structures affection, as mentioned by Brann et al. [28]. However, more information is needed to determine this.

There was clear evidence of CNS presence of SARS-CoV in brain tissues of patients with SARS in the early 2000s [19–21]. One of these studies found viral RNA of SARS-CoV infection in the brain of eight autopsied patients, and six of these had scattered red degeneration and edema in the cytoplasm of neurons from the hypothalamus and cortex [21].

In addition, Xu et al. reported a case of a 39-year-old doctor that was in contact with SARS-CoV patients and started to experience fever, chills, malaise, dizziness, and myalgia when he was admitted to the hospital [20]. After 35 days of illness onset, he died due to multiple organ failure and brain herniation. The autopsy revealed SARS-CoV RNA in the patient's brain tissue. While examining autopsied tissue samples from four SARS-CoV patients, Ding and colleagues found evidence of virus infection on the cerebrum and pituitary gland but not in the cerebellum of all four cases [19].

There is additional evidence that human coronaviruses can infect the human brain. For example, Arbour et al. described the presence of coronavirus RNA in autopsied brain samples from patients with multiple sclerosis and other neurological diseases, such as Alzheimer's, Parkinson's, schizophrenia, depression, and meningoencephalitis [34].

2.3 Neurological Manifestations of COVID-19

The manifestations of neurological symptoms in patients with COVID-19 involve the CNS, peripheral nervous system (PNS), and skeletal muscles. Severe patients commonly have neurological symptoms manifested as acute cerebrovascular diseases, consciousness impairment, and muscle injury, leading to a poor prognosis [35]. In a study carried out by Chen and colleagues, 22% of those who died from COVID-19 presented with impaired consciousness, compared with only 1% of the patients that survived [36].

CNS symptoms, such as headache, dizziness, impaired consciousness, ataxia, acute cerebrovascular disease, and epilepsy, were the main form of neurological injury in patients with the COVID-19 virus appearing in 53 out of 218 (24.8%) patients in a Chinese cohort [35]. Interestingly, patients presenting CNS involvement were associated with a more severe course of the disease [35]. On the other hand, PNS involvement occurred in 19 patients (8.9%), and hyposmia and dysgeusia were the most common symptoms, affecting 11 (5.1%) and 12 (5.6%), respectively. No differences in blood parameters were found in patients with or without PNS involvement [35].

There is still no clear evidence of severe neurological COVID-19 infection in children. However, despite the milder course of the disease, it is already known that children are susceptible to the virus with a prevalence of 1.7% in the United States [37]. To the best of our knowledge, Chacón-Aguilar et al. reported the first case of a febrile syndrome associated with neurological symptoms in early childhood [38]. It was a newborn (26 days) with two paroxysmal episodes, a 12 h fever, and nasal discharge and vomiting. On physical examination, the child was alert with mild hypertonia of the limbs and irritability and slightly increased tendon reflexes with normal tone. A nasopharyngeal swab sample tested positive for SARS-CoV-2. There were no changes in the cerebrospinal fluid (CSF). The patient was treated symptomatically and had a good outcome [38]. Considering the current epidemiological situation, fever and convulsive episodes should be suggestive of coronavirus infection and demanding early intervention and extra care of the clinical team.

In 2006, Hwang described a case of complete anosmia 3 weeks after the onset of the first symptoms of SARS-CoV infection [39]. The patient was a 27-year-old woman who presented with fever, cough, headache, myalgia, and diarrhea. Three weeks later, after upper respiratory tract improvement, the patient had complete anosmia for all kinds of odors on both sides of the nasal cavity. Although no abnormal findings that might cause anosmia were found on physical examination or via brain magnetic resonance imaging (MRI), this symptom persisted for the 2 years of follow-up without change [39]. As far as we know, this is the first case report of persisting anosmia after coronavirus infection. Further investigation and patient follow-up studies are necessary given the current reports of anosmia in COVID-19 patients. Moreover, a long period of anosmia may be linked to CNS lesions.

Hyposmia is also gaining the attention of the media and the medical community [40, 41]. A recent study on COVID-19 patients conducted by Leichien and colleagues described olfactory (85.6%) and gustatory (88%) dysfunctions of 417 patients with mild to moderate disease [42]. Among these patients that suffered from olfactory alterations, 12.6% had phantosmia and 32.4% had parosmia, and out of the 76 patients that did not suffer from nasal obstruction or rhinorrhea, 79.7% presented anosmia or hyposmia. This suggests that olfactory neuropathy may play a role in olfactory dysfunction. Smell and/or taste disorders in COVID-19 infections appear to have a variable prevalence between 5 and 48% [43], and the short-term recovery rate from anosmia or hyposmia in 59 recovered patients was 44% [42]. This factor may also be used with a small degree of diagnostic accuracy as sudden loss of smell has shown a specificity of 97% and sensitivity of 65% for COVID-19 infection [44]. Whether or not long-term persistence of anosmia will be observed should be studied further.

Patients with COVID-19 can also present with encephalopathy and other changes in their level of consciousness. Recently, three cases of encephalitis associated with SARS-CoV-2 were described. A study by the Beijing hospital was the first to find the SARS-CoV-2 in a patient's CSF [45]. In another case, the patient had encephalopathy and was positive for SARS-CoV-2 although no evidence of viral particles were detected in the CSF [46]. Therefore, if actual viral particles are present in CSF, it needs to be evaluated further.

Moriguchi and colleagues reported a case of meningitis/encephalitis in a 24-year-old man with no history of travel [47]. The suspicion of COVID-19 infection was made due to the patient's poor general condition and altered blood count, as well as a chest CT scan showing small ground-glass opacities on the right superior lobe and both sides of the inferior lobe. The disease was confirmed by means of a polymerase chain reaction (PCR) test for SARS-CoV-2 using a nasopharyngeal swab and CSF. The samples were negative for the swab and positive for the CSF [47]. Neurological findings of coronavirus infections also include cases of acute disseminated encephalomyelitis (ADEM) [48].

Chronic complications have already been described in SARS patients who presented with chronic myalgia and mood and sleep disorders [49]. However, organic neurological damage was not described in these patients. Chronic complications of coronavirus infection in the CNS have already been studied in murine models involving human coronavirus (HCoV-OC43) and mouse hepatitis virus [50, 51].

Thromboembolic and hemorrhagic phenomena can also be secondary to infection by SARS-CoV-2. In a retrospective study of 221 patients in Wuhan, China, Li et al. described 13 patients with acute cerebrovascular disease following COVID-19 infection [52]. The incidence was 5% for acute ischemic stroke, 0.5% for cerebral venous sinus thrombosis, and 0.5% for cerebral hemorrhage. Most of these patients were older (70–91 years old) and therefore had more cardiovascular and cerebrovascular risk factors [52].

Moreover, a new pattern of cerebrovascular condition known as large-vessel stroke appears to affect the young population (33–49 years old). Oxley and colleagues reported five young patients with signs of hemiplegia, dysarthria, and reduced levels of consciousness [53]. Technical imaging examinations of CT and CT angiography scans showed infarction and thrombosis in the right internal carotid artery, left middle cerebral artery, right middle cerebral artery, or right posterior cerebral artery. In addition, at the time of hospital admission of these five patients, the National Institutes of Health Stroke Scale (NIHSS) mean score was 17, consistent with a severe stroke of large vessels [53]. These events are probably related to the prothrombotic effect of the inflammatory response to viral infection [52] and may also justify the use of anticoagulant therapies such as heparin. In fact, this has been an important and critical care observation in COVID-19 patients.

Most recently, Zhao and colleagues described the first association of Guillain-Barré syndrome (GBS) with COVID-19 infection [54]. This was a 61-year-old woman with a complaint of acute weakness in both legs and severe fatigue. Despite the travel history for Wuhan, no respiratory symptoms were reported until 7 days after the onset of GBS symptoms. Oropharyngeal swabs were positive for SARS-CoV-2 by PCR assay [54]. Hence, COVID-19 appears to assume a parainfectious profile, in which GBS and viral infection occur concurrently, instead of the classic postinfectious profile. Curiously, a similar situation has already been described with GBS and Zika virus infection [55]. Cases of Miller Fisher syndrome and polyneuritis cranialis, both rare variants of GBS, have also been reported in patients with COVID-19 [56].

Another possible intriguing neurological association of COVID-19 infection is Takotsubo syndrome, which is characterized by transient left ventricular dysfunction and may be related to dysautonomia of the nervous system [57]. The case involved an 83-year-old woman who presented with typical chest pain and elevation of the ST segment in all precordial leads with deep T-wave inversions on electrocardiogram examination. The highly sensitive cardiac troponin T biomarker was elevated at 1142 ng/L, which is more than 100-fold over normal levels. Imaging tests, such as echocardiography and coronary angiography, were consistent with Takotsubo syndrome, ruling out acute coronary syndrome [58]. During hospitalization, the patient began to experience fever and bilateral opacity on lung radiography. The nasopharyngeal swab was negative for SARS-CoV-2, but the initial positive immunoglobin A and negative immunoglobulin G serology pattern proved acute infection [58]. Considering the association of Takotsubo syndrome with neurological disorders [57, 59], we believe that SARS-CoV-2 infection may induce autonomic dysfunctions. However, more reports are needed to exclude a stress-induced complication.

2.4 Conclusions

Although COVID-19 is mostly described as a lung disease, causing pneumonia and severe acute respiratory disease, several reports have indicated that patients may also display signs and symptoms related to effects on other organs. Here, we have discussed several CNS-related features, from mild symptoms such as headache, fatigue, and ataxia to more severe conditions as encephalitis, GBS, and stroke (Table 2.1 and Fig. 2.1). These have been observed in several patients, especially at the beginning of the disease. Although the virus has been detected in the CNS of patients, the mechanisms by which it reaches the brain are still unknown. It is possible that it directly infects the CNS through the olfactory epithelium or through the blood, either alone or transported by Trojan horse cells, as T lymphocytes may be infected by the virus with no productive replication (Fig. 2.1).

Due to the huge spread of the virus, and even if a small proportion of patients display neurological symptoms, it is important that the medical and research community be aware not only of these acute and critical problems like encephalitis and stroke but also of the possibility of further chronic or degenerative complications. It is important to be prepared to respond to such complications as occurred with the post-epidemic complications of the Zika virus [60–62] and von Economo's famous encephalitis lethargy followed by Parkinsonism symptoms in those who survived the Spanish flu pandemic [63].

Table 2.1 Summary of neurological manifestations found in coronaviruses infections. The first section shows the general findings of other coronaviruses that infect humans. The second section shows the specific findings of SARS-CoV-2, responsible for COVID-19, described in this paper

Neurological findings in human coronavirus infections			
Pathogens		Clinical manifestations	Reference
1.	Coronavirus (HCoV-229E, HCoV-OC43, SARS-CoV, and HCoV-OC43)	Acute: Febrile seizures, convulsions, loss of consciousness, ataxia, anosmia or hyposmia, encephalomyelitis, encephalitis, myelitis, neuritis, and acute disseminated encephalomyelitis (ADEM), headache	Bohmwald et al. 2018 [1]
		Chronic: Myalgia and mood and sleep disorders	Hwang 2006 [39]
			Ann et al. 2003 [48]
2.	COVID-19 (SARS-CoV-2) exclusively	Central nervous system: Headache, dizziness, impaired consciousness, ataxia, and epilepsy	Mao et al. 2020 [35]
		Peripheral nervous system: Dysgeusia, anosmia or hyposmia, peripheral neuropathy, and muscle injury	Chacón-Aguilar et al. 2020 [38]
			Lechien et al. 2020 [42]
		Encephalopathy with[1,3] and without[2] evidence of virus in central nervous system	1. Zhou L, et al. 2020 [45]
			2. Filatov, A. et al. 2020 [46]
			3.Moriguchi T, et al. 2020 [47]
		Stroke and venous thrombosis brain	Li et al. 2020 [14]
			Oxley et al. 2020 [53]
		Guillain-Barré syndrome including, Miller Fisher syndrome, polyneuritis cranialis, and Takotsubo syndrome	Zhao et al. 2020 [54]
			Gutiérrez-Ortiz et al. 2020 [56]
			Meyer et al. 2020 [58]

The involvement of the CNS was associated with more severe disease when compared to patients with no CNS involvement [35]. This neurotropism characteristic was observed with other human coronaviruses like SARS-CoV [19, 20] and HCoV OC43 [48, 50]. Thus, it is important to prioritize and to individualize treatment protocols based on the severity of the disease and the predominant organ systems involved. We recommend that in the presence of ataxia, loss of consciousness, convulsion, status epilepticus, encephalitis, myelitis, or neuritis [1, 35], differential diagnosis of COVID-19 should be considered, especially during the current pandemic.

The COVID-19 outbreak has spread worldwide, so careful surveillance is essential to monitor and control the disease. This is critical as clinical conditions of the patients can worsen rapidly, leading to respiratory failure. In addition, CNS-related symptoms may indicate a poor prognosis, and it is still not clear whether long-lasting impairments will be observed. Therefore, a fast and accurate diagnosis is necessary to allow the most effective interventions in a precision medicine approach. Such approaches will be more effective when new treatments that prevent or minimize the effects of COVID-19 become available.

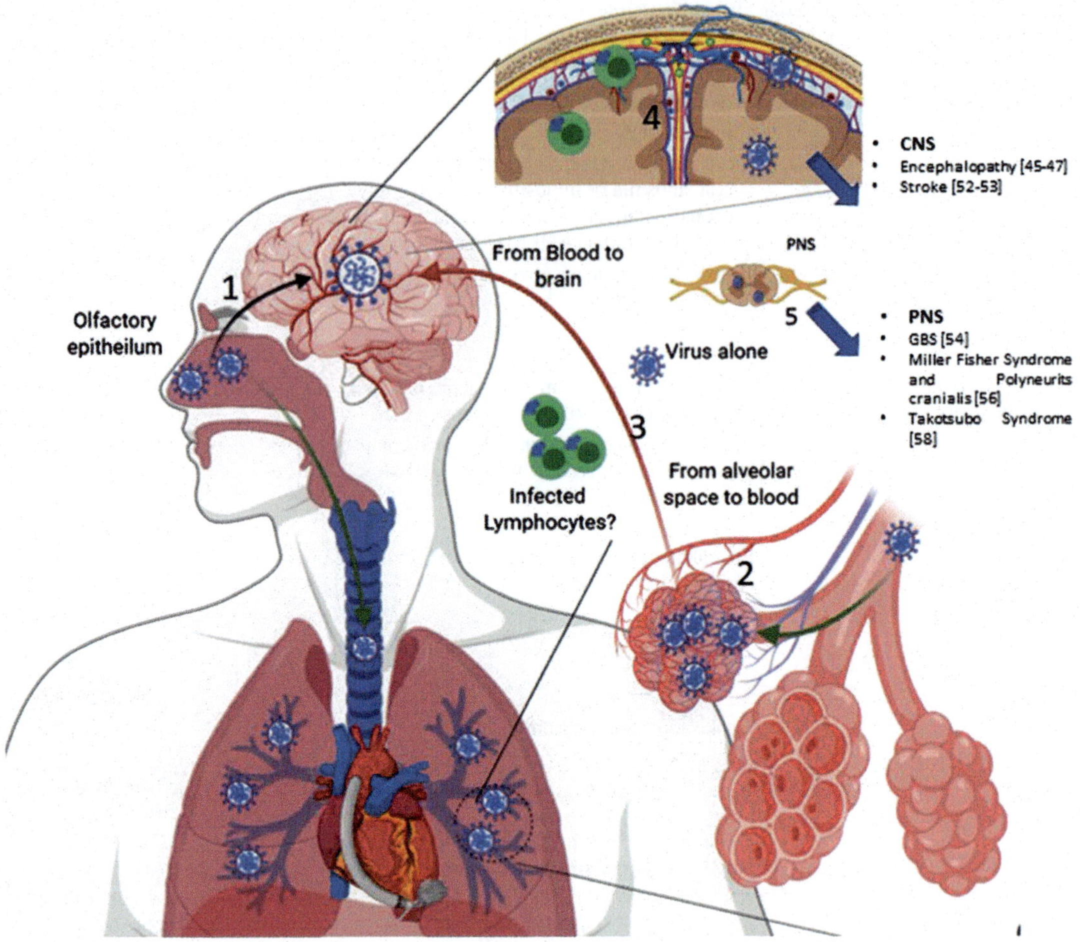

Fig. 2.1 Scheme of SARS-CoV-2 neurological complications. SARS-CoV-2 may reach the central nervous system either by direct infection and retrograde transportation from olfactory neurons (1) or by a hematogenous route (2). After replication in the lungs, the virus may reach the blood either alone or inside infected cells, such as lymphocytes (3). The virus then target the central nervous system by crossing the blood-brain barrier and reaching the meninges, brain parenchyma, and cerebrospinal fluid (4). The scheme was elaborated by the authors using www.biorender.com

Conflicts of Interest/Competing Interests The authors report no disclosures.

Availability of Data and Material We utilized publicly available data.

FundingJPSP is funded by FAPESP grants #2017/26170–0 and 2017/22504–1 and CNPq (301287/2016–3). No other targeted funding is reported.

References

1. Bohmwald K, Gálvez N, Ríos M, Kalergis A (2018) Neurologic alterations due to respiratory virus infections. Front Cell Neurosci 12:386. https://doi.org/10.3389/fncel.2018.00386
2. Guan W, Ni Z, Hu Y, Liang W, Ou C, He J et al (2020) Clinical characteristics of coronavirus disease 2019 in China. N Engl J Med 382(18):1708–1720
3. Huang C, Wang Y, Li X, Ren L, Zhao J, Hu Y et al (2020) Clinical features of patients infected with 2019 novel coronavirus in Wuhan, China. Lancet 395(10223):497–506
4. Baig A, Khaleeq A, Ali U, Syeda H (2020) Evidence of the COVID-19 virus targeting the CNS: tissue distribution, host–virus interaction, and proposed neurotropic mechanisms. ACS Chem Neurosci 11(7):995–998
5. Qin C, Zhou L, Hu Z, Zhang S, Yang S, Tao Y et al (2020) Dysregulation of immune response in patients with COVID-19 in Wuhan, China. Clin Infect Dis. pii: ciaa248. https://doi.org/10.1093/cid/ciaa248. [Epub ahead of print]
6. Sun D, Li H, Lu X, Xiao H, Ren J, Zhang F et al (2020) Clinical features of severe pediatric patients with coronavirus disease 2019 in Wuhan: a single center's observational study. World J Pediatr. https://doi.org/10.1007/s12519-020-00354-4. [Epub ahead of print]
7. Pinto B, Oliveira A, Singh Y, Jimenez L, Goncalves A, Ogava R et al (2020) ACE2 expression is increased in the lungs of patients with comorbidities associated with severe COVID-19. medRxiv. https://doi.org/10.1101/2020.03.21.20040261
8. Morichi S, Kawashima H, Ioi H, Ushio M, Yamanaka G, Kashiwagi Y et al (2009) Cerebrospinal fluid NOx (nitrite/nitrate) in RSV-infected children with CNS symptoms. J Infect 59(4):299–301
9. Schildgen O, Glatzel T, Geikowski T, Scheibner B, Simon A, Bindl L et al (2005) Human metapneumovirus RNA in encephalitis patient. Emerg Infect Dis 11(3):467–470
10. Hoffmann M, Kleine-Weber H, Schroeder S, Krüger N, Herrler T, Erichsen S et al (2020) SARS-CoV-2 cell entry depends on ACE2 and TMPRSS2 and is blocked by a clinically proven protease inhibitor. Cell 181(2):271–280.e8. https://doi.org/10.1016/j.cell.2020.02.052
11. Raj V, Mou H, Smits S, Dekkers D, Müller M, Dijkman R et al (2013) Dipeptidyl peptidase 4 is a functional receptor for the emerging human coronavirus-EMC. Nature 495(7440):251–254
12. Wang K, Chen W, Zhou Y, Lian J, Zhang Z, Du P et al (2020) SARS-CoV-2 invades host cells via a novel route: CD147-spike protein. medRxiv. https://doi.org/10.1101/2020.03.14.988345
13. Ziegler CGK, Allon SJ, Nyquist SK, Mbano IM, Miao VN, Tzouanas C et al (2020) SARS-CoV-2 receptor ACE2 is an interferon-stimulated gene in human airway epithelial cells and is detected in specific cell subsets across tissues. Cell (In Press, Journal Pre-proof). https://doi.org/10.1016/j.cell.2020.04.035
14. Li M, Li L, Zhang Y, Wang X (2020) Expression of the SARS-CoV-2 cell receptor gene ACE2 in a wide variety of human tissues. Infect Dis Poverty 9(1):45. https://doi.org/10.1186/s40249-020-00662-x
15. Leung W, To K, Chan P, Chan H, Wu A, Lee N et al (2003) Enteric involvement of severe acute respiratory syndrome-associated coronavirus infection. Gastroenterology 125(4):1011–1017
16. Dimitrov D (2003) The secret life of ACE2 as a receptor for the SARS virus. Cell 115(6):652–653
17. Gu J, Korteweg C (2007) Pathology and pathogenesis of severe acute respiratory syndrome. Am J Pathol 170(4):1136–1147
18. Oudit G, Kassiri Z, Jiang C, Liu P, Poutanen S, Penninger J et al (2009) SARS-coronavirus modulation of myocardial ACE2 expression and inflammation in patients with SARS. Eur J Clin Investig 39(7):618–625
19. Ding Y, He L, Zhang Q, Huang Z, Che X, Hou J et al (2004) Organ distribution of severe acute respiratory syndrome (SARS) associated coronavirus (SARS-CoV) in SARS patients: implications for pathogenesis and virus transmission pathways. J Pathol 203(2):622–630
20. Xu J, Zhong S, Liu J, Li L, Li Y, Wu X et al (2005) Detection of severe acute respiratory syndrome coronavirus in the brain: potential role of the chemokine mig in pathogenesis. Clin Infect Dis 41(8):1089–1096
21. Gu J, Gong E, Zhang B, Zheng J, Gao Z, Zhong Y et al (2005) Multiple organ infection and the pathogenesis of SARS. J Exp Med 202(3):415–424
22. Dubé M, Le Coupanec A, Wong A, Rini J, Desforges M, Talbot P (2018) Axonal transport enables neuron-to-neuron propagation of human coronavirus OC43. J Virol 92(17):pii: e00404–pii: e00418. https://doi.org/10.1128/JVI.00404-18
23. Bleau C, Filliol A, Samson M, Lamontagne L (2015) Brain invasion by mouse hepatitis virus depends on impairment of tight junctions and beta interferon

production in brain microvascular endothelial cells. J Virol 89(19):9896–9908
24. Netland J, Meyerholz DK, Moore S, Cassell M, Perlman S (2008) Severe acute respiratory syndrome coronavirus infection causes neuronal death in the absence of encephalitis in mice transgenic for human ACE2. J Virol 82(15):7264–7275
25. Perlman S, Evans G, Afifi A (1990) Effect of olfactory bulb ablation on spread of a neurotropic coronavirus into the mouse brain. J Exp Med 172(4):1127–1132
26. Li W, Zhang C, Sui J, Kuhn J, Moore M, Luo S et al (2005) Receptor and viral determinants of SARS-coronavirus adaptation to human ACE2. EMBO J 24(8):1634–1643
27. Hulswit R, de Haan C, Bosch B (2016) Coronavirus spike protein and tropism changes. Adv Virus Res 96:29–57
28. Brann D, Tsukahara T, Weinreb C, Logan D, Datta S (2020) Non-neural expression of SARS-CoV-2 entry genes in the olfactory epithelium suggests mechanisms underlying anosmia in COVID-19 patients. medRxiv. https://doi.org/10.1101/2020.03.25.009084
29. Choi R, Goldstein BJ (2018) Olfactory epithelium: cells, clinical disorders, and insights from an adult stem cell niche. Laryngoscope Investig Otolaryngol 3(1):35–42
30. Fletcher RB, Das D, Gadye L, Street KN, Baudhuin A, Wagner A et al (2017) Deconstructing olfactory stem cell trajectories at single-cell resolution. Cell Stem Cell 20(6):817–830.e8
31. Durrant DM, Ghosh S, Klein RS (2016) The olfactory bulb: an immunosensory effector organ during neurotropic viral infections. ACS Chem Neurosci 7(4):464–469
32. Doty RL (2019) Systemic diseases and disorders. Handb Clin Neurol 164:361–387
33. Wen P, Rao X, Xu L, Zhang Z, Jia F, He X et al (2019) Cortical organization of centrifugal afferents to the olfactory bulb: mono- and trans-synaptic tracing with recombinant neurotropic viral tracers. Neurosci Bull 35(4):709–723
34. Arbour N, Day R, Newcombe J, Talbot P (2000) Neuroinvasion by human respiratory coronaviruses. J Virol 74(19):8913–8921
35. Mao L, Wang M, Chen S, He Q, Chang J, Hong C et al (2020) Neurological manifestations of hospitalized patients with COVID-19 in Wuhan, China: a retrospective case series study. JAMA Neurol. https://doi.org/10.1001/jamaneurol.2020.1127. [Epub ahead of print]
36. Chen T, Wu D, Chen H, Yan W, Yang D, Chen G et al (2020) Clinical characteristics of 113 deceased patients with coronavirus disease 2019: retrospective study [published correction appears in BMJ. 2020 Mar 31;368:m1295]. BMJ 368:m1091. https://doi.org/10.1136/bmj.m1091
37. CDC COVID-19 Response Team (2020) Coronavirus disease 2019 in children – United States, February 12–April 2, 2020. MMWR Morb Mortal Wkly Rep 69(14):422–426
38. Chacón-Aguilar R, Osorio-Cámara J, Sanjurjo-Jimenez I, González-González C, López-Carnero J, Pérez-Moneo-Agapito B (2020) COVID-19: fever syndrome and neurological symptoms in a neonate. An Pediatr (Engl Ed) 27. https://doi.org/10.1016/j.anpede.2020.04.001. [Epub ahead of print]
39. Hwang C (2006) Olfactory neuropathy in severe acute respiratory syndrome: report of A case. Acta Neurol Taiwanica 15(1):26–28
40. Stone J (2020) There's an unexpected loss of smell and taste in coronavirus patients Forbes. https://www.forbes.com/sites/judystone/2020/03/20/theres-an-unexpected-loss-of-smell-and-taste-in-coronavirus-patients/#40567ff45101
41. Hopkins C, Kumar N (2020) Loss of sense of smell as a marker of COVID-19 infection. London: Ear, nose, and throat surgery professional membership in the United Kingdom (ENT UK). https://www.entuk.org/covid-19
42. Lechien J, Chiesa-Estomba C, De Siati D, Horoi M, LeBon S, Rodriguez A et al (2020) Olfactory and gustatory dysfunctions as a clinical presentation of mild-to-moderate forms of the coronavirus disease (COVID-19): a multicenter European study. Eur Arch Otorhinolaryngol. https://doi.org/10.1007/s00405-020-05965-1. [Epub ahead of print]
43. Beltrán-Corbellini Á, Chico-García J, Martínez-Poles J, Rodríguez-Jorge F, Natera-Villalba E, Gómez-Corral J et al Acute-onset smell and taste disorders in the context of Covid-19: a pilot multicenter PCR-based case-control study. Eur J Neurol. https://doi.org/10.1111/ene.14273. [Epub ahead of print]
44. Haehner A, Draf J, Draeger S, de With K, Hummel T (2020) Predictive value of sudden olfactory loss in the diagnosis of COVID-19. medRxiv. https://doi.org/10.1101/2020.04.27.20081356
45. Zhou L, Zhang M, Gao J, Wang J (2020) Sars-Cov-2: underestimated damage to nervous system. Travel Med Infect Dis:101642. https://doi.org/10.1016/j.tmaid.2020.101642. [Epub ahead of print]
46. Filatov A, Sharma P, Hindi F, Espinosa P (2020) Neurological complications of coronavirus disease (COVID-19): encephalopathy. Cureus 12(3):e7352. https://doi.org/10.7759/cureus.7352
47. Moriguchi T, Harii N, Goto J, Harada D, Sugawara H, Takamino J et al (2020) A first case of meningitis/encephalitis associated with SARS-Coronavirus-2. Int J Infect Dis 94:55–58
48. Ann Yeh E, Collins A, Cohen M, Duffner P, Faden H (2003) Detection of coronavirus in the central nervous system of a child with acute disseminated encephalomyelitis. Pediatrics 113(1):e73–e76
49. Moldofsky H, Patcai J (2011) Chronic widespread musculoskeletal pain, fatigue, depression and disordered sleep in chronic post-SARS syndrome; a case-controlled study. BMC Neurol 11:37. https://doi.org/10.1186/1471-2377-11-37
50. Jacomy H, Fragoso G, Almazan G, Mushynski W, Talbot P (2006) Human coronavirus OC43 infection

induces chronic encephalitis leading to disabilities in BALB/C mice. Virology 349(2):335–346
51. Hosking M, Lane T (2010) The pathogenesis of murine coronavirus infection of the central nervous system. Crit Rev Immunol 30(2):119–130
52. Li Y, Wang M, Zhou Y, Chang J, Xian Y, Mao L et al (2020) Acute cerebrovascular disease following COVID-19: a single center, retrospective, observational study. SSRN Electron J. https://doi.org/10.2139/ssrn.3550025
53. Oxley T, Mocco J, Majidi S, Kellner C, Shoirah H, Singh I et al (2020) Large-vessel stroke as a presenting feature of Covid-19 in the young. N Engl J Med 382(20):e60. https://doi.org/10.1056/NEJMc2009787
54. Zhao H, Shen D, Zhou H, Liu J, Chen S (2020) Guillain-Barré syndrome associated with SARS-CoV-2 infection: causality or coincidence? Lancet Neurol 19(5):383–384
55. Siu R, Bukhari W, Todd A, Gunn W, Huang QS, Timmings P (2016) Acute Zika infection with concurrent onset of Guillain-Barre syndrome. Neurology 87:1623–1624
56. Gutiérrez-Ortiz C, Méndez A, Rodrigo-Rey S, San Pedro-Murillo E, Bermejo-Guerrero L, Gordo-Mañas R et al (2020) Miller Fisher Syndrome and polyneuritis cranialis in COVID-19. Neurology. pii: 10.1212/WNL.0000000000009619. https://doi.org/10.1212/WNL.0000000000009619. [Epub ahead of print]
57. Ramos Ttito A, TipismanaBarbaran M, Najar Trujillo N, Escalaya Advincula A, Icumina Arevalo K, Moscol Ato G et al (2017) Takotsubo cardiomyopathy as a sequel of severe Dysautonomia from Guillain-Barré syndrome. J Neurol Sci 381:919–920
58. Meyer P, Degrauwe S, Delden C, Ghadri J, Templin C (2020) Typical takotsubo syndrome triggered by SARS-CoV-2 infection. Eur Heart J. pii: ehaa306. https://doi.org/10.1093/eurheartj/ehaa306. [Epub ahead of print]
59. Martins R, Barbarot N, Coquerel N, Baruteau A, Kolev I, Vérin M (2010) Takotsubo cardiomyopathy associated with Guillain–Barré syndrome: a differential diagnosis from dysautonomia not to be missed. J Neurol Sci 291(1–2):100–102
60. Nunes M, Carlini C, Marinovic D, Neto F, Fiori H, Scotta M et al (2016) Microcephaly and Zika virus: a clinical and epidemiological analysis of the current outbreak in Brazil. J Pediatr 92(3):230–240
61. Cugola F, Fernandes I, Russo F, Freitas B, Dias J, Guimarães K et al (2016) The Brazilian Zika virus strain causes birth defects in experimental models. Nature 534(7606):267–271
62. Figueiredo C, Barros-Aragão F, Neris R, Frost P, Soares C, Souza I et al (2019) Zika virus replicates in adult human brain tissue and impairs synapses and memory in mice. Nat Commun 10(1):3890. https://doi.org/10.1038/s41467-019-11866-7
63. Lutters B, Foley P, Koehler P (2018) The centennial lesson of encephalitis lethargica. Neurology 90(12):563–567

3 Endothelial Dysfunction as a Primary Consequence of SARS-CoV-2 Infection

Genevieve Mezoh and Nigel J. Crowther

Abstract

A number of different viral species are known to have effects on the endothelium. These include dengue, Ebola, Marburg, Lassa fever, yellow fever and influenza viruses, cytomegalovirus and coronaviruses. There are currently seven human endemic coronaviruses, all of which cause respiratory diseases and bind to receptors found within the endothelium. The severe acute respiratory syndrome coronavirus 2 (SARS-CoV-2) that causes the coronavirus disease 2019 (COVID-19) is highly infectious. Like its predecessor, SARS-CoV, it binds to angiotensin-converting enzyme-2 (ACE-2), which is expressed in many cell types, particularly in the lung, including endothelial cells. The initiation of a cytokine storm by the virus along with infection of endothelial cells leads to apoptosis and structural and functional changes that attenuate vascular integrity in many organs including the lungs, heart, liver and kidney. Endothelial damage also enhances the coagulation pathway leading to thrombus formation in major vessels and capillaries. Infection with SARS-CoV-2 has an adverse outcome for individuals with particular comorbid diseases, e.g. hypertension, obesity, type 2 diabetes and cardiovascular disease. It is possible that this is due to the presence of pre-existing endothelial dysfunction and systemic inflammation in subjects with these diseases. Therapies for COVID-19 that target the endothelium, the inflammatory response and the coagulation pathway are currently under trial.

Keywords

Endothelial dysfunction · Inflammation · COVID-19 · SARS-CoV-2

G. Mezoh (✉)
Department of Chemical Pathology, University of the Witwatersrand, Faculty of Health Sciences, Johannesburg, South Africa

N. J. Crowther
Department of Chemical Pathology, University of the Witwatersrand, Faculty of Health Sciences, Johannesburg, South Africa

Department of Chemical Pathology, National Health Laboratory Service, Johannesburg, South Africa

3.1 Introduction

Coronaviruses are a group of positive-sense single-stranded RNA viruses that were first identified over 50 years ago and known to infect both birds and mammals [1]. In humans, coronaviruses cause the common cold [2], severe acute respiratory syndrome (SARS),

P. C. Guest (ed.), *Clinical, Biological and Molecular Aspects of COVID-19*, Advances in Experimental Medicine and Biology 1321, https://doi.org/10.1007/978-3-030-59261-5_3

Middle East respiratory syndrome (MERS), gastroenteritis and hepatic and neurological disorders [3, 4]. More recently, a new strain of coronavirus, severe acute respiratory syndrome coronavirus 2 (SARS-CoV-2), was identified as the causative agent in the current pandemic of coronavirus disease 2019 (COVID-19) [5]. This is an atypical pneumonia in which patients present with shortness of breath and diarrhoea, as well as cold and flulike symptoms such as fatigue, fever, headaches, sore throat, cough, muscle aches, nausea, loss of taste or smell and runny nose [6, 7]. As of July 23, 2020, according to statistics from the World Health Organization (WHO), just over 15 million people had been infected with SARS-CoV-2 globally, of which the Americas accounted for 53% of confirmed cases, while 4% of confirmed cases were in Africa [8].

The SARS-CoV-2 is a spherical enveloped virus consisting of four structural proteins, spike (S), envelope (E), membrane (M) and nucleocapsid (N), eight accessory proteins (3a, 3b, 6, 7a, 7b, 8a, 8b and 9b) and 16 nonstructural proteins (nsp1–16) [9]. Stemming from the family of *Coronaviridae*, SARS-CoV-2 shares up to 80% sequence similarity to its predecessor, SARS-CoV [10]. Although SARS-CoV-2 is more infectious than SARS-CoV, it has a lower case fatality rate of 3.7%, compared to 10% for SARS-CoV [11]. The other five coronaviruses known to be found in humans are MERS-CoV, HCoV-229E, HCoV-NL63, HCoV-OC43 and HCoV-HKU1.

The endothelium consists of a layer of squamous cells that lines the internal surface of blood and lymph vessels. Acting as the interface between circulating blood and the wall of blood vessels, it has several functions and is regarded as an organ. Some of these functions include maintaining vascular integrity, inducing angiogenesis, controlling coagulation and reducing inflammation [12]. The endothelium equally plays a protective role, shielding the organs from damage, as it controls the movement of substances across the vessels into and out of the tissues. Damage to the endothelium or prolonged activation of endothelial cells via exposure to high cytokine levels, particularly of interleukin-6 (IL-6) and tumour necrosis factor (TNF), can cause endothelial dysfunction leading to activation of the coagulation pathways, enhanced inflammation and loss of vascular integrity. Recently, a study conducted by Varga et al. [13] showed that SARS-CoV-2 is able to infect endothelial cells resulting in endotheliitis in several organs including the lung, heart, kidney, liver and small intestine.

It is now known that SARS-CoV-2 and a number of other viruses can cause endothelial dysfunction by both direct and indirect methods, and this will be the focus of the current chapter.

3.2 Viruses and the Endothelium

Endothelial dysfunction is a common consequence of a number of different viral infections. One of the most intensely researched forms of virally induced endothelial damage is that associated with human immunodeficiency virus (HIV). It is thought to be a major cause of the increased prevalence of cardiovascular diseases (CVDs) observed in subjects infected with the virus [14]. This process is not thought to involve HIV entry into endothelial cells but may be due to the HIV accessory proteins, Nef and Tat. These viral proteins, particularly Nef, are thought to cause endothelial dysfunction via activation of the NF-*k*B pathway and by reactive oxygen species (ROS) generation within the endothelium [15]. Activation of endothelial cells results in the increased expression of adhesion molecules such as intercellular adhesion molecule-1 (ICAM-1), vascular cell adhesion molecule-1 (VCAM-1) and endothelial leukocyte adhesion molecule-1 (E-selectin) [16] and the propagation of atherosclerosis. The soluble versions of these molecules are used as biomarkers of endothelial dysfunction and have been shown to be present at high levels in the serum of subjects infected with HIV [17]. The causative association of HIV with endothelial dysfunction is further highlighted by the observation that antiretroviral therapy causes a reduction in the serum levels of markers of endothelial dysfunction [18, 19].

A number of other viruses have been associated with endothelial dysfunction and atherosclerosis [20]. The most prominent of these is cytomegalovirus (CMV), also known as human herpesvirus 5. This virus has been detected in multiple cell types within the vasculature, including the endothelium, causing vascular inflammation, endothelial dysfunction and the progression of atherosclerotic plaque formation [21]. Influenza viruses have also been implicated in atherosclerosis. The virus infects endothelial cells leading to apoptosis, and it also initiates a cytokine storm that activates the molecular pathways involved in plaque formation [22]. Furthermore, increased vascular permeability in the lungs is caused by influenza-induced endothelial dysfunction leading to pulmonary oedema.

Viral haemorrhagic fevers (VHFs) are a group of severe diseases caused by four families of RNA viruses, i.e. filoviruses, flaviviruses, arenaviruses, and bunyaviruses. These diseases include Ebola virus disease, Marburg disease, Lassa fever, yellow fever and dengue fever. A characteristic feature of many of these diseases is endothelial dysfunction and associated vascular damage and leakage. Haemorrhage is present but varies in severity across these viral types [23]. The mechanisms underlying the endothelial damaged caused by these viruses include infection of the endothelium and of monocytes, macrophages and dendritic cells leading to the production of inflammatory cytokines that induce endothelial dysfunction [24, 25]. A further mechanism specific to the flaviviruses involves production of the viral peptide non-structural protein 1 (NS1) [26]. This secreted peptide blocks the function of endothelial glycocalyx leading to reduced cell adhesion and increased permeability of the endothelium [26, 27]. It is interesting to note that the Zika virus is also a flavivirus, and studies of human foetal tissue have shown that Zika viral peptides can be found in brain endothelial cells [28].

It is therefore clear that a number of different viral species can negatively affect endothelial function, and recent studies have confirmed that SARS-CoV-2, and possibly other coronaviruses, are endothelial-tropic.

3.3 Infection of Endothelial Cells by SARS-CoV-2

Electron microscopies of lung and kidney tissue obtained from subjects who died from COVID-19 have demonstrated the presence of SARS-CoV-2 in endothelial cells from both tissue types [29]. The virus gains entry into cells by binding to angiotensin-converting enzyme-2 (ACE-2) via the S1 region of the viral S protein. The S protein is then cleaved by the host cell transmembrane serine protease 2 (TMPRSS2) at the boundary of the S1 and S2 subunits, with the latter then mediating fusion of the viral and host cell membranes [30]. The ACE-2 protein is expressed in several organs such as the kidney, intestine, heart and lungs, as well as on the surface of lung alveolar epithelial cells, enterocytes of the small intestine, smooth muscle cells and endothelial cells [31]. The ACE-2 protein is also used by SARS-CoV and HCoV-NL63 as a receptor for host cell entry [32]. The host cell protein used for viral entry by MERS-CoV is dipeptidyl peptidase-4 (DPP-4) [33], while aminopeptidase N (APN) is the receptor of HCoV-229E [34]. The receptor interactions of the coronavirus family are complex, and each virus can recognize multiple host cell proteins [35]. It is interesting to note that in humans, coronavirus receptors are often peptidases, but the biological significance of this is not fully understood. All the proteins used by endemic human coronaviruses to access host cells are widely expressed across various tissues, but, most importantly, each of these viral receptors is expressed in endothelial cells. This suggests that all of these viruses are capable of causing some degree of endothelial dysfunction, and it is interesting to note the similar symptomology of subjects infected with the different coronavirus species [36].

3.4 Signs of Endothelial Dysfunction in COVID-19 Cases

Histological studies have shown direct evidence of structural changes to, and SARS-CoV-2 infection of, endothelial cells [13, 29]. Clinical symptoms of infection with the virus also suggest endothelial involvement in COVID-19. Endothelial dysfunction can also be detected using blood-based biomarkers, but only a few studies have investigated these in subjects infected with SARS-CoV-2 [37, 38]. Such biomarkers include soluble forms of endothelial leukocyte adhesion molecule-1 (E-selectin), intercellular adhesion molecule-1 (ICAM-1), vascular cell adhesion molecule-1 (VCAM-1) and vascular adhesion protein-1 (VAP-1) and von Willebrand factor [37, 38]. The four adhesion molecules function in forming and stabilizing cell-to-cell interactions and also promote leukocyte transmigration through the endothelium [39], while von Willebrand factor plays an important role in haemostasis [40]. All these molecules are expressed by endothelial cells, and the levels of the adhesion molecules rise in the presence of inflammation [41]. Each of these biomarkers, with the exception of VAP-1, has been used to demonstrate the presence of endothelial dysfunction in subjects with HIV infection [39]. The plasma levels of soluble VCAM-1 and ICAM-1, as well as the numbers of circulating endothelial cells, have also been shown to be elevated in subjects with dengue virus infection [42, 43], a known cause of endothelial dysfunction.

Currently, only a small number of studies have analysed the levels of blood-based biomarkers of endothelial dysfunction in subjects with COVID-19. In an investigation involving 39 COVID-19 patients (9 with severe disease and 30 with mild disease) and 32 uninfected control participants, serum levels of fractalkine, ICAM-1, VCAM-1 and VAP-1 were significantly higher in cases than controls, with the levels of each biomarker increasing with disease severity and decreasing with disease recovery [37] (Fig. 3.1). Fractalkine (CX3CL1) is a chemokine and adhesion molecule that is also expressed on endothelial cells and has similar functions to ICAM-1 and VCAM-1 [44].

Two studies have investigated the level of circulating endothelial cells (CECs) in blood from COVID-19 patients [45, 46]. These cells are released from damaged endothelium and have been used as markers of vascular trauma. Their levels have been shown to be increased in a variety of disease conditions including CVD, inflammation and infection [47]. Blood levels of CECs have also been shown to be increased in the presence of untreated HIV infection [48]. In the first study, CECs were measured in 66 COVID-19 cases and 30 uninfected subjects [46] (Fig. 3.2), while in the second study, CECs were measured in blood taken from 30 COVID-19 patients and 6 healthy control subjects [45]. In both investigations, the CECs were found at significantly higher levels in the COVID-19 cases upon admission compared with uninfected subjects. Interestingly, in the first study, COVID-19 patients that had been treated with only anticoagulation therapy ($n = 10$) had lower CEC numbers than in untreated cases ($n = 47$), and COVID-19 patients treated with both an anticoagulation therapy and ACE inhibitors or angiotensin receptor blockers had the lowest CEC numbers [46].

Measuring these biomarkers of endothelial dysfunction in subjects with COVID-19 may aid in the assessment of disease severity and in monitoring the effectiveness of any treatment being administered, particularly in patients with pre-existing conditions associated with endothelial dysfunction.

The SARS-CoV-2 virus is able to infect endothelial cells, and studies show that markers of endothelial dysfunction are increased in subjects with COVID-19. It is important to understand how the virus is able to modulate endothelial function and how this may affect disease outcomes. A number of studies have investigated these topics and will be discussed in the following section of this chapter.

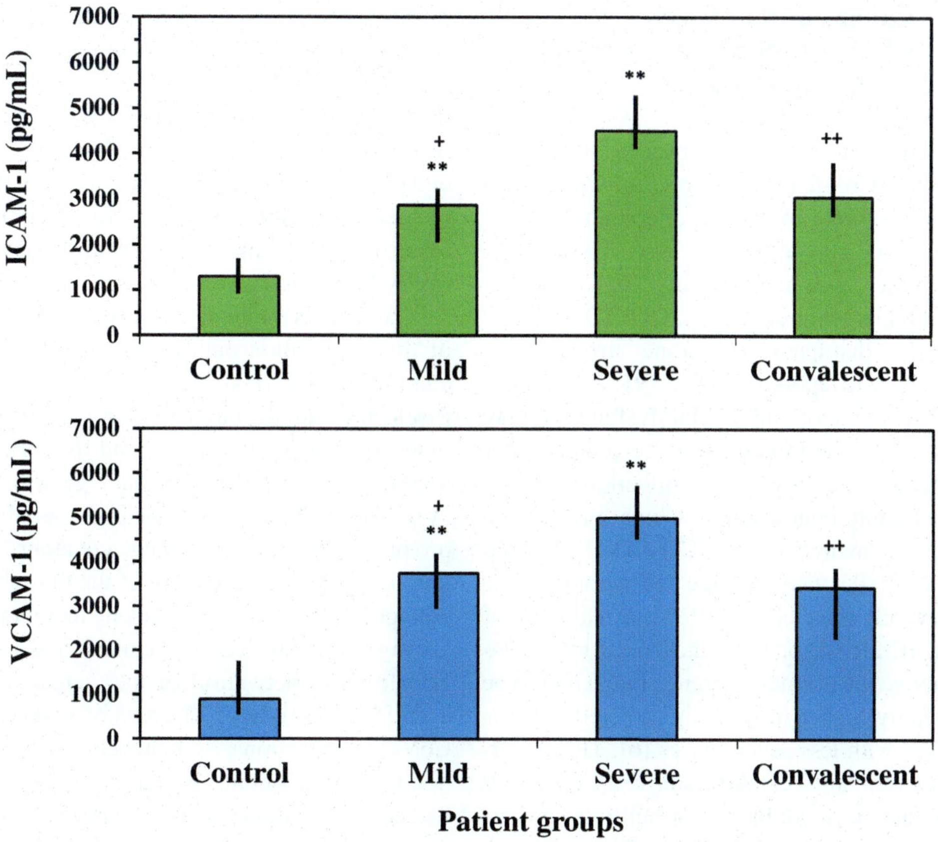

Fig. 3.1 Serum VCAM-1 and ICAM-1 levels in 32 non-infected control subjects and 30 mild and 9 severe COVID-19 cases. Data were also obtained from the severe cases of COVID-19 after convalescence. The data are given as median with interquartile range; **$p < 0.01$ vs. controls, $^{+}p < 0.05$ and $^{++}p < 0.01$ vs severe. (Data taken from [37])

Fig. 3.2 Blood levels of circulating endothelial cells (CECs) in 30 non-infected and 66 COVID-19 cases. The data are given as median with interquartile range;*$p = 0.008$ vs no COVID-19. (Data taken from [46])

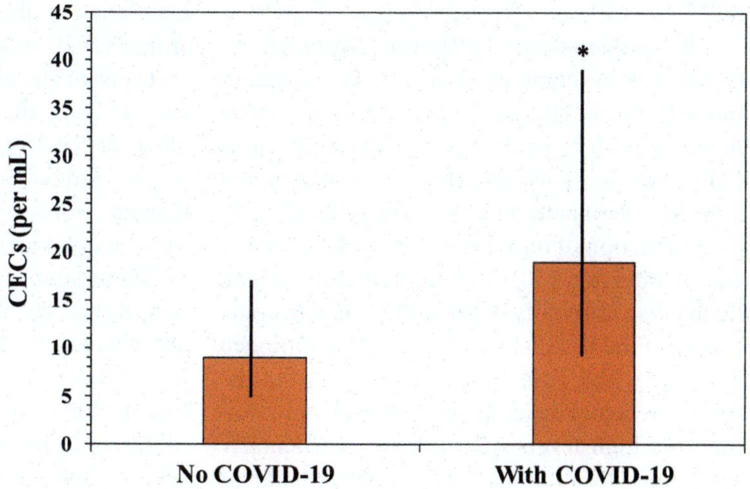

3.5 Causes and Results of Endothelial Dysfunction in COVID-19 Cases

The endothelial dysfunction observed in subjects with SARS-CoV-2 infection may be due to two main causes. Firstly, infection of endothelial cells by the virus may directly cause dysfunction. Microscopic examination of SARS-CoV-2-infected endothelial cells from COVID-19 cases has shown disrupted membrane structures in these cells and apoptosis [13, 29]. An equally important source of endothelial dysfunction may be induced by the inflammatory response that is characteristic of SARS-CoV-2 infection. A cross-sectional study conducted in China by Qin and colleagues involving 452 COVID-19-infected patients, 22–95 years of age, of whom 63.3% were severely infected, revealed plasma levels of TNF-α, IL-2R, IL-6, IL-8 and IL-10 to be elevated above the normal reference range and to be significantly higher in severely infected patients compared with less severe cases [6]. This observation of a so-called cytokine storm in COVID-19 patients has been confirmed in multiple reports. Thus, in a longitudinal study conducted by Liu et al. involving 40 COVID-19 patients, 34–62 years of age, the severely infected patients, who comprised approximately a third of the cohort, had higher serum levels of IL-2, IL-6, IL-10 and IFN-γ compared with the mild cases [49].

The inflammatory response observed in COVID-19 is complex and involves a major input from endothelial cells. The virus firstly interacts with type 2 alveolar epithelial cells (APCs) in the alveolar space of the lungs and activates the innate immune system, leading to the production of high levels of cytokines at the site of infection [50]. Endothelial cells, which lie in close proximity to the APCs, then become targets of the SARS-CoV-2 virus. As mentioned previously, the virus enters these cells via the ACE-2 receptor leading to apoptosis. In addition, the high levels of cytokines, particularly IL-6 and TNF, stimulate endothelial cells themselves to secrete more cytokines [IL-6, IL-8 and monocyte chemoattractant protein (MCP)-1] further enhancing the inflammatory milieu. This cytokine storm causes loosening of endothelial cell-to-cell interactions further increasing vascular permeability. The apoptosis of endothelial cells in combination with the loss of cell adhesion leads to vascular leakage resulting in oedema and in severe cases, respiratory failure [51].

The high cytokine levels also cause activation of the coagulation pathways by stimulating the endothelial cells to secrete von Willebrand factor, P-selectin and fibrinogen. These allow platelet binding to the endothelium, and the former also secrete vascular endothelial growth factor (VEGF) which stimulates the endothelial cells to produce tissue factor, a strong activator of coagulation. In addition, exposure of the thrombogenic collagen fibres of the basement membrane following endothelial cell detachment and apoptosis also leads to activation of the coagulation cascade [51]. A study of 77 cases of severe SARS-CoV-2 infections, matched with 145 non-COVID-19 controls (all cases and controls had acute respiratory distress syndrome), demonstrated that coagulation parameters in the cases were greater than those in the controls, and the cases had a sixfold higher risk of pulmonary embolism compared to the controls [52]. This activation of the coagulation cascade leads to thrombus formation in pulmonary arteries and capillaries as observed in autopsy tissue sections from COVID-19 cases [29].

Damage sustained to the pulmonary endothelium reduces the barrier function of this tissue, allowing SARS-CoV-2 to be transported to other organs. Studies have shown that endothelial cell damage is observed in multiple organs, including the kidneys, heart, liver and small intestine [13].

These studies suggest that SARS-CoV-2, by both direct and indirect mechanisms, can cause endothelial dysfunction which in turn leads to major vascular problems in multiple organs. Endothelial dysfunction is also a feature of many of the comorbid diseases that increase the risk of severe COVID-19, and this will be discussed below.

3.6 Comorbid Diseases, Endothelial Dysfunction and Severity of COVID-19

Reports from the United States and the United Kingdom have shown that individuals admitted to hospitals with COVID-19 have a high prevalence of comorbid diseases such as heart disease, hypertension, type 2 diabetes and obesity [53, 54]. Additionally, in a study comparing data from China and Italy, age and gender were associated with COVID-19 mortality, with individuals above 60 years of age and males having an increased risk [55]. In a retrospective study conducted in South Africa involving 22,308 COVID-19 patients, mortality was associated with diabetes, hypertension, male sex, increasing age and chronic kidney disease [56]. This study also showed that HIV and TB infection were associated with a 2.14- and a 2.70-fold increased risk of COVID-19-associated mortality, respectively, while diabetic subjects with poor glycaemic control (HbA1c $\geq$ 9.0%) had a 12.07-fold increased risk of death [56]. These studies, from both high- and low-to-middle-income countries, showed evidence of an association of both non-communicable and infectious diseases with higher mortality in subjects with COVID-19.

The reason for the higher risk of severe COVID-19 and mortality in subjects with comorbid diseases is not fully understood. However, it has been hypothesized that this may be due to the high level of endothelial dysfunction and systemic inflammation in subjects with cardiometabolic diseases [57, 58]. Many subjects with obesity, hypertension, type 2 diabetes or CVD have underlying endothelial dysfunction and systemic inflammation. Thus, the building blocks for vascular leakage and the cytokine storm are already in place, and infection with SARS-CoV-2 will build upon these pre-existing pathologies increasing the risk for severe COVID-19 (Fig. 3.3). Subjects with HIV infection also have underlying endothelial dysfunction and systemic inflammation [39], and this may partially explain the higher mortality rate observed for subjects coinfected with HIV and SARS-CoV-2 [56]. However, the immune suppression that is characteristic of HIV may also explain this finding.

The other endemic human coronavirus species, i.e. MERS-CoV, SARS-CoV, HCoV-229E, HCoV-NL63, HCoV-OC43 and HCoV-HKU1, have also been associated with more severe disease in subjects with CVDs [36]. Alongside the similar symptoms seen on infection with these viruses, this observation again suggests that all seven coronaviruses have similar effects on the endothelium and the inflammatory response.

Inflammation and endothelial damage and dysfunction are aetiological factors in atherosclerotic plaque formation [59]. It is therefore possible that coronavirus infection may enhance atherosclerosis [60] and that even after the infection has passed, plaque formation may progress, especially in those with pre-existing risk factors such as obesity, diabetes or dyslipidaemia. Long-term follow-up studies of survivors of SARS-CoV-2 infection are necessary to investigate if the incidence of atherosclerotic diseases such as coronary artery disease or subclinical atherosclerosis, as assessed using carotid intima-media thickness, is different from that in subjects who were never infected. A 12-year follow-up study comparing survivors of SARS-CoV-1 infection to those with no infection has been performed [61]. The number of study subjects was small (n = 25 per group), but it was shown that incident cardiovascular abnormalities were more common (44% vs 0%) in the infected group, although no details were given of these abnormalities.

Comorbid diseases obviously increase the risk of severe COVID-19 and of mortality and must therefore be monitored during infection and controlled as optimally as possible. Therapies for the treatment of COVID-19 itself are being intensively investigated, and the possible use of agents that improve endothelium function must be considered.

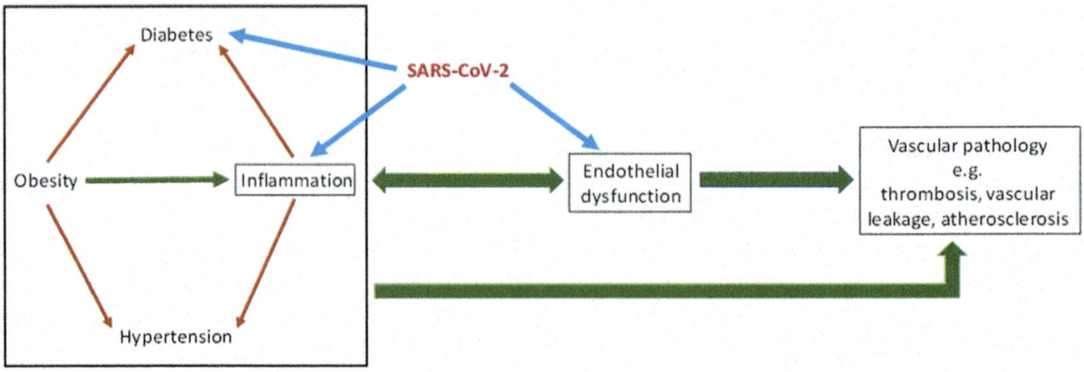

Fig. 3.3 The relationships between comorbid diseases, inflammation, vascular function and SARS-CoV-2. Obesity is an inflammatory state due to the secretion of cytokines such as IL-6 and TNF from adipose tissue. These cytokines cause insulin resistance and thus increase the risk of other comorbid diseases such as type 2 diabetes and hypertension. Obesity can also lead to diabetes and hypertension by mechanisms other than inflammation. Inflammation causes endothelial dysfunction as do obesity, hypertension and diabetes, and endothelial dysfunction can also enhance inflammation by the production of inflammatory cytokines from activated endothelial cells. Endothelial dysfunction is involved in vascular pathology including increased vascular permeability, enhanced coagulation leading to thrombosis and atherosclerotic disease. Comorbid diseases and inflammation can also affect the vasculature independently of their effects on endothelial dysfunction; for example, diabetes can have profound effects on major blood vessels and capillaries via chronic hyperglycaemia. The SARS-CoV-2 virus causes both inflammation and endothelial dysfunction enhancing the effects of the comorbid diseases. It should be noted that the virus has been associated with new-onset diabetes and worsening of glycaemic control in pre-diagnosed diabetics, and it has been suggested that the virus may also cause elevated blood pressure. Both of these effects are thought to be mediated by the ACE-2 protein

3.7 Targeting the Endothelium in Treatments for COVID-19

The endothelium is a major target of the cytokine response to SARS-CoV-2 infection, leading to dysfunction and weakening of the vascular barrier. Thus, attenuating this cytokine storm may protect the endothelium and reduce vascular leakage, and a number of anti-inflammatory agents are currently being assessed as therapeutic interventions for COVID-19. Dexamethasone, a strong anti-inflammatory agent, has already shown positive responses in severe COVID-19 cases [62]. It has been suggested that IL-6 should be the main focus of anti-cytokine therapy, because this molecule is one of the main drivers of the inflammatory response to SARS-CoV-2 infection and has major effects on endothelial function [49]. However, the use of anti-inflammatory therapies must be considered with caution as its usage is hindered by several factors such as correct timing of the treatment, secondary infections and cytokine measurement. It is thought that the use of IL-6 antagonists such as tocilizumab, which is a humanized monoclonal antibody targeting the IL-6 receptor, might only be beneficial in severely infected patients with elevated serum levels of IL-6 [63]. Moreover, COVID-19-infected patients do not share the same inflammatory profile. Significant fluctuation in serum levels of IL-6 in severely infected COVID-19 patients has been reported [49]. Despite these misgivings, small clinical trials without the use of a control group have shown both positive and negative effects of tocilizumab therapy in severe COVID-19 cases [64, 65]. These studies need to be replicated in larger populations set within more stringent clinical trial frameworks. Anti-TNF agents have also been suggested as therapy options for COVID-19 [66]. This cytokine is also a prominent role player in the cytokine storm with endothelial effects [51]. A small study of the anti-TNF monoclonal antibody infliximab in seven treated and non-treated COVID-19 cases did show clinical improvements with the therapy, but again, these results need to be confirmed in larger clinical trials.

Statins are known to have both anti-inflammatory and pro-endothelial effects, and it has been hypothesized that they may be an effective therapy for COVID-19. These drugs have been used to treat other viral infections such as influenza and Ebola, with some success [67, 68]. In fact, in the studies on Ebola, statins were used in conjunction with angiotensin receptor blockers (ARBs), as both agents have been reported to counteract endothelial dysfunction. The use of ARBs to treat COVID-19 is contentious as these drugs are known to upregulate expression of ACE2, the SARS-CoV-2 receptor [69].

Disruption of endothelial function by coronavirus leads to activation of the coagulation pathway [51]. Therapies directed at downregulation of this pathway have therefore been suggested for treatment of COVID-19, and it has been recommended that anticoagulants be used prophylactically in severe COVID-19 cases to reduce the risk of thrombosis [70].

There are a number of therapeutic agents which act directly or indirectly on the endothelium to attenuate inflammation, endothelial dysfunction and thrombotic events and hence improve vascular function. Many of these therapies have not yet completed testing in properly controlled clinical trials, although a number of such studies are currently in progress. The use of these drugs in combination must also be considered in future trials.

3.8 Conclusions

The endothelium has gained attention in recent years as a target for many different viral infections, and this has been strongly highlighted with the current SARS-CoV-2 pandemic. Its multiple functions have ensured that any pathological changes induced in this tissue will have profound effects on health. The targeting of therapies toward the endothelium is therefore essential, and the development of such agents has been augmented by studies showing that pre-existing, commonly used drugs do have positive effects on endothelial function. The outcome of current clinical trials on new therapies that modulate endothelial activity in the context of COVID-19 is eagerly awaited.

References

1. Estola T (1970) Coronaviruses, a new group of animal RNA viruses. Avian Dis 14:330–336
2. Mäkelä MJ, Puhakka T, Ruuskanen O, Leinonen M, Saikku P, Kimpimäki M et al (1998) Viruses and bacteria in the etiology of the common cold. J Clin Microbiol 36:539–542
3. Drosten C, Günther S, Preiser W, Van Der Werf S, Brodt HR, Becker S et al (2003) Identification of a novel coronavirus in patients with severe acute respiratory syndrome. N Engl J Med 348:1967–1976
4. Zaki AM, Van Boheemen S, Bestebroer TM, Osterhaus AD, Fouchier RA (2012) Isolation of a novel coronavirus from a man with pneumonia in Saudi Arabia. N Engl J Med 367:1814–1820
5. Chen N, Zhou M, Dong X, Qu J, Gong F, Han Y et al (2020) Epidemiological and clinical characteristics of 99 cases of 2019 novel coronavirus pneumonia in Wuhan, China: a descriptive study. Lancet 395:507–513
6. Qin C, Zhou L, Hu Z, Zhang S, Yang S, Tao Y et al (2020) Dysregulation of immune response in patients with coronavirus 2019 (COVID-19) in Wuhan, China. Clin Infect Dis 71:762–768
7. Jiang F, Deng L, Zhang L, Cai Y, Cheung CW, Xia Z (2020) Review of the clinical characteristics of coronavirus disease 2019 (COVID-19). J Gen Intern Med 35:1545–1549
8. WHO (2020) WHO Coronavirus Disease (COVID-19) Dashboard [Online]. Available: https://covid19.who.int/. Accessed 23rd July 2020
9. Perlman S, Dandekar AA (2005) Immunopathogenesis of coronavirus infections: implications for SARS. Nat Rev Immunol 5:917–927
10. Zhou P, Yang X-L, Wang X-G, Hu B, Zhang L, Zhang W et al (2020) A pneumonia outbreak associated with a new coronavirus of probable bat origin. Nature 579:270–273
11. Jiang S, Hillyer C, Du L (2020) Neutralizing antibodies against SARS-CoV-2 and other human coronaviruses. Trends Immunol 41:355–359
12. Brevetti G, Schiano V, Chiariello M (2008) Endothelial dysfunction: a key to the pathophysiology and natural history of peripheral arterial disease? Atherosclerosis 197:1–11
13. Varga Z, Flammer AJ, Steiger P, Haberecker M, Andermatt R, Zinkernagel AS et al (2020) Endothelial cell infection and endotheliitis in COVID-19. Lancet 395:1417–1418
14. Klein D, Hurley LB, Quesenberry JC (1999) Sidney S (2002) Do protease inhibitors increase the risk for

coronary heart disease in patients with HIV-1 infection? J Acquir Immune Defic Syndr 30:471–477
15. Wang T, Green LA, Gupta SK, Kim C, Wang L, Almodovar S et al (2014) Transfer of intracellular HIV Nef to endothelium causes endothelial dysfunction. PLoS One 9:e91063. https://doi.org/10.1371/journal.pone.0091063
16. Chang C-H, Huang Y, Issekutz AC, Griffith M, Lin K-H, Anderson R (2002) Interleukin-1α released from epithelial cells after adenovirus type 37 infection activates intercellular adhesion molecule 1 expression on human vascular endothelial cells. J Virol 76:427–431
17. Fourie C, Van Rooyen J, Pieters M, Conradie K, Hoekstra T, Schutte A (2011) Is HIV-1 infection associated with endothelial dysfunction in a population of African ancestry in South Africa? Cardiovasc J Afr 22:134–140
18. Arildsen H, Sorensen KE, Ingerslev JM, Ostergaard LJ, Laursen AL (2013) Endothelial dysfunction, increased inflammation, and activated coagulation in HIV-infected patients improve after initiation of highly active antiretroviral therapy. HIV Med 14:1–9
19. Van Vonderen MG, Hassink EA, Van Agtmael MA, Stehouwer CD, Danner SA, Reiss P et al (2009) Increase in carotid artery intima-media thickness and arterial stiffness but improvement in several markers of endothelial function after initiation of antiretroviral therapy. J Infect Dis 199:1186–1194
20. Hemmat N, Ebadi A, Badalzadeh R, Memar MY, Baghi HB (2018) Viral infection and atherosclerosis. J Clin Microbiol Infect Dis 37:2225–2233
21. Popović M, Smiljanić K, Dobutović B, Syrovets T, Simmet T, Isenović ERJJOT et al (2012) Human cytomegalovirus infection and atherothrombosis. J Thromb Thrombolysis 33:160–172
22. Peretz A, Azrad M, Blum A (2019) Influenza virus and atherosclerosis. QJM 112:749–755
23. Basler CF (2017) Molecular pathogenesis of viral hemorrhagic fever. Semin Immunopathol 39:551–561
24. Feldmann H, Bugany H, Mahner F, Klenk H-D, Drenckhahn D, Schnittler HJ (1996) Filovirus-induced endothelial leakage triggered by infected monocytes/macrophages. J Virol 70:2208–2214
25. Ter Meulen J, Sakho M, Koulemou K, Magassouba NF, Bah A, Preiser W et al (2004) Activation of the cytokine network and unfavorable outcome in patients with yellow fever. J Infect Dis 190:1821–1827
26. Rastogi M, Sharma N, Singh SK (2016) Flavivirus NS1: a multifaceted enigmatic viral protein. Virol J 13:131–131
27. Beatty PR, Puerta-Guardo H, Killingbeck SS, Glasner DR, Hopkins K, Harris E (2015) Dengue virus NS1 triggers endothelial permeability and vascular leak that is prevented by NS1 vaccination. Sci Transl Med 7:304ra141. https://doi.org/10.1126/scitranslmed.aaa3787
28. Noronha L, Zanluca C, Azevedo ML, Luz KG, Santos CN (2016) Zika virus damages the human placental barrier and presents marked fetal neurotropism. Mem Inst Oswaldo Cruz 111:287–293
29. Ackermann M, Verleden SE, Kuehnel M, Haverich A, Welte T, Laenger F et al (2020) Pulmonary vascular endothelialitis, thrombosis, and angiogenesis in Covid-19. N Engl J Med 383:120–128
30. Hoffmann M, Kleine-Weber H, Schroeder S, Krüger N, Herrler T, Erichsen S et al (2020) SARS-CoV-2 cell entry depends on ACE2 and TMPRSS2 and is blocked by a clinically proven protease inhibitor. Cell 181:271–280
31. Hamming I, Timens W, Bulthuis M, Lely A, Navis G, Van Goor H (2004) Tissue distribution of ACE2 protein, the functional receptor for SARS coronavirus. A first step in understanding SARS pathogenesis. J Pathol 203:631–637
32. Li W, Moore MJ, Vasilieva N, Sui J, Wong SK, Berne MA et al (2003) Angiotensin-converting enzyme 2 is a functional receptor for the SARS coronavirus. Nature 426:450–454
33. Raj VS, Mou H, Smits SL, Dekkers DH, Müller MA, Dijkman R et al (2013) Dipeptidyl peptidase 4 is a functional receptor for the emerging human coronavirus-EMC. Nature 495:251–254
34. Yeager CL, Ashmun RA, Williams RK, Cardellichio CB, Shapiro LH, Look AT et al (1992) Human aminopeptidase N is a receptor for human coronavirus 229E. Nature 357:420–422
35. Li F (2015) Receptor recognition mechanisms of coronaviruses: a decade of structural studies. J Virol 89:1954–1964
36. Zhao M, Wang M, Zhang J, Ye J, Xu Y, Wang Z et al (2020) Advances in the relationship between coronavirus infection and cardiovascular diseases. Biomed Pharmacother 127:110230. https://doi.org/10.1016/j.biopha.2020.110230
37. Tong M, Jiang Y, Xia D, Xiong Y, Zheng Q, Chen F et al (2020) Elevated serum endothelial cell adhesion molecules expression in COVID-19 patients. J Infect Dis:jiaa349. https://doi.org/10.1093/infdis/jiaa349
38. Escher R, Breakey N, Lämmle BJTR (2020) Severe COVID-19 infection associated with endothelial activation. Thromb Res 190:62. https://doi.org/10.1016/j.thromres.2020.04.014
39. Mezoh G, Crowther NJ (2019) Deciphering endothelial dysfunction in the HIV-infected population. Adv Exp Med Biol 1134:193–215
40. Sadler JE (1998) Biochemistry and genetics of von Willebrand factor. Annu Rev Biochem 67:395–424
41. Munro JM (1993) Endothelial-leukocyte adhesive interactions in inflammatory diseases. Eur Heart J 14:72–77
42. Koraka P, Murgue B, Deparis X, Van Gorp EC, Setiati TE, Osterhaus AD et al (2004) Elevation of soluble VCAM-1 plasma levels in children with acute dengue virus infection of varying severity. J Med Virol 72:445–450
43. Cardier JE, Rivas B, Romano E, Rothman AL, Perez-Perez C, Ochoa M et al (2006) Evidence of vascular damage in dengue disease: demonstration of high levels of soluble cell adhesion molecules and circulating endothelial cells. Endothelium 13:335–340

44. Umehara H, Bloom ET, Okazaki T, Nagano Y, Yoshie O, Imai T (2004) Fractalkine in vascular biology: from basic research to clinical disease. Arterioscler Thromb Vasc Biol 24:34–40
45. Nizzoli ME, Merati G, Tenore A, Picone C, Consensi E, Perotti L et al (2020) Circulating endothelial cells in COVID-19. Am J Hematol 95:E187–E188. https://doi.org/10.1002/ajh.25881
46. Khider L, Gendron N, Goudot G, Chocron R, Hauw-Berlemont C, Cheng C et al (2020) Curative anticoagulation prevents endothelial lesion in COVID-19 patients. J Thromb Haemost. https://doi.org/10.1111/jth.14968
47. Blann AD, Woywodt A, Bertolini F, Bull TM, Buyon JP, Clancy RM et al (2005) Circulating endothelial cells. Thromb Haemost 93:228–235
48. López M, Vispo E, San Román J, Herrero D, Peris A, Corral A et al (2012) High risk of endothelial dysfunction in HIV individuals may result from deregulation of circulating endothelial cells and endothelial progenitor cells. AIDS Res Hum Retrov 28:656–659
49. Liu J, Li S, Liu J, Liang B, Wang X, Wang H et al (2020) Longitudinal characteristics of lymphocyte responses and cytokine profiles in the peripheral blood of SARS-CoV-2 infected patients. EBioMedicine 55:102763. https://doi.org/10.1016/j.ebiom.2020.102763
50. Zhang C, Wu Z, Li J-W, Zhao H, Wang GQ (2020) The cytokine release syndrome (CRS) of severe COVID-19 and Interleukin-6 receptor (IL-6R) antagonist Tocilizumab may be the key to reduce the mortality. Int J Antimicrob Agents 55:105954. https://doi.org/10.1016/j.ijantimicag.2020.105954
51. Teuwen L-A, Geldhof V, Pasut A, Carmeliet P (2020) COVID-19: the vasculature unleashed. Nat Rev Immunol 20:389–391
52. Helms J, Tacquard C, Severac F, Leonard-Lorant I, Ohana M, Delabranche X et al (2020) High risk of thrombosis in patients with severe SARS-CoV-2 infection: a multicenter prospective cohort study. Intensive Care Med 46:1089–1098
53. Argenziano MG, Bruce SL, Slater CL, Tiao JR, Baldwin MR, Barr RG et al (2020) Characterization and clinical course of 1000 patients with coronavirus disease 2019 in New York: retrospective case series. BMJ 369:m1996. https://doi.org/10.1136/bmj.m1996
54. Atkins JL, Masoli JA, Delgado J, Pilling LC, Kuo C-L, Kuchel GA et al (2020) Preexisting comorbidities predicting COVID-19 and mortality in the UK biobank community cohort. J Gerontol A Biol Sci Med Sci:glaa183. https://doi.org/10.1093/gerona/glaa183
55. Onder G, Rezza G, Brusaferro S (2020) Case-fatality rate and characteristics of patients dying in relation to COVID-19 in Italy. JAMA. https://doi.org/10.1001/jama.2020.4683
56. Davies MA (2020) HIV and risk of COVID-19 death: a population cohort study from the Western Cape Province, South Africa. medRxiv:20145185. https://doi.org/10.1101/2020.07.02.20145185
57. Li X, Sun X, Carmeliet P (2019) Hallmarks of endothelial cell metabolism in health and disease. Cell Metab 30:414–433
58. Donath MY, Meier DT, Böni-Schnetzler M (2019) Inflammation in the pathophysiology and therapy of cardiometabolic disease. Endocr Rev 40:1080–1091
59. Zhu Y, Xian X, Wang Z, Bi Y, Chen Q, Han X et al (2018) Research progress on the relationship between atherosclerosis and inflammation. Biomol Ther 8:80. https://doi.org/10.3390/biom8030080
60. Vinciguerra M, Romiti S, Fattouch K, De Bellis A, Greco E (2020) Atherosclerosis as pathogenetic substrate for Sars-Cov-2 cytokine storm. J Clin Med 9:E2095. https://doi.org/10.3390/jcm9072095
61. Wu Q, Zhou L, Sun X, Yan Z, Hu C, Wu J et al (2017) Altered lipid metabolism in recovered SARS patients twelve years after infection. Sci Rep 7:9110. https://doi.org/10.1038/s41598-017-09536-z
62. Johnson RM, Vinetz JM (2020) Dexamethasone in the management of COVID-19. BMJ 370:m2648. https://doi.org/10.1136/bmj.m2648
63. Luo P, Liu Y, Qiu L, Liu X, Liu D, Li J (2020) Tocilizumab treatment in COVID-19: a single center experience. J Med Virol 92:814–818
64. Xu X, Han M, Li T, Sun W, Wang D, Fu B et al (2020) Effective treatment of severe COVID-19 patients with tocilizumab. Proc Natl Acad Sci U S A 117:10970–10975
65. Morena V, Milazzo L, Oreni L, Bestetti G, Fossali T, Bassoli C et al (2020) Off-label use of tocilizumab for the treatment of SARS-CoV-2 pneumonia in Milan, Italy. Eur J Intern Med 76:36–42
66. Feldmann M, Maini RN, Woody JN, Holgate ST, Winter G, Rowland M et al (2020) Trials of anti-tumour necrosis factor therapy for COVID-19 are urgently needed. Lancet 395:1407–1409
67. Lee KC, Sewa DW, Phua GC (2020) Potential role of statins in COVID-19. Int J Infect Dis 96:615–617
68. Fedson D, Opal S, Rordam OM (2020) Treating patients with severe COVID-19 infection. mBio 11:e00398–e00320. doi:https://doi.org/10.1128/mBio.00398-20
69. Danser AHJ, Epstein M, Batlle D (2020) Renin-angiotensin system blockers and the COVID-19 pandemic. At present there is no evidence to abandon renin-angiotensin system blockers. Hypertension 75:1382–1385
70. Pons S, Fodil S, Azoulay E, Zafrani L (2020) The vascular endothelium: the cornerstone of organ dysfunction in severe SARS-CoV-2 infection. Crit Care 24:353. https://doi.org/10.1186/s13054-020-03062-7

A Review Study on the Neonatal Outcomes of Pregnant Women with COVID-19

4

Somayeh Makvandi, Mitra Mahdavian, Goli Kazemi-Nia, Amir Vahedian-Azimi, Leila Karimi, and Amirhossein Sahebkar

Abstract

COVID-19 is a fatal respiratory disease caused by a novel coronavirus that quickly became a pandemic. Pregnant women and neonates are two vulnerable groups in COVID-19 infections because the immune system weakens during pregnancy. The present review study was conducted to investigate the rate of vertical transmission in infants born to women with COVID-19 infections and to describe the characteristics of the affected infants. We conducted a search of the various scientific databases using relevant keywords. All English-language studies involving neonates born to women who had COVID-19 infections were included. The main outcomes were rates of vertical transmission and the characteristics of the affected newborns. Out of 13 selected studies, 103 newborns were involved. The rate of vertical transmission was 5.4%. Of the five infected newborns, four were full-term and one was preterm. All were born by Cesarean section. The clinical symptoms were vomiting, fever, lethargy, shortness of breath, and cyanosis. In four newborns, a chest x-ray showed evidence of pneumonia. The most common laboratory finding was leukocytosis and elevated creatine kinase levels. One newborn needed mechanical ventilation. All newborns recovered and were discharged. The findings of this review study showed that the prognosis of newborns of infected mothers was satisfactory, and clinical symptoms of infected neonates did not differ from adults and were nonspecific. Due to the low amount of data regarding this field, further studies with higher sample sizes are required for more definitive conclusions.

S. Makvandi
Department of Midwifery, School of Nursing and Midwifery, Islamic Azad University, Ahvaz, Iran

M. Mahdavian
Department of Midwifery, School of Nursing and Midwifery, Islamic Azad University, Bojnourd, Iran

G. Kazemi-Nia
Sina Hospital, Ahvaz Jundishapur University of Medical Sciences, Ahvaz, Iran

A. Vahedian-Azimi
Trauma Research Center, Nursing Faculty, Baqiyatallah University of Medical Sciences, Tehran, Iran

L. Karimi (✉)
Behavioral Sciences Research Center, Life style institute, Nursing Faculty, Baqiyatallah University of Medical Sciences, Tehran, Iran
e-mail: leilakarimi1487@gmail.com

A. Sahebkar (✉)
Neurogenic Inflammation Research Center, Mashhad University of Medical Sciences, Mashhad, Iran

Biotechnology Research Center, Pharmaceutical Technology Institute, Mashhad University of Medical Sciences, Mashhad, Iran

Polish Mother's Memorial Hospital Research Institute (PMMHRI), Lodz, Poland

Halal Research Center of IRI, FDA, Tehran, Iran
e-mail: sahebkara@mums.ac.ir; amir_saheb2000@yahoo.com

P. C. Guest (ed.), *Clinical, Biological and Molecular Aspects of COVID-19*, Advances in Experimental Medicine and Biology 1321, https://doi.org/10.1007/978-3-030-59261-5_4

Keywords

COVID-19 · Novel coronavirus infection · Newborn · Vertical transmission

4.1 Introduction

The emergence of the novel SARS-CoV-2 (COVID-19) virus in December 2019 in Wuhan, China, has rapidly led to a global pandemic and has become one of the most important health threats in recent times [1]. SARS-CoV-2 is a member of the family of coronaviruses responsible for two dangerous diseases that occurred within the last two decades, SARS (severe acute respiratory syndrome) and MERS (Middle East respiratory syndrome) [2]. Such diseases can be fatal due to destruction of lung alveoli and progressive respiratory failure [3].

We are witnessing the daily growth of published research on different aspects of COVID-19 in different scientific databases, and researchers are trying to increase knowledge about different aspects of this disease. An important question that remains unanswered is whether or not COVID-19 can be transmitted from a pregnant woman to her fetus or neonate, a process called vertical transmission. If this turns out to be the case, it also remains to be determined what the severity and course of the disease will be in infants. Nissen et al. explained that the clinical symptoms of neonatal pneumonia are usually nonspecific, and it is therefore difficult to diagnose and treat [4]. Pregnant women and neonates are thought to be especially vulnerable to the novel coronavirus because the immune systems of both groups are weaker than others [5].

Studies of SARS, MERS, and other human coronavirus infections have suggested that such diseases can lead to adverse fetal and neonatal outcomes, such as intrauterine growth retardation, preterm labor, intensive care unit (ICU) hospitalization, spontaneous abortion, and perinatal mortality [6]. More scientific evidence regarding various aspects of COVID-19 infection is needed to develop effective strategies for prevention and clinical care. The present review was conducted to investigate the rate of vertical transmission in infants born to women who suffered from COVID-19 infection, and it describes the characteristics of the affected infants.

4.2 Methods

In this review, we conducted a search in the various scientific databases with varying combinations of the keywords "COVID-19," "COVID19," "2019 novel coronavirus infection," "COVID-19 pandemic," "coronavirus disease-19," "novel coronavirus disease," "pregnancy," "pregnancy outcomes," "neonate," "newborn," and "vertical transmission." Any type of English-language studies involving neonates born to women who suffered from COVID-19 infection was included. The main outcomes of our study were the rate of vertical transmission of novel coronavirus and the characteristics of the affected newborns. Two authors screened the titles and abstracts of resulting articles to exclude irrelevant studies. Then, they retrieved full text articles of seemingly relevant studies, examined these, and resolved any disagreement through discussion and final agreement. The same researchers designed a data extraction form that included the following information: first author's name, number of case(s), gestational age, method of birth, Apgar score [7], weight, result of throat swab, result of computerized tomography (CT) scan of lungs or chest x-ray, outcomes and clinical symptoms, diagnostic tests, and type of treatment in infected neonates.

In creating this study, in accordance with ethical principles, researchers refrained from data fabrication and never manipulated data for their own benefit. In all sections of the article, they also strived to avoid plagiarism.

4.3 Results

Out of 13 studies, 103 newborns were involved, ranging from 1 to 33 per study. The characteristics of the neonates studied are shown in

Table 4.1. Approximately one-fifth of the newborns were preterm and the rest were full-term. All of the studies were conducted in China, 83.5% of the newborns were born by Cesarean section, and 16.9% had low birth weight (LBW). A total of 93 tests were performed to detect coronavirus with five positive results (5.4%). One case of stillbirth [8] and one case of neonatal death [5] were reported.

The characteristics of the infected neonates are shown in Table 4.2. Of the five infected newborns, four were full-term and one was 31 weeks and 2 days old at birth. All were born by Cesarean section. The sex of four of these infants was male, but sex information was not provided in the study by Yu et al. [9]. Two cases of meconium-stained amniotic fluid [10, 11], one case of premature rupture of membranes, and one case of fetal distress were found [11]. The clinical symptoms were as follows: 30% of newborns had vomiting, 20% had fever, 20% had lethargy, 20% had shortness of breath, and 10% had cyanosis. In four newborns, a chest x-ray showed evidence of pneumonia. The most common laboratory finding was leukocytosis and elevated creatine kinase levels. One premature newborn needed mechanical ventilation. All newborns were cured and finally discharged from the neonatal intensive care unit (NICU).

4.4 Discussion

The findings of the present review study showed that the vertical transition rate of COVID-19 was 5.4%. As this only related to 5 out of 103 newborns, it was not possible to judge whether or not this finding is conclusive. Because four-fifths of the pregnant women whose neonates were included in the study had a full-term pregnancy at the time of developing of COVID-19 infection, the majority of newborns were also full-term. Therefore, it is not clear what the rate of transmission to the fetus would have been if the disease had occurred earlier in the pregnancy.

In general, the neonatal consequences in neonates born to mothers with COVID-19 are favorable. Of the five neonates who developed COVID-19 infections, only one appeared to be seriously ill. In addition to COVID-19, this neonate suffered from asphyxia, LBW, and other complications of prematurity.

According to the evidence obtained so far in adults, the most common clinical symptom of COVID-19 infection is fever. A report of 72,314 records in China showed that in patients with coronavirus, typical symptoms were fever, cough, and fatigue [12]. The findings of our study showed that vomiting was the most common symptom in infected neonates. Therefore, COVID-19 pneumonia in infants appears to have nonspecific clinical symptoms. In this regard, March et al. suggested that fever is not a good indicator of viral pneumonia in infants [13].

The findings of this study also showed that most neonates were born by Cesarean section, and the rate of vaginal delivery was only 16.5%. Also, the infected neonates were all born by Cesarean section. An expert consensus for managing pregnant women and neonates born to mothers with suspected or confirmed novel coronavirus infection stated that at present, there is no conclusive evidence of the best delivery method to reduce the risk of vertical transmission [14]. In other words, whether or not Cesarean section can reduce the risk of vertical transmission in COVID-19 remains to be determined. According to the evidence, the decision on the time and type of delivery in pregnant women suffering from COVID-19 infections requires a multidisciplinary teamwork approach and is influenced by several factors such as the patient's clinical condition and obstetrical factors [15].

Finally, it is important to note that, so far, there is little data about the impact of the 2019 novel coronavirus on neonatal outcomes. The papers reviewed above are mostly studies with a small sample size and may therefore have been of low quality. Thus, this factor may be limiting in interpreting the findings of this study. To achieve more realistic results, more studies with more detailed design are needed. We suggest that studies should be conducted to determine which factors can be used to predict the risk of pregnant women with COVID-19 infection, giving birth to neonates with viral infection. This may include a

Table 4.1 Characteristics of included studies

First author's name	N	GA (weeks)				Method of birth		APGAR score		Weight (g)		Result of throat swab			CT scan of lungs or chest x-ray			Outcome		
		<28	28–31	32–36	term	Vaginal delivery	Cesarean section	<7	7–10	≥2500	2500–4000	Positive	Negative	Not tested	Normal	Abnormal	Undone	Died	Cured and discharged	In hospital
Huaping Zhu [5]	10	–	2	4	4	3	7	–	10	7	3	–	9	1	3	7	–	1[a]	5	4
Yangli Liu [8]	10	–	–	5	5	–	10	1	9	NM	NM	–	9	1	–	–	–	1[b]	9	–
Xiaotong Wang [16]	1	–	1	–	–	–	1	–	1	1	0	–	1	–	–	–	–	–	1	–
Siyu Chen [17]	5	–	–	–	5	3	2	–	5	–	5	–	5	–	–	–	–	–	5	–
Suliman Khan [18]	3	–	–	1	2	3	–	–	3	–	3	–	3	–				–	3	–
Nan Yu [9]	7	–	–	–	7	–	7	–	7	–	7	1	2	4	–	1	–	–	7	–
Huijun Chen [19]	9	–	–	–	9	–	9	–	9	2	7	–	6	3	–	–	–	–	9	–
Cuifang Fan [20]	2	–	–	–	2	–	2	–	2	–	2	–	2	–	–	2	–	–	2	–
Yang Li [21]	1	–	–	1	–	–	1	–	1	NM	NM	–	1	–	–	–		–	1	–
Shaoshuai Wang [10]	1	–	–	–	1	–	1	–	1	–	1	1	–	–	–	1		–	1	–
Lingkong Zeng [11]	33	–	–	3	29	7	26	2	31	NM	NM	3	30	–	–	3	30	–	33	–
Rong Chen [22]	17	–	–	3	14	–	17	–	17	–	17	–	17	–	–	–	17	–	17	–
Yan Chen [23]	4	–	–	–	4	1	3	–	4	–	4	–	3	1	2	1	1	–	4	–
Total (%)	103	–	3 (2.9)	17 (16.5)	83 (80.6)	17 (16.5)	86 (83.5)	3 (2.9)	100 (97.1)	10 (16.9)	49 (83.1)	5 (4.9)	88 (85.4)	10 (9.7)	5 (7.3)	15 (22.1)	48 (70.6)	2 (2)	97 (94.2)	4 (3.8)

NM not mentioned

[a]Admitted 30 min after delivery due to shortness of breath and moaning. Died from multiple organ failure and DIC

[b]IUFD

Table 4.2 Characteristics of five infected newborns

		Neonate 1	Neonate 2	Neonate 3	Neonate 4	Neonate 5
First author's name		Nan Yu [9]	Shaoshuai Wang [10]	Lingkong Zeng [11]		
Gestational age		39 + 6	40	40	40 + 4	31 + 2
Sex		NM	Male	Male	Male	Male
History of chronic basic diseases		Hypothyroidism	No	NM	NM	NM
Pregnancy complications		No	Meconium-stained amniotic fluid	Premature rupture of membranes, meconium-stained amniotic fluid	No	Fetal distress
Method of birth		Cesarean	Cesarean	Cesarean	Cesarean	Cesarean
Weight (gr)		3250	3205	3250	3360	1580
Asphyxia		No	No	No	No	Yes
Symptoms		Mild shortness of breath	Vomiting once after feeding	Fever and lethargy	Fever, lethargy, and vomiting	Shortness of breath, cyanosis, and vomiting
Diagnostic tests	Chest x-ray	Mild pulmonary infection	Thickened lung texture with no abnormalities in heart and palate	Pneumonia	Pneumonia	Pneumonia and respiratory distress syndrome
	Laboratory data	NM	Lymphopenia, deranged liver function tests, and elevated creatine kinase level	Laboratory tests (except procalcitonin) were normal.	Leukocytosis, lymphopenia, and an elevated creatine kinase level	Enterobacter agglomerates: Positive blood culture, leukocytosis, and thrombocytopenia
Treatment	Mechanical ventilation	No	No	No	No	Yes
	Antibiotic	NM	Penicillin G	No	No	Yes
	Duration of neonatal ICU	14	17	NM	NM	NM
Discharged		Yes	Yes	Yes	Yes	Yes

NM not mentioned, *ICU* intensive care unit

combination of physiological, imaging, and blood-based molecular biomarker data.

4.5 Conclusions

The findings of this review study showed that the prognosis of newborns of infected mothers was satisfactory, and clinical symptoms of infected neonate differ from adults and are nonspecific. Due to the lack of data, the authors strongly recommend that more studies be performed on neonates of infected women to achieve more accurate and definitive results. Attempts should be made to identify risk factors of both vertical transmission and perinatal infection.

Acknowledgments Thanks to guidance and advice from the Clinical Research Development Unit of Baqiyatallah Hospital.

Conflict of Interests We declare no competing interests.

References

1. Huang C, Wang Y, Li X, Ren L, Zhao J, Hu Y et al (2020) Clinical features of patients infected with 2019 novel coronavirus in Wuhan, China. Lancet 395(10223):497–506
2. Qiao J (2020) What are the risks of COVID-19 infection in pregnant women? Lancet 395(10226):760–762
3. Xu Z, Shi L, Wang Y, Zhang J, Huang L, Zhang C et al (2020) Pathological findings of COVID-19 associated with acute respiratory distress syndrome. Lancet Respir Med 8(4):420–422
4. Nissen MD (2007) Congenital and neonatal pneumonia. Paediatr Respir Rev 8(3):195–203
5. Zhu H, Wang L, Fang C, Peng S, Zhang L, Chang G et al (2020) Clinical analysis of 10 neonates born to mothers with 2019-nCoV pneumonia. Transl Pediatr 9(1):51–60
6. Schwartz D, Graham A (2020) Potential maternal and infant outcomes from coronavirus 2019-nCoV (SARS-CoV-2) infecting pregnant women: lessons from SARS, MERS, and other human coronavirus infections. Viruses 12(2):pii: E194. https://doi.org/10.3390/v12020194
7. Colburn DW, Salzman M (1960) The Apgar scoring system in evaluation of the newborn infant. N Y State J Med 60:246–249
8. Liu Y, Chen H, Tang K, Guo Y (2020) Clinical manifestations and outcome of SARS-CoV-2 infection during pregnancy. J Infect. pii: S0163-4453(20)30109-2. https://doi.org/10.1016/j.jinf.2020.02.028. [Epub ahead of print]
9. Yu N, Li W, Kang Q, Xiong Z, Wang S, Lin X et al (2020) Clinical features and obstetric and neonatal outcomes of pregnant patients with COVID-19 in Wuhan, China: a retrospective, single-centre, descriptive study. Lancet Infect Dis 20(5):559–564
10. Wang S, Guo L, Chen L, Liu W, Cao Y, Zhang J et al (2020) A case report of neonatal COVID-19 infection in China. Clin Infect Dis. pii: ciaa225. https://doi.org/10.1093/cid/ciaa225. [Epub ahead of print]
11. Zeng L, Xia S, Yuan W, Yan K, Xiao F, Shao J et al (2020) Neonatal early-onset infection with SARS-CoV-2 in 33 neonates born to mothers with COVID-19 in Wuhan, China. JAMA Pediatr. https://doi.org/10.1001/jamapediatrics.2020.0878. [Epub ahead of print]
12. Tabatabaeizadeh SA, Avan A, Bahrami A, Khodashenas E, Esmaeili H, Ferns GA et al (2017) High-dose supplementation of vitamin D affects measures of systemic inflammation: reductions in high-sensitivity C-reactive protein level and neutrophil to lymphocyte ratio (NLR) distribution. J Cell Biochem 118(12):4317–4322
13. MFBP M, Sant'Anna CC (2005) Signs and symptoms indicative of community-acquired pneumonia in infants under six months. Braz J Infect Dis 9(2):150–155
14. Chen D, Yang H, Cao Y, Cheng W, Duan T, Fan C et al (2020) Expert consensus for managing pregnant women and neonates born to mothers with suspected or confirmed novel coronavirus (COVID-19) infection. Int J Gynaecol Obstet 149(2):130–136
15. Di Mascio D, Khalil A, Saccone G, Rizzo G, Buca D, Liberati M et al (2020) Outcome of coronavirus spectrum infections (SARS, MERS, COVID 1-19) during pregnancy: a systematic review and meta-analysis. Am J Obstet Gynecol MFM 25:100107. https://doi.org/10.1016/j.ajogmf.2020.100107
16. Wang X, Zhou Z, Zhang J, Zhu F, Tang Y, Shen X (2020) A case of 2019 Novel Coronavirus in a pregnant woman with preterm delivery. Clin Infect Dis. pii: ciaa200. https://doi.org/10.1093/cid/ciaa200. [Epub ahead of print]
17. Chen S, Liao E, Shao Y (2020) Clinical analysis of pregnant women with 2019 novel coronavirus pneumonia. J Med Virol. https://doi.org/10.1002/jmv.25789. [Epub ahead of print]
18. Khan S, Peng L, Siddique R, Nabi G, Xue M, Liu J et al (2020) Impact of COVID-19 infection on pregnancy outcomes and the risk of maternal-to-neonatal intrapartum transmission of COVID-19 during natural birth. Infect Control Hosp Epidemiol 19:1–3
19. Chen H, Guo J, Wang C, Luo F, Yu X, Zhang W et al (2020) Clinical characteristics and intrauterine vertical transmission potential of COVID-19 infection in nine pregnant women: a retrospective review of medical records. Lancet 395(10226):809–815
20. Fan C, Lei D, Fang C, Li C, Wang M, Liu Y et al (2020) Perinatal transmission of COVID-19 associ-

ated SARS-CoV-2: should we worry? Clin Infect Dis. pii: ciaa226. https://doi.org/10.1093/cid/ciaa226. [Epub ahead of print]
21. Li Y, Zhao R, Zheng S, Chen X, Wang J, Sheng X et al (2020) Lack of vertical transmission of severe acute respiratory syndrome coronavirus 2, China. Emerg Infect Dis 26:6. https://doi.org/10.3201/eid2606.200287. [Epub ahead of print]
22. Chen R, Zhang Y, Huang L, Cheng BH, Xia ZY, Meng QT (2020) Safety and efficacy of different anesthetic regimens for parturients with COVID-19 undergoing Cesarean delivery: a case series of 17 patients. Can J Anaesth 67(6):655–663
23. Chen Y, Peng H, Wang L, Zhao Y, Zeng L, Gao H et al (2020) Infants born to mothers with a new coronavirus (COVID-19). Front Pediatr 8:104. https://doi.org/10.3389/fped.2020.00104

Relationship Between COVID-19 and Angiotensin-Converting Enzyme 2: A Scoping Review

5

Asma Shojaee, Amir Vahedian-Azimi, Fakhrudin Faizi, Farshid Rahimi-Bashar, Alireza Shahriary, Hadi Esmaeili Gouvarchin Galeh, Batool Nehrir, Paul C. Guest, and Amirhossein Sahebkar

Abstract

Following the outbreaks of SARS-CoV in 2002 and MERS-CoV in 2012, the COVID-19 pandemic caused by the SARS-CoV-2 virus has become an increasing threat to human health around the world. Numerous studies have shown that SARS-CoV-2 appears similar to the SARS-CoV as it uses angiotensin converting enzyme 2 (ACE2) as a receptor to gain entry into cells. The main aims of this scoping review were to identify the primary hosts of coronaviruses, the relationship between the receptor binding domain of coronaviruses and ACE2, the organ specificity of ACE2 expression compared with clinical manifestations of

A. Shojaee
Student Research Committee, Nursing Faculty, Baqiyatallah University of Medical Sciences, Tehran, Iran

A. Vahedian-Azimi (✉)
Trauma Research Center, Nursing Faculty, Baqiyatallah University of Medical Sciences, Tehran, Iran
e-mail: amirvahedian63@gmail.com

F. Faizi
Atherosclerosis research center, Nursing Faculty, Baqiyatallah University of Medical Sciences, Tehran, Iran

F. Rahimi-Bashar
Anesthesia and Critical Care Department, Hamadan University of Medical Sciences, Hamadan, Iran

A. Shahriary
Chemical Injuries Research Center, Systems Biology and Poisonings Institute, Baqiyatallah University of Medical Sciences, Tehran, Iran

H. E. G. Galeh
Applied virology research center, Baqiyatallah University of medical science, Tehran, Iran

B. Nehrir
Health Management Research Center, Nursing Faculty, Baqiyatallah University of Medical Sciences, Tehran, Iran

P. C. Guest
Laboratory of Neuroproteomics, Department of Biochemistry and Tissue Biology, Institute of Biology, University of Campinas (UNICAMP), Campinas, Brazil

A. Sahebkar (✉)
Neurogenic Inflammation Research Center, Mashhad University of Medical Sciences, Mashhad, Iran

Biotechnology Research Center, Pharmaceutical Technology Institute, Mashhad University of Medical Sciences, Mashhad, Iran

Polish Mother's Memorial Hospital Research Institute (PMMHRI), Lodz, Poland

Halal Research Center of IRI, FDA, Tehran, Iran
e-mail: amir_saheb2000@yahoo.com

P. C. Guest (ed.), *Clinical, Biological and Molecular Aspects of COVID-19*, Advances in Experimental Medicine and Biology 1321, https://doi.org/10.1007/978-3-030-59261-5_5

the disease, and to determine if this information can be used in the development of novel treatment approaches for the COVID-19 pandemic.

Keywords

SARS-CoV · SARS-CoV-like · SARS-CoV-2 · COVID-19 · Angiotensin converting enzyme 2 · ACE2 · Spike protein · Receiver connection range · Bat-SARS-CoV

5.1 Introduction

Nidovirales encompasses three viral families known as Coronaviridae, Arteriviridae, and Roniviridae. Although these have common genomic characteristics and use the same strategy for replication inside hosts, they differ in morphology. The main pathogenic forms to humans involve two genera known as coronavirus and torovirus. Coronaviruses are spherical enveloped viruses with a diameter of 100–120 nm and contain a single-core RNA genome with positive polarity. They gained the "Corona" nomenclature due to their spike proteins having a similar appearance to a crown in electron micrographs (Fig. 5.1). These viruses also contain significantly more RNA than most other viruses at 27–32 kilobytes in length. The fast multiplicity of coronaviruses confers their high recombination capacity [1]. This makes them highly infectious in avian and mammalian species.

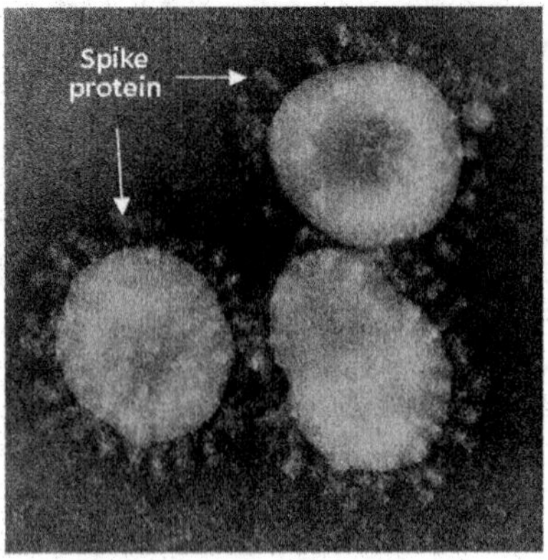

Fig. 5.1 Electron micrograph of SARS-CoV

Based on genomic sequences, coronaviruses can be divided into four groups known as, alpha (HCoV-229E, HCoV-NL63), beta (HCoV-OC43, HCoV-HKU1, MHV, SARS-COV, MERS-COV), gamma (IBV), and delta (pdCoV) [2–4]. The alpha and beta forms are infective in mammals, the gamma forms appear specific for birds, and the delta form is less defined. The Severe Acute Respiratory Syndrome epidemic of November 2002 to July 2003 was caused by a beta-coronavirus (SARS-CoV). This first erupted in the Guangdong province of China in November 2002 and spread to approximately 30 countries or territories, such as Hong Kong, Taiwan, Canada, Singapore, Vietnam, USA, and the Philippines [5]. Within 9 months, no new cases were reported but a total of 8098 people had been infected and, of these, 774 had died. Thus, the death rate of SARS-CoV was almost 10% of the infected population. An eruption of another beta-coronavirus known as Middle East Respiratory Syndrome (MERS-Cov) began in September 2012 with the majority of cases occurring in Saudi Arabia and some spreading to other countries, such as United Arab Emirates, Jordan, Qatar, and South Korea [6].

Late in 2019, a novel coronavirus erupted in the city of Wuhan of the Hui province of China. This virus was named SARS-CoV-2 and the World Health Organization (WHO) named the disease COVID-19 (for coronavirus disease 2019) [7, 8]. The recurrence and fulminant spreading of SARS-CoV-2 indicated that it was a potential threat to health around the world. The genome of SARS-CoV-2 is more similar to other beta-coronaviruses such as those from bats, as well as SARS-CoV and MERS-CoV. Early manifestations of the disease are fever, fatigue, dry coughing, myalgia, and dyspnea. Some patients may report headache, vertigo, stomach ache, diarrhea, nausea, and vomiting. In addition, some cases may progressively develop respiratory distress leading to alveolar injury and death [9].

The first step that occurs during a viral infection relies on the ability of the virus to enter the cells of the host via recognition and attachment to

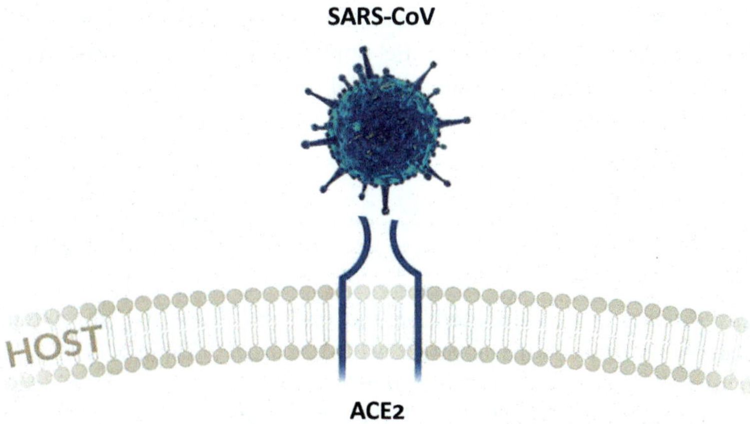

Fig. 5.2 Scheme showing the interaction between SARS-CoV and ACE2 to gain entry into host cells

a specific receptor [10]. Many studies have reported that the SARS-CoV receptor is angiotensin converting enzyme 2 (ACE2) (Fig. 5.2) [11, 12]. As the pandemic progressed, more studies were carried out on this topic and these confirmed that the novel coronavirus also uses ACE2 to gain entry into host cells [13–16]. ACE2 is homologous to ACE that regulates blood pressure, fluid and electrolyte balance, and systemic vascular resistance [17]. In this pathway, renin converts angiotensinogen to angiotensin 1 (AGT-I) and ACE converts AGT-I to AGT-II. In turn, AGT-II acts on the adrenal gland, causing it to release aldosterone. ACE2 converts AGT-I to AGT (1–9) and AGT-II to AGT (1–7) which bind to the mitochondrial assembly receptor (MAS), leading to antagonism of a wide variety of the effects of AGT-II. In general, ACE2 acts as a counter- regulatory enzyme that decreases the local concentration of AGT-II [18].

There are also two types of ACE2 with respect to functional characteristics. ACE2 contains a trans-membrane domain that connects its extra-cellular domain, which can act as a receptor for coronavirus spike proteins [11–16]. ACE2 is expressed in many cell types, especially pulmonary pneumocytes, myocardium cells, cholangiocytes, proximal tubules of the kidney, surface enterocytes of the intestines, cholecyst cells, lymphatic endothelial cells, epithelial cells of the bladder, corporeal cytotrophoblasts, and syncytiotrophoblasts, and it is also found in the eyes, epithelial cells of the mouth cavity, monocytes and macrophages, parietal cells of the stomach, the external layer of the adrenal glands, pancreatic islet cells, acidophilic cells of parathyroid glands, epithelial cells of sweat glands, and acidophilic cells of the pituitary [17, 18].

The spike proteins of SARS-CoV-2 provide the mechanism that allows it to enter cells in a manner similar to that used by the SARS coronavirus [13–16]. The spike protein contains two domains known as S1 and S2, and the receptor binding domain (RBD) is the main functional determinant within the S1 region that plays a crucial role in binding to ACE2 [19]. Species like civets, horseshoe bats, ferrets, golden Syrian hamsters, rabbits, turtles, monkeys, cows, sheep, pigs, weasels, and raccoon dogs are potential hosts for SARS-CoV-2 due to their inherent ACE2 receptors [20]. Studies of the RBD amino acid sequences of coronaviruses and the ACE2 attachment site have led to some information on severity of infections as well as the identity of potential intermediate hosts [11–16, 19, 20]. In general, a more comprehensive understanding of ACE2 expression regarding cells, tissues, organs and host species, as well as on the evolution and adaptability of the coronavirus spike proteins, may aid our development of effective treatments.

With this in mind, the aims of this review were to: 1) identify the primary reservoirs and intermediate hosts of coronaviruses; 2) explore the interaction between the coronavirus spike proteins and ACE2; 3) determine if any relationship exists between ACE2 tissue expression and the clinical manifestations of coronavirus infection; and 4) use this information to provide potential insights into novel treatment strategies against COVID-19.

5.2 Methods

This scoping review focused on the probable relationship between the novel COVID-19 coronavirus, SARS-CoV-2, and the ACE2 receptor. The selection process followed the Preferred Reporting Items for Systematic Reviews and Meta-Analyses (PRISMA) criteria (Fig. 5.3).

5.2.1 Data Sources and Research Strategies

All published and unpublished (gray literature) works up to the 21st of March 2020 were investigated. At first, suitable and related keywords were defined by the research team then the fields of title, abstract, keywords, topic, title/abstract were examined using the English language databases of Scopus, Web of science, ProQuest, Embase, and PubMed. Medical Subject Headings (MeSH) databases were also assessed and related synonyms were applied to increase the comprehensiveness of the study and minimize attrition. In addition unique Boolean syntax and operators related to each database were applied to extend the scope of the search (Table 5.1).

5.2.2 Study Selection

In the first stage of the search, all English-language studies were tracked considering title and abstracts, and papers addressing the key points were included. This included studies reporting on angiotensin converting enzyme 2 (ACE2) or SARS-like coronavirus in any hosts, studies covering any relation between the spike protein residues of coronaviruses and amino acid sequences of ACE2, as well as studies related to

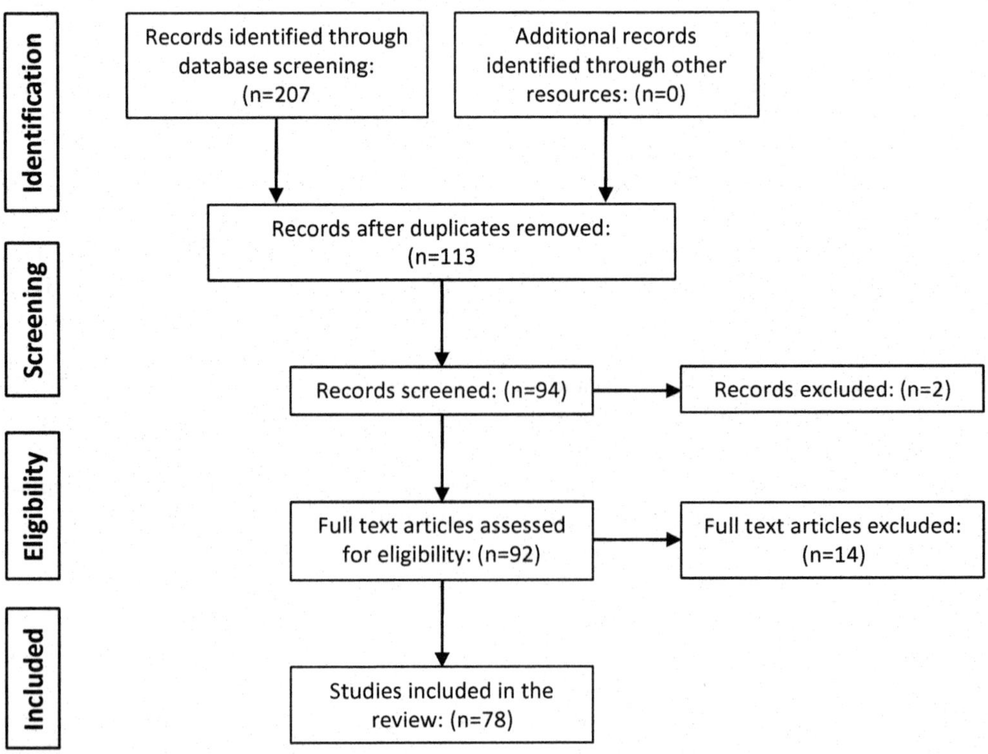

Fig. 5.3 Scheme showing the study selection procedure

Table 5.1 Search terms and databases

Scopus: 74 TITLE ("angiotensin converting enzyme 2" OR ace2) AND TITLE ("SARS CoV" OR coronavirus OR covid OR "SARSr CoV" OR "MERS CoV" OR ncov)
PubMed: 72 ("angiotensin converting enzyme 2" [title] OR ace2 "angiotensin converting enzyme 2" [title] OR ace2 [title]) AND ("SARS CoV" [title] OR coronavirus [title] OR covid [title] OR "SARSr CoV" [title] OR "MERS CoV" [title] OR ncov [title])
ProQuest: 171 ("angiotensin converting enzyme 2" OR ace2) AND ti ("SARS CoV" OR coronavirus OR covid OR "SARSr CoV" OR "MERS CoV" OR ncov)
Web of science: 65 ("angiotensin converting enzyme 2" OR ace2) AND TITLE: ("SARS CoV" OR coronavirus OR covid OR "SARSr CoV" OR "MERS CoV" OR ncov)
EMBASE: 79 ("angiotensin converting enzyme 2":Ti OR ace2:Ti) AND ("sars cov":Ti OR coronavirus:Ti OR covid:Ti OR "sarsr cov':Ti OR "mers cov":Ti OR ncov:Ti)

Table 5.2 Inclusion and exclusion criteria for selected articles

Inclusion criteria	Exclusion criteria
Published in English	Not written in English
Published between January 2003 and March 2020	Literature that did not include empirical data (letters, editorials, news, etc.)
Focus on relationship between ACE2 and SARS-coronaviruses	Articles found not be relevant

expression of ACE2 in cells, tissues, and body organs. The refining process was done by all the research team members to increase accuracy. The inclusion and exclusion criteria are summarized in Table 5.2.

5.2.3 Listing and Exploring Data and Analyzing the Studies

The format for listing data followed the Joanna Briggs Institute (JBI) approach as an accepted methodology for scoping reviews. The research team decided how to search questions. After discussions and investigations, the following topics were explored: 1) the country (or countries) in which the study was carried out, 2) study type, 3) the aims, and 4) the main findings. To increase accuracy, two external reviewers checked and explored the results separately. General conformity was obtained by discussions in cases of disagreement between the team members or between the team members and external reviewers. Kendall's coefficient of concordance was acceptable between the research team ($r = 0.95$; $p < 0.0001$) and between the team and reviewers ($r = 0.93$; $p < 0.0001$).

5.3 Results

5.3.1 Search Outcomes

The PRISMA flow chart was used to illustrate the study selection process and results (Fig. 5.3). Across the five databases a total of 207 studies were retrieved. After removal of duplicates, 94 titles and abstracts were screened for relevance and two were removed. The remaining 92 full-text articles were screened for eligibility and 78 articles were considered directly related to the research questions and included for the synthesis.

5.3.2 Article Information

Among the 78 studies, 73 used laboratory methods [22–31, 33–37, 38–42, 44–47, 48, 49, 50, 52–97], 2 were reviews [21, 98], 1 was a title of book [51], 1 was correspondence [99], and 1 was a perspective [32].

The included studies originated from different countries and, based on frequencies, 26 were from China [25, 27, 30, 31, 33, 39–42, 44–46, 53, 56, 57, 61, 62, 65, 73, 74, 76, 77, 82, 86, 94, 97], 18 from the USA [21, 22, 24, 26, 28, 36, 37, 38, 47, 49, 55, 58, 68, 69, 72, 79, 84, 99], 15 were carried out as multinational collaboration [29, 32, 35, 50, 51, 54, 59, 66, 70, 75, 88, 90–92, 98], 5 were from Japan [34, 48, 64, 78, 87], 4 from Germany [38, 71, 85, 96], 3 from Poland [52, 83, 89], 3 were from Taiwan [63, 66, 67], 2 Holland [23, 60], 1 Israel [81], and 1 from South Africa and Tunisia [80].

5.3.3 Narrative Summary of Studies

The main topics that the 78 studies focused on were: 1) the primary or intermediate reservoirs of coronaviruses; 2) the relationship of spike protein of the viruses and ACE2 as the related receptor; 3) the expression of ACE2 in various body organs; and 4) the recommended medical strategies based on the relationship of the spike protein and ACE2.

5.3.4 Studies Addressing the Primary Reservoir and Intermediate Hosts of Coronaviruses

Fifteen relevant studies are summarized below:

1. Li et al. (2006) addressed the following questions [98]:
   (a) If bats are a reservoir of SARS-CoV-like viruses, when and in which species did these viruses acquire a spike protein capable of using palm civet and human ACE2?
   (b) Are changes in the spike protein which enhanced human-to-human transmission a likely consequence of incubation in palm civets and other animals or is it a unique event not likely to recur?
   (c) Did SARS-CoV gain the use of ACE2 through recombination and, if so, with what virus?
   (d) What changes in other viral proteins were necessary for SARS-CoV to infect humans efficiently?

This paper described the emergence of dangerous variants of common pathogens including HCoVNL63 and animal equivalents and discussed coping strategies of viruses such as recombination.

2. Heller et al. reported that both mink and palm civet had 83/87 amino acid identity/similarity with human ACE2 [22]. This study suggested mink as a potential reservoir of SARS coronavirus in North America and established it as a suitable animal model to study this virus.
3. Zamoto et al. showed that ferret ACE2 acts as a SARS-CoV receptor with similar efficiency as human ACE2 and with greater efficiency than mouse ACE2 [48].
4. In 2007, a study by Fukushi et al. showed that SARS-CoV needs to bind via the RBD in the spike protein to ACE2 [78].
5. Chen et al. reported that, in comparison to human ACE2, 38 nonsynonymous changes exist in Chinese rhesus-ACE2, but this is just as effective as the human homolog in supporting viral entry [77]. The study also highlighted a natural mutation of tyrosine to asparagine at position 217 that can lead to downregulation of human-ACE2 and reduce viral entry.
6. Guo et al. reported that the number of amino acid differences between human-ACE2 and cat, civet, mouse, and rat ACE2 was 3, 8, 9, and 11, respectively [76]. Since there is no difference in the binding ability of cat ACE2 to the SARS-CoV spike protein, the possibility of zoonotic transmission of SARS-CoV from animals to humans is supported and, of the species tested, the cat ACE2 sequence was evolutionarily the closest.
7. Xu et al. stated there are six amino acid differences in raccoon dog ACE2 compared with human ACE2 and concluded that the raccoon dog may serve as a critical intermediate host for SARS-CoV and may have played a key role in SARS-CoV outbreaks [25].
8. In 2010, Hou et al. pointed out that two bat species, *Myotis daubentoni* and *Rhinolophus sinicus*, are likely to be susceptible to SARS-CoV and may be candidates as the natural host of the SARS-CoV progenitor or virus [75].
9. The study run by Demogines et al. reported that ACE2 utilization preceded the emergence of SARS-CoV-like viruses from bats [47]. Their results were consistent with a model in which an ACE2-utilizing bat coronavirus infected civets and/or other intermediate hosts, or possibly even humans directly.
10. Li et al. noted that human, civet, mouse, cat, golden Syrian hamster, and horseshoe bat support infection of SARS-CoV [45]. Therefore, comprehensive surveillance of these animals is suggested when SARS or SARS-like CoVs reemerge in the human

population in the future. This study also reported that rabbits and horseshoe bats are animal carriers of SARS-CoV.

11. Ge et al. identified two coronaviruses from Chinese horseshoe bats, RsSHC014 and Rs3367, which had the highest similarity to SARS-CoV, compared to other bat coronaviruses [29]. The similarity was highest in the RBD of the spike protein, supporting the case that these bat species as natural reservoirs of SARS-CoV.
12. Recently Cao et al. found that East Asian populations have higher allele frequencies in expression of quantitative trait loci variants associated with higher ACE2 expression in tissues [74]. This may indicate different susceptibilities or responses to SARS-CoV-2 infection in different populations.
13. Li et al. emphasized potential interspecies transmission of SARS-CoV-2 and the need for further surveillance in animal populations [46]. They found that the ACE2 amino acid positions 30–41, 82–84, and 353–357 are important in the interaction with SARS-CoV and amino acids 31, 35, 38, 82, and 353 are critical. As humans and nonhuman primates (gibbon, monkey, macaque, orangutan, and chimpanzee) showed identical sequences over these regions, this makes them potential hosts of SARS-CoV-2.
14. Another study by Liu et al. confirmed that other than pangolins and snakes, turtles are also potential intermediate hosts for transmission of SARS-CoV-2 to humans [73].

5.3.5 Studies Addressing the Interaction between Coronavirus Spike Proteins and ACE2

We found 27 studies which addressed this point:

1. Kuhn et al. reported ACE as a receptor for coronaviruses [21]. The paper stated that studying the receptor in detail is needed to progress in development of anti-viral drugs, vaccines, and animal models to survey pathogenesis of SARS-CoV. He concluded that the major questions that still need to be answered are the following: 1) Is ACE2 the only cellular factor permitting SARS-CoV cell entry or are co-receptors involved? 2) Does the inflammatory response to SARS-CoV infection lead to upregulation of ACE2 expression in lung tissue?
2. Hofmann et al. pointed out the central role of ACE2 in SARS-CoV infection and a minor contribution of the ACE2 cytoplasmic domain to receptor function [71].
3. Prabakaran et al. identified a deep channel on the top of the ACE2 molecule that contains the catalytic site and negatively charged ridges surrounding the channel that may provide a possible binding site for the positively charged receptor-binding domain of the spike protein [72]. He also noticed hydrophobic patches around the charges that could contribute to binding and the lack of carbohydrates at the top of the molecule could enable high-affinity binding.
4. Wong et al. stated that a 193-amino acid fragment of the spike protein (residues 318–510) bound to ACE2 more efficiently than did the full S1 domain (residues 12–672) [49]. In addition, smaller spike protein fragments, expressing residues 327–510 or 318–490, did not bind ACE2.
5. In their study, Zhang et al. reported that a SARS-CoV spike protein S1 residue (arginine 453) and an ACE2 residue (lysine 341) appear to be involved in the binding of SARS-CoV to ACE2 [50].
6. Li et al. carried out a study which found that the lower affinity of three SARS-CoV spike proteins from the less severe 2003–2004 outbreak could be enhanced by altering specific residues within the spike protein-binding site of human ACE2 to those of civet ACE2, or by altering spike protein residues 479 and 487 to those that were present in the more severe 2002–2003 outbreak. This study suggested that the reason for the low prevalence and intensity of SARS 2003–2004 outbreak was due to lower affinity of the spike protein of this coronavirus to bind ACE2 [93].

7. Lambert et al. showed that ADAM metallopeptidase domain 17 (ADAM17) is the protease responsible for ACE2 shedding [92].
8. Huang et al. reported on two coronaviruses (SARS-CoV, HCoV-NL63) that both utilize the ACE2 receptor, but enter cells through distinct mechanisms [24]. Specifically, only SARS-CoV utilized the enzymatic activity of the cysteine protease cathepsin L to infect ACE2-expressing cells.
9. Smith et al. reported that although the spike glycoprotein of HCoV-NL63 shares only 25% amino acid sequence identity with that of SARS-CoV, both viruses used ACE2 as a receptor [52]. This suggested that both viruses evolved separately to bind to the same receptor.
10. Pöhlmann et al. described how the ACE2 receptor was used for viral entry by CoV-NL63 despite little homology between this coronavirus and SARS-CoV [51].
11. The study run by Inoue et al. concluded that SARS-CoV mainly utilizes the clathrin-mediated endocytosis pathway for its entry into target cells and the cytoplasmic tail of ACE2 is not required for the penetration of SARS-CoV into cells [34].
12. The study of Li et al. noted that the spike proteins of SARS-CoV and HCoV-NL63 bind overlapping regions of ACE2 that include a critical loop between beta-strands IV and V [91]. In addition, changes to ACE2 residue 354, at the boundary of the SARS-CoV binding site, markedly inhibited utilization by HCoV-NL63 but not by SARS-CoV spike proteins.
13. Glende et al. in their study highlighted that cholesterol-rich micro-domains provide a platform facilitating efficient interaction of the SARS-CoV spike protein with ACE2 [90].
14. Mathewson et al. showed that the NL63 coronavirus spike protein has a weaker interaction with ACE-2 than the SARS-CoV spike protein [89].
15. Lin et al. reported that the NL63 coronavirus receptor binding domain binds to human ACE2 more efficiently than its full-length counterpart, with a binding efficiency comparable to the S1 or receptor binding domain of SARS-CoV [88].
16. Yoshikawa et al. reported that both AC70 and AC22 transgenic mice expressing the human ACE2 receptor were permissive to SARS-CoV infection, and caused elevated secretion of many inflammatory mediators within the lungs and brains, although infection was more intense with higher immunosuppression in AC70 than in AC22 mice, especially in the brain [26].
17. Haga et al. identified multiple ACE2-truncated variants that lost the SARS-CoV spike protein-induced shedding of ACE2 and TNF-α production in lung tissue [87].
18. A study by Chen et al. showed that the viral spike protein led to upregulation of fibrosis-associated chemokine ligand 2 (CCL2) and production of virus-like particles, and this was mediated by extracellular signal-regulated kinase 1 and 2 (ERK1/2) and the activator 1 protein (AP-1) transcription factor but not by the IκBα-NF-κB signaling pathway [86].
19. Glowacka et al. reported that SARS-CoV but not NL63 coronavirus replicated efficiently in ACE2-positive cells and reduced ACE2 expression [85].
20. The study of Wu et al. noted that binding to the same hot spot on human ACE2 was likely to be an outcome of convergent evolution by NL63-CoV and SARS-CoV [84].
21. Dijkman et al. showed that decreased ACE2 expression is dependent on the efficiency of NL63 coronavirus replication, and that NL63-CoV and SARS-CoV both affect cellular ACE2 expression during infection [83].
22. The study of Heurich et al. resulted in transmembrane protease serine 2 (TMPRSS2) but not ADAM17 protease promotion of SARS-CoV entry by two separate pathways: 1) ACE2 cleavage, which might promote viral uptake; and 2) SARS spike protein cleavage, which activates this protein for membrane fusion [96].
23. Song et al. showed that the spike glycoprotein retains the pre-fusion trimer structure

after trypsin cleavage and low-pH treatment [82]. Also, binding with the host cell receptor ACE2 promotes the release of S1 subunits from the S trimer and triggers the pre- to post-fusion conformational transition.

24. Brielle et al. described the evolution of coronaviruses (SARS-CoV, SARS-CoV-2, and NL63-CoV) towards host recognition [81].
25. Lan et al. suggested that SARS-CoV-2 is similar to SARS-CoV and reported that the similarities in structure and sequence of these two coronaviruses argue for convergent evolution towards improved binding to ACE2 [44].
26. Othman et al. reported that the interface segment of the spike protein RBD might have been acquired by SARS-CoV-2 via a complex evolutionary process rather than mutation accumulation [80].
27. Yan et al. showed that SARS-CoV-2 recognizes an ACE2 dimer that complexes with a membrane protein, and drugs which disrupt this interaction may be effective in reducing infection [31].

5.3.6 Studies Investigating the Relationship between ACE2 Expression and Clinical Manifestations of COVID-19 Infection

Seventeen studies addressed this topic:

1. To and Lo found that although ACE2 is expressed at high levels in pneumocytes and surface enterocytes of the small intestine, the tissue responses in these two organs are different [27]. They also found that the presence of ACE2 is not enough for coronavirus infection and that other receptors or cofactors may be required in some tissues.
2. Hamming et al. studied expression of the ACE2 protein on lung alveolar epithelial cells and enterocytes of the small intestine [60]. This revealed that ACE2 was present in arterial and venous endothelial cells, and arterial smooth muscle cells in all organs studied.
3. Mossel et al. reported that the human colon epithelial line CaCo-2 was the only human cell type out of 13 tested that supported efficient SARS-CoV replication [28].
4. The study by Jia et al. showed that ACE2 was more abundantly expressed on the apical surface of polarized epithelia, and well-differentiated cells support viral entry and replication [37].
5. Ren et al. showed that ACE2 is localized on the apical plasma membrane of polarized respiratory epithelial cells and mediates infection from the apical side of these cells [59].
6. Li et al. noted that both SARS-CoV receptors (ACE2 and CD209L) are expressed in organ/tissue-derived endothelial cells. The expression of the ACE2 receptor was highest in human lung microvascular endothelial cells, and expression of CD209L was higher in lymphatic endothelial cells [43].
7. Tseng et al. showed that pre-inflammatory mediators and viral titer were high in lung and brain of transgenic mice expressing ACE2 [58].
8. Yang et al. showed that SARS-CoV replicated more efficiently in lungs of ACE2 transgenic mice than in those of wild-type mice. Similar signs (vasculitis, degeneration, and necrosis) were also seen in other organs [57].
9. Dong et al. reported the mRNA of human ACE2 was expressed efficiently in normal lung tissue, but not in cartilage and cancellous bone under the weight-bearing area of the femoral head [56].
10. Netland et al. found that neurons are a susceptible target for SARS-CoV and that only the absence of host cell receptors prevents severe murine brain disease [55].
11. A study by Oudit et al. focused on myocardium showed that that SARS-CoV can mediate inflammation and damage associated with downregulation of the myocardial ACE2 system, which may be responsible for

the myocardial dysfunction and adverse cardiac outcomes in patients with SARS [54].
12. Chai et al. showed that SARS-CoV-2 might directly bind to ACE2 positive cholangiocytes but not necessarily to hepatocytes [53].
13. Deng et al. showed expression of ACE2 and TMPRSS2 in human kidney proximal tubules, indicating that the kidney is a potential target organ of SARS-CoV-2 infection [42].
14. Ji et al. showed that after triggering functional changes in ACE2, an imbalance in the steady-state cytokine regulatory axis involving the renin–angiotensin system and IP-10 leads to a cytokine storm [94].
15. Li et al. reported that the SARS-CoV-2 receptor ACE2 was widely spread in specific cell types of the maternal–fetal interface [41].
16. Lin et al. showed high ACE2 gene expression in all subtypes of kidney proximal tubule cells and low expression in bladder epithelial cells [39].
17. Xu et al. reported ACE2 expression on the mucosa of the oral cavity and epithelial cells of tongue [30].

5.3.7 Studies Investigating New Treatment Strategies for COVID-19 Infection

Twenty studies regarding new treatment approaches are summarized below:

1. Han et al. showed that a peptide derived from ACE2, which consisted of two discontinuous parts of ACE2 (amino acids. 22–44 and 351–357), was a good candidate for the treatment of coronary heart disease [69].
2. Li et al. described ACE2 as a functional receptor for SARS-CoV and showed that a solution form of ACE2 rather than ACE1 could block the spike S1 domain [36]. This suggested the potential use of ACE2 antibodies as a treatment for SARS infection, which may also be applicable to COVID-19 cases.
3. The findings of Moore et al. were in line with those of Li [70].
4. Batlle also recently reported that a soluble recombinant form of ACE2 appeared to neutralize SARS-CoV-2 in vitro [99].
5. Hoffmann et al. showed that SARS-CoV-2 uses ACE2 as a receptor and TMPRSS2 for spike protein priming [38]. This study supported that case that TMPRSS2 inhibitors might be a treatment option. The study also showed that sera from convalescent SARS-CoV patients cross-neutralized viral entry and could therefore provide a treatment and/or a vaccination strategy for patients with COVID-19.
6. Lei et al. generated a fusion protein containing the RBD of the SARS-CoV spike protein linked to the Fc portion of human IgG1 and found that this could be internalized into SARS-CoV-susceptible cells with ACE2 [61]. This may also have some implications for vaccine development [61].
7. Ho et al. also showed in their study that peptides derived from the spike protein, especially the use of amino acid residues 668–679, can compete with the ACE2-coronavirus interaction and prevent infection [67].
8. Kuba et al. found that recombinant spike IgG-Fc proteins can block coronary artery disease associated with SARS-CoV [35]. This study also introduced the idea of using ACE2 inhibitors as a way to reduce injury and pulmonary edema.
9. Zhang et al. showed that recombinant spike S1 subunit proteins (amino acid residues 388 to 496) can induce protective neutralizing antibodies against SARS-CoV [65].
10. Wang et al. also found that a SARS-CoV-RBD-IgG-Fc protein could bind to ACE2, again suggesting this as a potential vaccine approach [62].
11. de Lang et al. reported that the anti-inflammatory cytokines interferon-γ and interleukin (IL)-4 could reduce effects of coronary artery disease via reduced ACE2 expression [23].
12. He et al. showed that infection caused by coronaviruses can cause pro-inflammatory cytokines (MCP-1 and TGF-β1, TNF-α,

IL-1β, IL-6) in pneumocystis and macrophages of the lungs and bronchi, which can lead to acute lung damage [68]. This supports the use of anti-inflammatory cytokines as a therapeutic strategy.

13. Haga, S attributes the production of inflammatory cytokines, especially TNF-α, to the stimulation of the 2019-nCoV spike and the cytoplasmic tail of ACE2. This is a multifaceted interaction between the production of pre-inflammatory cytokines, protein spike SARS-CoV, and ACE2 [64].
14. Yan et al. showed that an siRNA approach can effectively prevent viral replication by targeting the ACE2 gene or viral nucleocapsid protein [66].
15. Lu et al. also showed that downregulation of ACE2 expression using an siRNA approach could effectively reduce the proliferation of SARS-CoV [63].
16. Wang et al. also showed that reducing expression of ACE2 by siRNA, makes ACE2 a therapeutic target [33].
17. Wu et al. suggested four potential treatment options for coronavirus infections: 1) the use of ACE2 recombinant proteins; 2) use of ACE2 inhibitors such as lisinopril; 3) use of ACE2 blockers such as losartan; and the use of angiotensin (7-1) to activate the MAS receptor for ACE2 neutralization [84].
18. Zhang et al. also provided treatment strategies for COVID-19 infection based on the role of ACE2, which included: 1) the use of vaccines against the spike protein; 2) the use of serum protease inhibitors against TMPRSS2; 3) blockade of ACE2 with small molecules; and 4) use of the ACE2 soluble form that binds competitively to the SARS-CoV spike protein [32].
19. Ho et al. reported on a number of small molecules that disrupted the SARS-CoV – ACE2 interaction and could therefore be promising leads for development of novel treatments for COVID-19 disease [67].

5.4 Discussion

To the best of our knowledge, this is the first scoping review on the SARS-CoV-2 which aims to integrate the existing knowledge on the primary hosts of coronaviruses, the relationship between the receptor binding domain of coronaviruses and the likely host cell receptor ACE2, the organ specificity of ACE2 expression compared with clinical manifestations of the disease, and whether or not this information can be used for development of novel treatment approaches.

In the case of the SARS-CoV, exotic marketplace animals were probably the immediate origin of the virus [100]. These animals included palm civets as the likely carriers since SARS-CoV could be isolated from these animals. In addition, the infections which occurred coincided with the preparation and consumption of palm civet meat products in restaurants. SARS-CoV infections of other marketplace species have also been observed such as the cat, red fox, and badger. Although these species may be an immediate source of SARS-CoV infections in humans, it is likely that they serve as a conduit of the virus from another reservoir species. The most likely of these reservoirs includes certain bat species such as the horseshoe bat [100].

For SARS-CoV-2, 6 amino acids in the RBD of spike protein amino have been found to be critical for ACE2 binding and host determination [101]. Interestingly, 5 of these amino acids differ between SARS-CoV-2 and SARS-CoV which seems to confer a higher affinity of SARS-CoV-2 to ACE2 in humans, cats, ferrets, and other species. As many early cases of SARS-CoV-2 infection were linked to the Huanan market in Wuhan, it is likely that bats served as the primary reservoir given the high genomic similarity of the RaTG13 bat coronavirus with SARS-CoV-2. In addition, illegally imported Malayan pangolins contain coronaviruses similar to SARS-CoV-2 especially within the RBD domain. This suggests that the SARS-CoV-2 spike protein was most likely optimized for binding to human-like ACE2 receptors by natural selection.

Taken together, this study provides insights into the spike protein of SARS-CoV-2 in relation to the probable host cell receptor, ACE2, in COVID-19 disease. Due to the diversity of coronavirus species transmission and the internal and intergenerational diversity of these viruses, the reservoir and intermediate host of SARS-CoV-2 is still not certain. However, as stated above, it is likely that the bat is the main animal reservoir and the results of a recent study are consistent with the pangolin being the intermediate host [102]. This latter study carried out molecular and phylogenetic analyses and showed that a pangolin coronavirus (pangolin-CoV-2020) is genetically related to SARS-CoV-2 and a group of bat coronaviruses and may therefore be natural hosts of betacoronaviruses. Thus, steps taken to minimize human exposure of humans to such wildlife will be important to reduce the risks of coronaviruses spreading from animals to humans.

In addition, it is still not clear if the interaction between ACE2 and the SARS-CoV-2 spike protein evolved separately or if they coevolved to permit the high infectivity of this coronavirus [103]. Recent studies have suggested that this could be due to the higher affinity of the SARS-CoV-2 spike protein receptor binding domain for ACE2 compared with other coronaviruses, such as SARS-CoV [104].

Although the clinical manifestations of COVID-19 disease are varied, at least some of these appear to be due to the targeting of ACE2 in different tissues and organs of the body. Although the virus likely enters the body at the level of the respiratory system due to the high levels of ACE2 expression there, the virus can spread out and cause damage to other vital organs and tissues expressing ACE2, triggering a wide spectrum of pathophysiological effects and symptoms, including digestive [105], neurological [106], and cardiovascular complications [107].

5.5 Conclusions and Future Perspectives

There is currently no proven effective treatment for COVID-19 disease and development of a safe and effective vaccine could take from 6 months to one and half years. However, since the virus gains access to the respiratory system through the cell surface ACE2 protein, a number of strategies are currently being explored to target this interaction [108–112]. One incredible feature of the COVID-19 pandemic has been the worldwide efforts to develop new treatments and vaccines to halt its spread and to raise our awareness of the dangers of pandemics due to such viruses and other pathogens. The emergence of COVID-19 highlights the critical importance of establishing a systematic coronavirus surveillance network. In addition, the current pandemic has instilled in all of us the value of setting in place a worldwide coronavirus surveillance network to prevent such events from reaching the dangerous levels that this one has and to manage outbreaks more effectively in the future.

Acknowledgements Thanks to guidance and advice from the Clinical Research Development Unit of Baqiyatallah Hospital.

Ethics Approval and Consent to Participate IR. BMSU.REC.1399.011.

Competing Interests The authors have no conflicts of interest.

References

1. Sanjuán R, Domingo-Calap P (2016) Mechanisms of viral mutation. Cell Mol Life Sci 73(23):4433–4448
2. Cui J, Li F, Shi ZL (2019) Origin and evolution of pathogenic coronaviruses. Nat Rev Microbiol 17(3):181–192
3. Lim YX, Ng YL, Tam JP, Liu DX (2016) Human coronaviruses: a review of virus-host interactions. Diseases 4(3):26. https://doi.org/10.3390/diseases4030026
4. https://www.cdc.gov/coronavirus/types.html
5. https://www.who.int/ith/diseases/sars/en/
6. https://www.who.int/emergencies/mers-cov/en/
7. Li H, Liu SM, Yu XH, Tang SL, Tang CK (2020) Coronavirus disease 2019 (COVID-19): current status and future perspectives. Int J Antimicrob Agents 55(5):105951. https://doi.org/10.1016/j.ijantimicag.2020.105951
8. https://www.who.int/emergencies/diseases/novel-coronavirus-2019/technical-guidance/naming-the-coronavirus-disease-(covid-2019)-and-the-virus-that-causes-it
9. Yuki K, Fujiogi M, Koutsogiannaki S (2020) COVID-19 pathophysiology: a review. Clin

Immunol 215:108427. https://doi.org/10.1016/j.clim.2020.108427
10. Belouzard S, Millet JK, Licitra BN, Whittaker GR (2012) Mechanisms of coronavirus cell entry mediated by the viral spike protein.Viruses 4(6):1011-1033
11. Turner AJ, Hiscox JA, Hooper NM (2004) ACE2: from Vasopeptidase to SARS virus receptor. Trends Pharmacol Sci 25(6):291–294
12. Hofmann H, Pöhlmann S (2004) Cellular entry of the SARS coronavirus. Trends Microbiol 12(10):466–472
13. Kruse RL (2020) Therapeutic strategies in an outbreak scenario to treat the novel coronavirus originating in Wuhan, China. F1000Res 9:72. https://doi.org/10.12688/f1000research.22211.2
14. Peng X, Xu X, Li Y, Cheng L, Zhou X, Ren B (2020) Transmission routes of 2019-nCoV and controls in dental practice. Int J Oral Sci 12(1):9. https://doi.org/10.1038/s41368-020-0075-9
15. Sun Z, Thilakavathy K, Kumar SS, He G, Liu SV (2020) Potential factors influencing repeated SARS outbreaks in China. Int J Environ Res Public Health 17(5):1633. https://doi.org/10.3390/ijerph17051633
16. Kannan S, Shaik Syed Ali P, Sheeza A, Hemalatha K (2020) COVID-19 (novel coronavirus 2019) - recent trends. Eur Rev Med Pharmacol Sci 24(4):2006–2011
17. Lambert DW, Clarke NE, Turner AJ (2010) Not just angiotensinases: new roles for the angiotensin-converting enzymes. Cell Mol Life Sci 67(1):89–98
18. Zisman LS (2005) ACE and ACE2: a tale of two enzymes. Eur Heart J 26(4):322–324
19. Magrone T, Magrone M, Jirillo E (2020) Focus on receptors for coronaviruses with special reference to angiotensin-converting enzyme 2 as a potential drug target - a perspective. Endocr Metab Immune Disord Drug Targets. https://doi.org/10.2174/1871530320666200427112902. Online ahead of print
20. Sutton TC, Subbarao K (2015) Development of animal models against emerging coronaviruses: from SARS to MERS coronavirus. Virology 479-480:247–258
21. Kuhn J, Li W, Choe H, Farzan M (2004) Angiotensin-converting enzyme 2: a functional receptor for SARS coronavirus. Cell Mol Life Sci 61(21):2738–2743
22. Heller LK, Gillim-Ross L, Olivieri ER (2006) Wentworth DE (2006) Mustela vison ACE2 functions as a receptor for SARS-coronavirus. Adv Exp Med Biol 581:507–510
23. de Lang A, Osterhaus AD, Haagmans BL (2006) Interferon-γ and interleukin-4 downregulate expression of the SARS coronavirus receptor ACE2 in Vero E6 cells. Virology 353(2):474–481
24. Huang I-C, Bosch BJ, Li F, Li W, Lee KH, Ghiran S et al (2006) SARS coronavirus, but not human coronavirus NL63, utilizes cathepsin L to infect ACE2-expressing cells. J Biol Chem 281(6):3198–3203
25. Xu L, Zhang Y, Liu Y, Chen Z, Deng H, Ma Z et al (2009) Angiotensin-converting enzyme 2 (ACE2) from raccoon dog can serve as an efficient receptor for the spike protein of severe acute respiratory syndrome coronavirus. J Gen Virol 90(11):2695–2703
26. Yoshikawa N, Yoshikawa T, Hill T, Huang C, Watts DM, Makino S et al (2009) Differential virological and immunological outcome of severe acute respiratory syndrome coronavirus infection in susceptible and resistant transgenic mice expressing human angiotensin-converting enzyme 2. J Virol 83(11):5451–5465
27. To K, Lo AW (2004) Exploring the pathogenesis of severe acute respiratory syndrome (SARS): the tissue distribution of the coronavirus (SARS-CoV) and its putative receptor, angiotensin-converting enzyme 2 (ACE2). J Pathol 203(3):740–743
28. Mossel EC, Huang C, Narayanan K, Makino S, Tesh RB, Peters C (2005) Exogenous ACE2 expression allows refractory cell lines to support severe acute respiratory syndrome coronavirus replication. J Virol 79(6):3846–3850
29. Ge XY, Li JL, Yang XL, Chmura AA, Zhu G, Epstein JH et al (2013) Isolation and characterization of a bat SARS-like coronavirus that uses the ACE2 receptor. Nature 503(7477):535–538
30. Xu H, Zhong L, Deng J, Peng J, Dan H, Zeng X et al (2020) High expression of ACE2 receptor of 2019-nCoV on the epithelial cells of oral mucosa. Int J Oral Sci 12(1):1–5
31. Yan R, Zhang Y, Guo Y, Xia L, Zhou Q (2020) Structural basis for the recognition of the 2019-nCoV by human ACE2. bioRxiv. https://doi.org/10.1101/2020.02.19.956946
32. Zhang H, Penninger JM, Li Y, Zhong N, Slutsky AS (2020) Angiotensin-converting enzyme 2 (ACE2) as a SARS-CoV-2 receptor: molecular mechanisms and potential therapeutic target. Intensive Care Med 46(4):586–590
33. Wang PH (2020) Increasing host cellular receptor—angiotensin-converting enzyme 2 (ACE2) expression by coronavirus may facilitate 2019-nCoV infection. bioRxiv. https://doi.org/10.1101/2020.02.24.963348
34. Inoue Y, Tanaka N, Tanaka Y, Inoue S, Morita K, Zhuang M et al (2007) Clathrin-dependent entry of severe acute respiratory syndrome coronavirus into target cells expressing ACE2 with the cytoplasmic tail deleted. J Virol 81(16):8722–8729
35. Kuba K, Imai Y, Rao S, Gao H, Guo F, Guan B et al (2005) A crucial role of angiotensin converting enzyme 2 (ACE2) in SARS coronavirus–induced lung injury. Nat Med 11(8):875–879
36. Li W, Moore MJ, Vasilieva N, Sui J, Wong SK, Berne MA et al (2003) Angiotensin-converting enzyme 2 is a functional receptor for the SARS coronavirus. Nature 426(6965):450–454
37. Jia HP, Look DC, Shi L, Hickey M, Pewe L, Netland J et al (2005) ACE2 receptor expression and severe acute respiratory syndrome coronavirus infection depend on differentiation of human airway epithelia. J Virol 79(23):14614–14621
38. Hoffmann M, Kleine-Weber H, Schroeder S, Krüger N, Herrler T, Erichsen S et al (2020) SARS-CoV-2 cell entry depends on ACE2 and TMPRSS2 and is

blocked by a clinically proven protease inhibitor. Cell 181(2):271–280.e8. https://doi.org/10.1016/j.cell.2020.02.052
39. Lin W, Hu L, Zhang Y, Ooi JD, Meng T, Jin P et al (2020) Single-cell analysis of ACE2 expression in human kidneys and bladders reveals a potential route of 2019-nCoV infection. bioRxiv. https://doi.org/10.1101/2020.02.08.939892
40. Zhao Y, Zhao Z, Wang Y, Zhou Y, Ma Y, Zuo W (2020) Single-cell RNA expression profiling of ACE2, the putative receptor of Wuhan 2019-nCov. bioRxiv. https://doi.org/10.1101/2020.01.26.919985
41. Li M, Chen L, Zhang J, Xiong C, Li X (2020) The SARS-CoV-2 receptor ACE2 expression of maternal-fetal interface and fetal organs by single-cell transcriptome study. PLoS One 15(4):e0230295. https://doi.org/10.1371/journal.pone.0230295
42. Deng YY, Zheng Y, Cai GY, Chen XM, Hong Q (2020) Single-cell RNA sequencing data suggest a role for angiotensin-converting enzyme 2 in kidney impairment in patients infected with 2019-nCoV. Chin Med J 133(9):1129–1131
43. Li GM, Li YG, Yamate M, Li S-M, Ikuta K (2007) Lipid rafts play an important role in the early stage of severe acute respiratory syndrome-coronavirus life cycle. Microbes Infect 9(1):96–102
44. Lan J, Ge J, Yu J, Shan S, Zhou H, Fan S et al (2020) Crystal structure of the 2019-nCoV spike receptor-binding domain bound with the ACE2 receptor. bioRxiv. https://doi.org/10.1101/2020.02.19.956235
45. Li K, Yip C, Hon C, Lam C, Zeng F, Leung F (2012) Characterisation of animal angiotensin-converting enzyme 2 receptors and use of pseudotyped virus to correlate receptor binding with susceptibility of SARS-CoV infection. Hong Kong Med J 18:35–38
46. Li R, Qiao S, Zhang G (2020) Analysis of angiotensin-converting enzyme 2 (ACE2) from different species sheds some light on cross-species receptor usage of a novel coronavirus 2019-nCoV. J Inf Secur 80(4):469–496
47. Demogines A, Farzan M, Sawyer SL (2012) Evidence for ACE2-utilizing coronaviruses (CoVs) related to severe acute respiratory syndrome CoV in bats. J Virol 86(11):6350–6353
48. Zamoto A, Taguchi F, Fukushi S, Morikawa S, Yamada YK (2006) Identification of ferret ACE2 and its receptor function for SARS-coronavirus. Adv Exp Med Biol 581:519–522
49. Wong SK, Li W, Moore MJ, Choe H, Farzan M (2004) A 193-amino acid fragment of the SARS coronavirus S protein efficiently binds angiotensin-converting enzyme 2. J Biol Chem 279(5):3197–3201
50. Zhang Y, Zheng N, Hao P, Cao Y, Zhong Y (2005) A molecular docking model of SARS-CoV S1 protein in complex with its receptor, human ACE2. Comput Biol Chem 29(3):254–257
51. Pöhlmann S, Gramberg T, Wegele A, Pyrc K, van der Hoek L, Berkhout B et al (2006) Interaction between the spike protein of human coronavirus NL63 and its cellular receptor ACE2. Adv Exp Med Biol 581:281–284
52. Smith M, Tusell S, Travanty EA, Berkhout B, van der Hoek L, Holmes KV (2006) Human angiotensin-converting enzyme 2 (ACE2) is a receptor for human respiratory coronavirus NL63. Adv Exp Med Biol 581:285–288
53. Chai X, Hu L, Zhang Y, Han W, Lu Z, Ke A et al (2020) Specific ACE2 expression in cholangiocytes may cause liver damage after 2019-nCoV infection. bioRxiv. https://doi.org/10.1101/2020.02.19.956235
54. Oudit G, Kassiri Z, Jiang C, Liu P, Poutanen S, Penninger J et al (2009) SARS-coronavirus modulation of myocardial ACE2 expression and inflammation in patients with SARS. Eur J Clin Investig 39(7):618–625
55. Netland J, Meyerholz DK, Moore S, Cassell M, Perlman S (2008) Severe acute respiratory syndrome coronavirus infection causes neuronal death in the absence of encephalitis in mice transgenic for human ACE2. J Virol 82(15):7264–7275
56. Dong W, Bai B, Lin Y, Gao J, Yu N (2008) Detection of the mRNA expression of human angiotensin-converting enzyme 2 as a SARS coronavirus functional receptor in human femoral head. Nan Fang Yi Ke Da Xue Xue Bao 28(3):441–443
57. Yang XH, Deng W, Tong Z, Liu YX, Zhang LF, Zhu H et al (2007) Mice transgenic for human angiotensin-converting enzyme 2 provide a model for SARS coronavirus infection. Comp Med 57(5):450–459
58. Tseng C-TK, Huang C, Newman P, Wang N, Narayanan K, Watts DM et al (2007) Severe acute respiratory syndrome coronavirus infection of mice transgenic for the human angiotensin-converting enzyme 2 virus receptor. J Virol 81(3):1162–1173
59. Ren X, Glende J, Al-Falah M, de Vries V, Schwegmann-Wessels C, Qu X et al (2006) Analysis of ACE2 in polarized epithelial cells: surface expression and function as receptor for severe acute respiratory syndrome-associated coronavirus. J Gen Virol 87(6):1691–1695
60. Hamming I, Timens W, Bulthuis M, Lely A, Navis G, van Goor H (2004) Tissue distribution of ACE2 protein, the functional receptor for SARS coronavirus. A first step in understanding SARS pathogenesis. J Pathol 203(2):631–637
61. Lei C, Fu W, Qian K, Li T, Zhang S, Ding M et al (2020) Potent neutralization of 2019 novel coronavirus by recombinant ACE2-Ig. bioRxiv. https://doi.org/10.1101/2020.02.01.929976
62. Wang S, Guo F, Liu K, Wang H, Rao S, Yang P et al (2008) Endocytosis of the receptor-binding domain of SARS-CoV spike protein together with virus receptor ACE2. Virus Res 136(1–2):8–15
63. Lu CY, Huang HY, Yang TH, Chang LY, Lee CY, Huang LM (2008) siRNA silencing of angiotensin-converting enzyme 2 reduced severe acute respiratory syndrome-associated coronavirus replications in Vero E6 cells. Eur J Clin Microbiol Infect Dis 27(8):709–715

64. Haga S, Yamamoto N, Nakai-Murakami C, Osawa Y, Tokunaga K, Sata T et al (2008) Modulation of TNF-α-converting enzyme by the spike protein of SARS-CoV and ACE2 induces TNF-α production and facilitates viral entry. Proc Natl Acad Sci USA105(22):7809-7814
65. Zhang Y, Wang W, Gao J-r, Ye L, X-n F, Y-c Z et al (2007) The functional motif of SARS-CoV S protein involved in the interaction with ACE2. Virol Sin 22(1):1–7
66. Yan X, Shen H, Feng Y, Wang J, Lou S, Wang L et al (2007) Potent inhibition of viral entry and replication of SARS-CoV by siRNAs targeting the genes encoding the cellular ACE2 receptor or the viral nucleocapsid protein. Antivir Res 74(3):A30–A31
67. Ho TY, Wu SL, Chen JC, Wei YC, Cheng E, Chang YH et al (2006) Design and biological activities of novel inhibitory peptides for SARS-CoV spike protein and angiotensin-converting enzyme 2 interaction. Antivir Res 69(2):70–76
68. He L, Ding Y, Zhang Q, Che X, He Y, Shen H et al (2006) Expression of elevated levels of pro-inflammatory cytokines in SARS-CoV-infected ACE2+ cells in SARS patients: relation to the acute lung injury and pathogenesis of SARS. J Pathol 210(3):288–297
69. Han DP, Penn-Nicholson A, Cho MW (2006) Identification of critical determinants on ACE2 for SARS-CoV entry and development of a potent entry inhibitor. Virology 350(1):15–25
70. Moore MJ, Dorfman T, Li W, Wong SK, Li Y, Kuhn JH et al (2004) Retroviruses pseudotyped with the severe acute respiratory syndrome coronavirus spike protein efficiently infect cells expressing angiotensin-converting enzyme 2. J Virol 78(19):10628–10635
71. Hofmann H, Geier M, Marzi A, Krumbiegel M, Peipp M, Fey GH et al (2004) Susceptibility to SARS coronavirus S protein-driven infection correlates with expression of angiotensin converting enzyme 2 and infection can be blocked by soluble receptor. Biochem Biophys Res 319(4):1216–1221
72. Prabakaran P, Xiao X, Dimitrov DS (2004) A model of the ACE2 structure and function as a SARS-CoV receptor. Biochem Biophys Res 314(1):235–241
73. Liu Z, Xiao X, Wei X, Li J, Yang J, Tan H et al (2020) Composition and divergence of coronavirus spike proteins and host ACE2 receptors predict potential intermediate hosts of SARS-CoV-2. J Med Virol 92(6):595–601
74. Cao Y, Li L, Feng Z, Wan S, Huang P, Sun X et al (2020) Comparative genetic analysis of the novel coronavirus (2019-nCoV/SARS-CoV-2) receptor ACE2 in different populations. Cell Discov 6(1):1–4
75. Hou Y, Peng C, Yu M, Li Y, Han Z, Li F et al (2010) Angiotensin-converting enzyme 2 (ACE2) proteins of different bat species confer variable susceptibility to SARS-CoV entry. Arch Virol 155(10):1563–1569
76. Guo H, Guo A, Wang C, Yan B, Lu H, Chen H (2008) Expression of feline angiotensin converting enzyme 2 and its interaction with SARS-CoV S1 protein. Res Vet Sci 84(3):494–496
77. Chen Y, Liu L, Wei Q, Zhu H, Jiang H, Tu X et al (2008) Rhesus angiotensin converting enzyme 2 supports entry of severe acute respiratory syndrome coronavirus in Chinese macaques. Virology 381(1):89–97
78. Fukushi S, Mizutani T, Sakai K, Saijo M, Taguchi F, Yokoyama M et al (2007) Amino acid substitutions in the s2 region enhance severe acute respiratory syndrome coronavirus infectivity in rat angiotensin-converting enzyme 2-expressing cells. J Virol 81(19):10831–10834
79. Li W, Greenough TC, Moore MJ, Vasilieva N, Somasundaran M, Sullivan JL et al (2004) Efficient replication of severe acute respiratory syndrome coronavirus in mouse cells is limited by murine angiotensin-converting enzyme 2. J Virol 78(20):11429–11433
80. Othman H, Bouslama Z, Brandenburg J-T, Da Rocha J, Hamdi Y, Ghedira K et al (2020) In silico study of the spike protein from SARS-CoV-2 interaction with ACE2: similarity with SARS-CoV, hot-spot analysis and effect of the receptor polymorphism. bioRxiv. https://doi.org/10.1101/2020.03.04.976027
81. Brielle ES, Schneidman D, Linial M (2020) The SARS-CoV-2 exerts a distinctive strategy for interacting with the ACE2 human receptor. bioRxiv. https://doi.org/10.1101/2020.03.10.986398
82. Song W, Gui M, Wang X, Xiang Y (2018) Cryo-EM structure of the SARS coronavirus spike glycoprotein in complex with its host cell receptor ACE2. PLoS Pathog 14(8):e1007236. https://doi.org/10.1371/journal.ppat.1007236
83. Dijkman R, Jebbink MF, Deijs M, Milewska A, Pyrc K, Buelow E et al (2012) Replication-dependent downregulation of cellular angiotensin-converting enzyme 2 protein expression by human coronavirus NL63. J Gen Virol 93(9):1924–1929
84. Wu K, Chen L, Peng G, Zhou W, Pennell CA, Mansky LM et al (2011) A virus-binding hot spot on human angiotensin-converting enzyme 2 is critical for binding of two different coronaviruses. J Virol 85(11):5331–5337
85. Glowacka I, Bertram S, Herzog P, Pfefferle S, Steffen I, Muench MO et al (2010) Differential downregulation of ACE2 by the spike proteins of severe acute respiratory syndrome coronavirus and human coronavirus NL63. J Virol 84(2):1198–1205
86. Chen IY, Chang SC, Wu HY, Yu TC, Wei WC, Lin S et al (2010) Upregulation of the chemokine (CC motif) ligand 2 via a severe acute respiratory syndrome coronavirus spike-ACE2 signaling pathway. J Virol 84(15):7703–7712
87. Haga S, Nagata N, Okamura T, Yamamoto N, Sata T, Yamamoto N et al (2010) TACE antagonists blocking ACE2 shedding caused by the spike protein of SARS-CoV are candidate antiviral compounds. Antivir Res 85(3):551–555
88. Lin HX, Feng Y, Wong G, Wang L, Li B, Zhao X et al (2008) Identification of residues in the receptor-binding domain (RBD) of the spike protein of human coronavirus NL63 that are critical for

the RBD–ACE2 receptor interaction. J Gen Virol 89(4):1015–1024
89. Mathewson AC, Bishop A, Yao Y, Kemp F, Ren J, Chen H et al (2008) Interaction of severe acute respiratory syndrome-coronavirus and NL63 coronavirus spike proteins with angiotensin converting enzyme-2. J Gen Virol 89(Pt 11):2741–2745
90. Glende J, Schwegmann-Wessels C, Al-Falah M, Pfefferle S, Qu X, Deng H et al (2008) Importance of cholesterol-rich membrane microdomains in the interaction of the S protein of SARS-coronavirus with the cellular receptor angiotensin-converting enzyme 2. Virology 381(2):215–221
91. Li W, Sui J, Huang I-C, Kuhn JH, Radoshitzky SR, Marasco WA et al (2007) The S proteins of human coronavirus NL63 and severe acute respiratory syndrome coronavirus bind overlapping regions of ACE2. Virology 367(2):367–374
92. Lambert DW, Yarski M, Warner FJ, Thornhill P, Parkin ET, Smith AI et al (2005) Tumor necrosis factor-α convertase (ADAM17) mediates regulated ectodomain shedding of the severe-acute respiratory syndrome-coronavirus (SARS-CoV) receptor, angiotensin-converting enzyme-2 (ACE2). J Biol Chem 280(34):30113–30119
93. Li W, Zhang C, Sui J, Kuhn JH, Moore MJ, Luo S et al (2005) Receptor and viral determinants of SARS-coronavirus adaptation to human ACE2. EMBO J 24(8):1634–1643
94. Ji X, Zhang C, Zhai Y, Zhang Z, Xue Y, Zhang C et al (2020) TWIRLS, an automated topic-wise inference method based on massive literature, suggests a possible mechanism via ACE2 for the pathological changes in the human host after coronavirus infection. medRxiv. https://doi.org/10.1101/2020.02.27.967588
95. Yan R, Zhang Y, Li Y, Xia L, Guo Y, Zhou Q (2020) Structural basis for the recognition of SARS-CoV-2 by full-length human ACE2. Science 367(6485):1444–1448
96. Heurich A, Hofmann-Winkler H, Gierer S, Liepold T, Jahn O, Pöhlmann S (2014) TMPRSS2 and ADAM17 cleave ACE2 differentially and only proteolysis by TMPRSS2 augments entry driven by the severe acute respiratory syndrome coronavirus spike protein. J Virol 88(2):1293–1307
97. Yang X, Yu Y, Xu J, Shu H, Liu H, Wu Y et al (2020) Clinical course and outcomes of critically ill patients with SARS-CoV-2 pneumonia in Wuhan, China: a single-centered, retrospective, observational study. Lancet Respir Med 8(5):475–481
98. Li W, Wong S-K, Li F, Kuhn JH, Huang I-C, Choe H et al (2006) Animal origins of the severe acute respiratory syndrome coronavirus: insight from ACE2-S-protein interactions. J Virol 80(9):4211–4219
99. Batlle D, Wysocki J, Satchell K (2020) Soluble angiotensin-converting enzyme 2: a potential approach for coronavirus infection therapy? Clin Sci 134(5):543–545
100. Li W, Shi Z, Yu M, Ren W, Smith C, Epstein JH et al (2005) Bats are natural reservoirs of SARS-like coronaviruses. Science 310(5748):676–679
101. Andersen KG, Rambaut A, Lipkin WI, Holmes EC, Garry RF (2020) The proximal origin of SARS-CoV-2. Nat Med 26(4):450–452
102. Liu P, Jiang JZ, Wan XF, Hua Y, Li L, Zhou J et al (2020) Are pangolins the intermediate host of the 2019 novel coronavirus (SARS-CoV-2)? PLoS Pathog 16(5):e1008421. https://doi.org/10.1371/journal.ppat.1008421
103. He J, Tao H, Yan Y, Huang SY, Xiao Y (2020) Molecular mechanism of evolution and human infection with SARS-CoV-2. Viruses 12(4):428. Published online 2020. https://doi.org/10.3390/v12040428
104. Wrapp D, Wang N, Corbett KS, Goldsmith JA, Hsieh CL, Abiona O et al (2020) Cryo-EM structure of the 2019-nCoV spike in the Prefusion conformation. Science 367(6483):1260–1263
105. Agarwal A, Chen A, Ravindran N, To C, Thuluvath PJ (2020) Gastrointestinal and liver manifestations of COVID-19. J Clin Exp Hepatol 10(3):263–265
106. Bostanciklioglu M (2020) Severe acute respiratory syndrome coronavirus 2 is penetrating to dementia research. Curr Neurovasc Res. https://doi.org/10.2174/1567202617666200522220509. Online ahead of print
107. Geng YJ, Wei ZY, Qian HY, Huang J, Lodato R, Castriotta RJ (2020) Pathophysiological characteristics and therapeutic approaches for pulmonary injury and cardiovascular complications of coronavirus disease 2019. Cardiovasc Pathol 47:107228. https://doi.org/10.1016/j.carpath.2020.107228
108. Gurwitz D (2020) Angiotensin receptor blockers as tentative SARS-CoV-2 therapeutics. Drug Dev Res. https://doi.org/10.1002/ddr.21656. Online ahead of print
109. Niu M, Wang RL, Wang ZX, Zhang P, Bai ZF, Jing J et al (2020) Rapid establishment of traditional Chinese medicine prevention and treatment of 2019-nCoV based on clinical experience and molecular docking. Zhongguo Zhong Yao Za Zhi 45(6):1213–1218
110. Seif F, Aazami H, Khoshmirsafa M, Kamali M, Mohsenzadegan M, Pornour M et al (2020) JAK inhibition as a new treatment strategy for patients with COVID-19. Int Arch Allergy Immunol 11:1–9. https://doi.org/10.1159/000508247. Online ahead of print
111. Saber-Ayad M, Saleh MA, Abu-Gharbieh E (2020) The rationale for potential pharmacotherapy of COVID-19. Pharmaceuticals (Basel) 13(5):E96. https://doi.org/10.3390/ph13050096
112. Ameratunga R, Lehnert K, Leung E, Comoletti D, Snell R, Woon ST et al (2020) Inhaled modified angiotensin converting enzyme 2 (ACE2) as a decoy to mitigate SARS-CoV-2 infection. N Z Med J 133(1515):112–118

Experience in Nutrition Management of Diabetes-Affected COVID-19 Patients

6

Behnaz Abiri, Paul C. Guest, and Mohammadreza Vafa

Abstract

When diabetic patients are ill, their bodies react by releasing hormones to combat the illness. These hormones can be triggered by some states, such as infections. Some illnesses that most likely have an influence on blood glucose levels include common cold or flu, COVID-19, bronchitis, or chest infections. So, it is important for diabetic patients affected by COVID-19 to eat a healthy balanced diet to maintain stable blood glucose levels and enhance their immune functions. The immune response has often been demonstrated to be attenuated by insufficient nutrition in many model systems as well as in human studies. We summarize and propose potential nutritional therapeutic options available for the treatment of this novel coronavirus in diabetic patients.

Keywords

Diabetes · Coronavirus · COVID-19 · SARS · MERS · Nutrition management

6.1 Introduction

Coronaviruses belong to the subfamily *Orthocoronavirinae* in the family of *Coronaviridae* in the order *Nidovirales*. This subfamily consists of alpha-, beta-, gamma-, and delta-coronavirus [1]. There are many coronaviruses ranging from the common cold to much more dangerous viruses, which caused diseases such as Severe Acute Respiratory Syndrome (SARS) and Middle East Respiratory Syndrome (MERS). Coronaviruses were first noted to cause enzootic infections in birds and mammals and, in the last decades, they have been reported to be able to infect humans as well [2]. The outbreaks of SARS in 2002 and MERS in 2012 demonstrated the lethality of coronaviruses when they cross the species barrier and infect humans [2]. SARS and MERS belong to the cluster of beta-coronavirus [3]. Recently, a novel flu-like coronavirus called SARS-CoV-2, related to

B. Abiri
Department of Nutrition, Faculty of Paramedicine, Ahvaz Jundishapur University of Medical Sciences, Ahvaz, Iran

P. C. Guest
Laboratory of Neuroproteomics, Department of Biochemistry and Tissue Biology, Institute of Biology, University of Campinas (UNICAMP), Campinas, Brazil

M. Vafa (✉)
Department of Nutrition, School of Public Health, Iran University of Medical Sciences, Tehran, Iran
e-mail: vafa.m@iums.ac.ir

P. C. Guest (ed.), *Clinical, Biological and Molecular Aspects of COVID-19*, Advances in Experimental Medicine and Biology 1321, https://doi.org/10.1007/978-3-030-59261-5_6

MERS-CoV and SARS-CoV, was detected at the end of 2019 in Wuhan, China [4], and the evidence of human-to-human transmission was established between close contacts [5]. The disease caused by the virus was called coronavirus 2019 (COVID-19). A comparison between SARS-CoV-2 and the other two coronaviruses, SARS-CoV and MERS-CoV, is indicated in Table 6.1 [6, 7]. The genome of SARS-CoV-2 is a single-stranded positive-sense RNA [8, 9]. The sequence analysis demonstrated that COVID-19 had a typical genome structure of coronavirus and also belongs to the beta-coronavirus family [9]. SARS-CoV-2 was more than 82% similar to SARS-CoV-2 at the nucleotide level [8, 9]. The single-stranded RNA genome of SARS-CoV-2 was 29,891 nucleotides in size, encoding 9860 amino acids. The G + C content was 38%. Similar to other beta-coronaviruses, the SARS-CoV-2 genome contains two flanking untranslated regions (UTRs) and a single long open reading frame encoding a polyprotein. The SARS-CoV-2 genome is arranged in the order: 5′-replicase (orf1/ab); structural proteins [spike (S), envelope (E), membrane (M), nucleocapsid (N) 3′ and lacks the hemagglutinin-esterase gene which is characteristically found in beta-coronaviruses (Fig. 6.1) [9]. The World Health Organization (WHO) has declared COVID-19 outbreak as a public health emergency of international challenge and has given it pandemic status. In some severe cases, coronaviruses can lead to infection in the lungs (pneumonia), kidney failure, and even death.

Table 6.1 Comparison between COVID-19, SARS, and MERS [6, 7]

	COVID-19	SARS	MERS
Site of origin	Wuhan, China	Guangdong, China	Arabian Peninsula
Animal reservoir	Bat	Bat	Bat
Intermediate host	Unknown[a]	Palm civet	Camel
Human-to-human transmission	Yes	Yes	Yes
Pneumonia	Yes	Yes	Yes
Organ failure	Yes	Yes	Yes
Requiring mechanical ventilation	Yes	Yes	Yes
Case fatality (%)	6.6[a]	9.5	34.4

[a]Studies ongoing. Present figure is taken from the Johns Hopkins University website; https://coronavirus.jhu.edu/map.html

Older people and those with preexisting medical states such as diabetes, asthma, and heart disease seem to be more susceptible to becoming severely ill with the SARS-CoV-2 virus. When people with diabetes are affected by a viral infection, treatment can become more difficult because of fluctuations in blood glucose levels and the likely presence of diabetes-related complications. There seems to be two reasons for this. Firstly, the immune system is compromised, making it harder to combat the virus and probably causing a longer recovery period. Secondly, the virus may expand in an environment of elevated blood glucose levels [10].

The data regarding the biology, epidemiology, and clinical characteristics of the SARS-CoV-2 virus have been accumulating on a daily basis. The virus genome was quickly sequenced, which allowed the development of diagnostic tests and initiation of research into potential vaccines and therapeutics. However, the clinical spectrum of the disease continues to be described (including the potential for asymptomatic spread) and clinical trials investigating potential treatments have begun. At present, there is no established treatment or vaccine for the disease. Therefore, there is an urgent requirement to find an alternative solution to prevent and control its replication and spread. Here, we summarize and propose nutritional therapeutic options available for the treatment of this novel coronaviruses in diabetic patients.

6.2 What Happens When Diabetic Patients Are Ill?

When diabetic patients are ill, their bodies react by releasing hormones to combat the illness. These hormones can be triggered by some conditions or environmental factors, such as infections.

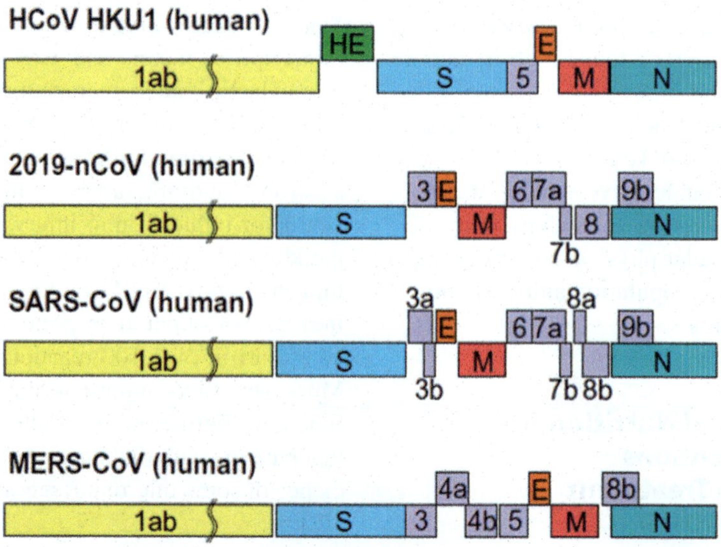

Fig. 6.1 Beta-coronavirus genome organization. The coronavirus genome consists of a 5′ untranslated region (5UTR) including 5′ leader sequence, open reading frame 1a/b (yellow box) encoding nonstructural proteins for replication, structural proteins including envelop (orange box), membrane proteins (red), nucleoproteins (cyan box), accessory proteins (purple boxes) of 2019-nCoV (HKU-SZ-005b) genome, and 3′ untranslated region (3UTR). Examples of each beta-coronavirus lineage are human coronavirus (HCoV) HKU1, 2019-nCoV (HKU-SZ-005b), SARS-CoV, and human MERS-CoV. The length of nonstructural proteins and open reading frames are not drawn to scale. (Adapted from Chan et al. with permission [9])

Some illnesses that most likely have an influence on blood glucose levels include the common cold or influenzas, viral infections, bronchitis, or respiratory conditions [11]. The hormones released throughout an illness raise blood glucose levels and subsequently make it more difficult for insulin to lower these. For individuals living with diabetes, even a minor illness can cause seriously high blood glucose levels. This may lead to life-threatening complications, including diabetic ketoacidosis or hyperosmolar hyperglycemic states [11].

6.3 Guidelines to Manage Diabetes During an Illness

Diabetic patients who are infected with the virus may see their glycemic control fail. They should practice the "sick day rules" recommended for any stressful condition to ameliorate the diabetic effects. Sick day rules for diabetic patients consist of: 1) keeping hydration; 2) monitoring levels of blood glucose; 3) monitoring body temperature; 4) if the patient is on insulin, also monitoring levels of ketone bodies; and 5) following the healthcare team recommendations [10].

Based on the above, if diabetic patients are infected with the SARS-CoV-2 virus, it is recommended that they:

1. pay extra attention to glucose control, as regular monitoring can help avoid complications resulting from high or low blood glucose,
2. ensure a sufficient supply of water as any infection is going to raise blood glucose levels and elevate the requirement for fluids,
3. ensure sufficient access to food,.
4. keep a regular schedule, avoid overwork, and ensure sufficient rest and sleep.

Healthy and balanced nutrition is an important component of diabetes management. Hence, it is important for diabetic patients affected by COVID-19 to eat a varied and healthy balanced diet to keep their blood sugar levels stable and enhance their immune system functions [10, 11]. It is recommended that patients:

1. give priority to foods with a low glycemic index (such as vegetables and whole grains),

2. avoid extreme consumption of fried foods,
3. limit consumption of foods high in carbohydrates and fat,
4. eat lean proteins (such as fish, eggs, milk, and beans after fully cooked),
5. eat cooked green, leafy vegetables, daily,
6. eat fruits in two or three servings, daily.

In addition, regular physical activity is important for the general population and even more so for diabetic patients.

6.4 Potential Nutritional Interventions for the Treatment of COVID-19

There is currently no established nutritional intervention for the disease. In the absence of a specific nutrition intervention for this novel virus, there is an urgent requirement to find alternative solutions to prevent and control the replication and spread of the virus and to manage the effects during infections. In addition, diabetes itself can affect the nutritional status [12]. Assuming that management of COVID-19 infection is not different from that of other viruses causing respiratory failure, we carried out a literature search of the PubMed database with the key words of SARS, MERS, and coronavirus. Some of the findings which described nutrient effects were related to viral pneumonias causing respiratory failure and immune function. These nutrients and the specific mechanisms by which they can affect coronaviruses and enhance immunity are discussed in the following sections.

6.4.1 B Vitamins

B vitamins are water-soluble vitamins and act as a part of coenzymes. Each B vitamin has its specific functions. Keil et al. showed that vitamin B2 (riboflavin) and the use of ultraviolet light effectively decreased the titer of the MERS virus in human plasma [13]. Administration of vitamin B3 (nicotinamide) increased killing of *Staphylococcus aureus* via a myeloid-specific transcription factor, and was effective in both prophylactic and therapeutic settings [14]. Furthermore, vitamin B3 therapy significantly reduced neutrophil infiltration into the lungs with a strong antiinflammatory impact throughout ventilator-induced lung injury. However, it also paradoxically resulted in the development of significant hypoxemia [15]. Vitamin B6 (pyridoxine) is also required in protein metabolism and takes part in over 100 reactions in body tissues. Moreover, it has important roles in body immune function. Furthermore, studies have suggested that patients with diabetes are susceptible to deficiency of some micronutrients such as B vitamins [12].

6.4.2 Vitamin C

Vitamin C is another water-soluble vitamin and it is also named ascorbic acid, which means "no-scurvy acid." Vitamin C is best known for its role in the synthesis of collagen in connective tissues and works as an antioxidant. Vitamin C also supports immune functions and defends against coronavirus infection [16]. For example, Atherton et al. demonstrated that vitamin C elevated the resistance of chick embryo tracheal organ cultures to avian coronavirus infection [17]. Vitamin C may also work as a weak antihistamine-like compound to provide relief from influenza-like symptoms including sneezing, runny nose, and swollen sinuses [18]. Three human controlled clinical trials have been conducted which revealed a significantly lower incidence of pneumonia in vitamin C-supplemented groups, suggesting that vitamin C supplementation might reduce vulnerability to lower respiratory tract infections under certain conditions [19]. In addition, vitamin C has antioxidant properties with a potent inhibitory impact on peroxidation of polyunsaturated lipids in plasma [20], and can regenerate lipid-soluble antioxidants, such as vitamin E [20]. Findings have shown that vitamin C levels may be significantly reduced in some patients with diabetes [20, 21]. Although decreased dietary intake or uncontrolled excretion may have a role, the most probable causes of this are the following:

1. Elevated consumption (oxidation) of vitamin C due to increased free radical activity in diabetes (evidenced as increased production of the oxidation product, dehydroascorbic acid).
2. Failed regeneration of vitamin C from dehydroascorbic acid due to competitive inhibition of its transport across the cell membrane by glucose (a structurally similar molecule).

6.4.3 Vitamin A

Vitamin A is a fat-soluble vitamin derived from beta-carotene. There are three active forms of vitamin A in the body known as retinol, retinal, and retinoic acid. It is also known as the "anti-infective" vitamin as many of the body's mechanisms against infection rely on a sufficient supply of this vitamin [22]. Vitamin A deficiency is known to occur in measles, which can become more severe in children with vitamin A deficiency [23]. Furthermore, Semba et al. found that vitamin A supplementation diminished morbidity and mortality in various infectious diseases, such as diarrheal disease, measles-associated pneumonia, human immunodeficiency virus (HIV) infection, and malaria [24, 25]. Jee et al. showed that diets low in vitamin A might lead to compromised efficacy of inactivated bovine coronavirus vaccines and render calves more vulnerable to infectious disease [26]. The impact of infectious bronchitis virus (IBV), one type of coronavirus, was more noticeable in chickens fed a diet deficient in vitamin A compared to those fed a vitamin A sufficient diet [27]. The mechanism by which vitamin A and retinoids impede measles replication is via upregulating elements of the innate immune response in uninfected bystander cells, making these resistant to infection in subsequent rounds of viral replication [28]. Notably, some studies have demonstrated that vitamin A levels are significantly decreased in patients with diabetes [29, 30].

6.4.4 Vitamin D

Vitamin D acts as both a micronutrient and a hormone. In addition to its effects on maintaining bone integrity, it also stimulates maturation of many cells such as immune cells. Many adults have been shown to have vitamin D deficiency, especially at the end of the winter season [31]. Moreover, individuals who are housebound or institutionalized, as well as night workers, may become vitamin D deficient, as do many older people, due to limited exposure to sunlight [32]. It is interesting in this regard that the COVID-19 outbreak was first recognized during the winter of 2019 and this mostly affected the elderly population [5–7, 9]. Vitamin D levels have also been shown to be significantly lower in diabetic patients compared to nondiabetic controls [33, 34]. It Also, studies have suggested that vitamin D deficiency may predispose individuals to type 1 diabetes and type 2 diabetes, and may contribute to the pathogenesis of both forms of the disease [34, 35]. Furthermore, lower vitamin D levels in calves have been noted to increase their susceptibility to infection by bovine coronavirus [36].

6.4.5 Vitamin E

Vitamin E is another lipid-soluble vitamin which includes the tocopherols and tocotrienols. It plays a major role in reducing oxidative stress via acting as an antioxidant to damaging free radicals [37]. Vitamin E deficiency has been demonstrated to augment the myocardial damage [38] and augment the virulence of coxsackievirus B3 viral infection in mice [39]. In addition, deficiency of both vitamin D and E in calves rendered them more susceptible to infection by bovine coronavirus [36]. The antioxidant activity of vitamin E helps minimize damage to lipids caused by the free oxygen radical-mediated tissue damage in diabetes [40], and it works as the first line of defense against lipid peroxidation of cell membranes [41]. A study by Halliwell showed that vitamin E administration can also delay the onset of insulin resistance in diabetic rat models [42].

This is consistent with findings of a negative correlation between vitamin E and fasting blood glucose, low density lipoprotein-cholesterol (LDL-C) and triglyceride levels [43].

6.4.6 Zinc

Zinc is a dietary trace mineral and is necessary for maintenance and enhancement of immune cells of both the innate and adaptive immune system [44]. Zinc deficiency leads to dysfunction of both humoral and cell-mediated immunity and increases vulnerability to infectious diseases [45]. For example, zinc supplementation in zinc-deficient children has been found to decrease measles-associated morbidity and mortality, resulting from lower respiratory tract infections [46]. Studies have shown that elevating the levels of intracellular zinc with zinc-ionophores like pyrithione can effectively disrupt replication of different RNA viruses [47]. Moreover, the combination of zinc and pyrithione at low concentrations impedes the replication of the SARS coronavirus [47]. It has also been proposed that zinc deficiency is an important risk factor of type 2 diabetes and low levels of zinc have been observed in diabetic patients compared to healthy controls [48–50]. Another study found that zinc plasma concentrations were inversely associated with glycemia [hemoglobin A1c (HbA1c) levels] in diabetes mellitus [51]. An association between inadequate zinc intake and raised insulin levels in blood has also been reported in adolescents [52]. In addition, recent studies demonstrated that zinc supplementation ameliorated glucose metabolism and insulin sensitivity deficiencies in diabetic patients [53, 54].

6.4.7 Iron

Iron is needed for pathogen defense and an iron deficient state can lead to an impaired immune response. Conversely, an oversupply of iron can lead to an oxidative stress state [55]. Iron deficiency has been demonstrated as a risk factor for the development of recurrent acute respiratory tract infections [56]. Studies by El-Agouza et al. [57] and Coban et al. [58] demonstrated that HbA1c levels were higher in patients with iron deficiency anemia and reduced significantly by iron administration.

6.4.8 Selenium

Selenium is an important trace element for mammalian oxidative-reduction biology [59]. This is because some micronutrients can provide protection against infectious diseases [60]. Conversely, nutritional deficiencies influence not only the host immune functions but also the viral pathogens themselves [22]. Dietary selenium deficiency that leads to oxidative stress in the host can change the viral genome, so that a virus that is normally mildly pathogenic can become more virulent. Beck et al. noted that selenium deficiency could not only elevate the pathology of an influenza virus infection [61] but also drive alterations in the genome of coxsackievirus, allowing an avirulent virus to shift to virulence via genetic mutation [62]. This appears to be due to the fact that selenium is a cofactor in many enzymatic functions and, in cooperation with vitamin E, acts to prevent the formation of free radicals and minimize the oxidative harm to cells and tissues [63]. One study found a synergistic impact of selenium with ginseng stem-leaf saponins that could enhance the immune response to a live bivalent infectious bronchitis coronavirus vaccine in chickens [64]. Although numerous epidemiological studies have investigated the relationship between selenium and diabetes, their findings have been inconsistent. Some researchers have proposed that high selenium levels could reduce the prevalence of diabetes [65], whereas others have reported that a high level of serum selenium could be associated with increased prevalence of the disease [66, 67]. In addition, other studies have proposed that selenium supplementation in patients with type 2 diabetes may have an adverse impact on blood glucose homeostasis [68, 69]. It is possible that these discrepancies might reflect the use of heterogeneous study populations or different detection methods.

Finally, it may be that the relationship between selenium and cardiometabolic outcomes is U-shaped, with potential damage occurring both below and above an optimal physiological range of this micronutrient [70]. Therefore, further studies are required to determine if selenium supplementation is an effective choice for the treatment of pre-diabetes and diabetes-affected COVID-19 patients.

6.4.9 Omega-3 Polyunsaturated Fatty Acids

The long-chain omega-3 polyunsaturated fatty acids (PUFAs) eicosapentaenoic acid (EPA, 20:5n-3) and docosahexaenoic acid (DHA, 22:6n-3) from fish and other seafood have received substantial interest, because of their association with reduced risk of cardiovascular diseases and for several risk factors for diabetes, such as inflammation, adiposity, hypertension, and dyslipidemia [71]. In animal models, long-chain omega-3 PUFAs have also been found to reduce insulin resistance [72], but the results from randomized human trials have generally reported few advantages on glucose or insulin homeostasis [73, 74]. No relationships have been found in such studies from Europe, whereas those from Australasia have found that fish or EPA/DHA intake was related to lower risk of type 2 diabetes and studies from the USA found that they were associated with higher risk [75–78]. These geographical variations in the effects of omega-3 PUFAs on risk may reflect genetic differences, gene–diet interactions or variations in the type of fish consumed (fatty fish vs. lean fish) or in the fish preparation methods (raw/steamed/boiled vs. deep-fried) [71].

The positive effects of omega-3 PUFAs on glucose homeostasis and type 2 diabetes could be due to beneficial effects on adiposity, hypertension, and dyslipidemia [71]. The means of achieving higher serum omega-3 PUFA levels is likely to be due to higher intake of fish in place of red meat and lower intake of saturated fatty acids. Processed red meat intake has been related to a higher risk of type 2 diabetes [79], and saturated fatty acids have been found to be linked to increased insulin resistance [80]. In addition, PUFAs are important mediators in the antiinflammatory and adaptive immune responses [81]. Furthermore, the omega-3 PUFA-derived lipid mediator protectin D1 has been shown to decrease influenza virus replication through effects on the RNA export machinery and may protect against influenza mortality [81]. However, the contradictory findings of some of the above studies necessitate further studies on the effects of the omega-3 PUFAs in diabetes in relation to viral infectivity.

6.4.10 Natural Polyphenols and Isoflavones

Antioxidants have been extensively evaluated in disease prevention and health maintenance due to their effects on suppressing factors associated with oxidative stress. Polyphenols have been established as potent antioxidants that can aid in the prevention of type 2 diabetes via antiinflammatory, antimicrobial, and immunomodulatory mechanisms [82]. Most polyphenols impede amylase and glycosidase activity and, therefore, block glucose absorption in the intestine. In addition, polyphenols activate peroxisome proliferator-activated receptor (PPAR) and enhance adiponectin production, thereby ameliorating insulin resistance [82].

Flavonoids are an important class of natural products, which include chalcones, flavones, flavonols, and isoflavones [83]. Flavonoids have many functions in addition to antioxidant effects, including antiviral properties. Jo et al. proposed that the anti-coronavirus activity of some flavonoids (herbacetin, rhoifolin, and pectolinarin) resulted from inhibition of the SARS 3C-like viral protease (3CLpro) [84]. Other flavonoids (herbacetin, isobavachalcone, quercetin3-β-d-glucoside, and helichrysetin) have been found to inhibit the enzymatic activity of the MERS 3CLpro protease [85]. Moreover, Ryuet al. showed that biflavonoids from *Torreya nucifera* also inhibited SARS 3CLpro protease [86].

One study investigated phytocompounds for activity against the SARS virus by measuring the virus-induced cytopathogenic impact on Vero E6 cells [87]. Ten diterpenoids, two sesquiterpenoids, two triterpenoids, five lignoids, curcumin, and the reference controls niclosamide and valinomycin were all found to be potent inhibitors at concentrations between 3.3 and 10 μM, with one of the triterpenoids (betulinic acid) and one of the lignoids (savinin) having the strongest impacts. Some protein molecules encoded by the SARS viral genome are potential targets for chemotherapeutic suppression of viral infection and replication. These consist of the spike protein, which mediates the targeting and entry of the virus into host cells, the SARS 3CLpro protease, the NTPase/helicase, the RNA-dependent RNA polymerase, the membrane protein needed for virus budding, the envelope protein involved in virus assembly, and the nucleocapsid phosphoprotein that associates with the viral RNA inside the virion [87]. Thus, some of the compounds mentioned above may have efficacy in targeting some of these proteins. In the diterpenoid group of compounds, pinusolidic acid is also known as a platelet-activating factor inhibitor [88], and forskolin has been reported to activate adenylate cyclase and elevate cyclic AMP levels in several cell types [89]. Both of these compounds have shown potent anti-SARS virus activity [87]. The anti-SARS activity of these two compounds could be due to a combination of two antiviral mechanisms. One of these is protease inhibition, as indicated in the study by Wen and colleagues [87]. Additionally, studies have shown that betulinic acid derivatives could effectually interfere with HIV-1 virus entry in test cells at a post-bonding, envelope-dependent step apparently associated with fusion of virus with the host cell membrane. Because of the similarity between the gp41 of the retrovirus HIV-1 and the S2 subunit of the spike protein of the SARS virus, it was speculated that another anti-SARS mechanism might be the blocking of viral entry at the post-binding step during the fusion of virus particle to host cell membrane [87].

Some studies have also noted that abietane-type diterpenes exhibited antiviral activities against influenza [90] and HIV-1 [91] viruses. The activity of lignans against some types of viruses, such as HIV-1, has also been demonstrated [92–94]. In addition, the five lignoids investigated by Wen et al. possessed notable anti-SARS activities [87].

Curcumin (20), a known phytocompound from turmeric (Curcuma longa), has been noted to exhibit antiinflammatory, antioxidant, anticarcinogenic, and anti-HIV activities [95]. In the study by Wen and colleagues, mild activity against SARS viral replication and inhibition of the 3CLpro protease were reported.

Finally, 20 phytocompounds, including the abietane-type and labdane-type diterpenes, lupane-type triterpenes, liganoids, and curcumin, were shown to exhibit significant and specific anti-SARS activity and hence may provide a new direction for development of anti-COVID-19 agents. Alpha-lipoic acid (ALA), a naturally occurring disulfide compound, works as a cellular coenzyme and has been used for the treatment of polyneuropathies and hepatic disorders for many years [96]. ALA has antioxidant effects in scavenging free radicals to protect against oxidative damage in some diseases [97]. In addition, ALA appears to promote intracellular levels of the antioxidant glutathione [97] and normalize the oxidative stress induced by dexamethasone administration in chickens [98]. Wu et al. also showed that the oxidative stress, as well as glucose-6-phosphate dehydrogenase (G6PD) deficiency, in host cells was an important factor in infectivity of the human coronavirus 229E, and infectivity was reduced following addition of ALA [99]. In addition, Baur et al. reported that ALA administration was effective in blocking the replication of HIV-1 [100].

6.5 Conclusions

In conclusion, it is important for diabetic patients affected by COVID-19 infection to eat a varied and healthy balanced diet to maintain their blood glucose levels and immune functions. In this review, we summarized potential nutritional interventions for diabetic patients affected by COVID-19 infection, according to previous treatments of SARS and MERS. The immune response has often been demonstrated to be impaired by insufficient nutrition in many model systems as well as in human studies. On the other hand, diabetes itself can alter nutritional status, and experiments have shown that patients with diabetes are prone to deficiencies of micronutrients and antioxidants. However, the nutritional state of the host has not been considered until recently as a potential contributing factor in the propagation of viral infectious diseases. Therefore, we suggest that it is important to verify the nutritional state of COVID-19 infected patients before the administration of general therapies. Moreover, we described coronavirus-specific and antiviral managements that were advantageous for treatment of the SARS and MERS viruses. Therefore further attention should be paid to these as potential nutritional interventions for COVID-19 infection.

Conflict of Interest None.

Funding This study did not receive any financial support.

References

1. Banerjee A, Kulcsar K, Misra V, Frieman M, Mossman K (2019) Bats and coronaviruses. Viruses 11(1):41. https://doi.org/10.3390/v11010041
2. Schoeman D, Fielding BC (2019) Coronavirus envelope protein: current knowledge. Virol J 16:69. https://doi.org/10.1186/s12985-019-1182-0
3. Zumla A, Hui DS, Perlman S (2015) Middle East respiratory syndrome. Lancet 386(9997):995–1007
4. del Rio C, Malani PN (2020) COVID-19-new insights on a rapidly changing epidemic. JAMA. https://doi.org/10.1001/jama.2020.3072. Online ahead of print
5. Li Q, Guan X, Wu P, Wang X, Zhou L, Tong Y et al (2020) Early transmission dynamics in Wuhan, China, of novel coronavirus-infected pneumonia. N Engl J Med 382(13):1199–1207
6. Arabi YM, Murthy S, Webb S (2020) Correction to: COVID-19: a novel coronavirus and a novel challenge for critical care. Intensive Care Med 46(5):1087–1088
7. Petrosillo N, Viceconte G, Ergonul O, Ippolito G, Petersen E (2020) COVID-19, SARS and MERS: are they closely related? Clin Microbiol Infect 28. https://doi.org/10.1016/j.cmi.2020.03.026. Online ahead of print
8. Zhang N, Wang L, Deng X, Liang R, Su M, He C et al (2020) Recent advances in the detection of respiratory virus infection in humans. J Med Virol 92(4):408–417
9. Chan JF, Kok KH, Zhu Z, Chu H, To KK, Yuan S et al (2020) Genomic characterization of the 2019 novel human-pathogenic coronavirus isolated from a patient with atypical pneumonia after visiting Wuhan. Emerg Microbes Infect 9(1):221–236
10. https://www.idf.org/covid-19
11. www.gmc-uk.org/news/news-archive/how-we-will-continue-to-regulate-in-light-of-novel-coronavirus
12. Satyanarayana A, Balakrishna N, Pitla S, Reddy PY, Mudili S, Lopamudra P et al (2011) Status of B-vitamins and homocysteine in diabetic retinopathy: association with vitamin-b12 deficiency and hyperhomocysteinemia. PLoS One 6(11):e26747. https://doi.org/10.1371/journal.pone.0026747
13. Keil SD, Bowen R, Marschner S (2016) Inactivation of Middle East respiratory syndrome coronavirus (MERS-CoV) in plasma products using a riboflavin-based and ultraviolet light-based photochemical treatment. Transfusion 56:2948–2952
14. Kyme P, Thoennissen NH, Tseng CW, Thoennissen GB, Wolf AJ, Shimada K et al (2012) C/EBPepsilon mediates nicotinamide-enhanced clearance of Staphylococcus aureus in mice. J Clin Invest 122:3316–3329
15. Jones HD, Yoo J, Crother TR, Kyme P, Ben-Shlomo A, Khalafi R et al (2015) Nicotinamide exacerbates hypoxemia in ventilator-induced lung injury independent of neutrophil infiltration. PLoS One 10:e0123460. https://doi.org/10.1371/journal.pone.0123460
16. Hemila H (2003) Vitamin C and SARS coronavirus. J Antimicrob Chemother 52:1049–1050
17. Atherton JG, Kratzing CC, Fisher A (1978) The effect of ascorbic acid on infection chick-embryo ciliated tracheal organ cultures by coronavirus. Arch Virol 56:195–199. https://doi.org/10.1007/bf01317848
18. Field CJ, Johnson IR, Schley PD (2002) Nutrients and their role in host resistance to infection. J Leukoc Biol 71:16–32
19. Hemila H (1997) Vitamin C intake and susceptibility to pneumonia. Pediatr Infect Dis J 16:836–837
20. Maxwell SR, Thomason H, Sandler D, Leguen C, Baxter MA, Thorpe GH et al (1997) Antioxidant status in patients with uncomplicated insulin-dependent and non-insulin-dependent diabetes mellitus. Eur J Clin Investig 27(6):484–490
21. Sinclair AJ, Girling AJ, Gray L, Le Guen C, Lunec J, Barnett AH (1991) Disturbed handling of ascorbic acid in diabetic patients with and without microangiopathy during high dose ascorbate supplementation. Diabetologia 34(3):171–175
22. Guillin OM, Vindry C, Ohlmann T, Chavatte L (2019) Selenium, selenoproteins and viral infection. Nutrients 11(9):2101. https://doi.org/10.3390/nu11092101

23. Kantoch M, Litwinska B, Szkoda M, Siennicka J (2002) Importance of vitamin A deficiency in pathology and immunology of viral infections. Rocz Panstw Zakl Hig 5394:385–392
24. Semba RD (1999) Vitamin A and immunity to viral, bacterial and protozoan infections. Proc Nutr Soc 58:719–727
25. Villamor E, Mbise R, Spiegelman D, Hertzmark E, Fataki M, Peterson KE et al (2002) Vitamin A supplements ameliorate the adverse effect of HIV-1, malaria, and diarrheal infections on child growth. Pediatrics 109(1):E6. https://doi.org/10.1542/peds.109.1.e6
26. Jee J, Hoet AE, Azevedo MP, Vlasova AN, Loerch SC, Pickworth CL et al (2013) Effects of dietary vitamin A content on antibody responses of feedlot calves inoculated intramuscularly with an inactivated bovine coronavirus vaccine. Am J Vet Res 74(10):1353–1362
27. West CE, Sijtsma SR, Kouwenhoven B, Rombout JH, van der Zijpp AJ (1992) Epithelia-damaging virus infections affect vitamin A status in chickens. J Nutr 122:333–339
28. Trottier C, Colombo M, Mann KK, Miller WH Jr, Ward BJ (2009) Retinoids inhibit measles virus through a type I IFN-dependent bystander effect. FASEB J 23(9):3203–3212
29. Tsai EC, Hirsch IB, Brunzell JD, Chait A (1994) Reduced plasma peroxyl radical trapping capacity and increased susceptibility of LDL to oxidation in poorly controlled IDDM. Diabetes 43(8):1010–1014
30. Krempf M, Ranganathan S, Ritz P, Morin M, Charbonnel B (1991) Plasma vitamin A and E in type 1 (insulin-dependent) and type 2 (non-insulin-dependent) adult diabetic patients. Int J Vitam Nutr Res 61(1):38–42
31. Tangpricha V, Pearce EN, Chen TC, Holick MF (2002) Vitamin D insufficiency among free-living healthy young adults. Am J Med 112:659–662
32. Holick MF (2004) Sunlight and vitamin D for bone health and prevention of autoimmune diseases, cancers, and cardiovascular disease. Am J Clin Nutr 80:1678S–1688S
33. Yoho RM, Frerichs J, Dodson NB, Greenhagen R, Geletta S (2009) A comparison of vitamin D levels in nondiabetic and diabetic patient populations. J Am Podiatr Med Assoc 99:35–41
34. Mathieu C, Gysemans C, Giulietti A, Bouillon R (2005) Vitamin D and diabetes. Diabetologia 48:1247–1257
35. Palomer X, González-Clemente JM, Blanco-Vaca F, Mauricio D (2008) Role of vitamin D in the pathogenesis of type 2 diabetes mellitus. Diabetes Obes Metab 10:185–197
36. Nonnecke BJ, McGill JL, Ridpath JF, Sacco RE, Lippolis JD, Reinhardt TA et al (2014) Acute phase response elicited by experimental bovine diarrhea virus (BVDV) infection is associated with decreased vitamin D and E status of vitamin-replete preruminant calves. J Dairy Sci 97:5566–5579
37. Galmes S, Serra F, Palou A (2018) Vitamin E metabolic effects and genetic variants: a challenge for precision nutrition in obesity and associated disturbances. Nutrients 10(12):1919. https://doi.org/10.3390/nu10121919
38. Beck MA, Kolbeck PC, Rohr LH, Shi Q, Morris VC, Levander OA (1994) Vitamin E deficiency intensifies the myocardial injury of coxsackievirus B3 infection of mice. J Nutr 124:345–358
39. Beck MA (1997) Increased virulence of coxsackievirus B3 in mice due to vitamin E or selenium deficiency. J Nutr 127:966S–970S
40. Maxwell SR (1995) Prospects for the use of antioxidant therapies. Drugs 49(3):345–361
41. Rizvi S, Raza ST, Ahmed F, Ahmad A, Abbas S, Mahdi F (2014) The role of vitamin E in human health and some diseases. Sultan Qaboos Univ Med J 14(2):e157–e165
42. Halliwell B (2002) Vitamin E and the treatment and prevention of diabetes: a case for a controlled clinical trial. Singap Med J 43(9):479–484
43. Said NS, Hadhoud KM, Nada WM, El Tarhouny SA (2013) Superoxide dismutase, glutathione peroxidase and vitamin E in patients with diabetic retinopathy. Life Sci J 10(1):1851–1856
44. Maares M, Haase H (2016) Zinc and immunity: an essential interrelation. Arch Biochem Biophys 611:58–65. https://doi.org/10.1016/j.abb.2016.03.022
45. Tuerk MJ, Fazel N (2009) Zinc deficiency. Curr Opin Gastroenterol 25:136–143
46. Awotiwon AA, Oduwole O, Sinha A, Okwundu CI (2017) Zinc supplementation for the treatment of measles in children. Cochrane Database Syst Rev (6):CD011177. https://doi.org/10.1002/14651858.CD011177.pub3
47. te Velthuis AJ, van den Worm SH, Sims AC, Baric RS, Snijder EJ, van Hemert MJ (2010) Zn(2+) inhibits coronavirus and arterivirus RNA polymerase activity in vitro and zinc ionophores block the replication of these viruses in cell culture. PLoS Pathog 6:e1001176. https://doi.org/10.1371/journal.ppat.1001176
48. Devi TR, Hijam D, Dubey A, Debnath S, Oinam P, Devi NGT et al (2016) Study of serum zinc and copper levels in type 2 diabetes mellitus. Int J Contem Med Res 3(4):2454–7379
49. Kaur J, Singh T (2015) Estimation of serum magnesium and zinc levels in type-2 diabetes mellitus. Int J Bioassays 4(1):3654–3656
50. Jyothirmayi B, Vasantha M (2015) Study of zinc and glycated Hb levels in diabetic complications. Int J Pharm Clin Res 7(5):360–363
51. Sinha S, Sen S (2014) Status of zinc and magnesium levels in type 2 diabetes mellitus and its relationship with glycemic status. Int J Diabetes Dev Ctries 34(4):220–223
52. Ho M, Baur LA, Cowell CT, Samman S, Garnett SP (2016) Zinc status, dietary zinc intake and metabolic

risk in Australian children and adolescents; Nepean Longitudinal Study. Eur J Nutr 30:1–8
53. El-Ashmony SMA, Morsi HK, Abdelhafez AM (2012) Effect of zinc supplementation on glycemic control, lipid profile, and renal functions in patients with type II diabetes: a single blinded, randomized, placebo-controlled, trial. J Biol Agric Health 2(6):33. ISSN 2225-093X (Online)
54. Islam MR, Attia J, Ali L, McEvoy M, Selim S, Sibbritt D et al (2016) Zinc supplementation for improving glucose handling in pre-diabetes: a double-blind randomized placebo controlled pilot study. Diabetes Res Clin Pract 115:39–46
55. Wessling-Resnick M (2018) Crossing the Iron Gate: why and how transferrin receptors mediate viral entry. Annu Rev Nutr 38:431–458
56. Jayaweera J, Reyes M, Joseph A (2019) Childhood iron deficiency anemia leads to recurrent respiratory tract infections and gastroenteritis. Sci Rep 9:12637. https://doi.org/10.1038/s41598-019-49122-z
57. El-Agouza I, Abu Shohla A, Sirdah M (2002) The effect of iron deficiency anaemia on the levels of haemoglobin subtypes: possible consequences for clinical diagnosis. Clin Lab Haematol 24:285–289
58. Coban E, Ozdogan M, Timuragaoglu A (2004) Effect of iron deficiency anemia on the levels of hemoglobin A1c in nondiabetic patients. Acta Haematol 112:126–859
59. Rayman MP (2012) Selenium and human health. Lancet 379(9822):1256–1268
60. Beck MA, Matthews CC (2000) Micronutrients and host resistance to viral infection. Proc Nutr Soc 59:581–585
61. Beck MA, Nelson HK, Shi Q, Van Dael P, Schiffrin EJ, Blum S et al (2001) Selenium deficiency increases the pathology of an influenza virus infection. FASEB J 15(8):1481–1483
62. Beck MA, Shi Q, Morris VC, Levander OA (1995) Rapid genomic evolution of a non-virulent coxsackievirus B3 in selenium-deficient mice results in selection of identical virulent isolates. Nat Med 1:433–436
63. Harthill M (2011) Review: micronutrient selenium deficiency influences evolution of some viral infectious diseases. Biol Trace Elem Res 143:1325–1336
64. Ma X, Bi S, Wang Y, Chi X, Hu S (2019) Combined adjuvant effect of ginseng stem-leaf saponins and selenium on immune responses to a live bivalent vaccine of Newcastle disease virus and infectious bronchitis virus in chickens. Poult Sci 98:3548–3556
65. Navarro-Alarcon M, Lopez GSH, Perez-Valero V, Lopez-Martinez C (1999) Serum and urine selenium concentrations as indicators of body status in patients with diabetes mellitus. Sci Total Environ 228(1):79–85
66. Yuan Y, Xiao Y, Yu Y, Liu Y, Feng W, Qiu G et al (2018) Associations of multiple plasma metals with incident type 2 diabetes in Chinese adults: the Dongfeng- Tongji cohort. Environ Pollut 237:917–925
67. Lu CW, Chang HH, Yang KC, Kuo CS, Lee LT, Huang KC (2016) High serum selenium levels are associated with increased risk for diabetes mellitus independent of central obesity and insulin resistance. BMJ Open Diabetes Res Care 4(1):e000253. https://doi.org/10.1136/bmjdrc-2016-000253
68. Faghihi T, Radfar M, Barmal M, Amini P, Qorbani M, Abdollahi M et al (2014) A randomized, placebo-controlled trial of selenium supplementation in patients with type 2 diabetes: effects on glucose homeostasis, oxidative stress, and lipid profile. Am J Ther 21(6):491–495
69. Kohler LN, Florea A, Kelley CP, Chow S, Hsu P, Batai K et al (2018) Higher plasma selenium concentrations are associated with increased odds of prevalent type 2 diabetes. J Nutr 148(8):1333–1340
70. Rayman MP, Stranges S (2013) Epidemiology of selenium and type 2 diabetes: can we make sense of it? Free Radic Biol Med 65:1557–1564
71. Virtanen JK, Mursu J, Voutilainen S, Uusitupa M, Tuomainen TP (2014) Serum omega-3 polyunsaturated fatty acids and risk of incident type 2 diabetes in men: the Kuopio Ischemic Heart Disease Risk Factor study. Diabetes Care 37(1):189–196
72. Storlien LH, Kraegen EW, Chisholm DJ, Ford GL, Bruce DG, Pascoe WS (1987) Fish oil prevents insulin resistance induced by high-fat feeding in rats. Science 237(4817):885–888
73. Akinkuolie AO, Ngwa JS, Meigs JB, Djousse L (2011) Omega-3 polyunsaturated fatty acid and insulin sensitivity: A meta-analysis of randomized controlled trials. Clin Nutr 30:702–707
74. Hartweg J, Perera R, Montori V, Dinneen S, Neil HA, Farmer A (2008) Omega-3 polyunsaturated fatty acids (PUFA) for type 2 diabetes mellitus. Cochrane Database Syst Rev 1:CD003205. https://doi.org/10.1002/14651858.CD003205.pub2
75. Wu JH, Micha R, Imamura F, Pan A, Biggs ML, Ajaz O et al (2012) Omega-3 fatty acids and incident type 2 diabetes: a systematic review and metaanalysis. Br J Nutr 107(Suppl 2):S214–S227
76. Wallin A, Di Giuseppe D, Orsini N, Patel PS, Forouhi NG, Wolk A (2012) Fish consumption, dietary long-chain n-3 fatty acids, and risk of type 2 diabetes: systematic review and meta-analysis of prospective studies. Diabetes Care 35:918–929
77. Zheng JS, Huang T, Yang J, Fu YQ, Li D (2012) Marine N-3 polyunsaturated fatty acids are inversely associated with risk of type 2 diabetes in Asians: a systematic review and meta analysis. PLoS One 7(9):e44525. https://doi.org/10.1371/journal.pone.0044525
78. Xun P, He K (2012) Fish consumption and incidence of diabetes: meta-analysis of data from 438,000 individuals in 12 independent prospective cohorts with an average 11-year follow-up. Diabetes Care 35:930–938
79. Pan A, Sun Q, Bernstein AM, Schulze MB, Manson JE, Willett WC et al (2011) Red meat consumption and risk of type 2 diabetes: 3 cohorts of US

adults and an updated metaanalysis. Am J Clin Nutr 94:1088–1096
80. Riserus U, Willett WC, Hu FB (2009) Dietary fats and prevention of type 2 diabetes. Prog Lipid Res 48:44–51
81. Morita M, Kuba K, Ichikawa A, Nakayama M, Katahira J, Iwamoto R et al (2013) The lipid mediator protectin D1 inhibits influenza virus replication and improves severe influenza. Cell 153(1):112–125
82. Umeno A, Horie M, Murotomi K, Nakajima Y, Yoshida Y (2016)Antioxidative and antidiabetic effects of natural polyphenols and isoflavones. Molecules:21(6). pii: E708.doi: https://doi.org/10.3390/molecules21060708
83. Panche AN, Diwan AD, Chandra SR (2016) Flavonoids: an overview. J Nutr Sci 5:e47. https://doi.org/10.1017/jns.2016.41
84. Jo S, Kim S, Shin DH, Kim MS (2020) Inhibition of SARS-CoV 3CL protease by flavonoids. J Enzyme Inhib Med Chem 35:145–151
85. Jo S, Kim H, Kim S, Shin DH, Kim MS (2019) Characteristics of flavonoids as potent MERS-CoV 3C-like protease inhibitors. Chem Biol Drug Des 94:2023–2030
86. Ryu YB, Jeong HJ, Kim JH, Kim YM, Park JY, Kim D et al (2010) Biflavonoids from Torreyanucifera displaying SARS-CoV3CL(pro) inhibition. Bioorg Med Chem 18:7940–7947
87. Wen CC, Kuo YH, Jan JT, Liang PH, Wang SY, Liu HG et al (2007) Specific plant terpenoids and lignoids possess potent antiviral activities against severe acute respiratory syndrome coronavirus. J Med Chem 50(17):4087–4095
88. Yang HO, Han BH (1998) Pinusolidic acid: a platelet-activating factor inhibitor from Biota orientalis. Planta Med 64:72–74
89. Insel PA, Ostrom RS (2003) Forskolin as a tool for examining adenylyl cyclase expression, regulation, and G protein signaling. Cell Mol Neurobiol 23:305–314
90. Staschke KA, Hatch SD, Tang JC, Hornback WJ, Munroe JE, Colacino JM et al (1998) Inhibition of influenza virus hemagglutinin-mediated membrane fusion by a compound related to podocarpic acid. Virology 248:264–274
91. Paris A, Strukelj B, Renko M, Turk V, Pukl M, Umek A et al (1993) Inhibitory effect of carnosic acid on HIV-1 protease in cell-free assays. J Nat Prod 56:1426–1430
92. Charlton JL (1998) Antiviral activity of lignans. J Nat Prod 61:1447–1451
93. Eich E, Pertz H, Kaloga M, Schulz J, Fesen MR, Mazumder A et al (1996) (−)-Arctigenin as a lead structure for inhibitors of human immunodeficiency virus type-1 integrase. J Med Chem 39:86–95
94. Hara H, Fujihashi T, Sakata T, Kaji A, Kaji H (1997) Tetrahydronaphthalenelignan compounds as potent anti-HIV type 1 agents. AIDS Res Hum Retrovir 13:695–705
95. Joe B, Vijaykumar M, Lokesh BR (2004) Biological properties of curcumin-cellular and molecular mechanisms of action. Crit Rev Food Sci Nutr 44:97–111
96. Sachse G, Willms B (1980) Efficacy of thioctic acid in the therapy of peripheral diabetic neuropathy. Horm Metab Res Suppl 9:105–107
97. Tibullo D, Li Volti G, Giallongo C, Grasso S, Tomassoni D, Anfuso CD et al (2017) Biochemical and clinical relevance of alpha lipoic acid: antioxidant and anti-inflammatory activity, molecular pathways and therapeutic potential. Inflamm Res 66:947–959
98. El-Senousey HK, Chen B, Wang JY, Atta AM, Mohamed FR, Nie QH (2018) Effects of dietary vitamin C, vitamin E, and alpha-lipoic acid supplementation on the antioxidant defense system and immune-related gene expression in broilers exposed to oxidative stress by dexamethasone. Poult Sci 97:30–38
99. Wu YH, Tseng CP, Cheng ML, Ho HY, Shih SR, Chiu DT (2008) Glucose-6-phosphate dehydrogenase deficiency enhances human coronavirus 229E infection. J Infect Dis 197:812–816
100. Baur A, Harrer T, Peukert M, Jahn G, Kalden JR, Fleckenstein B (1991) Alpha-lipoic acid is an effective inhibitor of human immuno-deficiency virus (HIV-1) replication. Klin Wochenschr 69:722–724

Emerging Technologies for the Treatment of COVID-19

7

Hossein Aghamollaei, Rahim Sarvestani, Hamid Bakherad, Hamed Zare, Paul C. Guest, Reza Ranjbar, and Amirhossein Sahebkar

Abstract

The new coronavirus, named severe acute respiratory syndrome coronavirus 2 (SARS-CoV-2), turned into a pandemic affecting more than 200 countries. Due to the high rate of transmission and mortality, finding specific and effective treatment options for this infection is currently of urgent importance. Emerging technologies have created a promising platform for developing novel treatment options for various viral diseases such as the SARS-CoV-2 virus. Here, we have described potential novel therapeutic options based on the structure and pathophysiological mechanism of the SARS-CoV-2 virus, as well as the results of previous studies on similar viruses such as SARS and MERS. Many of these approaches can be used for controlling viral infection by reducing the viral damage or by increasing the potency of the host response. Owing to their high sensitivity, specificity, and reproducibility, siRNAs, aptamers, nanobodies, neutralizing antibodies, and different types of peptides can be used for interference with viral replication or for blocking internal-

H. Aghamollaei
Chemical Injuries Research Center, Systems biology and Poisonings Institute, Baqiyatallah University of Medical Sciences, Tehran, Iran

R. Sarvestani
Research and Development Department, PersisGen Par Biopharma Accelerator, Tehran, Iran

H. Bakherad
Department of Pharmaceutical Biotechnology, School of Pharmacy and Pharmaceutical Sciences, Isfahan University of Medical Sciences, Isfahan, Iran

H. Zare
Cellular and Molecular Research Center, Birjand University of Medical Sciences, Birjand, Iran

P. C. Guest
Laboratory of Neuroproteomics, Department of Biochemistry and Tissue Biology, Institute of Biology, University of Campinas (UNICAMP), Campinas, Brazil

R. Ranjbar (✉)
Molecular Biology Research Center, Systems Biology and Poisonings Institute, Baqiyatallah University of Medical Sciences, Tehran, Iran
e-mail: ranjbarre@gmail.com

A. Sahebkar (✉)
Biotechnology Research Center, Pharmaceutical Technology Institute, Mashhad University of Medical Sciences, Mashhad, Iran

Neurogenic Inflammation Research Center, Mashhad University of Medical Sciences, Mashhad, Iran

Halal Research Center of IRI, FDA, Tehran, Iran

Polish Mother's Memorial Hospital Research Institute (PMMHRI), Lodz, Poland
e-mail: sahebkara@mums.ac.ir; amir_saheb2000@yahoo.com

P. C. Guest (ed.), *Clinical, Biological and Molecular Aspects of COVID-19*, Advances in Experimental Medicine and Biology 1321, https://doi.org/10.1007/978-3-030-59261-5_7

ization. Receptor agonists and interferon-inducing agents are also potential options to balance and enhance the innate immune response against SARS-CoV-2. Solid evidence on the efficacy and safety of such novel technologies is yet to be established although many well-designed clinical trials are underway to address these issues.

Keywords

COVID-19 · SARS-CoV-2 · Coronavirus · Treatment · Biotechnology

7.1 Introduction

The novel coronavirus, which is known as SARS-CoV-2, was first identified in the city of Wuhan, the People's Republic of China, which has spread globally. Its fast outbreak resulted in the 2019–2020 coronavirus pandemic of what has been termed COVID-19 disease. The primary symptoms of COVID-19 infection are fever, dry cough, sputum production, fatigue, and shortness of breath. In severe cases, other symptoms including persistent chest pain or pressure, confusion, anosmia, and gastrointestinal symptoms are seen.

Like other coronaviruses, such as severe acute respiratory syndrome (SARS) (SARS-CoV) and Middle East respiratory syndrome (MERS-CoV), the new coronavirus carries a single-positive stranded RNA. The genome size for SARS-CoV-2 is 29,891 nucleotides which encodes 9860 amino acids. This genome has 82% nucleotide identity with human SARS-CoV [1]. The organization of genes in the coronaviruses shares the same order, coding for polyproteins 1a and 1b and the spike (S), membrane (M), envelope (E), and nucleocapsid (N) proteins.

Of the several common drugs currently used to treat COVID-19 infection, chloroquine, remdesivir, lopinavir (LPV), and ritonavir (RTV) have gained the most attention. Despite the relative effectiveness of these medications, they also have side effects. The most hopeful antiviral to combat SARS-CoV-2 is remdesivir. Remdesivir was effective against MERS-CoV and SARS-CoV by acting as an inhibitor of RNA-dependent RNA polymerases in the RNA replication process [2–5]. LPV is an HIV-1 protease inhibitor. This drug has been used in combination with RTV to improve its half-life and found to be effective against SARS-CoV in tissue culture and in patients with HIV-1 [6]. However, the antiviral property of LPV against MERS-CoV remains ambiguous. There are some safety concerns to the use of this drug, including risk of cardiac arrhythmia, caution in patients with hepatic disease, and significant drug interactions [6]. Chloroquine is an antimalarial drug which has been shown to have in vitro activity against SARS-CoV-2. Its mechanism of action may include inhibition of viral enzymes or processes such as viral DNA and RNA polymerase. Using chloroquine may have some limitations including the risk of cardiac arrhythmia and risk of retinal injury, with cautions in patients with diabetes and those with glucose 6-phosphate dehydrogenase (G6PD) deficiency [7–10]. In addition, it has significant drug interactions.

Another possible treatment option is to use the serum from patients infected with the SARS-CoV-2. However, it is not yet clear whether a sufficient set of potential donors is possible. Studies on MERS-CoV have shown that the sera from patients recovering from infection do not appear to contain adequate antibody titers for therapeutic use [11].

Based on the recommendation of the World Health Organization (WHO) and the Centers for Disease Control and Prevention (CDC), there are currently no approved drugs or vaccines for the treatment or prevention of COVID-19. In recent years, with the development and integration of different scientific branches, several techniques have been proposed for the treatment of viral infections. These methods, which have emerged from a successful combination of medical sciences, biotechnology, chemistry, and bioinformatics, have shown promising results in the treatment of viral infections. Here, we have aimed to review the potency of novel and emerging techniques for the treatment of COVID-19.

7.2 Small Interfering RNA

The discovery of RNA interference (RNAi) provided a new approach for silencing the expression of specific genes in order to treat a wide range of human disorders. RNA interference as an antiviral mechanism was originally discovered in plants. Later, it was also observed in other organisms, including nematodes, Drosophila, and vertebrates [12, 13]. As a tool, siRNAs is a powerful approach specifically designed to reduce and prevent the synthesis of the target protein. This has brought the opportunity to develop a new generation of drugs for several diseases. Currently, many pharmaceutical companies are developing RNA-based therapeutics to specifically regulate various disease-causing genes. A detailed discussion of RNA therapies and their advantages, disadvantages, and challenges has been reviewed [see 14]. Several research teams have successfully used siRNA technology for developing various antiviral treatments (Table 7.1). Their effects have been evaluated on cell lines and animal models, and positive results have been reported. The great potential of siRNA for the management of serious human and animal respiratory viruses including respiratory syncytial virus (RSV), SARS-CoV, influenza, adenovirus, avian metapneumovirus, and the porcine respiratory virus has also been reviewed [see 15, 16].

Table 7.1 Reported studies on siRNA against respiratory viruses

Target disease	Target gene/genes	Reference
Respiratory syncytial virus (RSV)	Viral fusion (F) phosphoprotein (P) proteins	[105]
Influenza virus	RNA-dependent RNA polymerase (RdRP)	[17, 18, 105]
Influenza virus	Ran-binding protein 5	[106]
SARS	Spike protein	[23, 24]
SARS	RdRP protein	[19–21]
SARS	Envelop protein	[21]
SARS	Leader sequence	[107]
Human CoronavirusNL63	Spike protein	[22]
Adenoviruses	Adenoviral E1A	[108]
Influenza A1 virus	Nucleocapsid protein; polymerase acidic protein	[109–111]
SARS	Spike protein	[112]
MERS	Spike protein	[113]
SARS	Leader, TRS, 3′-UTR and spike	[114]

Using siRNA technology to specifically target the key mRNAs for SARS-Cov-2 infection and assembly could be a valuable tool for the treatment of COVID-19. Similar to MERS and SARS, SARS-Cov-2 belongs to the coronavirus family, and its genome structure, host infection, and assembly likely share a common pattern. Therefore, reports on the application of siRNA for SARS and MERS could be informative for designing siRNA treatments for SARS-Cov-2. One of the most important proteins in the coronavirus family is the RNA-dependent RNA polymerase (RdRP), which is responsible for genome replication once the host cell is infected by the virus. One strategy to prevent the viral amplification and spread of the disease is to decrease the corresponding mRNA for this protein. Several studies have shown that targeting of RdRP using an siRNA approach is an effective strategy for controlling influenza and SARS disease in cell lines and animal models, by resulting in an 80–90% reduction of virus replication [17–21]. Other potential targets for using siRNA against SARS-CoV-2 are the structural proteins. These proteins are involved in virus assembly and binding to the host cell. Several reports have shown that siRNA against structural proteins, S, E, and M protein, could reduce the progression of coronavirus infections, including SARS-CoV, MERS-CoV, and HCoV-NL63 [19, 21–24].

7.3 Neutralizing Antibodies

Since the 1980s, when the first therapeutic monoclonal antibody (mAb) was approved, dozens of monoclonal antibodies have been used in the treatment of various diseases. Today, the majority of the biotherapeutic product market is occupied by monoclonal antibodies [25].

During recent years, monoclonal antibodies are increasingly being considered as agents to fight severe viral diseases. In this section, we highlight the potential targets for neutralizing antibodies against SARS-CoV-2 inspired by those mAbs developed for combating of SARS-CoV or MERS-CoV.

Once coronavirus binds to the cell surface receptors via the spike protein, its replication begins [26]. Specific interaction between S1 subunit of the spike protein and its receptor creates a conformational change in the S2 subunit, which causes the viral envelope to fuse with the cellular membrane and release of the nucleocapsid into the cytoplasm [27]. The cell surface enzymes are used as a specific receptor for most of the human coronaviruses. For example, angiotensin-converting enzyme 2 (ACE2) works as a receptor for HCoV-NL63, SARS-CoV, and SARS-CoV-2 coronaviruses, while MERS-CoV attaches via dipeptidyl peptidase 4 (DPP4) [28] (Fig. 7.1). Thus, an effective treatment against SARS-CoV-2 might be developed based on the use of neutralizing or blocking monoclonal antibodies targeting either the viral spike protein or the host receptor [29]. Monoclonal antibodies against the spike protein in coronaviruses have shown promising results both in vitro and in vivo. Coughlin et al. generated dozens of mAbs against the SARS-CoV spike protein, and some of them were effective in in vitro studies [30]. Recently, a comprehensive review on the possibility of applying monoclonal antibody-based treatment for SARS-Cov-2 was published by Shanmugaraj et al. [29]. One potential source of mAbs for SARS-Cov-2 is via identification and isolation from either an antibody human phage library or memory B cells from infected and recovered patients.

Interleukin 6 (IL-6) acts as a pro-inflammatory cytokine by stimulating the acute phase response. Research has shown that the levels of cytokines such as IL-6 in COVID-19 cases can increase dramatically and the use of drugs that can inhibit this cytokine improves patient recovery. There are different kinds of FDA-approved antibodies that block IL-6 or IL-6 receptors such as siltuximab (Sylvant), sarilumab (Kevzara), and tocilizumab (Actemra). Studies have shown that the administration of these antibodies in COVID-19 patients with high levels of IL-6 can greatly improve the severity of this disease [31]. Currently, both intravenous and subcutaneous administration of RoActemra and subcutaneous administration of Kevzara are considered in phase 2 clinical trials as a treatment for COVID-19 [32]. The results are expected in June 2021.

7.4 Aptamer-Based Viral Treatment

Aptamers are non-coding single-stranded nucleotide sequences that specifically bind to their targets. Aptamers are synthesized by an in vitro process called systematic evolution of ligands by exponential enrichment (SELEX). Compared to mAbs, aptamers are easily synthe-

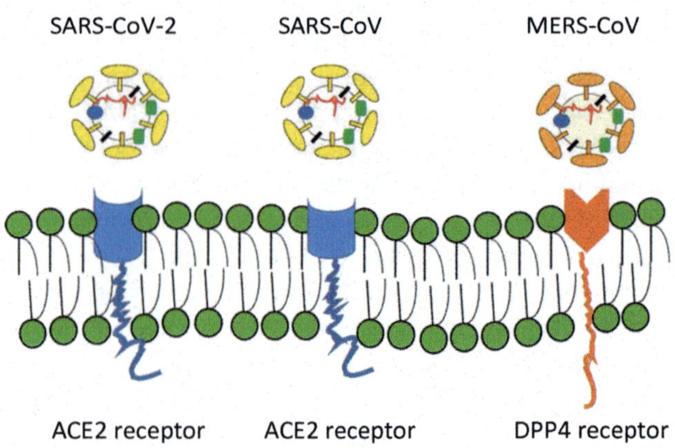

Fig. 7.1 Schematic of SARS-CoV, MERS-CoV, SARS-CoV-2, and their cellular receptors

sized, and their targets include a wide range of biomolecules. Aptamer-based therapeutics has the potential to create a revolution in the development of antiviral drugs. Nowadays, several approved aptamer-based drug for various diseases is available on the market. There are several advantages over using aptamer-based treatments including high specificity, rapid selection process, no need for the complex process of protein expression and purification, and the simple process needed for large-scale manufacturing. These advantages make aptamer technology well-suited for treatment of viral infections [33].

Because of the high specificity and affinity for their targets, aptamers are being increasingly applied in research and therapeutics. Many researchers have studied the development of aptamer-based antiviral treatment especially with respect to HIV and influenza viruses. In various studies, many proteins and enzymes from the HIV virus, including reverse transcriptase, integrase, and transactivation responsive protein, have been targeted for aptamer development [34–37]. Several studies have also focused on the hemagglutinin protein in the influenza virus structure for aptamer selection [38–41]. In addition, in two separate studies, whole influenza virus was targeted for aptamer selection, and the results showed more than 90% inhibition of receptor binding in the presence of aptamers [38, 39]. Shum et al. produced a comprehensive overview of aptamer-based therapies and their challenges in the treatment of various viruses [33].

Aptamer technology has also been used to combat the previous SARS-CoV outbreak. The SARS-CoV helicase contains a functional domain with double-stranded nucleic acid unwinding and ATPase activities. A study showed that the aptamer might bind to the nucleic acid binding site of the helicase and block the unwinding and subsequent helicase activities [42].

The number of patients infected by SARS-CoV-2 is increasing rapidly, and in these circumstances, research plans need to be pushed forward in the right direction to find effective treatments. Using aptamer technology could potentially lead to an effective treatment against COVID-19 disease in a short time and at a relatively low cost. For COVID-19, it is recommended to consider one of the strategies below. One suitable target is blocking the viral fusion with the target cell. The spike protein is responsible for cell attachment and entry, and blockade by an aptamer could be an effective way to inhibit infection. Proteins and enzymes involved in the viral replication cycle including polymerases and nucleocapsid protein are other potential targets for inhibitory aptamers. Another promising target for aptamer treatment is RNA-dependent RNA polymerase, because of its importance in virus replication. In addition, specific regions of the viral genome interact with various proteins responsible for transcription initiation, translation, and replication, or viral assembly. These proteins are also promising targets for the generation of aptamers with selective affinity to these regions [43].

Although the intrinsic physicochemical features of aptamers pose serious challenges for their transport to infected organs or cells, they may be well-suited for respiratory viruses because the upper airways and lungs are relatively accessible as target organs.

7.5 Nanobodies

Nanobodies are a new class of recombinant antibody derived from heavy-chain antibodies in camels and sharks. Unlike traditional antibodies, the variable domain of these types of antibodies is made from a single region [44]. These types of antibodies have specific properties that distinguish them from others, including smaller size, higher affinity, more solubility and resistance to denaturation, stability in intolerable condition (high and low pH, high temperature), a broad diversity of epitopes recognition, faster tissue permeability, high sequence homology with human antibodies, and cost-effective production [44–49].

Nanobodies are now used and tested in the treatment of many diseases, including viruses such as hepatitis B, influenza, polio, rabies, HIV, RSV, FMDV, and rotavirus [50]. One of these nanobodies called ALX-0171 inhibits RSV infection by binding to the F-protein of virus and blocking uptake into the cells [51]. It is important to note that due to the specific properties of this type of antibody, ALX-0171 was used as an inhaled form. This method of administration at the site of infection has many advantages, including an increase in the efficacy of the drug and reduced dose compared to systemic injections [52].

The receptor-binding domain (RBD) of the spike protein is the priority target against the coronavirus family as this allows binding to the host cell surface receptor. In a recent study, scientists isolated nanobodies against the RBD domain of the MERS-CoV which potently neutralized MERS-CoV infection [53]. Therefore, there is some hope that a similar approach can be used against SARS-Cov-2.

7.6 Peptide Inhibitors

Peptides are short chains of amino acids that are usually composed of less than 50 amino acids. Peptides have many advantages over proteins and antibodies, such as being small in size, easy to synthesize, as well as efficient in cell and tissue penetration [54]. They can also have high affinity, specificity, and activity and do not accumulate in a particular tissue, resulting in low toxicity.

In the treatment of viral diseases, peptides have two important applications. By studying viral antigens and selecting the appropriate peptides, they can be used as vaccines, and subsequently, the immune system can detect and eliminate the virus. They can also be used competitively against viral proteins and thereby prevent viral entry into cells. Guo et al. attempted to identify the most potent peptides to stimulate the humoral immune system as a SARS vaccine [55]. They synthesized 4942 overlapping peptides from all proteins of the SARS genome and evaluated these against serum from patients recovering from the virus. Peptides recognized by antibodies in the serum samples were selected for potential use as a polyvalent immunogen [55]. In order to investigate the possibility of using these peptides against SARS-Cov-2, we conducted a blast analysis of the same peptides against SARS-Cov-2 proteins. The results of this analysis showed that among 24 peptides presented in the SARS-CoV study, 13 are highly conserved to regions SARS-Cov-2 proteins and might therefore be used as vaccine candidates against the virus (Table 7.2).

Wang et al. analyzed various peptides of the SARS-CoV spike protein using a bioinformatics approach and synthesized the most promising candidates [24]. Next, they screened these

Table 7.2 Blast of peptides recognized in SARS convalescent sera against NCBI databases which gave high sequence identities with SARS-Cov-2 proteins

Category	SARS peptide	Covid-19 peptide	Homology (%)	Antibody classes
Orf1a	NQDVNLHSSRLS	NQDVNLHSSRLS	100	IgA, IgM
Nucleocapsid (N)-protein	QLPQGTTLPKGFYA	QLPQGTTLPKGFYA	100	IgG, IgA
	TVTLLPAADMDDF	TVTLLPAADLDDF	92	IgG, IgM
	YKTFPPTEPKKD	YKTFPPTEPKKD	100	IgA
	GGSQASSRSSSR	GGSQASSRSSSR	100	IgG, IgM
	IRQGTDYKHWPQ	IRQGTDYKHWPQ	100	IgG, IgM
Spike (S)-protein	CPFGEVFNATKF	CPFGEVFNATRF	91	IgA
	PIGAGICASYHT	PIGAGICASYQT	91	IgG, IgA, IgM
	QYGSFCTQLNRA	QYGSFCTQLNRA	100	IgG, IgM
	PFAMQMAYRFNG	PFAMQMAYRFNG	100	IgM
Membrane (M)-protein	KEITVATSRTLS	KEITVATSRTLS	100	IgG, IgA, IgM
	GTITVEELKQLL	GTITVEELKKLL	91	IgG, IgA, IgM
E-protein	YVYSRVKNLNSS	YVYSRVKNLNSS	100	IgG, IgA, IgM

peptides using T cells from individuals who had recovered from the disease. They found that two peptides (FIAGLIAIV and LITGRLQSL) were immunogenic and effectively stimulated a T-cell immune response against this virus. To investigate the possibility of using these two peptides as immunogens against COVID-19, we compared the sequences against those in the SARS-CoV-2 spike protein. This revealed 100% identity, lending support to their potential use as a SARS-Cov-2 vaccine. Another study targeted MHC-I and MHC-II epitopes within the spike protein of the SARS-CoV-2 virus in an informatics-based approach to identify the most promising peptide vaccine candidates [56]. They identified 29 peptides within the MHC-I region and 8 within the MHC-II region, which they used to synthesize a single vaccine complex.

Zheng et al. synthesized 24 peptides against the SARS-COV spike protein and tested these as inhibitors of viral entry into cells. They found that SARS-CoV infection was completely inhibited by two peptides [57]. The sequence of one of these peptides (IQKEIDRLNEVAKNLNESLI) is identical to a sequence in the S2 subunit of SARS-CoV-2, suggesting that it might be a suitable candidate for the treatment of COVID-19 disease.

7.7 Fusion Inhibitors

Fusion inhibitors are a class of drugs that were first introduced in HIV infection, and their mechanism of action is to prevent and interfere with the binding, fusion, and entry of the virus into the target cells [58]. Various strategies can be proposed to produce a fusion inhibitor against SARS-CoV-2. The first is the production of small molecules that can bind to the virus target receptor and prevent its binding and entry, such as 1-thia-4- azaspiro[4.5]decan-3-one derivatives [59]. An alternative approach is the use of drugs developed for other coronaviruses such as Nafamostat, Griffithsin, and Dihydrotanshinone E-64-C and E-64-D [5, 60–64].

Xia et al. found that a peptide derived from the heptad repeat 2 (HR2) domain of human coronaviruses has a pan inhibitory function against several members of this viral family [65]. In vivo studies showed that the inhalation of this peptide had a high potency in suppressing viral infection and good safety profile [65]. Subsequently, the same research group developed lipopeptides derived from the same region and showed that one of these was more than 100-fold more potent than the original peptide in preventing infection with the SARS-CoV-2 virus [66].

The use of recombinant proteins can also be an effective way of inhibiting the virus from entering the cell. Wong et al. reported that the RBD domain of SARS-Cov S protein potently binds to ACE2 and prevents infection [67]. The main advantage of methods that disrupt the virus host interaction is that the host receptor (ACE2) does not undergo rapid mutation [68].

Li et al. demonstrated that administration of recombinant ACE2 effectively bound to the SARS-CoV virus and inhibited infection of cells in culture [69]. Recently, Monteil et al. reported that treatment of Vero E6 cells with recombinant ACE2 in the early stage of infection can reduce the SARS-CoV-2 growth rate by more than 1000-fold [70]. However, this study only examined the effects of this protein in the early stages of infection, and its effectiveness in the later stages of COVID-19 infection has yet to be determined.

7.8 Antimicrobial Peptides (AMPs)

The development of antimicrobial peptides during the late 1990s and 2000s led to first marketing approvals in 2012 for 6 peptides [68]. Peptides are an important part of the drug industry, and about 140 peptides are currently being tested in various clinical trials [71]. The use of peptides in treating infections has three advantages, including the shorter market time, inhibition of protein-protein interactions, and the availability of methods to increase the peptide half-lives. Through the creation of a pore and eliciting changes in the structure of bacterial cell membranes, peptides have broad-spectrum activ-

Table 7.3 The peptide sequence of the seven selected AMPs

Peptide	Sequence
AP00225	ACYCRIGACVSGERLTGACGLNGRIYRLCCR
AP00180	ATCYCRTGRCATRESLSGVCEISGRLYRLCCR
AP00549	GFGCNGPWDEDDMQCHNHCKSIKGYKGGYCAKGGFVCKCY
AP00744	GLPQDCERRGGFCSHKSCPPGIGRIGLCSKEDFCCRSRWYS
AP00729	GLPVCGETCVGGTCNTPGCTCSWPVCTRN
AP00764	GLRSKIWLWVLLMIWQESNKFKKM
AP00223	VTCYCRSTRCGFRERLSGACGYRGRIYRLCCR

ity against several microorganisms [72–74], with fewer side effects compared to chemical drugs [72, 75]. Many peptides are also available which can inhibit viral activities. For example, a peptide called RVFV-6, which originates from the Rift Valley FeverVirus (RVFV) glycoprotein, is an inhibitor of viral fusion [75, 76]. Kn2-7, a new derivative of a scorpion venom peptide, has inhibitory activity against HIV-1, with a weak cytotoxic effect in mammalian cells [77]. Numerous studies have shown that AMPs are good candidates for the development of new therapeutic agents against coronaviruses [78–81]. Antiviral AMPs function in different ways, including prevention of viral entry through particular receptors, viral fusion blockage through interaction with the viral envelope and membrane, and stopping viral entry through interaction with heparansulfate [78].

Zhao and colleagues examined the antiviral activity of 11 mouse defensin-derived peptides. Among them, one peptide (NGAICWGPCPTA FRQIGNCGHFKVRCCKIR) showed strong and wide-ranging antiviral effects on several respiratory viruses including MERS-CoV, SARS-CoV, and influenza A H1N1 virus [80]. This peptide interrupts the RBD interaction [81].

In addition to the above approaches, prevention of viral replication is one of the strategies to control viral infections [82]. Mucropin-M1 (LFRLIKSLIKRLVSAFK) is a derivative from mucropin AMP (LFGLIPSLIGGLVSAFK). In this case, the proline (P) was replaced by arginine (R), and glycine (G) was changed to lysine (K). Mucropin-M1 demonstrated activity against SARS-CoV and influenza A virus H5N1, by preventing viral replication [83]. The original peptide mucropin showed no antiviral activity against any of these viruses.

In another study, Mustafa et al. developed several AMPs which bind to the MERS-CoV spike protein [84]. These peptides belong to the defensin family and may be very important in providing inhibitory activity. The results of the study showed that seven peptides had a high affinity for MERS-CoV spike protein at its active site, suggesting their potential use in the treatment of COVID-19 (Table 7.3).

Zhou et al. recognized that the glycopeptide antibiotic teicoplanin could inhibit the entrance of Ebola viruses into the cell cytoplasm [85]. This was carried out by high-throughput screening of FDA-approved drugs. Further analysis confirmed that teicoplanin was also capable of blocking the entry of SARS-CoV and MERS-CoV viruses. This antibiotic has been shown to have an inhibitory effect on viral replication and transcription.

Evaluation of the AMP rhesus theta-defensin 1 (RTD-1) showed 100% survival and a moderate decrease in lung injury in a mouse model of SARS-CoV infection [86]. The mechanism appeared via an effect on the inflammatory system as the cytokine responses in the treated animals were altered compared to the untreated group.

7.9 Interferon-Inducing Agents

Another way of modulating the body's protection system against SARS infection is through treatment with interferons or the use of agents that induce interferon production [87–89]. IFN inducers have several advantages compared to exogenous IFN. They motivate the production of the body's own IFN, which has no antigenic properties, unlike recombinant forms of IFN [90]. IFN inducers can be mixed with IFN and other antiviral drugs, a strategy that could have both immunomodulating and etiotropic effects [91].

Bao and colleagues developed a method based on CpG oligodeoxynucleotides (ODNs) for the treatment and prevention of SARS-CoV disease [92]. They found a new CpG ODN called BW001 which could stimulate human peripheral blood mononuclear cells (PBMCs) to protect Vero cells against SARS-CoV. In addition, BW001 stimulated human dendritic cells and PBMCs to secrete high levels of IFN-α and stimulated B cell and PBMC proliferation. Additionally, BW001 can increase the secretion of IFN-γ and natural killer cell cytotoxicity. In another study, Barnard et al. used a mismatched double-stranded (ds)-RNA called Ampligen® (poly I: poly C124) as an interferon inducer and a hybrid human interferon (IFN-α B/D) against SARS-CoV infection [93]. In this study, Ampligen was injected intraperitoneally 4 h before the mice were exposed to SARS-CoV. As a result, the titers of the lung viruses decreased below the detectable level.

Kumaki and colleagues used polyriboinosinic-polyribocytidylic acid stabilized with poly-l-lysine and carboxymethyl cellulose (poly-ICLC) as an interferon inducer in SARS-CoV-infected mice [94]. Treatment with poly-ICLC (5 mg/kg) was initiated 24 h after infection with SARS- Cov and continued 2 times a day for 5 days. All treated mice were protected against lethal viral infection, and virus titers were reduced in the lungs.

7.10 Peptidomimetics

Any compound that can mimic the biological activities and structural properties of a peptide are referred to as a peptidomimetic. Changes in peptide structure in antimicrobial research include side-chain and backbone modifications with the use of unnatural amino acids (such as D-amino acids), peptoids, β-peptides, and lipidation [95]. The main protease (Mpro) is responsible for proteolytic processing of polyproteins 1a and 1ab, causing the release of 15 proteins involved in the viral replication process [96]. As Mpro is vital at the beginning of coronavirus replication, it is a promising target against infection [97, 98].

Kumar et al. designed and synthesized three peptidomimetic inhibitors that inhibit 3CLpro of SARS-CoV and MERS-CoV with IC_{50} values of 0.2–0.7 μM and 1.7–4.7 μM, respectively [98]. These agents demonstrated a desirable selectivity index and could potentially lead to the discovery of wide-spectrum antiviral drugs against newly emerging coronaviruses. In addition, Arun et al. designed and synthesized several peptidomimetic SARS-CoV protease inhibitors with good SARS-CoV 3CLpro inhibitory activity [97]. In another study, Kankanamalage et al. designed and evaluated a new compound which inhibits the 3CLpro of the MERS-CoV [99]. These compounds effectively prevented MERS-CoV replication.

Finally, Chuck et al. investigated the inhibitory effects of several numbers of nitrile-based peptidomimetic inhibitors with various peptide lengths and N-terminal protective groups, on the enzymatic activity of 3CLpro of SARS-CoV [100]. Three nitrile-based inhibitors with carboxybenzyl (Cbz), tert-butyloxycarbonyl (Boc), and 5-methylisoxazole-3-carboxyl (Mic) protective groups were synthesized containing the SARS-CoV auto-cleavage sequence AVLQ. Protease activity was measured in the presence of inhibitors, and the IC_{50} values of Cbz-AVLQ-CN, Boc-AVLQ-CN, and Mic-AVLQ-CN were 4.6 ± 0.2, 49 ± 2, and 49 ± 2 μM, respectively. Thus, the inhibitory effect of components with Cbz group was 10 times stronger than the others. This demonstrated that the nitrile cap could efficiently deactivate the 3CLpro activity. Further studies showed that Cbz-AVLQ-CN is a wide-spectrum inhibitor against several coronavirus strains (e.g., OC43, NL63, 229E, and HKU1), suggesting that this approach may have promise for treatment of COVID-19.

7.11 Toll-like Receptor Agonists

Toll-like receptors (TLRs) are a group of proteins that allow the immune system to discriminate between "self" and "non-self" [34, 35]. Consequently, TLR antagonists and agonists have been suggested as antiviral or adjuvant compounds [101, 102].

A study by Totura et al. showed that TLR signaling via the TIR-domain-containing adapter-inducing interferon-β (TRIF) protein protects mice from SARS-CoV disease lethality [102]. Their findings showed a balanced immune response that operates via both MyD88 adapter-driven and TRIF-driven pathways. Since the TLR3−/−, TLR4−/−, and TRAM−/− mice are more sensitive to SARS-CoV than normal mice, using TLR agonists can be effective in the treatment of MERS-CoV and SARS-CoV infection [101, 102].

Zhao and colleagues used intranasal poly(I·C), lipopolysaccharide, R848, or CpG (TLR3, TLR4, TLR7/8, or TLR9 agonists) in mice infected with SARS-CoV [103]. After treatment, approximately 95% survival was found for poly(I·C) against SARS-CoV. Pretreatment with poly (I·C) led to upregulation of IFN-γ, IFN-β, tumor necrosis factor alpha (TNFα), and IL-1β gene expression in the lungs. Their investigation also showed that treatment with poly(I·C) repressed viral replication in human host cells. These findings suggest that TLR adapters are crucial in producing a balanced innate immune response to COVID-19 infection.

7.12 Conclusions

Emerging techniques can be used for controlling viral infection by reducing the damage or increasing the potency of the host response. The development of siRNAs or aptamers for targeting genes coding for critical structural (i.e., S, E, and M) and nonstructural (e.g., RdRP, 3CL protease) proteins can be used to block the effects of SARS-CoV-2 infection. Also, the sensitivity, specificity, reproducibility, and ease of use make mAbs an attractive option for the treatment of COVID-19. However, this strategy might be time-consuming and costly compared to other treatments. Future studies for mAb development against SARS-CoV-2 may be focused on the identification and use of S1 epitopes as a key target for inhibition of viral entry into the cells.

Peptides are one of the most promising options for the development of anti-COVID-19 drugs as they can be used as antigens for vaccine production or as inhibitors for preventing viral infection. Due to the homology of SARS-CoV and SARS-Cov-2 protein sequences, several peptides proposed for use in the former could be applicable for the treatment of COVID-19. Based on our blast results, we propose 13 peptides with high homology for consideration as a target for vaccine development (Table 7.2). Peptidomimetics can also help to improve peptide effectiveness as antiviral agents. Unique features of nanobodies such as the small size, low immunogenicity, and capacity for conjugation with other agents make them ideal candidates for viral detection and therapy.

In addition, Toll-like receptor agonists can protect against SARS-CoV, and IFN inducers stimulate the natural production of the IFN by the host, thereby improving the host response against viral infection. Production of different inhibitors by genetic engineering and recombinant protein expression is another approach which may be promising as viral therapies. Specifically, the use of ACE2 recombinant proteins for inhibition of viral entry may also work against future coronavirus infections given that it this protein is an endogenous factor.

The approaches mentioned in this review prove that it is possible to quickly start well-designed randomized controlled studies even in the middle of a global emergency such as the COVID-19 pandemic. Table 7.4 shows the potential drugs in different phases of clinical trials for treatment of COVID-19, which highlights this capacity. However, there is a need for novel platforms for the development and

Table 7.4 Potential biological-derived drug in different phase of clinical trial for treatment of COVID-19 (clinicaltrials.gov)

Number	Drug/ Molecule	Type	Effect	Status
1	TJ003234	mAb	Anti-GM-CSF monoclonal antibody	Phase I
2	Lopinavir	Protease inhibitory	–	Phase II
3	Ritonavir	Protease inhibitory	Cytochrome P450-3A4	Phase II
4	Fludase (DAS181)	Fusion inhibitor	Preventing of viral entry by removing sialic receptor	–
5	Sarilumab	Immunosuppressive	Blocking of interleukin-6 receptor	Phase II
6	Tocilizumab	Immunosuppressive	Blocking of interleukin-6 receptor	Phase II
7	Sargramostim	Leukocyte growth factor	Recombinant granulocyte macrophage colony-stimulating factor	Phase IIII
8	Mavrilimumab	mAb	Granulocyte macrophage colony-stimulating factor receptor inhibitor	Phase II
9	bacTRL-Spike1	Vaccine	Stimulation of antibody production against SARS-CoV-2 Spike protein	Phase I
10	Ad5-nCoV	Vaccine	Stimulation of antibody production against SARS-CoV-2 Spike protein	Phase II
11	Emapalumab	mAb	Anti-interferon-gamma (IFNγ)	Phase II
12	IFN-α2β	Interferon	Recombinant human interferon α1β	Early Phase I
13	rhIFNα	Interferon	Recombinant human interferon Alpha-1b	Phase III
14	INO-4800	DNA vaccine	Stimulation of antibody production against SARS-CoV-2 Spike protein	Phase III
15	mRNA-1273	Vaccine	Stimulation of antibody production against SARS-CoV-2 Spike protein	Phase I

manufacturing of therapeutic agents and vaccines that can be readily adapted to new viral agents in line with the National Institute of Allergy and Infectious Diseases initiative [104]. Such a platform would facilitate the development of therapeutic agents and vaccines to enter clinical trials in less than 16 weeks and fast-track large-scale manufacturing if a given drug proves to be effective [115]. Such approaches are now essential given that the continuance of the current pandemic and the likely eruption of future coronavirus outbreaks. Finally, the authors of this article believe that both traditional and emerging approaches are essential for the prevention and treatment of COVID-19 [116, 117].

Acknowledgments Thanks to guidance and advice from the Clinical Research Development Unit of Baqiyatallah Hospital.

References

1. Chan JF, Kok KH, Zhu Z, Chu H, To KK, Yuan S et al (2020) Genomic characterization of the 2019 novel human-pathogenic coronavirus isolated from a patient with atypical pneumonia after visiting Wuhan. Emerg Microb Infect 9(1):221–236
2. Agostini ML, Andres EL, Sims AC, Graham RL, Sheahan TP, Lu X et al (2018) Coronavirus susceptibility to the antiviral remdesivir (GS-5734) is mediated by the viral polymerase and the proofreading exoribonuclease. MBio 9(2):e00221–e00218. https://doi.org/10.1128/mBio.00221-18
3. Brown AJ, Won JJ, Graham RL, Dinnon KH III, Sims AC, Feng JY et al (2019) Broad spectrum antiviral remdesivir inhibits human endemic and zoonotic deltacoronaviruses with a highly divergent RNA dependent RNA polymerase. Antivir Res 169:104541. https://doi.org/10.1016/j.antiviral.2019.104541
4. de Wit E, Feldmann F, Cronin J, Jordan R, Okumura A, Thomas T et al (2020) Prophylactic and therapeutic remdesivir (GS-5734) treatment in the rhesus

macaque model of MERS-CoV infection. Proc Natl Acad Sci USA 117(12):6771–6776
5. Wang M, Cao R, Zhang L, Yang X, Liu J, Xu M et al (2020) Remdesivir and chloroquine effectively inhibit the recently emerged novel coronavirus (2019-nCoV) in vitro. Cell Res 30(3):269–271
6. Chu C, Cheng V, Hung I, Wong M, Chan K, Chan K et al (2004) Role of lopinavir/ritonavir in the treatment of SARS: initial virological and clinical findings. Thorax 59(3):252–256
7. Fox R (1996) Anti-malarial drugs: possible mechanisms of action in autoimmune disease and prospects for drug development. Lupus 5(Suppl 1):4–10
8. Gao J, Tian Z, Yang X (2020) Breakthrough: chloroquine phosphate has shown apparent efficacy in treatment of COVID-19 associated pneumonia in clinical studies. Biosci Trends 14(1):72–73
9. Yao X, Ye F, Zhang M, Cui C, Huang B, Niu P et al (2020) In vitro antiviral activity and projection of optimized dosing design of hydroxychloroquine for the treatment of severe acute respiratory syndrome coronavirus 2 (SARS-CoV-2). Clin Infect Dis Mar 9:ciaa237. https://doi.org/10.1093/cid/ciaa237. Online ahead of print
10. Smith T, Prosser T (2020) COVID-19 drug therapy-potential options. https://www.lusiadas.pt/pt/Covid19%20Public%20Resource%20Center/Clinical%20and%20Research%20Articles/18032020/COVID-19-Drug-Therapy_Mar-2020.pdf
11. Arabi Y, Balkhy H, Hajeer AH, Bouchama A, Hayden FG, Al-Omari A et al (2015) Feasibility, safety, clinical, and laboratory effects of convalescent plasma therapy for patients with Middle East respiratory syndrome coronavirus infection: a study protocol. Springerplus 4:709. https://doi.org/10.1186/s40064-015-1490-9
12. Agami R (2002) RNAi and related mechanisms and their potential use for therapy. Curr Opin Chem Biol 6(6):829–834
13. Cerutti H (2003) RNA interference: traveling in the cell and gaining functions? Trends Genet 19(1):39–46
14. Bajan S, Hutvagner G (2020) RNA-based therapeutics: from antisense oligonucleotides to miRNAs. Cell 9(1):137. https://doi.org/10.3390/cells9010137
15. Bitko V, Barik S (2007) Respiratory viral diseases: access to RNA interference therapy. Drug Discov Today Ther Strateg 4(4):273–276
16. Wu CJ, Chan YL (2006) Antiviral applications of RNAi for coronavirus. Expert Opin Investig Drugs 15(2):89–97
17. Gao Y, Sun L, Dong JX, Xingye X, Shu Y, Chen M et al (2006) Rapid identification of small interfering RNA that can effectively inhibit the replication of multiple influenza B virus strains. Antivir Ther 11(4):431–438
18. Ge Q, McManus MT, Nguyen T, Shen C, Sharp PA, Eisen HN et al (2003) RNA interference of influenza virus production by directly targeting mRNA for degradation and indirectly inhibiting all viral RNA transcription. Proc Natl Acad Sci U S A 100(5):2718–2723
19. He ML, Zheng BJ, Chen Y, Wong KL, Huang JD, Lin MC et al (2006) Kinetics and synergistic effects of siRNAs targeting structural and replicase genes of SARS-associated coronavirus. FEBS Lett 580(10):2414–2420
20. He ML, Zheng B, Peng Y, Peiris JSM, Poon LLM, Yuen KY et al (2003) Inhibition of SARS-associated coronavirus infection and replication by RNA interference. JAMA 290(20):2665–2666
21. Meng B, Lui YW, Meng S, Cao C, Hu Y (2006) Identification of effective siRNA blocking the expression of SARS viral envelope E and RDRP genes. Mol Biotechnol 33(2):141–148
22. Pyrc K, Bosch BJ, Berkhout B, Jebbink MF, Dijkman R, Rottier P et al (2006) Inhibition of human coronavirus NL63 infection at early stages of the replication cycle. Antimicrob Agents Chemother 50(6):2000–2008
23. Qin ZL, Zhao P, Zhang XL, Yu JG, Cao MM, Zhao LJ et al (2004) Silencing of SARS-CoV spike gene by small interfering RNA in HEK 293T cells. Biochem Biophys Res Commun 324(4):1186–1193
24. Zhang Y, Li T, Fu L, Yu C, Li Y, Xu X et al (2004) Silencing SARS-CoV spike protein expression in cultured cells by RNA interference. FEBS Lett 560(1–3):141–146
25. Ecker DM, Jones SD, Levine HL (2015) The therapeutic monoclonal antibody market. MAbs 7(1):9–14
26. Fung TS, Liu DX (2019) Human coronavirus: host-pathogen interaction. Annu Rev Microbiol 73:529–557
27. Masters PS (2006) The molecular biology of coronaviruses. Adv Virus Res 66:193–292
28. Lim YX, Ng YL, Tam JP, Liu DX (2016) Human coronaviruses: a review of virus-host interactions. Diseases 4(3):26. https://doi.org/10.3390/diseases4030026
29. Shanmugaraj B, Siriwattananon K, Wangkanont K, Phoolcharoen W (2020) Perspectives on monoclonal antibody therapy as potential therapeutic intervention for Coronavirus disease-19 (COVID-19). Asian Pac J Allergy Immunol 38(1):10–18
30. Coughlin M, Lou G, Martinez O, Masterman SK, Olsen OA, Moksa AA et al (2007) Generation and characterization of human monoclonal neutralizing antibodies with distinct binding and sequence features against SARS coronavirus using XenoMouse. Virol 361(1):93–102
31. Zhang C, Wu Z, Li JW, Zhao H, Wang GQ (2020) The cytokine release syndrome (CRS) of severe COVID-19 and Interleukin-6 receptor (IL-6R) antagonist tocilizumab may be the key to reduce the mortality. Int J Antimicrob Agents 55(5):105954. https://doi.org/10.1016/j.ijantimicag.2020.105954
32. Coomes EA, Haghbayan H (2020) Interleukin-6 in COVID-19: a systematic

review and meta-analysis. medRxiv. https://doi.org/10.1101/2020.03.30.20048058
33. Shum KT, Zhou J, Rossi JJ (2013) Aptamer-based therapeutics: new approaches to combat human viral diseases. Pharmaceuticals (Basel) 6(12):1507–1542
34. DeStefano JJ, Nair GR (2008) Novel aptamer inhibitors of human immunodeficiency virus reverse transcriptase. Oligonucleotides 18(2):133–144
35. Kim SJ, Kim MY, Lee JH, You JC, Jeong S (2002) Selection and stabilization of the RNA aptamers against the human immunodeficiency virus type-1 nucleocapsid protein. Biochem Biophys Res Commun 291(4):925–931
36. Kissel JD, Held DM, Hardy RW, Burke DH (2007) Single-stranded DNA aptamer RT1t49 inhibits RT polymerase and RNase H functions of HIV type 1, HIV type 2, and SIVCPZ RTs. AIDS Res Hum Retrovir 23(5):699–708
37. Watrin M, Von Pelchrzim F, Dausse E, Schroeder R, Toulmé JJ (2009) In vitro selection of RNA aptamers derived from a genomic human library against the TAR RNA element of HIV-1. Biochemistry 48(26):6278–6284
38. Cheng C, Dong J, Yao L, Chen A, Jia R, Huan L et al (2008) Potent inhibition of human influenza H5N1 virus by oligonucleotides derived by SELEX. Biochem Biophys Res Commun 366(3):670–674
39. Gopinath SCB, Sakamaki Y, Kawasaki K, Kumar PKR (2006) An efficient RNA aptamer against human influenza B virus hemagglutinin. J Biochem 139(5):837–846
40. Park SY, Kim S, Yoon H, Kim KB, Kalme SS, Oh S et al (2011) Selection of an antiviral RNA aptamer against hemagglutinin of the subtype H5 avian influenza virus. Nucleic Acid Ther 21(6):395–402
41. Wongphatcharachai M, Wang P, Enomoto S, Webby RJ, Gramer MR, Amonsin A et al (2013) Neutralizing DNA aptamers against swine influenza H3N2 viruses. J Clin Microbiol 51(1):46–54
42. Jang KJ, Lee NR, Yeo WS, Jeong YJ, Kim DE (2008) Isolation of inhibitory RNA aptamers against severe acute respiratory syndrome (SARS) coronavirus NTPase/helicase. Biochem Biophys Res Commun 366(3):738–744
43. Wandtke T, Wozniak J, Kopinski P (2015) Aptamers in diagnostics and treatment of viral infections. Viruses 7(2):751–780
44. Hamers-Casterman C, Atarhouch T, Muyldermans S, Robinson G, Hammers C, Songa EB et al (1993) Naturally occurring antibodies devoid of light chains. Nature 363(6428):446–448
45. Arbabi Ghahroudi M, Desmyter A, Wyns L, Hamers R, Muyldermans S (1997) Selection and identification of single domain antibody fragments from camel heavy-chain antibodies. FEBS Lett 414(3):521–526
46. Bakherad H, Gargari SLM, Rasooli I, RajabiBazl M, Mohammadi M, Ebrahimizadeh W et al (2013) In vivo neutralization of botulinum neurotoxins serotype E with heavy-chain camelid antibodies (VHH). Mol Biotechnol 55(2):159–167
47. Ebrahimizadeh W, Gargari SM, Rajabibazl M, Ardekani LS, Zare H, Bakherad H (2013) Isolation and characterization of protective anti-LPS nanobody against V. cholerae O1 recognizing Inaba and Ogawa serotypes. Appl Microbiol Biotechnol 97(10):4457–4466
48. Harmsen M, De Haard H (2007) Properties, production, and applications of camelid single-domain antibody fragments. Appl Microbiol Biotechnol 77(1):13–22
49. Aghamolaei H, Gargari SLM, Rasaee MJ, Ghanei M (2017) Camelid variable fragments of heavy chain antibodies (Nanobody): its applications in research, diagnosis and therapy. Minerva Biotecnol 29(2):89–100
50. Vanlandschoot P, Stortelers C, Beirnaert E, Ibañez LI, Schepens B, Depla E et al (2011) Nanobodies®: new ammunition to battle viruses. Antivir Res 92(3):389–407
51. Detalle L, Stohr T, Palomo C, Piedra PA, Gilbert BE, Mas V et al (2016) Generation and characterization of ALX-0171, a potent novel therapeutic nanobody for the treatment of respiratory syncytial virus infection. Antimicrob Agents Chemother 60(1):6–13
52. De Bruyn S, De Smedt T, Allosery K, Crabbe P, De Brabandere V, Detalle L et al (2015) ALX-0171: safety and therapeutic potential of an inhaled anti-RSV Nanobody. RDD Europe 1:37–48
53. Zhao G, He L, Sun S, Qiu H, Tai W, Chen J et al (2018) A novel nanobody targeting Middle East respiratory syndrome coronavirus (MERS-CoV) receptor-binding domain has potent cross-neutralizing activity and protective efficacy against MERS-CoV. J Virol 92(18):e00837–e00818
54. Marqus S, Pirogova E, Piva TJ (2017) Evaluation of the use of therapeutic peptides for cancer treatment. J Biomed Sci 24(1):21. https://doi.org/10.1186/s12929-017-0328-x
55. Guo JP, Petric M, Campbell W, McGeer PL (2004) SARS corona virus peptides recognized by antibodies in the sera of convalescent cases. Virol 324(2):251–256
56. Bhattacharya M, Sharma AR, Patra P, Ghosh P, Sharma G, Patra BC et al (2020) Development of epitope-based peptide vaccine against novel coronavirus 2019 (SARS-COV-2): immunoinformatics approach. J Med Virol 92(6):618–631
57. Zheng BJ, Guan Y, Hez M, Sun H, Du L, Zheng Y et al (2005) Synthetic peptides outside the spike protein heptad repeat regions as potent inhibitors of SARS-associated coronavirus. Antivir Ther 10(3):393–403
58. Schneider SE, Bray BL, Mader CJ, Friedrich PE, Anderson MW, Taylor TS et al (2005) Development of HIV fusion inhibitors. J Pept Sci 11(11):744–753
59. Apaydın ÇB, Cesur N, Stevaert A, Naesens L, Cesur Z (2019) Synthesis and anti-coronavirus activity of a series of 1-thia-4-azaspiro [4.5]

decan-3-one derivatives. Arch Pharm (Weinheim) 352(6):e1800330. https://doi.org/10.1002/ardp.201800330
60. Barton C, Kouokam JC, Lasnik AB, Foreman O, Cambon A, Brock G et al (2014) Activity of and effect of subcutaneous treatment with the broad-spectrum antiviral lectin griffithsin in two laboratory rodent models. Antimicrob Agents Chemother 58(1):120–127
61. Dyall J, Coleman CM, Hart BJ, Venkataraman T, Holbrook MR, Kindrachuk J et al (2014) Repurposing of clinically developed drugs for treatment of Middle East respiratory syndrome coronavirus infection. Antimicrob Agents Chemother 58(8):4885–4893
62. Ito K, Yotsuyanagi H, Sugiyama M, Yatsuhashi H, Karino Y, Takikawa Y et al (2016) Geographic distribution and characteristics of genotype A hepatitis B virus infection in acute and chronic hepatitis B patients in Japan. J Gastroenterol Hepatol 31(1):180–189
63. O'Keefe BR, Giomarelli B, Barnard DL, Shenoy SR, Chan PK, McMahon JB et al (2010) Broad-spectrum in vitro activity and in vivo efficacy of the antiviral protein griffithsin against emerging viruses of the family Coronaviridae. J Virol 84(5):2511–2521
64. Wang C, Hua C, Xia S, Li W, Lu L, Jiang S (2019) Combining a fusion inhibitory peptide targeting the MERS-CoV S2 protein HR1 domain and a neutralizing antibody specific for the S1 protein receptor-binding domain (RBD) showed potent synergism against pseudotyped MERS-CoV with or without mutations in RBD. Viruses 11(1):31. https://doi.org/10.3390/v11010031
65. Xia S, Yan L, Xu W, Agrawal AS, Algaissi A, Tseng CTK et al (2019) A pan-coronavirus fusion inhibitor targeting the HR1 domain of human coronavirus spike. Sci Adv 5(4):eaav4580. https://doi.org/10.1126/sciadv.aav4580
66. Xia S, Liu M, Wang C, Xu W, Lan Q, Feng S et al (2020) Inhibition of SARS-CoV-2 infection (previously 2019-nCoV) by a highly potent pan-coronavirus fusion inhibitor targeting its spike protein that harbors a high capacity to mediate membrane fusion. Cell Res 30(4):343–355
67. Wong SK, Li W, Moore MJ, Choe H, Farzan M (2004) A 193-amino acid fragment of the SARS coronavirus S protein efficiently binds angiotensin-converting enzyme 2. J Biol Chem 279(5):3197–3201
68. Kruse RL (2020) Therapeutic strategies in an outbreak scenario to treat the novel coronavirus originating in Wuhan, China. F1000Res 9:72. https://doi.org/10.12688/f1000research.22211.2
69. Li W, Moore MJ, Vasilieva N, Sui J, Wong SK, Berne MA et al (2003) Angiotensin-converting enzyme 2 is a functional receptor for the SARS coronavirus. Nature 426(6965):450–454
70. Monteil V, Kwon H, Prado P, Hagelkrüys A, Wimmer RA, Stahl M et al (2020) Inhibition of SARS-CoV-2 infections in engineered human tissues using clinical-grade soluble human ACE2. Cell 181(4):905–913
71. Mustafa S, Balkhy H, Gabere MN (2018) Current treatment options and the role of peptides as potential therapeutic components for Middle East Respiratory Syndrome (MERS): a review. J Infect Public Health 11(1):9–17
72. Moghaddam MM, Aghamollaei H, Kooshki H, Barjini KA, Mirnejad R, Choopani A (2015) The development of antimicrobial peptides as an approach to prevention of antibiotic resistance. Rev Med Microbiol 26(3):98–110
73. Heiat M, Aghamollaei H, Moghaddam MM, Kooshki H (2014) Using CM11 peptide as a cell permeable agent for the improvement of conventional plasmid transformation methods in Escherichia coli and Bacillus subtilis. Minerva Biotecnol 26:149–157
74. Kaspar AA, Reichert JM (2013) Future directions for peptide therapeutics development. Drug Discov Today 18(17–18):807–817
75. Koehler JW, Smith JM, Ripoll DR, Spik KW, Taylor SL, Badger CV et al (2013) A fusion-inhibiting peptide against Rift Valley fever virus inhibits multiple, diverse viruses. PLoS Negl Trop Dis 7(9):e2430. https://doi.org/10.1371/journal.pntd.0002430
76. Vigant F, Santos NC, Lee B (2015) Broad-spectrum antivirals against viral fusion. Nat Rev Microbiol 13(7):426–437
77. Chen Y, Cao L, Zhong M, Zhang Y, Han C, Li Q et al (2012) Anti-HIV-1 activity of a new scorpion venom peptide derivative Kn2-7. PLoS One 7(4):e34947. https://doi.org/10.1371/journal.pone.0034947
78. Jenssen H, Hamill P, Hancock RE (2006) Peptide antimicrobial agents. Clin Microbiol Rev 19(3):491–511
79. Taguchi F, Shimazaki YK (2000) Functional analysis of an epitope in the S2 subunit of the murine coronavirus spike protein: involvement in fusion activity. J Gen Virol 81(12):2867–2871
80. Zhao H, Zhou J, Zhang K, Chu H, Liu D, Poon VKM et al (2016) A novel peptide with potent and broad-spectrum antiviral activities against multiple respiratory viruses. Sci Rep 6:22008. https://doi.org/10.1038/srep22008
81. Zumla A, Hui DS, Perlman S (2015) Middle East respiratory syndrome. Lancet 386(9997):995–1007
82. de Wit E, van Doremalen N, Falzarano D, Munster VJ (2016) SARS and MERS: recent insights into emerging coronaviruses. Nat Rev Microbiol 14(8):523–534
83. Li Q, Zhao Z, Zhou D, Chen Y, Hong W, Cao L et al (2011) Virucidal activity of a scorpion venom peptide variant mucroporin-M1 against measles, SARS-CoV and influenza H5N1 viruses. Peptides 32(7):1518–1525
84. Mustafa S, Balkhy H, Gabere M (2019) Peptide-protein interaction studies of antimicrobial peptides targeting middle east respiratory syndrome

coronavirus spike protein: an in silico approach. Adv Bioinforma 2019:6815105. https://doi.org/10.1155/2019/6815105
85. Zhou N, Pan T, Zhang J, Li Q, Zhang X, Bai C et al (2016) Glycopeptide antibiotics potently inhibit cathepsin L in the late endosome/lysosome and block the entry of Ebola virus, Middle East respiratory syndrome coronavirus (MERS-CoV), and severe acute respiratory syndrome coronavirus (SARS-CoV). J Biol Chem 291(17):9218–9232
86. Wohlford-Lenane CL, Meyerholz DK, Perlman S, Zhou H, Tran D, Selsted ME et al (2009) Rhesus theta-defensin prevents death in a mouse model of severe acute respiratory syndrome coronavirus pulmonary disease. J Virol 83(21):11385–11390
87. Padalko E, Nuyens D, De Palma A, Verbeken E, Aerts JL, De Clercq E et al (2004) The interferon inducer ampligen [poly (I)-poly (C12U)] markedly protects mice against coxsackie B3 virus-induced myocarditis. Antimicrob Agents Chemother 48(1):267–274
88. Barnard DL, Kumaki Y (2011) Recent developments in anti-severe acute respiratory syndrome coronavirus chemotherapy. Future Virol 6(5):615–631
89. Tong TR (2006) SARS coronavirus anti-infectives. Recent Pat Antiinfect Drug Discov 1(3):297–308
90. De Clercq E (2006) Interferon and its inducers–a never-ending story: "old" and "new" data in a new perspective. J Infect Dis 194(Suppl 1):S19–S26
91. Ershov F, Tazulakhova E (1999) Interferon inducers: new generation of immunomodulators. Vestn Ross Akad Med Nauk (4):52–56
92. Bao M, Zhang Y, Wan M, Dai L, Hu X, Wu X et al (2006) Anti-SARS-CoV immunity induced by a novel CpG oligodeoxynucleotide. Clin Immunol 118(2–3):180–187
93. Barnard DL, Day CW, Bailey K, Heiner M, Montgomery R, Lauridsen L et al (2006) Evaluation of immunomodulators, interferons and known in vitro SARS-coV inhibitors for inhibition of SARS-coV replication in BALB/c mice. Antivir Chem Chemother 17(5):275–284
94. Kumaki Y, Day CW, Bailey KW, Wandersee MK, Wong MH, Madsen JR et al (2010) Induction of interferon-γ-inducible protein 10 by SARS-CoV infection, interferon alfacon 1 and interferon inducer in human bronchial epithelial Calu-3 cells and BALB/c mice. Antivir Chem Chemother 20(4):169–177
95. Mojsoska B, Jenssen H (2015) Peptides and peptidomimetics for antimicrobial drug design. Pharmaceuticals 8(3):366–415
96. Chuck C, Ke Z, Chen C, Wan D, Chow H, Wong K (2014) Profiling of substrate-specificity and rational design of broad-spectrum peptidomimetic inhibitors for main proteases of coronaviruses. Hong Kong Med J 20(Suppl 4):22–25
97. Ghosh AK, Xi K, Grum-Tokars V, Xu X, Ratia K, Fu W et al (2007) Structure-based design, synthesis, and biological evaluation of peptidomimetic SARS-CoV 3CLpro inhibitors. Bioorg Med Chem Lett 17(21):5876–5880
98. Kumar V, Shin JS, Shie JJ, Ku KB, Kim C, Go YY et al (2017) Identification and evaluation of potent Middle East respiratory syndrome coronavirus (MERS-CoV) 3CLPro inhibitors. Antivir Res 141:101–106
99. Kankanamalage ACG, Kim Y, Damalanka VC, Rathnayake AD, Fehr AR, Mehzabeen N et al (2018) Structure-guided design of potent and permeable inhibitors of MERS coronavirus 3CL protease that utilize a piperidine moiety as a novel design element. Eur J Med Chem 150:334–346
100. Chuck CP, Chen C, Ke Z, Wan DCC, Chow HF, Wong KB (2013) Design, synthesis and crystallographic analysis of nitrile-based broad-spectrum peptidomimetic inhibitors for coronavirus 3C-like proteases. Eur J Med Chem 59:1–6
101. Li SW, Wang CY, Jou YJ, Huang SH, Hsiao LH, Wan L et al (2016) SARS coronavirus papain-like protease inhibits the TLR7 signaling pathway through removing Lys63-linked polyubiquitination of TRAF3 and TRAF6. Int J Mol Sci 17(5):678. https://doi.org/10.3390/ijms17050678
102. Totura AL, Whitmore A, Agnihothram S, Schäfer A, Katze MG, Heise MT et al (2015) Toll-like receptor 3 signaling via TRIF contributes to a protective innate immune response to severe acute respiratory syndrome coronavirus infection. MBio 6(3):e00638–e00615
103. Zhao J, Wohlford-Lenane C, Zhao J, Fleming E, Lane TE, McCray PB et al (2012) Intranasal treatment with poly (I• C) protects aged mice from lethal respiratory virus infections. J Virol 86(21):11416–11424
104. Marston HD, Paules CI, Fauci AS (2017) The critical role of biomedical research in pandemic preparedness. JAMA 318(18):1757–1758
105. Bitko V, Barik S (2001) Phenotypic silencing of cytoplasmic genes using sequence-specific double-stranded short interfering RNA and its application in the reverse genetics of wild type negative-strand RNA viruses. BMC Microbiol 1(34):1–11
106. Deng T, Engelhardt OG, Thomas B, Akoulitchev AV, Brownlee GG, Fodor E (2006) Role of ran binding protein 5 in nuclear import and assembly of the influenza virus RNA polymerase complex. J Virol 80(24):11911–11919
107. Li T, Zhang Y, Fu L, Yu C, Li X, Li Y et al (2005) siRNA targeting the leader sequence of SARS-CoV inhibits virus replication. Gene Ther 12:751–761
108. Chung YS, Kim MK, Lee WJ, Kang C (2007) Silencing E1A mRNA by RNA interference inhibits adenovirus replication. Arch Virol 152(7):1305–1314
109. Ge Q, Filip L, Bai A, Nguyen T, Eisen HN, Chen J (2004) Inhibition of influenza virus production in virus-infected mice by RNA interference. Proc Natl Acad Sci U S A 101(23):8676–8681
110. Tompkins SM, Lo CY, Tumpey TM, Epstein SL (2004) Protection against lethal influenza virus chal-

lenge by RNA interference in vivo. Proc Natl Acad Sci U S A 101(23):8682–8686
111. Zhou H, Jin M, Yu Z, Xu X, Peng Y, Wu H et al (2007) Effective small interfering RNAs targeting matrix and nucleocapsid protein gene inhibit influenza A virus replication in cells and mice. Antivir Res 76(2):186–193
112. Li BJ, Tang Q, Cheng D, Qin C, Xie FY, Wei Q et al (2005) Using siRNA in prophylactic and therapeutic regimens against SARS coronavirus in rhesus macaque. Nat Med 11(9):944–951
113. Millet JK, Whittaker GR (2014) Host cell entry of Middle East respiratory syndrome coronavirus after two-step, furin-mediated activation of the spike protein. Proc Natl Acad Sci U S A 111(42):15214–15152
114. Wu CJ, Huang HW, Liu CY, Hong CF, Chan YL (2005) Inhibition of SARS-CoV replication by siRNA. Antivir Res 65(1):45–48
115. Sheikhshahrokh A, Ranjbar R, Saeidi E, DEHKORDI FS, Heiat M, Ghasemi-Dehkordi P et al (2020) Frontier therapeutics and vaccine strategies for sars-cov-2 (COVID-19): A review. Iran J Public Health 49:18–29
116. Heiat M, Hashemi-Aghdam MR, Heiat F, Rastegar Shariat Panahi M, Aghamollaei H, Moosazadeh Moghaddam M et al (2020) Integrative role of traditional and modern technologies to combat COVID-19. Expert Rev Anti Infect Ther 19:1–1
117. Mirzaie A, Halaji M, Dehkordi FS, Ranjbar R, Noorbazargan H (2020) A narrative literature review on traditional medicine options for treatment of corona virus disease 2019 (COVID-19). Complement Ther Clin Pract 17:101214

Obesity and Risk of COVID-19 Infection and Severity: Available Evidence and Mechanisms

8

Behnaz Abiri, Paul C. Guest, and Mohammadreza Vafa

Abstract

The coronavirus disease 2019 (COVID-19) pandemic has resulted in worldwide research efforts to recognize people at greatest risk of developing critical illness and dying. Growing numbers of reports have connected obesity to more severe COVID-19 illness and death. Although the exact mechanism by which obesity may lead to severe COVID-19 outcomes has not yet been determined, the mechanisms appear to be multifactorial. These include mechanical changes of the airways and lung parenchyma, systemic and airway inflammation, and general metabolic dysfunction that adversely affect pulmonary function and/or response to treatment. As COVID-19 continues to spread worldwide, clinicians should carefully monitor and manage obese patients for prompt and targeted treatment.

Keywords

Coronavirus · SARS-CoV-2 · COVID-19 · Obesity · Overweight · Metabolism

8.1 Introduction

In late December 2019, a group of pneumonia patients with unknown origin was reported in Wuhan, China [1]. Since then, COVID-19, caused by severe acute respiratory syndrome coronavirus-2 (SARS-CoV-2), has taken the world by storm and was officially announced as a pandemic by the World Health Organization (WHO) on March 11, 2020. SARS-CoV-2 is reported to be a member of the betacoronavirus family, associated with the severe acute respiratory syndrome (SARS) virus SARS-CoV [2]. Clinical signs of COVID-19 disease range from asymptomatic or mild infection to severe manifestations that are life-threatening. In China, those over 65 years old and/or with comorbidities were found to be at higher risk of a more severe course of SARS-CoV-2 infection. Among the comorbidities, the highest fatality rates were observed for individuals with cardiovascular dis-

B. Abiri
Department of Nutrition, Faculty of Paramedicine, Ahvaz Jundishapur University of Medical Sciences, Ahvaz, Iran

P. C. Guest
Laboratory of Neuroproteomics, Department of Biochemistry and Tissue Biology, Institute of Biology, University of Campinas (UNICAMP), Campinas, Brazil

M. Vafa (✉)
Department of Nutrition, School of Public Health, Iran University of Medical Sciences, Tehran, Iran
e-mail: vafa.m@iums.ac.ir

P. C. Guest (ed.), *Clinical, Biological and Molecular Aspects of COVID-19*, Advances in Experimental Medicine and Biology 1321, https://doi.org/10.1007/978-3-030-59261-5_8

ease (CVD) and diabetes mellitus, followed by chronic respiratory diseases, hypertension, cancer, immunosuppressive disorders, and organ failure [3, 4]. As both CVD and diabetes mellitus are linked with increased adipose tissue mass [5], a high body mass index (BMI) and other anthropometric indices associated with obesity might be key risk factors indicative of a more severe course of disease, including development of pneumonia. It has been established that both under- and overnutrition are related to a worse prognosis of viral infections, as occurred in the case of the devastating 1918 influenza pandemic [6]. In addition, recent evidence has shown that obesity and overweight are associated with more severe COVID-19 outcomes [7–9]. The parameters mediating this high risk are thought to be due to an impaired respiratory system in obese persons, mediated by elevated airway resistance, disrupted respiratory gas exchange, as well as low respiratory muscle strength and reduced lung volumes [10]. Furthermore, other studies have proposed that obesity is related to an impaired immune response which is a critical factor in COVID-19 disease [11, 12].

The obesity-related effects on the immune system play a key role in the pathogenesis and outcome of most viral infections such as COVID-19 disease, and obesity is also linked to an increased inflammation response in adipose tissue. In turn, the inflammatory response in adipose tissue can lead to metabolic dysfunction, potentially resulting in dyslipidemia, insulin resistance, diabetes mellitus, hypertension, and CVD [13]. In addition, anthropometric studies have shown that abdominal obesity can cause impaired mechanical ventilation at the base of the lungs, leading to decreased oxygenation of vital tissues [14]. The abnormal secretion of adipokines and cytokines such as tumor necrosis factor-alpha (TNF-α) and interferon (INF) is indicative of a chronic low-grade inflammation characteristic of abdominal-centered obesity, and this may further impair immune responses [15] and negatively impact lung physiology [16]. Figure 8.1 highlights some of the obesity-associated comorbidities related to COVID-19 disease severity.

The COVID-19 pandemic is now spreading all over the world, especially in Europe and the Americas, where obesity has a high prevalence [16]. Although this is only suggestive and does not necessarily imply causation, the links mentioned make a strong case for more thorough investigations into the potential associations between obesity and severity of COVID-19 disease. Obesity has already been identified as a risk factor for individuals experiencing a more severe course of infection during the 2009 influenza A H1N1 epidemic [15, 17, 18]. Taken together, these findings suggest that obesity is an independent risk factor for SARS-CoV-2 infection [19].

Since rapid diagnosis and early treatment appear to produce the best patient outcomes in many disease areas, recognition of risk factors for morbidity and mortality is important to protect the most vulnerable individuals in the society and to guide the most appropriate treatment response in a precision medicine manner. The aim of the present chapter was to investigate the hypothesis that having a higher BMI is a risk factor for COVID-19 infection and its progression to a more severe disease course, independent of other common risk factors. We also discuss potential key mechanisms by which obesity affects COVID-19 disease and may suggest potential therapeutic avenues.

8.2 Association Between COVID-19 and Obesity: Early Data

Much has been learned from influenza in patients with obesity. The Centers for Disease Control and Prevention (CDC) suggests that individuals with a BMI ≥ 40 kg/m^2 have a higher risk of influenza complications [20]. During the H1N1 influenza pandemic, obesity was recognized as an independent risk factor for increasing disease severity [21]. A study also showed that individuals with obesity have reduced protection from influenza immunization [22]. Therefore, it has been deemed likely that obesity is an independent risk factor for COVID-19 severity.

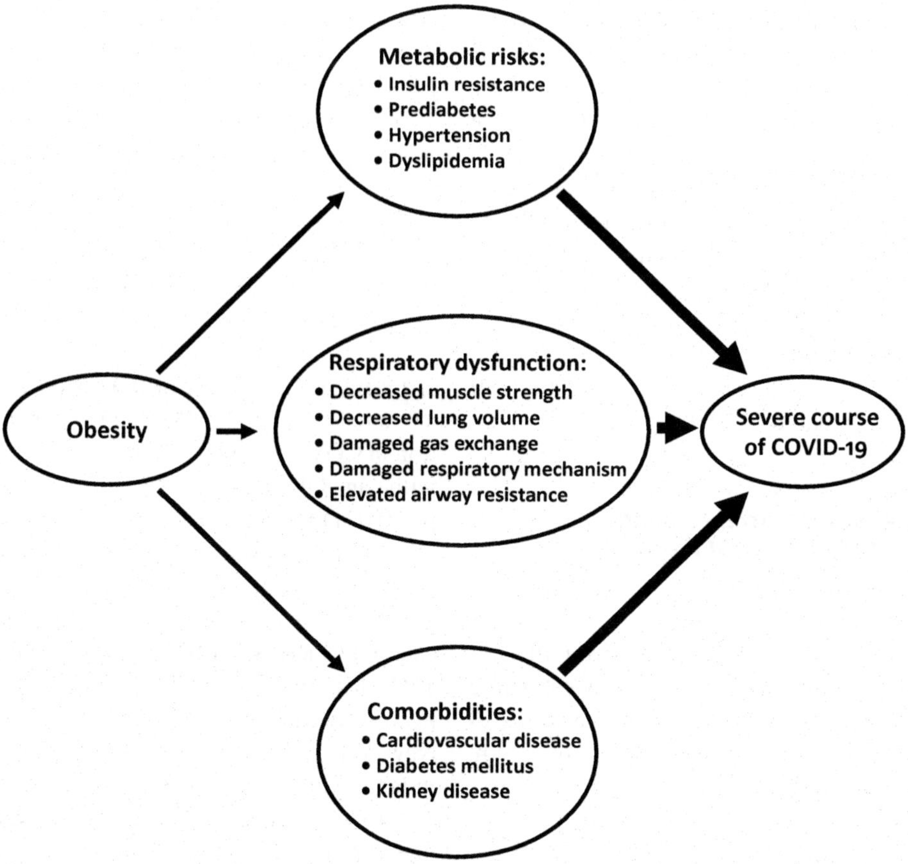

Fig. 8.1 Obesity-associated comorbidities and mechanisms of COVID-19 severity

A descriptive study between a small sample of 24 critically ill patients diagnosed with COVID-19 in the Seattle region was one of the first studies to report BMI data with 20 of the patients being either overweight or obese [23]. Although the numbers were too small for statistical analyses, 20 of the patients needed mechanical ventilation and 15 died. In a study in China, older age (≥65 years) and the presence of comorbidities were found to be related to a more severe course in patients infected with SARS-CoV-2 [3]. Across the recorded comorbidities, the highest fatality rates were reported for CVD (10.5%) and diabetes mellitus (7.3%), followed by chronic respiratory diseases (6.3%), hypertension (6.0%), and cancer (5.6%) [3].

Concern about the impacts of BMI was further demonstrated by preliminary data from Shenzhen, China [24], and New York City (NYC), USA [25]. In the Chinese study, it was found that obesity, particularly in men, significantly elevated the risk of developing severe COVID-19 [24]. In addition, the obese patients tended to have upper respiratory tract infection symptoms, including fever and cough, although no significant differences were observed in terms of disease duration or progression. In the New York study, having a BMI >40 kg/m^2 was the next strongest independent predictor of hospitalization, after old age [25]. In addition, a small study at the University Hospital in Lille, France, reported that the need for mechanical ventilation in 124 COVID-19 patients was greater for those with a BMI ≥35 kg/m^2, independent of other risk factors [26].

In California, USA, between April and August 2009, 1088 individuals were either hospitalized or died from H1N1 influenza infection [27]. Of these individuals, 58% were obese (BMI >30 kg/

m^2), and 67% had severe obesity (BMI >40 kg/m^2). Approximately two-thirds of the patients with obesity had comorbidities, including chronic lung disease, asthma, cardiac problems, or diabetes.

In another investigation, Peng et al. conducted a retrospective analysis of 112 patients infected with SARS-CoV-2 who had been admitted to the Union Hospital in Wuhan, from January 20 to February 15, 2020 [28]. The BMI of the critical patients was higher than that of the general population, and 88% of the non-survivors had a BMI >25 kg/m^2 compared to 18.9% of the survivors.

8.3 Association Between COVID-19 and Obesity: Recent Data

A number of more recent studies over the last 2 months have now been carried out investigating the effects of COVID-19 outcomes in terms of infection, disease severity, and risk of death. These are highlighted in the following sections.

8.3.1 Association of Obesity with a SARS-CoV-2-Positive Test

A study was carried out by the Oxford Royal College of General Practitioners (RCGP) Research, and Surveillance Center primary care network analyzed routinely collected data from patients tested for SARS-CoV-2 between Jan 28 and April 4, 2020 [29]. They used multivariable logistic regression models to identify risk factors associated with a positive test result. This showed that 20.9% of people with obesity tested positive for the disease compared with only 13.2% of normal-weight people [odds ratio (OR): 1.41, 95% confidence interval (CI): 1.04–1.91). Another study from the UK found that both BMI and waist circumference were positively associated with a positive test for COVID-19 [30]. This investigation also showed a dose-response-like relationship between BMI and a positive test for COVID-19, with odds ratios for overweight (BMI: 25–<30 kg/m^2), obese (BMI: 30- < 35 kg/m^2), and severely obese (BMI: ≥35 kg/m^2) subjects of 1.31 (1.05–1.62), 1.55 (1.19–2.02), and 1.57 (1.14–2.17), respectively, compared to normal weight controls (BMI: 18.5–<25 kg/m^2).

8.3.2 Association of Obesity with COVID-19 Disease Severity

A recent meta-analysis of three studies reported an increased need of invasive mechanical ventilation in COVID-19 patients with a BMI >35 kg/m^2 with an OR of 7.36 (95% CI: 1.63–33.14, p = 0.021) [31]. A study in China which investigated the association between obesity and COVID-19 illness severity among patients with confirmed SARS-CoV-2 infection found that each unit increase in BMI was associated with a 12% increase in the risk of severe COVID-19 [32]. A study by Cummings et al. carried out in New York (NY), USA, found that of 257 individuals who were critically ill with COVID-19, 171 (67%) were males, 212 (82%) had at least one chronic illness, and 119 (46%) were obese [33]. Hur et al. analyzed data from 10 Chicago Illinois hospitals in the USA and found that among patients who required intubation, those who were older or more obese required longer intubation times [34].

8.3.3 Association of Obesity with Increased Risk of Death from COVID-19

A retrospective study of 13 young patients who died of COVID-19 and 40 matched survivors found that the deceased patients had higher BMIs (p = 0.010), increased C-reactive protein (CRP) inflammation biomarker (p = 0.014), increased troponin I (TPNI) cardiac biomarker (p = 0.005), and elevated D-dimer coagulation activity biomarker (p = 0.047) [35]. Klang et al. carried out a retrospective analysis of data from COVID-19 patients hospitalized in New York between March 1 and May 17, 2020, using multivariable logistic regression models and found that among the younger patients that died (<50 years old), having a BMI >40 kg/m^2 was independently associated with mortality (OR: 5.1, 95% CI: 2.3–11.1) [36]. Another study in New York of 770 patients found that those who were obese were more likely to present with fever, cough, and shortness of breath, with a significantly higher rate of intensive care unit (ICU) admission or death ($p = 0.002$) [37].

A prospective study of 20,133 patients in the UK carried out by the International Severe Acute Respiratory and emerging infections consortium (ISARIC) World Health Organization (WHO) Clinical Characterization Protocol UK (CCP-UK) found that increasing age, male sex, and comorbidities, such as obesity and chronic cardiac, pulmonary, kidney, and liver diseases, were associated with higher mortality outcomes [38]. Bello-Chavolla and co-workers carried out a study in Mexico of 51,633 people with SARS-CoV-2 infection, which evaluated risk factors and proposed a lethality score for the disease [39]. In this study, 5332 of the individuals died, and it appeared that obesity increased risk for the need for ICU admission and intubation and was associated with 49.5% of the lethality. Another investigation assessed the obesity prevalence of the top 20 countries ranked according to total COVID-19-related deaths as of May 20, 2020 [40]. This showed that the USA had the highest obesity (36.2%) and overweight (31.7%) prevalence, as well as the highest number of total deaths. In addition, correlation analysis showed that the number of total deaths was significantly correlated with the obesity prevalence in each country ($r = 0.464$, $p = 0.039$).

8.4 Mechanisms

8.4.1 Inflammation

Obesity alters the innate and adaptive immune responses, which cause a state of chronic low-grade inflammation (Fig. 8.2) [41, 42]. This state is characterized by higher levels of pro-inflammatory cytokines such as TNF-α, macrophage chemoattractant protein I (MCP-1), and interleukin-6 (IL-6), which are mainly secreted from visceral and subcutaneous adipose tissue [43]. However, presentation of an antigen such as a virus results in decreased macrophage activation and blunted pro-inflammatory cytokine production, as well as exacerbation of viral symptoms [44]. This may explain the poorer vaccination response in obese individuals [45]. In addition, B- and T-cell responses are disrupted in obesity which elevates susceptibility to viral infections and a delay in their resolution. A study by Zhang et al. suggested that leptin resistance was a cofactor in the H1N1 influenza pandemic, as this hormone is an important regulator of B cell maturation, development, and performance [46]. In addition, obese patients may have impaired memory T-cell and antibody responses, which could also explain vaccine ineffectiveness [47].

A disturbed pro-inflammatory response is the likely cause of lung lesions observed in victims of influenza pandemics. In line with this, a study on influenza A virus infection in obese ob/ob mice showed elevated disease severity, increased secondary bacterial infections, and decreased vaccine efficacy [48]. A study by the same research group showed that serial passage of a human H1N1 influenza virus through diet-induced and genetic (ob/ob) models of obesity in mice leads to a more severe disease with elevated virulence and morbidity, which may br related to disruption of the INF response [49].

8.4.2 Impaired Insulin Signaling

Patients with obesity consume a higher than normal percentage of oxygen during respiratory work [50, 51]. Obesity is also linked with respiratory conditions, such as exertional dyspnea, obesity hypoventilation syndrome, chronic obstructive pulmonary disease (COPD), asthma, and aspiration pneumonia [52]. Obesity is a known risk factor for diabetes, metabolic syndrome, and CVD, which may also contribute to higher mortality in COVID-19 cases. Insulin resistance is a major feature of these conditions and can be caused by obesity [53]. Under normal

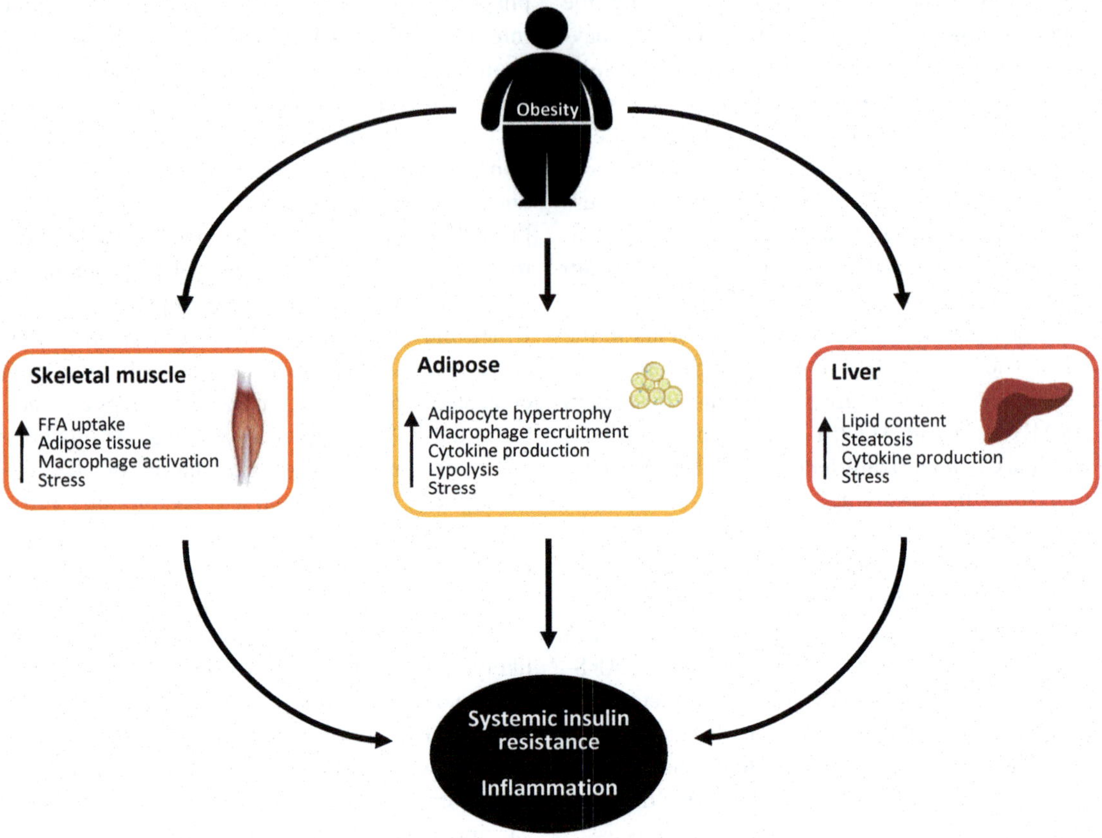

Fig. 8.2 Effects of obesity on inflammation and insulin resistance, predicted to lead to a more severe form of COVID-19 disease. The effects are shown in adipose tissue, muscle, and liver, although they are likely to be systemic

conditions, binding of insulin to the insulin receptor in target tissues results in parallel signaling via the PI3K-Akt and Ras-MAPK networks in the regulation of metabolism and growth pathways (Fig. 8.3). Insulin-resistant states like obesity are characterized by specific impairments in the PI3K-Akt pathway, leading to compensatory hyperinsulinemia in order to maintain normal glycemia. This leads to excessive signaling via the growth pathway, contributing to increased inflammation, proliferation, and hypertrophy.

8.4.3 Other Mechanisms

Another factor might also lead to the elevated risk from COVID-19 for patients with obesity that was highlighted by a previous study which showed that adipose tissue can serve as a reservoir for human viruses [54]. More studies should be performed to determine if adipose tissue also serves as a focal point of SARS-CoV-2 infection and spreading to other organs.

8.5 Potential Treatment Avenues

8.5.1 Biomarkers

As described above, the presence of metabolic diseases such as obesity, hypertension, diabetes, and CVD is likely to contribute to a poorer prognosis in COVID-19 patients. Since these conditions are marked by insulin resistance and a latent

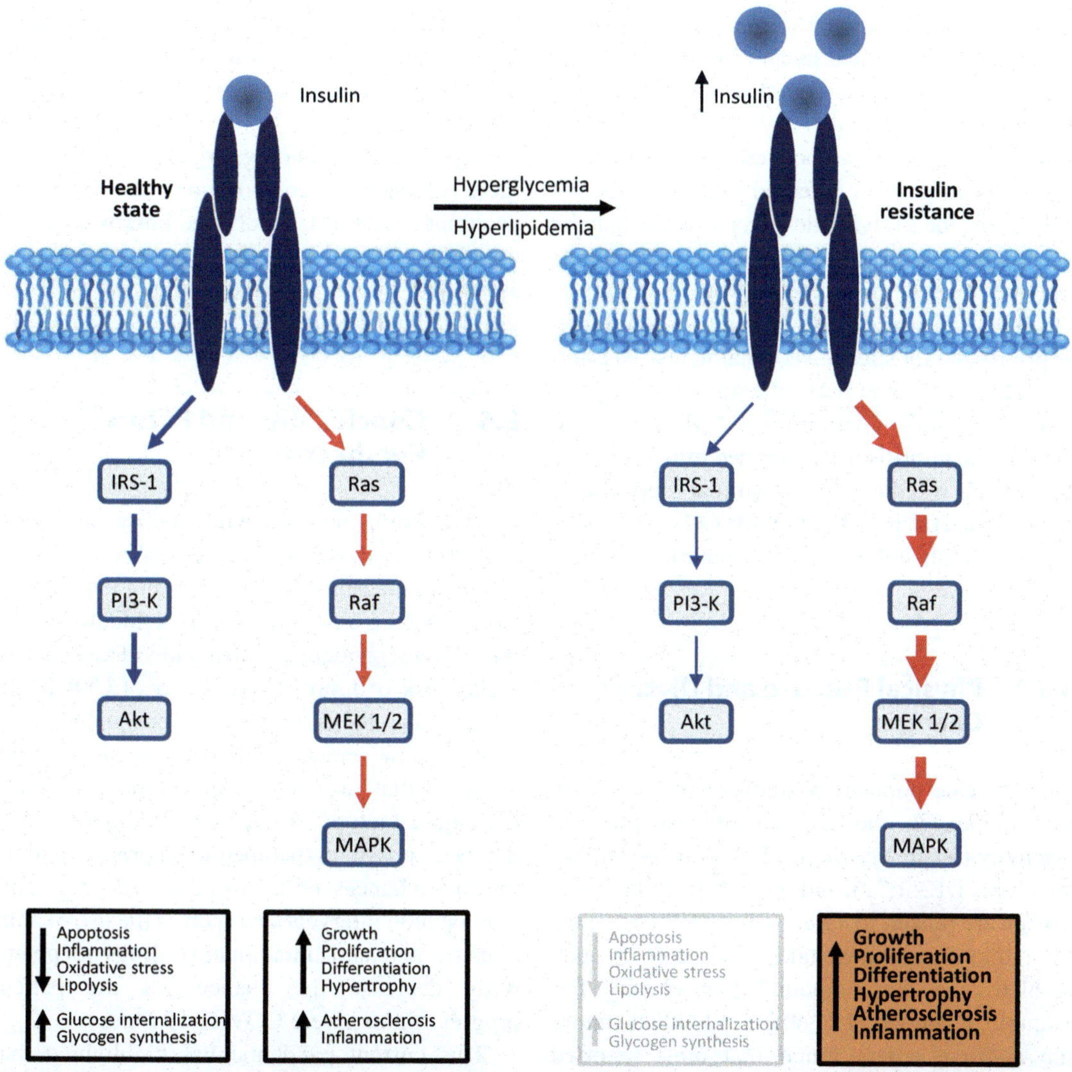

Fig. 8.3 Insulin signaling in healthy and insulin-resistant states. Akt, protein kinase b; ERK, extracellular receptor kinase; IRS-1, insulin substrate receptor-1; MAPK, mitogen-activated protein kinase; MEK, mitogen-activated protein kinase/ extracellular receptor kinase; p38, p38 mitogen-activated protein kinase; PI3-K, phosphatidylinositol (PI)3-kinase; RAS, a small GTPase involved in signal transduction; RAF, proto-oncogene c-RAF

chronic inflammatory state, it is possible that the application of approved antidiabetic drugs such as pioglitazone could lead to a better outcome for patients with such comorbidities [55]. As a preventative measure, obtaining a higher level of cardiorespiratory fitness by prior physical exercise may offer some innate immune-protection against SARS-CoV-2 infection by improving insulin signaling and attenuating the "cytokine storm syndrome"that can occur in high-risk individuals [56, 57]. The term cytokine storm describes an excessive and uncontrolled release of pro-inflammatory cytokines which can cause damage to the lungs and other tissues. High-risk patients could be identified by biomarker tests for insulin resistance, such as an oral glucose tolerance test or the measurement of the triglyceride and glucose index. Ren et al. showed that the latter index was closely associated with severity and morbidity in patients infected with SAR-CoV-2 [58]. Thus, the triglyceride-glucose index may be a useful marker for identification of those patients who are likely to experience a worse outcome of COVID-19 disease. These individuals could then be prioritized for specialized treatments. The successful use of some anti-inflammatory drugs in other hyperinflammation-related diseases like rheumatoid arthritis has generated much speculation about whether or not similar approaches could be useful in patients with COVID-19 disease and high inflammatory biomarker profiles [59, 60].

8.5.2 Physical Exercise and Dietary Changes

A large-scale population study of 387,109 men and women in the UK found that physical inactivity (relative risk = 1.32, 1.10–1.58), smoking (1.42, 1.12–1.79), and high BMI (2.05, 1.68–2.49) were related to cases of COVID-19 serious enough to warrant hospital admission [61]. Such problems may be compounded by obesity. For example, a study of 123 obese individuals under stay-at-home orders found that most reported increased anxiety and depression, increased stress eating, increased difficultly in achieving weight loss goals, and reduced exercise time and intensity [62]. This problem may have been exacerbated during the imposed lockdown in many countries due to negative effects on eating behaviors and dietary habits [63]. Physical inactivity is known to increase symptom severity and death outcomes in individuals with chronic diseases due to blunting of the immune response and macrophage activation, caused by the associated increased insulin resistance [64]. In contrast, exercise is known to reduce the risk of mortality from metabolic diseases through an increase in physiological reserve and enhanced immunological benefits.

Together, these findings argue for the development of specialized programs to encourage healthier lifestyles involving improved nutritional quality and increased physical activity to assist with disease management during and after the COVID-19 pandemic.

8.6 Conclusions and Future Perspectives

In conclusion, patients with obesity and, most importantly, those with severe obesity should take extra measures to avoid coming into contact with SARS-CoV-2-infected individuals during the current pandemic. Such individuals have a higher risk of more severe forms of COVID-19 disease due to impaired insulin signaling and chronic low grade inflammation. It is now accepted that researchers and clinicians should take these factors into account in order to offer the best possible therapeutic approach and to improve chances of a favorable outcome. This may include interventions such as the use of anti-diabetic and anti-inflammatory drugs to potentially decrease the chances of the patient progressing to severe COVID-19 illness.

This current pandemic has highlighted that more should be done at the individual level to reduce the effects of obesity in our societies to minimize the effects of the current and future pandemics. As a preventative measure, policies should be adopted worldwide which encourage individuals to adopt a healthier lifestyle, involving improved nutrition and increased physical activity. This will also have the added benefit of decreasing the effects of other communicable and noncommunicable diseases on society and relieve the ever increasing burden on healthcare at a global level.

References

1. Zhu N, Zhang D, Wang W, Li X, Yang B, Song J et al (2020) A novel coronavirus from patients with Pneumonia in China, 2019. N Engl J Med 382(8):727–733
2. Epidemiology Working Group for NCIP Epidemic Response, Chinese Center for Disease Control and Prevention (2020) The epidemiological characteristics of an outbreak of 2019 novel coronavirus diseases (COVID-19) in China. Zhonghua Liu Xing Bing XueZaZhi 41(2):145–151
3. Wu Z, McGoogan JM (2020) Characteristics of and important lessons from the coronavirus disease 2019 (COVID-19) outbreak in China: summary of a report of 72,314 cases from the Chinese Center for Disease Control and Prevention. JAMA 323:1239–1242
4. Wu Z, McGoogan JM (2020) Characteristics of and important lessons from the coronavirus disease 2019 (COVID-19) outbreak in China: summary of a report of 72,314 cases from the Chinese Center for Disease Control and Prevention. JAMA 323:1239–1242
5. Stefan N, Schick F, Häring HU (2017) Causes, characteristics, and consequences of metabolically unhealthy normal weight in humans. Cell Metab 26(2):292–300
6. Short KR, Kedziereska K, van de Sandt CE (2018) Back to the future: lessons learned from the 1918 influenza pandemic. Front Cell Infect Microbiol 8(8):343. https://doi.org/10.3389/fcimb.2018.00343
7. Dietz W, Santos-Burgoa C (2020) Obesity and its implications for COVID-19 mortality. Obesity (Silver Spring) 28(6):1005. https://doi.org/10.1002/oby.22818
8. Lighter J, Phillips M, Hochman S, Sterling S, Johnson D, Francois F et al (2020) Obesity in patients younger than 60 years is a risk factor for Covid-19 hospital admission. Clin Infect Dis:ciaa415. https://doi.org/10.1093/cid/ciaa415. Online ahead of print
9. Simonnet A, Chetboun M, Poissy J, Raverdy V, Noulette J, Duhamel A et al (2020) High prevalence of obesity in severe acute respiratory syndrome coronavirus-2 (SARS-CoV-2) requiring invasive mechanical ventilation. Obesity (Silver Spring). https://doi.org/10.1002/oby.22831. Online ahead of print
10. Murugan AT, Sharma G (2008) Obesity and respiratory diseases. Chron Respir Dis 5(4):233–242
11. Sheridan PA, Paich HA, Handy J, Karlsson EA, Hudgens MG, Sammon AB et al (2012) Obesity is associated with impaired immune response to influenza vaccination in humans. Int J Obes 36(8):1072–1077
12. Green WD, Beck MA (2017) Obesity impairs the adaptive immune response to influenza virus. Ann Am Thorac Soc 14(Suppl 5):S406–S409
13. Kassir R (2020) Risk of COVID-19 for patients with obesity. Obes Rev 21(6):e13034. https://doi.org/10.1111/obr.13034
14. Dixon AE, Peters U (2018) The effect of obesity on lung function. Expert Rev Respir Med 12(9):755–767
15. Huttunen R, Syrjänen J (2013) Obesity and the risk and outcome of infection. Int J Obes 37(3):333–340
16. Simonnet A, Chetboun M, Poissy J, Raverdy V, Noulette J, Duhamel A et al (2020) High prevalence of obesity in severe acute respiratory syndrome Coronavirus-2 (SARS-CoV-2) requiring invasive mechanical ventilation. Obesity (Silver Spring). https://doi.org/10.1002/oby.22831. Online ahead of print
17. Wang W, Chen H, Li Q, Qiu B, Wang J, Sun X et al (2011) Fasting plasma glucose is an independent predictor for severity of H1N1 pneumonia. BMC Infect Dis 11:104. https://doi.org/10.1186/1471-2334-11-104
18. Honce R, Schultz-Cherry S (2019) Impact of obesity on influenza A virus pathogenesis, immune response, and evolution. Front Immunol 10:1071. https://doi.org/10.3389/fimmu.2019.01071
19. Ryan DH, Ravussin E, Heymsfield S (2020) COVID 19 and the patient with obesity – the editors speak out. Obesity (Silver Spring) 28(5):847. https://doi.org/10.1002/oby.22808
20. Centers for Disease Control and Prevention. People at high risk for flu complications. https://www.cdc.gov/flu/highr isk/index.htm. Updated August 27, 2018. Accessed 19 Mar 2020
21. Louie JK, Acosta M, Winter K, Jean C, Gavali S, Schechter R et al (2009) Factors associated with death or hospitalization due to pandemic 2009 influenza A(H1N1) infection in California. JAMA 302(17):1896–1902
22. Neidich SD, Green WD, Rebeles J, Karlsson EA, Schultz-Cherry S, Noah TL et al (2017) Increased risk of influenza among vaccinated adults who are obese. Int J Obes 41(9):1324–1330
23. Bhatraju PK, Ghassemieh BJ, Nichols M, Kim R, Jerome KR, Nalla AK et al (2020) Covid-19 in critically ill patients in the Seattle region- case series. N Engl J Med 382:2012–2022
24. Cai Q, Chen F, Wang T, Luo F, Liu X, Wu Q et al (2020) Obesity and COVID-19 severity in a designated hospital in Shenzhen, China. Diabetes Care:dc200576. https://doi.org/10.2337/dc20-0576. Online ahead of print
25. Petrilli CM, Jones SA, Yang J, Rajagopalan H, O'Donnell LF, Chernyak Y et al (2020) Factors associated with hospitalization and critical illness among 4,103 patients with COVID-19 disease in New York City. medRxiv. https://doi.org/10.1101/2020.04.08.20057794

26. Simonnet A, Chetboun M, Poissy J, Raverdy V, Noulette J, Duhamel A et al (2020) High prevalence of obesity in severe acute respiratory syndrome Coronavirus-2 (SARS-CoV-2) requiring invasive mechanical ventilation. Obesity (Silver Spring). https://doi.org/10.1002/oby.22831. Online ahead of print
27. Venkata C, Sampathkumar P, Afessa B (2010) Hospitalized patients with 2009 H1N1 influenza infection: the Mayo Clinic experience. Mayo Clin Proc 85(9):798–805
28. Peng YD, Meng K, Guan HQ, Leng L, Zhu RR, Wang BY et al (2020) Clinical characteristics and outcomes of 112 cardiovascular disease patients infected by 2019-nCoV. Zhonghuaxinxue Guan Bing Za Zhi 48(0):E004. https://doi.org/10.3760/cma.j.cn112148-20200220-00105
29. de Lusignan S, Dorward J, Correa A, Jones N, Akinyemi O, Amirthalingam G et al (2020) Risk factors for SARS-CoV-2 among patients in the Oxford Royal College of general practitioners research and surveillance Centre primary care network: a cross-sectional study. Lancet Infect Dis. https://doi.org/10.1016/S1473-3099(20)30371-6
30. Yates T, Razieh C, Zaccardi F, Davies MJ, Khunti K (2020) Obesity and risk of COVID-19: analysis of UK biobank. Prim Care Diabetes:S1751-9918(20)30196-0. https://doi.org/10.1016/j.pcd.2020.05.011. Online ahead of print
31. Tamara A, Tahapary DL (2020) Obesity as a predictor for a poor prognosis of COVID-19: a systematic review. Diabetes Metab Syndr 14(4):655–659
32. Gao F, Zheng KI, Wang XB, Sun QF, Pan KH, Wang TY et al (2020) Obesity is a risk factor for greater COVID-19 severity. Diabetes Care:dc200682. https://doi.org/10.2337/dc20-0682. Online ahead of print
33. Cummings MJ, Baldwin MR, Abrams D, Jacobson SD, Meyer BJ, Balough EM et al (2020) Epidemiology, clinical course, and outcomes of critically ill adults with COVID-19 in New York City: a prospective cohort study. Lancet 395(10239):1763–1770
34. Hur K, Price CPE, Gray EL, Gulati RK, Maksimoski M, Racette SD et al (2020) Factors associated with intubation and prolonged intubation in hospitalized patients with COVID-19. Otolaryngol Head Neck Surg. https://doi.org/10.1177/0194599820929640. Online ahead of print
35. Zhang F, Xiong Y, Wei Y, Hu Y, Wang F, Li G et al (2020) Obesity predisposes to the risk of higher mortality in young COVID-19 patients. J Med Virol. https://doi.org/10.1002/jmv.26039. Online ahead of print
36. Klang E, Kassim G, Soffer S, Freeman R, Levin MA, Reich DL et al (2020) Morbid obesity as an independent risk factor for COVID-19 mortality in hospitalized patients younger than 50. Obesity (Silver Spring). https://doi.org/10.1002/oby.22913. Online ahead of print
37. Hajifathalian K, Kumar S, Newberry C, Shah S, Fortune B, Krisko T et al (2020) Obesity is associated with worse outcomes in COVID-19: analysis of early data from new York City. Obesity (Silver Spring). https://doi.org/10.1002/oby.22923. Online ahead of print
38. Docherty AB, Harrison EM, Green CA, Hardwick HE, Pius R, ISARIC4C Investigators et al (2020) Features of 20 133 UK patients in hospital with covid-19 using the ISARIC WHO Clinical Characterisation Protocol: prospective observational cohort study. BMJ 369:m1985. https://doi.org/10.1136/bmj.m1985
39. Bello-Chavolla OY, Bahena-López JP, Antonio-Villa NE, Vargas-Vázquez A, González-Díaz A, Márquez-Salinas A et al (2020) Predicting mortality due to SARS-CoV-2: A mechanistic score relating obesity and diabetes to COVID-19 outcomes in Mexico. J Clin Endocrinol Metab:dgaa346. https://doi.org/10.1210/clinem/dgaa346. Online ahead of print
40. Ekiz T, Pazarlı AC (2020) Relationship between COVID-19 and obesity. Diabetes Metab Syndr 14(5):761–763
41. Anderson CJ, Murphy KE, Fernandez ML (2016) Impact of obesity and metabolic syndrome on immunity. Adv Nutr 7:66–77
42. Ouchi N, Parker JL, Lugus JJ, Walsh K (2011) Adipokinesin inflammation and metabolic disease. Nat Rev Immunol 11(2):85–97
43. Richard C, Wadowski M, GorukS CL, Sharma AM, Field CJ (2017) Individuals with obesity and type 2 diabetes have additional immune dysfunction compared with obese individuals who are metabolically healthy. BMJ Open Diabetes Res Care 5(1):e000379. https://doi.org/10.1136/bmjdrc-2016-000379
44. Ahn SY, Sohn SH, Lee SY, Park HL, Park YW, Kim H et al (2015) The effect of lipopolysaccharide-induced obesity and its chronic inflammation on influenza virus-related pathology. Environ Toxicol Pharmacol 40(3):924–930
45. Karlsson EA, Hertz T, Johnson C, Mehle A, Krammer F, Schultz-Cherry S (2016) Obesity outweighs protection conferred by adjuvanted influenza vaccination. mBio 7(4):e01144-16. https://doi.org/10.1128/mBio.01144-16
46. Zhang AJ, To KK, Li C, Lau CC, Poon VK, Chan CC et al (2013) Leptin mediates the pathogenesisi of severe 2009 pandemic influenza A (H1N1) infection associated with cytokine dysregulation in mice with diet-induced obesity. J Infect Dis 207(8):1270–1280
47. Honce R, Schultz-Cherry S (2019) Impact of obesity on influenza, a virus pathogenesis, immune response, and evolution. Front Immunol 10:1071. https://doi.org/10.3389/fimmu.2019.01071
48. O'Brien KB, Vogel P, Duan S, Govorkova EA, Webby RJ, McCullers JA et al (2012) Impaired wound healing predisposes obese mice to severe influenza virus infection. J Infect Dis 205(2):252–261
49. Honce R, Karlsson EA, Wohlgemuth N, Estrada LD, Meliopoulos VA, Yao J et al (2020) Obesity related microenvironment promotes emergence of virulent influenza virus strains. mBio 11(2):e03341-19. https://doi.org/10.1128/mBio.03341-19

50. Baik I, Curhan GC, Rimm EB, Bendich A, Willett WC, Fawzi WW (2000) A prospective study of age and lifestyle factors in relation to community acquired pneumonia in US men and women. Arch Intern Med 160(20):3082–3088
51. Dixon AE, Peters U (2018) The effect of obesity on lung function. Expert Rev Respir Med 12(9):755–767
52. Zammit C, Liddicoat H, Moonsie I, Makker H (2010) Obesity and respiratory diseases. Int J Gen Med 3:335–343
53. de Luca C, Olefsky JM (2006) Stressed out about obesity and insulin resistance. Nat Med 12(1):41–42. discussion 42
54. Bourgeois C, Gorwood J, Barrail-Tran A, Lagathu C, Capeau J, Desjardins D et al (2019) Specific biological features of adipose tissue, and their impact on HIV persistence. Front Microbiol 10:2837. https://doi.org/10.3389/fmicb.2019.02837
55. Carboni E, Carta AR, Carboni E (2020) Can pioglitazone be potentially useful therapeutically in treating patients with COVID-19? Med Hypotheses 140:109776. https://doi.org/10.1016/j.mehy.2020.109776
56. Zbinden-Foncea H, Francaux M, Deldicque L, Hawley JA (2020) Does high cardiorespiratory fitness confer some protection against pro-inflammatory responses after infection by SARS-CoV-2? Obesity (Silver Spring). https://doi.org/10.1002/oby.22849. Online ahead of print
57. Kim J, Nam JH (2020) Insight into the relationship between obesity-induced low-level chronic inflammation and COVID-19 infection. Int J Obes (Lond) 22:1–2. https://doi.org/10.1038/s41366-020-0602-y. Online ahead of print
58. Ren H, Yang Y, Wang F, Yan Y, Shi X, Dong K et al (2020) Association of the insulin resistance marker TyG index with the severity and mortality of COVID-19. Version 2. Cardiovasc Diabetol 19(1):58. https://doi.org/10.1186/s12933-020-01035-2
59. Korakas E, Ikonomidis I, Kousathana F, Balampanis K, Kountouri A, Raptis A et al (2020) Obesity and COVID-19: immune and metabolic derangement as a possible link to adverse clinical outcomes. Am J Physiol Endocrinol Metab. https://doi.org/10.1152/ajpendo.00198.2020. Online ahead of print
60. Almerie MQ, Kerrigan DD (2020) The association between obesity and poor outcome after COVID-19 indicates a potential therapeutic role for montelukast. Med Hypotheses 143:109883. https://doi.org/10.1016/j.mehy.2020.109883
61. Hamer M, Kivimäki M, Gale CR, Batty GD (2020) Lifestyle risk factors, inflammatory mechanisms, and COVID-19 hospitalization: a community-based cohort study of 387,109 adults in UK. Brain Behav Immun:S0889-1591(20)30996-X. https://doi.org/10.1016/j.bbi.2020.05.059. Online ahead of print
62. Almandoz JP, Xie L, Schellinger JN, Mathew MS, Gazda C, Ofori A et al (2020) Impact of COVID-19 stay-at-home orders on weight-related behaviors among patients with obesity. Clin Obes Jun 9:e12386. https://doi.org/10.1111/cob.12386. Online ahead of print
63. Sidor A, Rzymski P (2020) Dietary choices and habits during COVID-19 lockdown: experience from Poland. Nutrients 12(6):E1657. https://doi.org/10.3390/nu12061657
64. Hudson GM, Sprow K (2020) Promoting physical activity during the COVID-19 pandemic: implications for obesity and chronic disease management. J Phys Act Health:1–3. https://doi.org/10.1123/jpah.2020-0318. Online ahead of print

COVID-19 Is an Endothelial Disease: Implications of Nitric Oxide

9

Jason Kidde, Armita Mahdavi Gorabi, Tannaz Jamialahmadi, and Amirhossein Sahebkar

Abstract

Endothelial cells are a clinically important infection site for COVID-19, both as a mechanism for disease pathogenesis and as a therapeutic target. People with dysfunctional endothelium, defined by nitric oxide deficiency, appear to have a more severe disease course. As such, nitric oxide has therapeutic potential to mitigate COVID-19 severity. Inhaled nitric oxide appears to improve outcomes, although this strategy neglects systemic endothelium. Meanwhile, early studies have documented that endothelial protective medications, such as the administration of statins and ACE-inhibitors, are associated with less severe disease and reduced mortality. Importantly, these medications augment endothelial sources of nitric oxide, which may explain this effect.

Keywords

COVID-19 · Sars-CoV-2 · Endothelium · Endothelial dysfunction · Nitric oxide · Statins · ACE inhibitors

J. Kidde
Department of Physical Therapy, University of Utah, Salt Lake City, UT, United States

A. M. Gorabi
Research Center for Advanced Technologies in Cardiovascular Medicine, Tehran Heart Center, Tehran University of Medical Sciences, Tehran, Iran

T. Jamialahmadi
Department of Food Science and Technology, Quchan Branch, Islamic Azad University, Quchan, Iran

Department of Nutrition, Faculty of Medicine, Mashhad University of Medical Sciences, Mashhad, Iran

A. Sahebkar (✉)
Biotechnology Research Center, Pharmaceutical Technology Institute, Mashhad University of Medical Sciences, Mashhad, Iran

Neurogenic Inflammation Research Center, Mashhad University of Medical Sciences, Mashhad, Iran

Polish Mother's Memorial Hospital Research Institute (PMMHRI), Lodz, Poland

Halal Research Center of IRI, FDA, Tehran, Iran
e-mail: sahebkara@mums.ac.ir; amir_saheb2000@yahoo.com

9.1 Introduction

Our vascular endothelium is under attack by Sars-CoV-2. Researchers from China believe that while initial infection occurs within the respiratory epithelium, subsequent viremia results in multiorgan infection as well as infection of the distal vasculature [1]. This was recently corroborated by Swiss researchers who demonstrated

P. C. Guest (ed.), *Clinical, Biological and Molecular Aspects of COVID-19*, Advances in Experimental Medicine and Biology 1321, https://doi.org/10.1007/978-3-030-59261-5_9

infection of the endothelium by autopsy [2]. In response to these results, they went as far as to call COVID-19 an "endothelial disease." Both groups independently attribute the coagulopathies associated with COVID-19 to endothelial damage and acknowledged that those who are most at risk for severe disease appear to have underlying dysfunctional endothelium at baseline [1, 2]. This arguably makes endothelial dysfunction a significant risk factor for mortality from COVID-19, making it a likely therapeutic target.

The endothelium is a single layer of cells that lines all 60,000 miles of our blood vessels which perfuse every organ in the body and has such diverse functions that it can be considered as an organ in its own right. It regulates blood flow by controlling vascular smooth muscle tone and checks vessel narrowing by preventing cell and platelet aggregation along its surface. Additionally, it is also an important immune system activator. Instrumental to a healthy endothelium is its ability to produce nitric oxide, which is the linchpin for these functions [3]. Therefore, decreased nitric oxide production is central to the pathology of dysfunctional endothelium, and deficiency is ubiquitous among people with hypertension, cardiovascular disease, diabetes, and chronic kidney disease [3]. These are all notably on the top tier of risk factors for severe COVID-19, which is likely not a coincidence. Therefore, nitric oxide may have therapeutic potential to mitigate the disease course, particularly in the most at-risk populations.

The idea that nitric oxide has immunogenic and antiviral activity is not entirely novel. It has known antimicrobial activity against a wide range of organisms including bacteria, fungus, protozoa, helminths, and an assortment of viruses [4]. Jan Martel and colleagues recently published a medical theory paper on the importance of nasal nitric oxide regarding natural immunity, which suggested that variations in nasal nitric oxide levels may explain variable susceptibility to COVID-19 [4]. The paper also suggested that people with higher basal expired nitric oxide are less symptomatic to the common cold. Alternatively, animal studies have shown that mice deficient in nitric oxide are more prone to respiratory viral infections [4]. Additionally, prior research has demonstrated that administering nitric oxide donors to mice suffering from coxsackievirus-induced myocarditis improves outcomes [5] and that inhibition of nitric oxide synthase increases viral load [6].

9.2 Nitric Oxide, Demographics, and COVID-19

While several aforementioned chronic disease states associated with nitric oxide deficiency are considered to be high risk for severe COVID-19 illness, disease risk can also be stratified by demographics defined by race and sex in the absence of chronic disease. These factors also seem to correlate well with population-specific nitric oxide levels. For instance, estrogen and progesterone have a stimulating effect on nitric oxide synthase, which results in comparatively greater nitric oxide production in women compared with men [7, 8], and pregnant women have a surge in nitric oxide levels associated with elevated sex hormones [8]. This presents a dose-like protective effect of nitric oxide with women conferring an ~50% reduction in risk of death. Statistics suggests additional protection in pregnancy with ~96% of symptomatic pregnant women having mild symptoms [9] and 87% of pregnant women screening positive for COVID-19 being asymptomatic altogether [10].

Black adults are at increased risk of both incidence and severity of COVID-19 illness, and while these observations are largely attributed to social factors, it has been recently suggested there is likely to be a concomitant biological component. A possible explanation is that this population is comparatively deficient in nitric oxide compared to white adults [11]. This relative deficiency, and subsequent endothelial dysfunction, is recognized as a primary cause of increased risk of hypertension, cardiovascular disease, and kidney disease in this population [11]. Furthermore, reduced endothelial function is observed across the lifespan and apparent even in healthy, young Black adults [12].

9.3 Evidence for Inhaled Nitric Oxide and Systemic Therapy

Trials are underway investigating the utility of supplemental nitric oxide. These are largely based on a study that came from the SARS outbreak in 2003, in which nitric oxide given via mechanical ventilation had improved oxygenation and earlier hospital discharge [13]. Another study during this same period demonstrated specific viricidal effects of nitric oxide on Sars-CoV [14]. Additionally, nitric oxide appears to remove palmitic acid from the spike protein (depalmytoilation), which decreases the virus's ability to bind to the ACE-2 receptor [14]. Recently, Gilly Regev of SaNOtize (Vancouver, Canada) showed that nitric oxide is viricidal against Sars-CoV-2 in vitro. SaNOtize is now exploring the application of a nitric oxide solution to the nasopharynx in a multicenter prevention and efficacy trial against COVID-19 [15]. While inhaled or topical nitric oxide applications are promising therapies, they fail to address deficient vascular sources of NO and therefore the associated systemic consequences that define illness severity.

The benefits of nitric oxide are also indicated in the action of some drugs. Statins are mostly known for their cholesterol-lowering effects, but many lipid-independent actions have also been discovered for these drugs [16–21]. Among these so-called pleiotropic effects, statins increase endothelial nitric oxide via multiple pathways, and some studies have theorized that many of the cholesterol-independent effects of statins are mediated by this gas [22, 23]. Such effects may be at play in a study carried out by Zhang and colleagues, which demonstrated reduced mortality in hospitalized patients with COVID-19 receiving in-hospital statins [24]. The risk reduction was apparent before matching for age and comorbid conditions, with the statin group being older and more burdened by chronic disease. Specifically, the higher-risk statin-treated group had a mortality rate of 5.5% compared with 6.8% in the younger, healthier cohort. After matching the groups for age and comorbidities, the improvement was predictably more profound with a 5.2% death rate in the statin-treated group and a 9.4% mortality rate in the non-statin group [24]. The study authors hypothesized that the anti-inflammatory properties and immune modulating effects of statins likely explain the survival benefit; however, it should not be overlooked that these pleiotropic effects of statins are possibly mediated by nitric oxide [22, 23].

Angiotensin-converting enzyme (ACE) inhibitors and angiotensin receptor blockers also lead to increased nitric oxide [25–27], and Zhang et al. similarly demonstrated mortality benefit of both types of compounds in hospitalized COVID-19 patients [28]. The mortality rate was less than 5% in the treated group, while those not taking the medications had a mortality rate of greater than 10%. Unfortunately, this study did not evaluate the medication classes separately to examine superiority, as ACE inhibitors have a more robust nitric oxide response than angiotensin receptor blockers [26]. Independent studies have shown an increase in nitiric oxide production of 64% to 110% in response to ACE inhibitors compared with a peak increase of 30% observed with angiotensin receptor blockers [26, 27]. Therefore, it may be informative that a study currently in press associates ACE inhibitors with a 40% reduction in hospitalization of older adults, but no benefit was observed with angiotensin receptor blockers [29].

9.4 Conclusions

Given the antiviral, immunologic, vasodilatory, and antithrombotic properties of nitric oxide (Fig. 9.1), a deficiency may create an environment in which the endothelium is both more susceptible to infection and more prone to severe consequences. After infection ensues, subsequent inflammation results in further decreases in bioavailability of nitric oxide. Several inflammatory cytokines inhibit nitric oxide synthase, thereby limiting production, and reactive oxygen species scavenge existing nitric oxide further decreasing bioavailability [3]. Pro-inflammatory cytokines have a direct effect on vasoconstriction and hypercoagulation, usually countered by processes including nitric oxide [3]. Theoretically, a

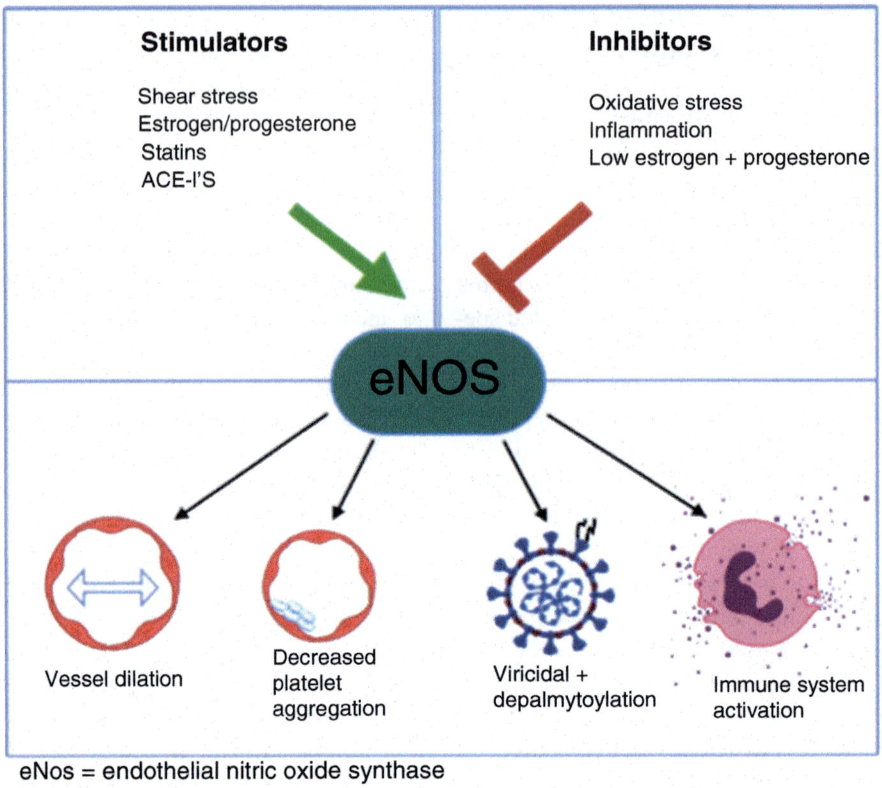

Fig. 9.1 Relevant inhibitors and stimulators of eNOS regarding COVID-19 and subsequent physiological responses

person who is already deficient in nitric oxide with preexisting endothelial inflammation will have less reserve to address the cytokine storm associated with COVID-19, and the scales begin to be unfairly tipped in the direction of dysfunction.

The variability of nitric oxide production across populations in both health and disease appears to accurately predict protection from and susceptibility to COVID-19 disease in a dose-like fashion. Furthermore, there are rational mechanisms to explain the conferred protection from elevated levels of endothelial nitric oxide. There are some studies underway exploring this relationship, although most are limited to inhaled nitric oxide. Meanwhile, some have investigated endothelial stabilizing medications with evidence of benefit. While it remains unclear as to what mechanism explains the protective effect of these medications, increased nitric oxide is a plausible answer. Although these studies are encouraging, a clearer understanding of the role of systemic nitric oxide on disease course is needed to define the path to potentially important therapeutic discoveries.

References

1. Cao W, Li T (2020) COVID-19: towards understanding of pathogenesis. Cell Res 30:367–369
2. Varga Z, Flammer AJ, Steiger P, Haberecker M, Andermatt R, Zinkernagel AS et al (2020) Endothelial cell infection and Endotheliitis in COVID-19. Lancet 395(10234):1417–1418
3. Rajendran P, Rengarajan T, Thangavel J, Nishigaki Y, Sakthisekaran D, Sethi G et al (2013) The vascular endothelium in human diseases. Int J Biol Sci 9(10):1057–1069
4. Martel J, Ko YF, Yong J, Ojcius D (2020) Could nasal nitric oxide help to mitigate the severity of COVID-19? Microbes Infect 22(4):168–171
5. Roland Z, Markgraf R, Schmidtke M, Corlach SA, Henke A et al (2004) Nitric Oxide donors inhibit the coxsackievirus B3 proteinases 2A and 3C in vitro, virus production in cells, and signs of myocardi-

tis in virus-infected mice. Med Microbiol Immunol 193:91–100
6. Hiraoka Y, Kishimoto C, Takada H, Nakamura M, Kurokawa M, Ochiai H et al (1996) Nitric oxide and murine coxsackievirus B3 myocarditis: aggravation of myocarditis by inhibition of nitric oxide synthase. J Am Coll Cardiol 28(6):1610–1615
7. Cattanep MG, Vanetti C, Decimo I, Di Chio M, Martano G, Garrone G et al (2017) Sex-specific eNOS activity and function in human endothelial cells. Sci Rep 7(1):9612. https://doi.org/10.1038/s41598-017-10139-x
8. Sutton E, Gemmel M, Powers R (2020) Nitric oxide signaling in pregnancy and preeclampsia. Nitric Oxide 95:55–62
9. Elshafeey F, Magdi R, Hindi N, Elshebiny M, Farrag N, Mahdy S et al (2020) A systematic scoping review of COVID-19 during pregnancy and childbirth. Int J Gynaecol Obstet 150(1):47–52
10. Sutton D, Fuchs K, D'Alton M, Goffman D (2020) Universal Screening for SARS-CoV-2 in women admitted for delivery. N Engl J Med 382:2163–2164
11. Brothers RM, Fadel PJ, Keller DM (2019) Racial disparities in cardiovascular disease risk: mechanisms of vascular dysfunction. Am J Physiol Heart Circ Physiol 317(4):H777–H789
12. Kim K, Herr C, Patik JC, Brothers RM (2018) Attenuated cutaneous microvascular function in healthy young African Americans: Role of intradermal l-arginine supplementation. Microvasc Res 116:1–6
13. Chen L, Liu P, Gao H, Sun B, Chao D, Wang F et al (2004) Inhalation of nitric oxide in the treatment of severe acute respiratory syndrome: a rescue trial in Beijing. Clin Infect Dis 39(10):1531–1535
14. Akerström S, Gunalan V, Keng CT, Tan YJ, Mirazimi A (2009) Dual effect of nitric oxide on SARS-CoV replication: viral RNA production and palmitoylation of the S protein are affected. Virology 395(1):1–9
15. Could a simple gas produced by our bodies be used to treat COVID-19? Canadian trials underway. https://www.ctvnews.ca/health/coronavirus/could-a-simple-gas-produced-by-our-bodies-be-used-to-treat-covid-19-canadian-trials-underway-1.4962611
16. Chruściel P, Sahebkar A, Rembek-Wieliczko M, Serban MC, Ursoniu S, Mikhailidis DP et al (2016) Impact of statin therapy on plasma adiponectin concentrations: A systematic review and meta-analysis of 43 randomized controlled trial arms. Atherosclerosis 253:194–208
17. Parizadeh SMR, Azarpazhooh MR, Moohebati M, Nematy M, Ghayour-Mobarhan M, Tavallaie S et al (2011) Simvastatin therapy reduces prooxidant-antioxidant balance: Results of a placebo-controlled cross-over trial. Lipids 46(4):333–340
18. Sahebkar A, Serban C, Mikhailidis DP, Undas A, Lip GYH, Muntner P et al (2015) Association between statin use and plasma d-dimer levels: A systematic review and meta-analysis of randomised controlled trials. Thromb Haemost 114(3):546–557
19. Serban C, Sahebkar A, Ursoniu S, Mikhailidis DP, Rizzo M, Lip GYH et al (2015) A systematic review and meta-analysis of the effect of statins on plasma asymmetric dimethylarginine concentrations. Sci Rep 5:9902. https://doi.org/10.1038/srep09902
20. Bagheri H, Ghasemi F, Barreto GE, Sathyapalan T, Jamialahmadi T, Sahebkar A (2020) The effects of statins on microglial cells to protect against neurodegenerative disorders: a mechanistic review. BioFactors 46(3):309–325. https://doi.org/10.1002/biof.1597
21. Kouhpeikar H, Delbari Z, Sathyapalan T, Simental-Mendía LE, Jamialahmadi T, Sahebkar A (2020) The effect of statins through mast cells in the pathophysiology of atherosclerosis: a review. Curr Atheroscler Rep 22(5): art. no. 19. https://doi.org/10.1007/s11883-020-00837-9
22. Gorabi AM, Kiaie N, Hajighasemi S, Banach M, Penson P, Jamialahmadi T et al (2019) Statin-Induced Nitric Oxide Signaling: Mechanisms and Therapeutic Implications. J Clin Med 8(12):2051. https://doi.org/10.3390/jcm8122051
23. Heeba G, Moselhy ME, Hassan M, Khalifa M, Gryglewski R, Malinski T (2009) Anti-atherogenic effect of statins: role of nitric oxide, peroxynitrite and haem oxygenase-1. Br J Pharmacol 156(8):1256–1266
24. Zhang XJ, Yibin W, Cai J, Guo J, Li H (2020) In hospital use of statin is associated with a reduced risk of mortality among individuals with COVID-19. Cell Metab:S1550-4131(20)30316-8. https://doi.org/10.1016/j.cmet.2020.06.015. Online ahead of print
25. Gauthier KM, Cepura CJ, Campbell WB (2013) ACE inhibition enhances bradykinin relaxations through nitric oxide and B1 receptor activation in bovine coronary arteries. Biol Chem 394(9):1205–1212
26. Mason PR, Jacob RF, Kubant R, Jacoby A, Louka F, Corbalan JJ et al (2012) Effects of angiotensin receptor blockers on endothelial nitric oxide release: the role of eNOS variants. Br J Pharmacol 74(1):141–146
27. Desideri G,Grassi D,Croce G,Bocale R, Tiberti S., Evangelista Set al (2008) Different effects of angiotensin converting enzyme inhibitors on endothelin-1 and nitric oxide balance in human vascular endothelial cells: evidence of an oxidant-sensitive pathway. Mediators Inflamm 2008:305087. doi:https://doi.org/10.1155/2008/305087
28. Zhang XJ, Yibin W, Cai J, Guo J, Li H (2020) Association of inpatient use of angiotensin-converting enzyme inhibitors and angiotensin II receptor blockers with mortality among patients with hypertension hospitalized With COVID-19. Circulation 126:1671–1681
29. Seniors with COVID-19 taking ACE inhibitors have lower hospitalization risk. https://news.yale.edu/2020/05/27/seniors-covid-19-taking-ace-inhibitors-have-lower-hospitalization-risk

The Immune Response and Effectiveness of COVID-19 Therapies

10

Fataneh Tavasolian, Gholam Reza Hatam, Sayed Hussain Mosawi, Mahdiyar Iravani Saadi, Elham Abdollahi, Tannaz Jamialahmadi, Thozhukat Sathyapalan, and Amirhossein Sahebkar

Abstract

Coronavirus disease 2019 (COVID-19) caused by severe acute respiratory syndrome coronavirus 2 (SARS-CoV-2) is highly pathogenic with relatively high mortality and morbidity. In addition to pneumonia, acute respiratory distress syndrome, and microembolic disorder, a high proportion of patients with SARS-CoV-2 develop lymphopenia and cytokine storm disorder. This review explores the underlying mechanisms behind the pathogen-

F. Tavasolian
Department of Immunology, Faculty of Medical Sciences, Tarbiat Modares University, Tehran, Iran

G. R. Hatam
Basic Sciences in Infectious Diseases Research Center, School of Medicine, Shiraz University of Medical Sciences, Shiraz, Iran

S. H. Mosawi
Medical Sciences Research Center, Ghalib University, Kabul, Afghanistan

M. I. Saadi
Hematology Research Center, Department of Hematology, Medical Oncology and Stem Cell Transplantation, Shiraz University of Medical Sciences, Shiraz, Iran

E. Abdollahi
Department of Medical Immunology and Allergy, School of Medicine, Mashhad University of Medical Sciences, Mashhad, Iran

T. Jamialahmadi
Biotechnology Research Center, Pharmaceutical Technology Institute, Mashhad University of Medical Sciences, Mashhad, Iran

Department of Food Science and Technology, Quchan Branch, Islamic Azad University, Quchan, Iran

Department of Nutrition, Faculty of Medicine, Mashhad University of Medical Sciences, Mashhad, Iran

T. Sathyapalan
Academic Diabetes, Endocrinology and Metabolism, Hull York Medical School, University of Hull, Hull, UK

A. Sahebkar (✉)
Biotechnology Research Center, Pharmaceutical Technology Institute, Mashhad University of Medical Sciences, Mashhad, Iran

Neurogenic Inflammation Research Center, Mashhad University of Medical Sciences, Mashhad, Iran

Halal Research Center of IRI, FDA, Tehran, Iran

Polish Mother's Memorial Hospital Research Institute (PMMHRI), Lodz, Poland
e-mail: amir_saheb2000@yahoo.com; sahebkara@mums.ac.ir

P. C. Guest (ed.), *Clinical, Biological and Molecular Aspects of COVID-19*, Advances in Experimental Medicine and Biology 1321, https://doi.org/10.1007/978-3-030-59261-5_10

esis of SARS-CoV-2, especially the immune mechanisms, which could be potentially used as therapeutic targets for the management of COVID-19.

Keywords

COVID-19 · Cytokine storm syndrome · Immune response · Lymphopenia

10.1 Introduction

Infection of SARS-CoV-2 in the lungs results in stimulation of macrophages and monocytes, the release of cytokines, and adaptive immune responses [1, 2]. In certain instances, this immune response can overcome viral infection and the patient recovers. However, in some instances, a dysfunctional immune response will result in pneumonia and multiorgan failure [3]. Dysfunctional immune response in some patients activates a cytokine storm which results in widespread inflammation of the lungs. There is some evidence to show that lymphopenia and the cytokine storm result in a worse prognosis. Managing the inflammatory response due to dysfunctional immune response is found to be as crucial as controlling the infection. Medications that prevent viral infection as well as those which regulate defective immune defenses can potentially prevent development of multiorgan failure [4].

10.2 Virus Entry

SARS-CoV-2 is a solitary strand RNA virus with four main basic proteins: spike (S), envelope (E), nucleocapsid (N), and membrane (M) proteins which infect human respiratory tract cells. The virus enters cells through binding of the S protein to the angiotensin-converting enzyme 2 (ACE2) receptor following S protein priming by the host transmembrane protease, serine 2 (TMPRSS2). Another receptor, CD147, is also associated with SARS-CoV-2 entry into the host cells [5]. After these processes, the virus enters the cells by endocytosis, and the viral RNA is discharged into the cytosol. The virus utilizes the cell hardware for multiplication and then erupts from the cell through exocytosis. Patients with relatively high viral burdens tend to develop more severe COVID-19 disease. Furthermore, downregulation and shedding of ACE2 by the viral S protein may disrupt the renin-angiotensin framework and increase vascular penetrability resulting in more severe lung injury [6].

10.3 Immune Responses in COVID-19 Disease

The immune responses induced by SARS-CoV-2 have two principal stages. The initial stage is the protective response, and the second stage is the inflammatory response. COVID-19 causes an imbalance of the immune system and hyperactivation of the immune response. The adaptive immune response is needed during the asymptomatic phase to get rid of infection [7]. Thus, strategies related to improving the immune system are essential at this point. Patients must be in good health and have a favorable genetic background (e.g., HLA) that could contribute to the first line of defense against the virus [8, 9]. However, if this response is not adequate, the virus will spread mainly to tissues with high ACE2 expression, such as the intestines and kidneys. The infected cells can cause latent pulmonary inflammation which is principally mediated by pro-inflammatory macrophages and granulocytes. At this stage, strategies to reduce inflammation could be potentially helpful. Effective intervention at this stage will bring down the virus load and prevent hyperinflammation. In this regard, type I interferon (IFN) is crucial for early viral clearance to minimize viral replication, T-cell exhaustion, and cytokine storms (Fig. 10.1) [10].

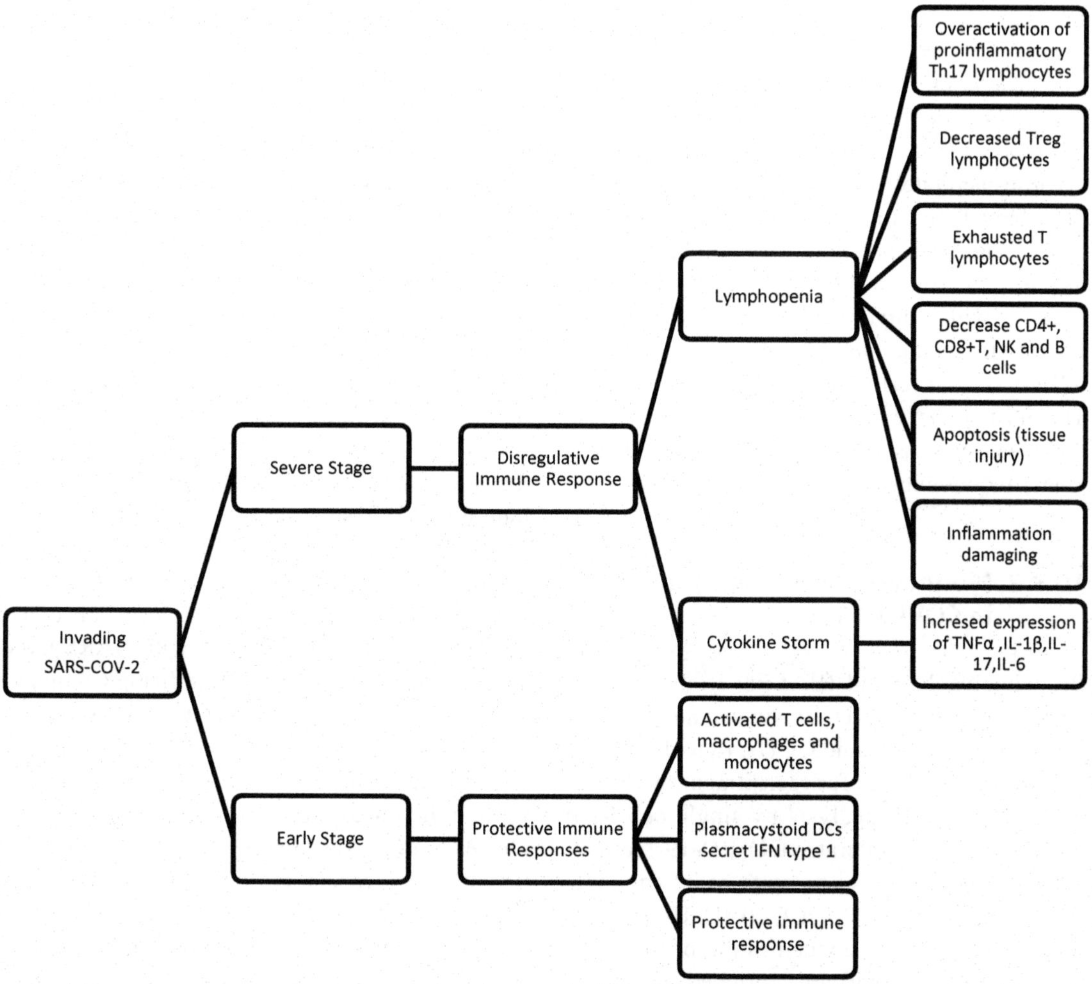

Fig. 10.1 The invading SARS-CoV-2 virus induces non-serious symptoms during the incubation period, which elicits defensive immune responses. Successful clearance of viruses relies on the state of safety. If the affected individual's general health condition does not remove the infection, then the patient reaches the severe stage of an intense and harmful inflammatory reaction, particularly in the lungs

10.3.1 Innate Immune Responses in COVID-19: Role of Cytokines and Chemokines

SARS-CoV-2 stimulates expression of numerous IFN-stimulated genes (ISGs) [11, 12]. These have an immunopathogenic capacity through overexpression of the inflammatory genes. Type I IFN is essential for protection against viral diseases as it facilitates intracellular destruction of RNA and recovery from viral infections, induces tissue repair, and activates a continuous adaptive immune response [13]. Type I IFN is delivered by plasmacytoid dendritic cells (pDCs) because they are less susceptible to active viral infection and virus-mediated cytotoxicity. They also release inflammatory cytokines, including tumor necrosis factor (TNF)-α and interleukin (IL)-6 to control the T-cell response. PDCs disperse immune cells that serve as guardians and are activated after physical contact within infected cells as part of a process called the interferogenic synapse. This results in transition of pathogen-associated molecular pattern molecules (PAMPs) to toll-like receptor 7 (TLR7) receptors in pDCs [14]. This synapse facilitates development of type I IFN at the infected area, therefore restricting viral replication and the potentially harmful systemic response. A reduced IFN type I response is associated with higher COVID-19 severity. Hypercytokinemia in COVID-19 patients is related to the severity of COVID-19 disease [15]. The most crucial cytokines in this regard are chemokines, such as neutrophil-recruiting chemokines, and monocyte attractants [16].

10.3.2 Monocytes and Macrophages in COVID-19

Bronchoalveolar fluid (BALFs) from individuals with severe COVID-19 showed an increased expression of CCL2 and CCL7, two of the most essential chemokines for recruitment of CCR2+ monocytes. BALF analysis by single-cell RNA sequencing of moderate COVID-19 patients reported increased concentrations of mononuclear phagocytes [11]. In COVID-19 patients, there is an increased concentration of the group of macrophages that are enriched in tissue-repaired genes and promote generation of fibrosis, as found in liver cirrhosis. This suggests that the pathogenicity of invading macrophages may go farther than acute inflammation to fibrosis in ventilated patients [11, 17]. Park et al. have referenced macrophages as a Trojan horse in COVID-19. ACE2-expressive CD68+CD169+ macrophages were found in the splenic marginal zone and marginal sinuses of the lymph node which expresses nucleoprotein antigen SARS-CoV-2 and produces a significant rise in IL-6 concentrations. This suggests CD169+ macrophages can facilitate viral spread during SARS-CoV-2 disease, heighten inflammation, and activation-induced lymphocytic cell death [18].

10.3.3 Role of Complement in COVID-19

Complement is one of the essential factors helpful in shielding from pathogens. However, the excessive and deregulated response of complement can trigger injury to the tissue. Complement is both an integral part of the innate immune system of the pathogens and a pro-inflammatory reaction orchestrator. C3-lacking mice infected with SARS-CoV show reduced pulmonary injury, lower neutrophil and monocyte infiltration, and diminished cytokine and chemokine levels in both the lungs and the sera [19]. This suggests that the inactivation of C3 in the inflammatory lung may likewise reduce the severity of SARS-CoV-2 injury in tissues. The reduction in lung-invading neutrophils and the reduced intrapulmonary and plasma IL-6 levels observed in C3-deficient mice infected with SARS-CoV suggests the opportunities for utilizing C3 blockers with anti-IL-6 agents [19]. C3 inhibition can simultaneously block the development of C3a and C5a, as well as intrapulmonary activation of C3 and the release of IL-6 from alveolar macrophages or other cells expressing C3a (C3aR) and C5a (C5aR) receptors, thereby limiting lung injury. Ex vivo experiments of whole blood infection with the AMY-101 C3 inhibitor have demonstrated that it will reduce IL-6 levels. The lung biopsy specimens from individuals with extreme COVID-19 revealed extensive activation of complement, characterized by the production of C3a and deposition of C3 fragment. There was also a rise in the serum C5a levels [20]. Patients with an anti-C5a antibody showed better lung oxygenation and diminished inflammatory responses [21, 22].

10.3.4 B-Cell Immunity

In patients with COVID-19, B-cell reactions emerge around the same time as T follicular helper cell responses, beginning about 1 week after the inception of symptoms. A B-cell response mainly occurs in patients with SARS-CoV disease for the most part against the viral N protein [23]. Antibody responses to the S protein were seen within 4–8 days of the beginning of symptoms. Neutralizing responses of antibodies against the S antigen begin to increase by week 2, and by week 3, most patients develop neutralizing antibodies. However, this does not appear to result in durable SARS-CoV-2 antibodies [23]. The neutralization of the virus is viewed as a fundamental mode of action for antibodies although the specific titer of antibodies remains unresolved [10, 23].

10.3.5 T-Cell Immunity

CD8+ T cells are expected to attack and destroy virus-infected cells specifically, while CD4+ T cells are essential in the activation of both CD8+ T cells and B cells. CD4+ T cells likewise produce cytokines to activate immune cells [10]. It seems that SARS-CoV-2 can cause a protective immune response mediated by the T lymphocyte, in comparison to other CoVs. Patients with COVID-19 have increased monocytes and T cells in the lungs and a significant reduction in the amounts of CD4+ and CD8+ T cells in the peripheral blood due to insufficient activation as seen by elevated HLA-DR and CD38 double-positive fractions [9]. Such outcomes demonstrate that T cells are attracted to monitor virus infection away from the blood and toward the affected region. Likewise, the intense stage response in patients with SARS-CoV is associated with a significant reduction in CD4+ T and CD8+ T cells [8]. Albeit additional precautionary measures ought to be taken in patients determined to have SARS-CoV-2 who are hospitalized with lymphopenia, cellular immune reactions also appear to be reduced. A cellular immune response efficiently destroys SARS-CoV-2 in the safest-case scenario without any (or mild) clinical signs of infection. However, this is not always the case as the virus also induces immunosuppression that reduces and sometimes overcomes the host's defenses [8].

10.4 Pro-inflammatory Th17 Lymphocytes and Disease Progression

Xu et al. observed that patients with severe COVID-19 infection had high concentrations of CCR4+ CCR6+ TH17 cells in the peripheral blood, thereby indicating a TH17-type cytokine storm. This research demonstrated a crucial role of Th17 inflammatory response in the pathogenesis of COVID-19 pneumonia [24]. This involves releasing essential cytokines such as IL-17 and other factors to intensify viral immunopathogenesis by downregulating Treg cells, facilitating neutrophil relocation, and simultaneously inciting Th2 reactions. IL-17 can induce severe eosinophilic reactions, allergic disease, and, to some degree, extravasation into the lungs [24]. Most recent outcomes show that the N protein is a potential inducer of IL-6 reactions that could intervene in coronavirus lung pathology [24, 25].

10.5 Lymphopenia and COVID-19

Lymphopenia is one of the most noteworthy markers of COVID-19. All lymphocyte subsets, which incorporate CD4+ and CD8+ cytotoxic T cells, natural killer (NK) cells, memory and Treg cells along with B cells, have been shown to be diminished in COVID-19 disease. This is critical as lower levels of lymphocytes are strongly related to the seriousness of disease [26]. T-cell numbers are inversely related with serum levels of IL-6, IL-10, and TNF-α and elevated levels of programmed cell death protein 1(PD-1) or T-cell immunoglobulin (Ig) and mucin domain-containing molecule-3 (TIM-3) [27].

The evidence suggests that SARS-CoV-2 targets T cells by receptor-dependent, S-protein-mediated membrane fusion. T cells have a low level of ACE2 expression, suggesting both an alternate receptor and a strong S-sensitivity to the protein. Invasion of T cells is abortive, showing that replication of SARS-CoV-2 inside T cells is not possible yet causes cell death instead [28].

Second, impaired lung macrophages or epithelial cells (in the first stage of hypercytokinemia) build up a variety of inhibitory cytokines, particularly TNF-α causing T-cell apoptosis, IL-10-restraining T-cell expansion, and type I IFN in the regulation of lymphocyte distribution. Thirdly, lymphopenia has been accepted to be the result of immune cell redistribution, with lymphocytes proliferating in the lungs or lymphoid glands [28]. Immunohistochemical staining of spleen and lymph nodes has shown decreased levels of CD4+ and CD8+ T cells. Finally, metabolic molecules generated by metabolic disorders such as lactic acid can block lymphocytes. Severe cases of COVID-19 patients have been found to have high blood lactic acid levels which can limit lymphocyte expansion [29].

10.6 Exhausted T Lymphocytes Associated with COVID-19 Disease Severity

Immune system homeostasis represents a vital role in preventing COVID-19 pneumonia [28]. Yong-Tang Zheng provides substantial differences in the levels of exhaustion modules between the three target groups (healthy, mild, severe), in particular PD-1, cytotoxic T-lymphocyte-associated protein 4 (CTLA-4), and T-cell immunoglobulin and ITIM domain (TIGIT), and functional modules, such as IFN-γ, TNF-α, and IL-2. In the severe group, the amount of multifunctional CD4+ T cells declined significantly relative to the levels in healthy controls and a mild group, whereas the number of non-functional subgroups (IFN − TNF-α − IL-2) increased. The disturbance of CD4+ T cells could also have predisposed COVID-19 patients to severe diseases [28]. Prior research from Guang Chen et al. demonstrated that the Treg (CD4+ CD25+ CD127low+) and CD45RA+ Treg rates were lower in practically every severe and moderate case, with CD45RA+ Treg cells falling more significantly in severe cases than in moderate ones. It should be remembered that in certain patients with extreme and moderate COVID-19, CD4+T, CD8+T, and NK cells, the levels of IFN-γ secretion are reduced [30]. An early IFN response is fundamental for an effective T-cell reaction as a delayed IFN response may reduce T-cell expansion or T-cell departure from lymphoid organs. It might also bring about T-cell depletion and cell death. In patients with extreme COVID-19, lung damage was found to be correlated with cytokine release syndrome, suggesting an expected failure to trigger opportune immunosuppressive systems. Even so, Treg cell counts have been suggested to be associated inversely with the seriousness of the disease in COVID-19 patients [15]. IFNs are viewed as significant controllers for Treg cell development. Zheng et al. showed that the total amount of NK and CD8+ T cells diminished significantly in patients with SARS-CoV-2 disease. With the increased production of NKG2A in patients with COVID-19, the activity of the NK and CD8+ T cells was found to be reduced. Interestingly, the amount of NK and CD8+ T cells with decreased NKG2A expression has been shown to be increased in patients convalescing following treatment [31]. Consequently, infection with SARS-CoV-2 can destroy antiviral immunity rapidly. Hence, expression of SARS-CoV-2 induced NKG2A in COVID-19 patients with serious pulmonary inflammation associated with an initial phase of functional exhaustion of cytotoxic lymphocytes, which may lead to progression of the disease. Persistent infection and cancer have been found to have immune-inhibitory "checkpoint" receptors that lead to dysfunction of NK and T cells. It is important to take note that checkpoint inhibitors, such as anti-PD-1 and anti-TIGIT, help in the case of chronic infection and cancer and revitalize depleted T- or NK-cell responses [32]. NKG2A is believed to be another inhibitory molecule in the immune-checkpoint blockade. Such results show the importance of improving the immune response of NK cells and CTLs underlying SARS-CoV-2 infection and maintaining strategies for avoiding cytotoxic lymphocyte exhaustion [33].

10.7 Cytokine Storm, a Lethal Phase

Cytokine release syndrome (CRS) tends to involve severely affected individuals with COVID-19 disease. Considering that lymphocytopenia is frequently seen in severe COVID-19 patients, leukocytes other than T cells will mediate the CRS induced by the SARS-CoV-2 virus. Cytokines are highly significant for COVID-19 pathophysiology [4]. Although some are protective (type I interferon, IL-7), others seem hazardous (IL-1β, IL-6, and TNF-α) and result in cytokine storms. This cytokine storm appears to be more likely to occur by a combination of the defective or delayed first line of protection, accompanied by chronic hypercytokinemia and an abnormal T-cell response [4]. This tends to result in incomplete removal of apoptotic cells or affected macrophages, an increase in viral proliferation and expansion, accompanied by an IL-18-/IFN-γ-activating macrophages, and which leads to massive secretion of cytokines, hemophagocytosis, coagulopathy, and acute respiratory distress syndrome (ARDS). By examining the immunopathology of SARS-CoV-2-related ARDS more closely, two mechanisms of immune failure have been identified as COVID-19 worsens: (1) an IL-1β-driven macrophage activation syndrome and (2) an immune dysregulation process guided by IL-6 [34]. The latter has been depicted by a combination of hypercytokinemia, immunoparalysis (CD14 monocytes with reduced HLA-DR molecules), and generalized lymphopenia (such as CD4+ and NK cells). Nonetheless, the blockade of IL-6 tocilizumab preserved HLA-DR expression on CD14 monocytes and increased the number of circulating lymphocytes. Cytokine storms could cause ARDS, severe cardiac attack, and secondary infection, culminating in systemic sepsis, and multisystem failure that may result in death (Fig. 10.2) [17, 35].

10.8 Coronavirus Treatments: Which Therapies Could Be Effective for COVID-19?

10.8.1 Apoptosis Inhibitors

Currently, there are no definite treatment protocols developed for COVID-19. Traditional therapies are mostly directed at the symptom level. Evidence shows that lymphopenia is generally identical in SARS-COV-2, respiratory syncytial virus infection, measles, and sepsis. The main triggers of sepsis and measles is apoptosis, which is expected to promote lymphopenia. For example, apoptosis inhibitors ameliorate inflammation and reduce mortality in sepsis models. These results have given us valuable insights concerning SARS-CoV-2 patients [36]. The proliferation of lymphocytes or targeting drugs for apoptosis (PD1/PD-L1 inhibitors) may forestall lymphopenia or recuperate lymphocyte in severe COVID-19 patients [37]. Restoration of the leukomonocyte populations of COVID-19-hospitalized patients appears to be associated with viral clearance. By comparing the numbers of leukomonocytes in COVID-19 patients at different periods of the sicknesses, research showed that CD3+, CD4+, and CD8+ T cells and B cells appear to play significant roles in viral clearance. It has been proposed that stabilization of leukomonocyte levels may be used as a guide for releasing and discharging patients in the COVID-19 diagnostic guidelines [37].

10.8.2 Convalescent Plasma and COVID-19

Immune or convalescent plasma is plasma obtained from patients after recovering from infection and antibody production. As can be shown, there are also many issues concerning convalescent plasma or immunoglobulins regard-

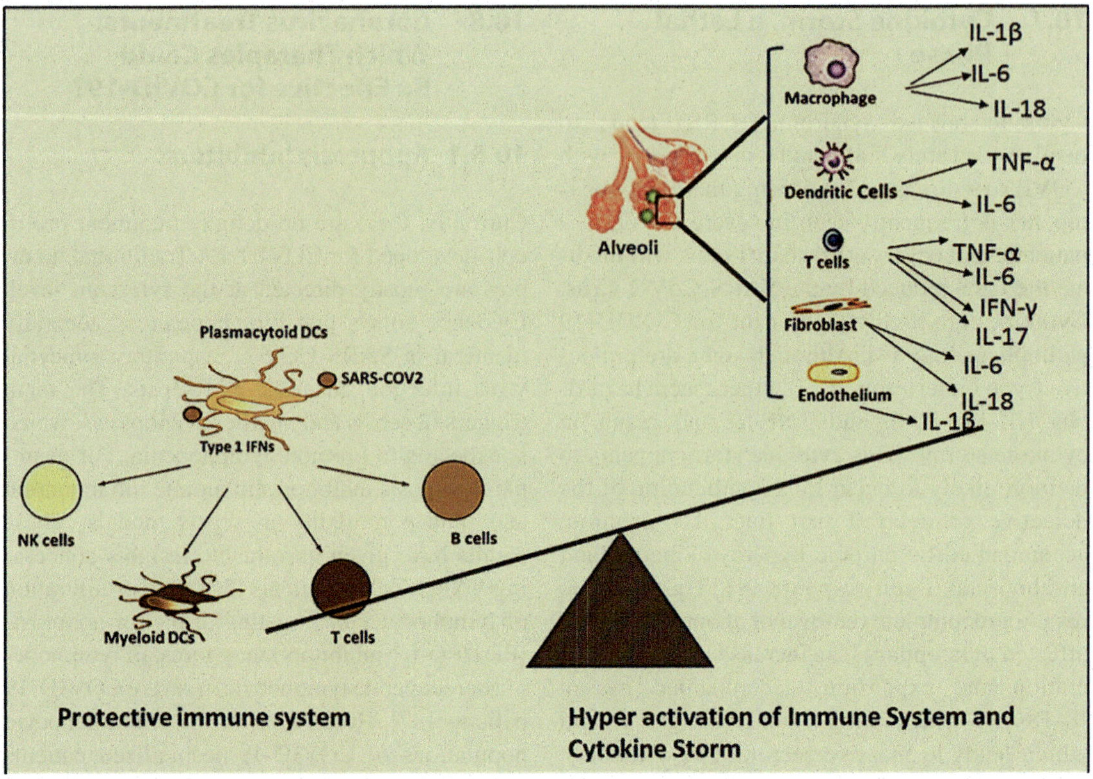

Fig. 10.2 The immune responses mediated by SARS-CoV-2 are two main phases: the protective immune step and the second damage processing step done by inflammation and the cytokine storm

less of their wide approval. They had been used to increase the recovery rate of patients with SARS, human immunodeficiency virus (HIV), and severe H1N1 influenza whose conditions worsened despite the treatment with pulsed methylprednisolone [38]. One possible rationale behind the viability of convalescent plasma therapy might be that the antibodies could forestall viremia and provide passive immunity [39]. An in vivo examination likewise found that these neutralizer activities were not only restrictive for virus clearance and prevented new invasion but also increased infected cell clearance. Convalescent plasma, acquired from recovered COVID-19 patients with humoral immunity to the virus, has a large amount of neutralizing antibodies which could neutralize SARS-CoV-2 and eliminate the pathogen from the blood circulation and pulmonary tissues [40]. The outcome might be particularly advantageous for individuals with severe or life-threatening COVID-19 disease and, by using this medication, reduce the length or extent of the illness. The neutralization antibody titers correlate with the numbers of virus-specific T cells. Given the unavailability of data of SARS-CoV-2 fundamental biology, particularly virus heterogeneity and mutation, locally acquired plasma that better represents the circulating virus in the community may be a viable treatment choice. However, for this to work, we need appropriate donor selection with significantly higher serum titers of antibody that are neutralizing [40].

10.8.3 Intravenous Immunoglobulin (IVIG)

Individuals with debilitated immune systems, in general, have a greater danger of the related complications of COVID-19. Coupled with antiviral medications, IVIG-utilizing immunotherapy can be utilized to control or eliminate COVID-19 and improve the immune response to this virus. IVIG antibodies have two fundamental parts. These are the F (ab′) 2 piece, which is essential for the recognition of antigens, and the crystallizable fragment (Fc), which is essential for activating the immune response by communicating with B cells as well as other innate human immune cells with Fcγ receptors [41]. The Fc section additionally plays a pivotal role in enacting complement and evacuating the microorganisms. Elective treatment may, for the most part, be given with safe IVIG as an adjunctive prescription combined with antivirals. Tolerant IVIG antibodies acquired from healing patients would be successful against COVID-19 by reinforcing the immune response reaction in recently infected patients. However, no COVID-19 immunization is authoritatively accessible, and the mix of insusceptible IVIG antibodies with antiviral medications only gives short- and medium-term protection against COVID-19 [41, 42].

10.8.4 Cytokine-Based Interventions

10.8.4.1 Type I Interferon

Patient immune status will establish the efficacy of the COVID-19 treatments. Emerging data indicates that SARS-CoV-2 can activate a range of immune processes, allowing immunosuppressive agents in clinical trials to be beneficial in certain patients but dangerous to others [43]. IFN-α and IFN-β have recognized as potentially beneficial anti-SARS-CoV-2 medications. Type I IFN should be given as early as possible following diagnosis (ideally before the initiation of manifestations), however not in the late stage because of likely disturbance to tissues. Until peak viral replication, early IFN therapy was found to save mice from fatal SARS-CoV or MERS-CoV challenge, while late IFN therapy disrupted viral clearance and exacerbated immunopathology [14].

10.8.4.2 Granulocyte-Macrophage Colony-Stimulating Factor (GM-CSF)

GM-CSF performs a primary controlling function in cytokine production and myeloid cell-induced hyperinflammation. In addition, the late phases of COVID-19 disease will most likely be controlled not by forceful viral replication and cell lysis but by immunopathology, especially myeloid cell immunopathology [44]. In this manner, the putative pathogenic function of GM-CSF in immune hyperactivation opens up the possibility for starting continuous clinical investigations using GM-CSF focusing on monoclonal antibodies (mAbs) for treating COVID-19 patients. The reasoning and the risks for both therapeutic administration and inhibition of GM-CSF in COVID-19 have been established. The use of GM-CSF in COVID-19 patients may improve lung capacity by supporting the alveolar wall and enhancing viral removal and thereby offer a positive advantage in early COVID-19 phases. Conversely, GM-CSF or GM-CSFR inhibition can be a protective treatment for the cytokine storm and myeloid cell tissue invasion. As it may influence the release of several pro-inflammatory cytokines and chemokines by myeloid cells, the GM-CSF approach may have noteworthy immunomodulatory implications [44].

10.8.5 Anti-Cytokine Interventions

10.8.5.1 IL-6 Inhibition

Inflammation caused by SARS-CoV-2 results in a dose-dependent release of IL-6 from bronchial epithelial cells [45]. In patients with SARS-CoV-2 infections, alterations in IL-6 levels could potentially be a crucial mediator when severe systemic inflammatory reactions occur. IL-6 is involved in two particular pathways in SARS-CoV-2 pathology. The first of these is thought to be anti-inflammatory (induced by the membrane-bound variant of the IL-6 receptor, mIL6R) and the second is the trans-signaling pathway (induced by the soluble IL6R form) which is presumed to have a pro-inflammatory role [25].

Specific blockers of these pathways should be investigated to determine which is the most effective as a treatment for COVID-19 disease.

10.8.5.2 TNF Inhibitors

As mentioned above, SARS-CoV-2 infection is related to downregulation of ACE2 expression combined with activation of the renin-angiotensin system liable for the lung injury [6]. In addition, the viral spike protein will cause a TNF-α-converting enzyme (TACE)-dependent shedding of the ectodomain ACE2, which is fundamental for the viral entry into the cell. For these reasons, it has been speculated that the use of TNF inhibitors may be efficient in lowering both SARS-CoV-2 infection and the resulting organ damage. Subsequently, the Chinese Clinical Trial Registry (ChiCTR2000030089) announced an investigation of adalimumab in patients with COVID-19 infection [46].

10.8.5.3 Targeting Chemokine Receptors

In patients with COVID-19, a significant rise of CCL2 and CCL3 expression in macrophages has been seen alongside diminished expression of CCR1, the receptor for both chemokines. Since binding of CCL2 or CCL3 to CCR1, CCR2, or CCR5 causes monocyte recruitment into the lung parenchyma with eventual differentiation into inflammatory macrophages and resulting recruitment and activation of multiple immune cells, and epithelial injury, CCR1, CCR2, and CCR5 could be potent anti-inflammatory targets in COVID-19 [11]. HIV and other viral diseases target the CCR2/CCL2 axis. However, the evidence has not verified the production of CCR2 in the respiratory tract of COVID-19 patients (possibly due to its accelerated inhibition in monocytes as they leave the circulation and reach tissues). This leaves CCR1 and CCR5 as potential targets, which should be investigated further [11, 16].

10.8.6 Nontargeted Therapies

10.8.6.1 Corticosteroids

Corticosteroids are effective cytokine inhibitors that act by various pathways but essentially through inhibition of the transcription factor NF-κB. These drugs are the foundation of therapies for autoimmune and autoinflammatory disorders with cytokine storms. Dexamethasone is a medication that has been utilized in a variety of treatments since the 1960s to minimize inflammation involving autoimmune diseases and certain cancers [47]. According to early findings in a clinical trial conducted in the United Kingdom, dexamethasone was found to reduce mortality by about one-third for patients on ventilators, and mortality was decreased by around one-fifth for patients needing only oxygen [47].

10.8.6.2 Remdesivir

Remdesivir is a nucleotide-analog prodrug that prevents polymerases of viral RNA, which has been shown to have efficacy against SARS-CoV-2 in vitro [48]. Remdesivir is intracellularly metabolized to an analog of adenosine triphosphate that suppresses viral RNA polymerases. Remdesivir has broad actions on a variety of viral agents, including filoviruses (Ebola) and coronaviruses (SARS-CoV and MERS-CoV). The US Food and Drug Administration has provided an urgent usage permit for the investigational antiviral medication remdesivir for the care of suspected or laboratory-affirmed COVID-19 in hospitalized adults and children with serious illness. One study showed that remdesivir reduced the period of rehabilitation in some instances of COVID-19 [49].

10.9 Conclusions

The occurrence and development of SARS-CoV-2 depend on the interaction between virus infection and the immune system. Dysregulation of the immune system in COVID-19 patients can contribute to serious illness. Dysregulation of the immune system such as lymphopenia and cytokine storm could be a crucial factor related to the severity of COVID-19. Decreased T lymphocytes and elevated cytokines could potentially serve as COVID-19 prognostic markers. Antiviral or immunomodulatory therapies have not been shown to be useful for treatment of COVID-19 to date. In clinical trials, interventions could be timed based on immune response. For example, antivirals and immune boosters should be started soon after the start of symptoms, whereas immunosuppressants should be delivered at the beginning of the cytokine storm.

References

1. Shoenfeld Y (2020) Corona (COVID-19) time musings: our involvement in COVID-19 pathogenesis, diagnosis, treatment and vaccine planning. Autoimmun Rev 19(6):102538. https://doi.org/10.1016/j.autrev.2020.102538
2. Fauci AS, Lane HC, Redfield RR (2020) Covid-19 – navigating the uncharted. N Engl J 382(13):1268–1269
3. Li X, Geng M, Peng Y, Meng L, Lu S (2020) Molecular immune pathogenesis and diagnosis of COVID-19. J Pharm Anal 10(2):102–108
4. Mehta P, McAuley DF, Brown M, Sanchez E, Tattersall RS, Manson JJ et al (2020) COVID-19: consider cytokine storm syndromes and immunosuppression. Lancet 395(10229):1033–1034
5. Ulrich H, Pillat MM (2020) CD147 as a target for COVID-19 treatment: suggested effects of azithromycin and stem cell engagement. Stem Cell Rev Rep 16(3):434–440
6. Acosta FASAD (2020) Entry of SARS-CoV2 through the basal surface of alveolar endothelial cells–A proposed mechanism mediated by CD147 in COVID-19. Preprints 2020050359. https://doi.org/10.20944/preprints202005.0359.v1
7. Tufan A, GÜLER AA, Matucci-Cerinic M (2020) COVID-19, immune system response, hyperinflammation and repurposing antirheumatic drugs. Turk J Med Sci 50(SI-1):620–632
8. Shi Y, Wang Y, Shao C, Huang J, Gan J, Huang X et al (2020) COVID-19 infection: the perspectives on immune responses. Cell Death Differ 27(5):1451–1454
9. Nguyen A, David JK, Maden SK, Wood MA, Weeder BR, Nellore A et al (2020) Human leukocyte antigen susceptibility map for SARS-CoV-2. J Virol 94(13):e00510–e00520. https://doi.org/10.1128/JVI.00510-20
10. Giamarellos-Bourboulis EJ, Netea MG, Rovina N, Akinosoglou K, Antoniadou A, Antonakos N et al (2020) Complex immune dysregulation in COVID-19 patients with severe respiratory failure. Cell Host Microbe 27(6):992–1000.e3
11. Merad M, Martin JC (2020) Pathological inflammation in patients with COVID-19: a key role for monocytes and macrophages. Nat Rev Immunol 20(6):355–362
12. Zhou Z, Ren L, Zhang L, Zhong J, Xiao Y, Jia Z et al (2020) Heightened innate immune responses in the respiratory tract of COVID-19 patients. Cell Host Microbe 27(6):883–890.e2
13. Trouillet-Assant S, Viel S, Gaymard A, Pons S, Richard JC, Perret M et al (2020) Type I IFN immunoprofiling in COVID-19 patients. J Allergy Clin Immunol 146(1):206–208.e2
14. Acharya D, Liu G, Gack MU (2020) Dysregulation of type I interferon responses in COVID-19. Nat Rev Immunol 20(7):397–398
15. Park A, Iwasaki A (2020) Type I and type III interferons–induction, signaling, evasion, and application to combat COVID-19. Cell Host Microbe 27(6):870–878
16. Coperchini F, Chiovato L, Croce L, Magri F, Rotondi M (2020) The cytokine storm in COVID-19: an overview of the involvement of the chemokine/chemokine-receptor system. Cytokine Growth Factor Rev 53:25–32
17. McGonagle D, Sharif K, O'Regan A, Bridgewood C (2020) Interleukin-6 use in COVID-19 pneumonia related macrophage activation syndrome. Autoimmun Rev 19(6):102537. https://doi.org/10.1016/j.autrev.2020.102537
18. Park MD (2020) Macrophages: a Trojan horse in COVID-19? Nat Rev Immunol Apr 17:1. https://doi.org/10.1038/s41577-020-0317-2. [Epub ahead of print]
19. Campbell CM, Kahwash R (2020) Will complement inhibition be the new target in treating COVID-19–related systemic thrombosis? Circulation 141(22):1739–1741
20. Mastaglio S, Ruggeri A, Risitano AM, Angelillo P, Yancopoulou D, Mastellos DC et al (2020) The first case of COVID-19 treated with the complement C3 inhibitor AMY-101. Clin Immunol 215:108450. doi: https://doi.org/10.1016/j.clim.2020.108450
21. Risitano AM, Mastellos DC, Huber-Lang M, Yancopoulou D, Garlanda C, Ciceri F et al (2020) Complement as a target in COVID-19? Nat Rev Immunol 20(6):343–344
22. Magro C, Mulvey JJ, Berlin D, Nuovo G, Salvatore S, Harp J et al (2020) Complement associated microvascular injury and thrombosis in the pathogenesis of severe COVID-19 infection: a report of five cases. Transl Res 220:1–13
23. Tay MZ, Poh CM, Rénia L, MacAry PA, Ng LF (2020) The trinity of COVID-19: immunity, inflammation and intervention. Nat Rev Immunol 20:1–12
24. Wu D, Yang XO (2020) TH17 responses in cytokine storm of COVID-19: an emerging target of JAK2 inhibitor Fedratinib. J Microbiol Immunol Infect 53(3):368–370
25. Zhang C, Wu Z, Li JW, Zhao H, Wang GQ (2020) The cytokine release syndrome (CRS) of severe COVID-19 and Interleukin-6 receptor (IL-6R) antagonist Tocilizumab may be the key to reduce the mortality. Int J Antimicrob Agents 55(5):105954. https://doi.org/10.1016/j.ijantimicag.2020.105954
26. Tan L, Wang Q, Zhang D, Ding J, Huang Q, Tang YQ et al (2020) Lymphopenia predicts disease severity of COVID-19: a descriptive and predictive study. Signal Transduct Target Ther 5(1):1–3
27. Diao B, Wang C, Tan Y, Chen X, Liu Y, Ning L et al (2020) Reduction and functional exhaustion of T cells in patients with coronavirus disease 2019

(COVID-19). Front Immunol 11:827. https://doi.org/10.3389/fimmu.2020.00827

28. Zheng HY, Zhang M, Yang CX, Zhang N, Wang XC, Yang XP et al (2020) Elevated exhaustion levels and reduced functional diversity of T cells in peripheral blood may predict severe progression in COVID-19 patients. Cell Mol Immunol 17(5):541–543
29. Wilk AJ, Rustagi A, Zhao NQ, Roque J, Martínez-Colón GJ, McKechnie JL et al (2020) A single-cell atlas of the peripheral immune response in patients with severe COVID-19. Nat Med 26(7):1070–1076
30. Chen G, Wu D, Guo W, Cao Y, Huang D, Wang H et al (2020) Clinical and immunologic features in severe and moderate forms of coronavirus disease 2019. J Clin Invest 130(5):2620–2629
31. Zheng M, Gao Y, Wang G, Song G, Liu S, Sun D et al (2020) Functional exhaustion of antiviral lymphocytes in COVID-19 patients. Cell Mol Immunol 17(5):533–535
32. Pickles OJ, Lee LY, Starkey T, Freeman-Mills L, Olsson-Brown A, Cheng V et al (2020) Immune checkpoint blockade: releasing the breaks or a protective barrier to COVID-19 severe acute respiratory syndrome? Br J Cancer Jun 16;1–3. https://doi.org/10.1038/s41416-020-0930-7. Online ahead of print
33. Bersanelli M (2020) Controversies about COVID-19 and anticancer treatment with immune checkpoint inhibitors. Immunotherapy 12(5):269–273
34. Ye Q, Wang B, Mao J (2020) The pathogenesis and treatment of the 'Cytokine Storm' in COVID-19. J Inf 80(6):607–613
35. Henderson LA, Canna SW, Schulert GS, Volpi S, Lee PY, Kernan KF et al (2020) On the alert for cytokine storm: immunopathology in COVID-19. Arthritis Rheum 72:1059–1063
36. Cao X (2020) COVID-19: immunopathology and its implications for therapy. Nat Rev Immunol 20(5):269–270
37. Bermejo-Martin JF, Almansa R, Menéndez R, Mendez R, Kelvin DJ, Torres A (2020) Lymphopenic community acquired pneumonia as signature of severe COVID-19 infection. J Inf Secur 80(5):e23–e24
38. Chen L, Xiong J, Bao L, Shi Y (2020) Convalescent plasma as a potential therapy for COVID-19. Lancet Infect Dis 20(4):398–400
39. Duan K, Liu B, Li C, Zhang H, Yu T, Qu J et al (2020) Effectiveness of convalescent plasma therapy in severe COVID-19 patients. Proc Natl Acad Sci U S A 117(17):9490–9496
40. Rojas M, Rodríguez Y, Monsalve DM, Acosta-Ampudia Y, Camacho B, Gallo JE et al (2020) Convalescent plasma in Covid-19: possible mechanisms of action. Autoimmun Rev 19(7):102554. https://doi.org/10.1016/j.autrev.2020.102554
41. Jawhara S (2020) Could intravenous immunoglobulin collected from recovered coronavirus patients protect against COVID-19 and strengthen the immune system of new patients? Int J Mol Sci 21(7):2272. https://doi.org/10.3390/ijms21072272
42. Xie Y, Cao S, Dong H, Li Q, Chen E, Zhang W et al (2020) Effect of regular intravenous immunoglobulin therapy on prognosis of severe pneumonia in patients with COVID-19. J Infect Apr 10;S0163-4453(20):30172–30179. https://doi.org/10.1016/j.jinf.2020.03.044. Online ahead of print
43. Ritchie AI, Singanayagam A (2020) Immunosuppression for hyperinflammation in COVID-19: a double-edged sword? Lancet 395(10230):1111. https://doi.org/10.1016/S0140-6736(20)30691-7
44. Lang FM, Lee KMC, Teijaro JR, Becher B, Hamilton JA (2020) GM-CSF-based treatments in COVID-19: reconciling opposing therapeutic approaches. Nat Rev Immunol Jun 23;1–8. https://doi.org/10.1038/s41577-020-0357-7. Online ahead of print
45. Liu B, Li M, Zhou Z, Guan X, Xiang Y (2020) Can we use interleukin-6 (IL-6) blockade for coronavirus disease 2019 (COVID-19)-induced cytokine release syndrome (CRS)? J Autoimmun 111:102452. https://doi.org/10.1016/j.jaut.2020.102452
46. Feldmann M, Maini RN, Woody JN, Holgate ST, Winter G, Rowland M et al (2020) Trials of anti-tumour necrosis factor therapy for COVID-19 are urgently needed. Lancet 395(10234):1407–1409
47. Theoharides T, Conti P (2020) Dexamethasone for COVID-19? Not so fast. J Biol Regul Homeost Agents Jun 4;34(3). https://doi.org/10.23812/20--EDITORIAL_1-5. Online ahead of print
48. Grein J, Ohmagari N, Shin D, Diaz G, Asperges E, Castagna A et al (2020) Compassionate use of remdesivir for patients with severe Covid-19. N Engl J Med 382(24):2327–2336
49. Wang Y, Zhang D, Du G, Du R, Zhao J, Jin Y et al (2020) Remdesivir in adults with severe COVID-19: a randomised, double-blind, placebo-controlled, multicentre trial. Lancet 395(10236):1569–1578

The Novel Coronavirus and Inflammation

11

J. A. George and E. S. Mayne

Abstract

The SARS-CoV-2 virus which causes COVID-19 disease was initially described in the Hubei Province of China and has since spread to more than 200 countries and territories of the world. Severe cases of the disease are characterised by release of high levels of pro-inflammatory cytokines and other inflammatory mediators in a process characterised as a cytokine storm. These inflammatory mediators are associated with pathological leukocyte activation states with tissue damage. Here, we review these effects with a focus on their potential use in diagnosis, patient stratification and prognosis, as well as new drug targets.

Keywords

Biomarker · COVID-19 · Cytokine storm · Inflammation · Pandemic · SARS-CoV-2 · Therapeutic target

J. A. George
Department of Chemical Pathology, National Health Laboratory Services and Faculty of Health Sciences, University of the Witwatersrand, Johannesburg, South Africa

E. S. Mayne (✉)
Department of Immunology, National Health Laboratory Services and Faculty of Health Sciences, University of the Witwatersrand, Johannesburg, South Africa
e-mail: Elizabeth.mayne@nhls.ac.za

11.1 Introduction

The novel coronavirus, severe acute respiratory syndrome-coronavirus-2 (SARS-CoV-2 or COVID-19), was initially described in a cohort of patients presenting in Wuhan, Hubei Province, China. The virus belongs to a family of small RNA viruses which cause illness in mammals and birds. Viruses from the family are considered responsible primarily for upper respiratory tract infections. SARS-CoV-2 is considered a zoonotic infection (with possible intermediary hosts including bats and pangolins) [1]. The virus is transmitted in respiratory droplets and infects the pulmonary tissue, gaining access through the angiotensin-converting enzyme 2 (ACE2) receptor (expressed on airway epithelial cells, alveolar epithelial cells, vascular endothelial cells and macrophages in the lung) [2–4]. It is estimated that a majority of individuals (up to 80% in some studies) are asymptomatic or mildly symptomatic (with predominantly upper respiratory tract infection symptoms), ~10% are moderately to severely symptomatic and ~10% require admission [5, 6]. The mortality rates of the virus have varied geographically between approximately 2 and approximately 16% although these data may be skewed by incomplete reporting and case finding. It has become apparent that certain individuals are more likely to suffer from severe disease – specifically patients with severe underlying diseases (especially morbid obesity and

P. C. Guest (ed.), *Clinical, Biological and Molecular Aspects of COVID-19*, Advances in Experimental Medicine and Biology 1321, https://doi.org/10.1007/978-3-030-59261-5_11

type 2 diabetes mellitus) and patients over the age of 60 years [7–10].

11.2 The Inflammatory Response

Inflammation is the body's response to injury and tissue damage which could result from trauma or infection. It is characterised by the secretion of inflammatory mediators and activation by innate and adaptive immune effector cells which are targeted to resolve infection and to initiate wound repair and healing [11, 12]. The inflammatory response to infections aims to restrict the spread of the infectious agent through the body and to activate cells to eliminate the agent. Infection is initiated when the mucosal or skin surface is breached, and pattern recognition receptors on cells recognise specific pathogen-associated molecular patterns (including lipopolysaccharide and peptidoglycan or non-mammalian nucleic acid signals). Cells are stimulated to secrete intercellular cytokines which have an autocrine, paracrine and endocrine effect including the type 1 interferons, interleukin (IL)-1, IL-6 and tumour necrosis factor-alpha (TNF-α). Levels of IL-1, IL-6 and TNF-α are often elevated in viral infections and exert pleiotropic effects. These include:

- Increased apoptosis of endothelial cells with a shift to a procoagulant state [11, 13–15]
- Activation of natural killer cells, B cells and T cells [13, 15]
- Tissue cytotoxicity with effects on glucose metabolism, cardiac function and lung function

Damaged cells also produce arachidonic acid mediators of inflammation (also called eicosanoids) [16]. These are small lipid molecules derived from the action of phospholipase-A2 on the phospholipid cell membrane bilayer and include prostaglandins, prostacyclins and leukotrienes [17, 18]. These have slightly different effects which, in some cases, are tissue dependent but include smooth muscle constriction, increased coagulation and platelet activation and increased endothelial permeability with leukocyte recruitment [17].

Cells which play an acute role in inflammation include phagocytes, innate-like lymphocytes and, in more chronic processes, conventional T and B lymphocytes. Neutrophils, cells of the acute inflammatory process, are essential in the initial phagocytosis of infectious agents [19]. Neutrophils are responsible for oxygen-dependent and independent killing of pathogens and also extrude their cellular DNA as neutrophil extracellular traps (NETs). In overwhelming infection, neutrophils die and form pus. Monocyte (or macrophages in the tissue resident form) and myeloid dendritic cells are responsible for phagocytosing pathogens and presenting peptide antigen to CD4+ T cells triggering an adaptive response. The innate-like lymphoid cells have a number of functions including killing virally infected cells (natural killer cells), recognising non-peptide antigen (natural killer T cells) and producing thymic-independent B-cell responses (B1-B cells).

The conclusion of an inflammatory response should be a compensatory anti-inflammatory response characterised by secretion of cytokines like transforming growth factor beta (TGF-β). This allows for wound healing and proliferation of tissues with accompanying angiogenesis. Activated lymphocytes undergo apoptosis, leaving a small residual population of memory cells who can then respond to any future insults [16, 20].

11.3 The Role of Inflammation in SARS-CoV-2-Associated Morbidity and Mortality

From early on in the pandemic, it was recognised that inflammation plays a role in SARS-CoV-2-associated severe disease which is characterised by release of high levels of pro-inflammatory cytokines and inflammatory mediators described as a cytokine storm (Table 11.1) [11, 21–23]. This has been compared to similar conditions noted in, for example, the use of chimeric-antigen receptor T cells [24] and processes of abnormal

Table 11.1 Soluble inflammatory mediators with potential pathogenic effects in SARS-CoV-2-associated disease

Inflammatory mediator	Source (s) of secretion in COVID-19	Potential benefit in COVID infection resolution	Potential pathogenic role in COVID-19 disease	References
Interleukin 1β	Primarily macrophages	1. Promotes T-cell proliferation and maturation 2. Recruitment of monocytes and lymphocytes	1. Pyrogen 2. Cardiac hyperstimulation and arrhythmias 3. Stimulation of hepatic acute-phase response 4. Stimulation of pain and CNS receptors 5. Macrophage activation syndrome	[14, 28, 30–32]
Interleukin 4 and 5 (T-helper 2 cytokines)	T cells and B cells	1. Promotes antibody formation, B-cell maturation, proliferation and class switch 2. Recruits basophils and eosinophils	1. Eosinophil recruitment with bronchoconstriction and dyspnoea 2. Viral exacerbation of allergic lung disease 3. Upper respiratory tract symptoms including rhinorrhoea 4. May predispose to more severe disease in certain groups, e.g., in pregnant patients	[33–36]
Interleukin 6	Alveolar macrophages, stromal cells, T cells, monocyte/macrophages and endothelial cells	1. Recruitment of monocytes and lymphocytes 2. Promotes CD4+ T-cell maturation 3. Plasma cell differentiation	1. Increased arterial and venous thrombosis with endothelial cell activation and dysfunction 2. Macrophage activation syndrome 3. Stimulation of release of toll-like receptor agonists with tissue damage in lung 4. Prolongs neutrophil half-life in lungs 5. Reduced numbers of T cells 6. Stimulation of hepatic acute-phase response 7. Stimulation of chemotaxis 8. Control of metabolism and cachexia 9. Induction of secretion of other cytokines 10. Stimulation of pain and CNS receptors 11. Reduced myocardial contractility	[13, 25, 29, 37–43]

(continued)

Table 11.1 (continued)

Inflammatory mediator	Source (s) of secretion in COVID-19	Potential benefit in COVID infection resolution	Potential pathogenic role in COVID-19 disease	References
Interleukin 8	Chemokine released by a number of cells including neutrophils	1. Phagocyte activation	1. Granulocyte chemotaxis 2. Production of neutrophil extracellular traps 3. Phagocyte activation	[30, 44]
Interleukin 17	T-helper 17 cells	1. Promotes antiviral immune response 2. Enhances mucosal barrier function 3. Promotes endothelial repair	1. Increased arterial and venous thrombosis with endothelial cell activation and dysfunction 2. Reduced type 1 interferon production with viral persistence 3. Stimulates interleukin 8 production with neutrophil recruitment 4. Stimulates triggering receptor expressed on myeloid cells (TREM-1) with pro-inflammatory cytokine production 5. Neutrophil extracellular trap formation	[45–50]
Tumour necrosis factor (TNF)-α	Damaged epithelial cells, granulocytes, monocytes and other leukocytes	1. Recruitment of monocytes and lymphocytes 2. T-cell maturation and activation	1. Increased arterial and venous thrombosis with endothelial cell activation and dysfunction 2. Stimulation of secretion of pro-inflammatory cytokines including interleukin 6 3. Stimulation of production of pro-inflammatory arachidonic acid mediators 4. Stimulation of liver production of acute-phase response 5. Quantitative and qualitative effects on sleep 6. Stimulation of pain receptors 7. Suppression of appetite 8. Chemotaxis of leukocytes including neutrophils and monocytes 9. Increased endothelial permeability with capillary leak 10. Cell damage and apoptosis 11. T-cell overactivation and exhaustion	[11, 12, 34, 51–58]

(continued)

Table 11.1 (continued)

Inflammatory mediator	Source (s) of secretion in COVID-19	Potential benefit in COVID infection resolution	Potential pathogenic role in COVID-19 disease	References
Pro-inflammatory arachidonic acid pathway mediators/ eicosanoids including prostaglandins	Damaged epithelial cells, monocytes, macrophages and neutrophils	1. Upregulates antiviral immune responses	1. Promote acute-phase response	[16–18, 59]
		2. Disruption of viral envelop integrity	2. Stimulate pain receptors	
		3. Promote tissue healing	3. Stimulate fever response	
			4. Stimulate secretion of pro-inflammatory cytokines including IL-6 and TNF-α	
			5. Activate platelets with increased arterial and venous thrombosis	
			6. Bronchoconstriction	
			7. Increased endothelial permeability and dysfunction	

inflammation seen in autoimmune pathology [22]. It has also been described in other coronavirus outbreaks including the SARS-CoV and Middle East respiratory syndrome (MERS-CoV) epidemics [25]. Potential risk factors for developing the increased levels of pro-inflammatory mediators include rapid viral replication and delayed viral clearance (with reduced or delayed interferon responses and T-cell lymphopaenia) [15, 26, 27]. Of note, levels of type 1 interferons (interferon α and β) which act both as autocrine and paracrine viral restriction cytokines are often reduced especially in patients requiring admission for severe SARS-CoV-2 disease.

Elevated levels of cytokines are also associated with abnormal activation of immune effector cells including macrophage activation syndrome (also known as secondary lymphohistiocytosis) [28]. The presence of these pro-inflammatory mediators appears to correlate with increased pulmonary tissue damage with features of acute respiratory distress syndrome: proteinaceous exudate, hyperplasia of type 2 pneumocytes and a characteristic 'ground-glass' appearance on chest radiography [29]. If untreated, the cytokine release syndrome seen in severe SARS-CoV-2 disease is associated with multiorgan (and particularly pulmonary) damage and high mortality [11].

In addition to the elevated pro-inflammatory cytokines produced during severe disease, there are disruptions of the cellular immune system as well. Monocyte/macrophages express the ACE-2 receptor and are, therefore, permissive for SARS-CoV-2 infection [52]. These cells show abnormal levels of activation and are responsible for secretion of additional pro-inflammatory cytokines [52]. Neutrophils also proliferate and track to the lungs [60]. In the lungs, there is evidence that neutrophils produce NETs. NETs are pro-inflammatory and prothrombotic [19, 60–66]. Other granulocyte subclasses (eosinophils, basophils and mast cells) appear to be elevated primarily with recovery rather than acute infection and may be important in controlling the pro-inflammatory response. Eosinopaenia has been described in acute severe disease, and rises in eosinophil counts are seen with enhanced secretion of the antiviral interferon-γ [67–69]. Basophils are hypothesised to help to coordinate an effective antibody response to the infection [67].

Lymphopaenia remains a marker for severe disease. Reduced numbers of NK cells as well as T cells and B cells are documented in patients who have required admission for disease, and predictive models combining lymphopaenia and cytokine enumeration have been developed for prognostication [52, 60, 70–72]. In addition to quantitative changes, functional changes have been described with increased expression of markers of cellular exhaustion (like FAS) particularly for CD4+ T cells [34, 50, 73, 74]. Morphological changes seen in infection include increased numbers of plasmacytoid B cells and atypical lymphocytes [75, 76].

11.4 Role of Suppressing Inflammation in SARS-CoV-2

11.4.1 Targeting Viral Entry

One potential mechanism to suppress inflammation is to target the viral entry process through the ACE2 receptor and the serine protease, TMPRSS2, which is a cofactor required for host entry [77]. Loss of pulmonary ACE2 function is associated with acute lung injury, and the SARS-CoV-2 infection downregulates ACE2 expression in the lung [78].

Drugs that target the ACE2 receptor or TMPRSS2 have been proposed as possible therapeutic options. One such drug is baricitinib, a Janus kinase pathway (JAK-STAT) inhibitor, that has been approved for the treatment of rheumatoid arthritis [79]. Concern has been expressed because downregulation of the JAK-STAT pathway may result in suppression of viral killing [80, 81].

Another potential strategy is the use of human recombinant soluble ACE2 (hsACE2) to inhibit viral entry. HsACE2 has been shown to be well tolerated in healthy adults [82]. A phase 2 clinical trial of hsACE2 in intensive care unit (ICU) patients with acute respiratory distress syndrome was, however, terminated following a number of adverse events [83]. A recent in vitro study has shown that hsACE2 blocks viral entry; however, at this time, the authors are not aware of any trials using it for SARS-CoV-2 [84].

11.4.2 Targeting Inflammation

As a cytokine storm is common in critically ill patients and leads to clinical deterioration, a number of anti-inflammatory therapies have been used to reduce inflammation (Table 11.2). A preliminary report of the RECOVERY trial, a randomised, controlled, open-label, adaptive, platform trial comparing a range of possible treatments with standard of care in patients hospitalised with COVID-19, showed that dexamethasone reduced 28-day mortality among those receiving invasive mechanical ventilation or oxygen at randomisation [85].

Table 11.2 Anti-inflammatory agents utilised in severe SARS-CoV-2 infection

Drug	Effect	References
Tocilizumab	IL-6 monoclonal antibody	[87–90]
Gimsilumab	Inhibition of GM-CSF	[94]
Lenzilumab		
Namilumab		
Sagromostin	GM-CSF	[93]
Cytosorb®	Adsorb cytokines	[104]
Thalidomide	Inhibits TNFα	[101]
Chloroquine or hydroxychloroquine	Uncertain	[98–100]
Colchicine	Inhibits NLRP3 inflammasome	[103, 105]

In the pathogenesis of severe COVID-19, the cytokine IL-6 seems to play a dominant role. Plasma levels of IL-6 are higher than usually seen in severe (bacterial) sepsis. In addition, increased IL-6 levels are a strong predictor of mortality, and IL-6 levels were found to be related to more severe lung injury [41]. Tocilizumab (TCZ) is a recombinant human IL-6 monoclonal antibody, which specifically binds to soluble and membrane-bound IL-6 receptors (IL-6R), thus blocking IL-6 signalling and its mediated inflammatory response. TCZ was initially used in rheumatoid arthritis and was later approved for the treatment of cytokine storm following cancer treatment [86]. Preliminary studies suggest that TCZ may be beneficial with reduced mortality and improved laboratory parameters in patients with SARS-CoV-2 who received the drug [87–90].

Anakinra is a recombinant IL-1 receptor antagonist that may be useful in the cytokine storm. Currently, anakinra is used in the treatment of rheumatoid arthritis and in neonatal-onset multisystem inflammatory disease [91].

Granulocyte-macrophage colony-stimulating factor (GM-CSF) is a myelopoietic growth factor and pro-inflammatory cytokine which plays a key role in the maintenance of lung macrophage homeostasis and inflammation in the lung [92]. Both the administration and the inhibition of GM-CSF are currently under investigation in SARS-CoV-2 clinical trials [93–97].

In an attempt to reduce mortality, several commonly used drugs have been tested to reduce inflammation. Chloroquine (CQ) and hydroxychloroquine (HCQ) are 4-aminoquinoline derivatives that are approved by the US Food and Drug Administration (FDA) for the treatment of malaria, systemic lupus erythematosus and rheumatoid arthritis (RA). HCQ has a better safety profile than CQ. The exact mechanism of action of HCQ and effects on the immune system are largely unknown. It is thought that they may have some antiviral effect by blocking acidification in endosomes, lysosomes and Golgi bodies where they inhibit proteases and thereby viral release. Currently, there is limited evidence of efficacy of HCQ in patients with SARS-CoV-2 infection, and its use is associated with adverse events [98–100].

Thalidomide, which is an anti-inflammatory agent used to treat autoimmune disorders, was effective in the cytokine surge in a single patient in Wenzhou City, China [101]. Currently, there are phase 2 trials investigating its use [34, 102].

Colchicine, used in the treatment of gout and acute pericarditis, is undergoing clinical trials for its anti-inflammatory effects [103]. It is a non-specific inhibitor of the NOD-, LRR- and pyrin domain-containing protein 3 (NLRP3) inflammasome.

11.5 Diagnostic Application of Measuring Inflammation in SARS-CoV-2

The gold standard for the diagnosis of SARS-CoV-2 is nucleic acid testing. Serological assays for the detection of antibodies have the potential to be used in certain situations; however, as antibody responses take a few days to weeks to develop, their utility during the acute infection is limited [106].

Given the rapid rise in SARS-CoV-2 infections and the shortage of ICU beds, it is necessary to triage patients early. The evidence from the literature suggests that a number of laboratory markers may indicate more severe disease.

Infected patients show increased white cell counts and decreased lymphocytes counts. The neutrophil lymphocyte ratio (NLR), a measure of the inflammatory response, is easily measured. Patients with severe SARS-CoV-2 have higher NLR than non-severe patients, and a raised NLR can be an indicator of disease progression and increased mortality [72, 107]. Both neutrophil and lymphocyte counts are inexpensive, routinely requested tests, and the utility of which should be investigated in resource-poor countries.

Procalcitonin (PCT), a marker of severe inflammation, is produced by the parafollicular cells of the thyroid gland as well as by neuroendocrine cells of the lungs and intestine. It rises mainly in response to bacterial sepsis and can be used to monitor response to antibiotic therapy [108]. A review of risk factors for severe SARS-CoV-2 showed that a PCT value greater than 0.5 ng/mL was associated with increased severity and mortality [108]. PCT levels are higher in severely ill patients compared to moderately ill patients and highest in patients who subsequently die [109].

C-reactive protein (CRP) is produced by the liver in response to inflammation. It is an acute-phase protein induced by IL-6, and levels tend to track with the degree of inflammation. In patients with SARS-CoV-2, significantly higher CRP levels are seen in severe cases [110, 111]. Ferritin, another acute phase protein, is raised in SARS-CoV-2 patients, and albumin, a negative acute-phase reactant, drops with infection. Elevated levels of IL-6 and IL-10 are associated with increased disease severity along with other cytokines and may be produced earlier than acute-phase reactants [28]. IL-6 levels can be used to monitor response to TCZ therapy in clinical applications where this drug is available.

The exponential increase in SARS-CoV-2 patients has resulted in an increase in attempts to identify patients at risk of severe disease and to identify better therapeutics. This has resulted in repurposing old drugs such as attempts to use HCQ and CQ for infected patients, as well as the search for newer therapies. Unlike during the Spanish flu in 1918, we have better diagnostic platforms and increased computing ability which may be of use. Two such examples are the use of machine learning algorithms using clinical symptoms and signs along with CRP and haematological changes to predict the need for ICU admission and the use of proteomic signatures as disease classifiers [112].

11.6 Conclusions

The inflammatory response to SARS-CoV-2 appears to contribute significantly to morbidity and mortality. This includes production of high levels of pro-inflammatory cytokines and lipid mediators as well as derangements in cellular immunity. Some of these changes show promise as potential therapeutic targets, possible diagnostic or prognostic biomarkers or both. With additional clinical trials, these will be further evaluated for potential use in this evolving pandemic.

References

1. Zhang T, Wu Q, Zhang Z (2020) Probable pangolin origin of SARS-CoV-2 associated with the COVID-19 outbreak. Curr Biol 30(7):1346–1351.e2
2. Ahmad T, Khan M, Haroon, Musa TH, Nasir S, Hui J et al (2020) COVID-19: zoonotic aspects. Travel Med Infect Dis Feb 27:101607. doi: https://doi.org/10.1016/j.tmaid.2020.101607. Online ahead of print
3. Arshad Ali S, Baloch M, Ahmed N, Arshad Ali A, Iqbal A (2020) The outbreak of Coronavirus Disease 2019 (COVID-19)-an emerging global health threat. J Infect Public Health 13(4):644–646
4. Boldog P, Tekeli T, Vizi Z, Denes A, Bartha FA, Rost G (2020) Risk assessment of novel coronavirus COVID-19 outbreaks outside China. J Clin Med 9(2):571. https://doi.org/10.3390/jcm9020571
5. Al-Tawfiq JA (2020) Asymptomatic coronavirus infection: MERS-CoV and SARS-CoV-2 (COVID-19). Travel Med Infect Dis 35:101608. https://doi.org/10.1016/j.tmaid.2020.101608
6. Lai CC, Shih TP, Ko WC, Tang HJ, Hsueh PR (2020) Severe acute respiratory syndrome coronavirus 2 (SARS-CoV-2) and coronavirus disease-2019 (COVID-19): the epidemic and the challenges. Int J Antimicrob Agents 55(3):105924. https://doi.org/10.1016/j.ijantimicag.2020.105924
7. Li B, Yang J, Zhao F, Zhi L, Wang X, Liu L et al (2020) Prevalence and impact of cardiovascular metabolic diseases on COVID-19 in China. Clin Res Cardiol 109(5):531–538
8. Wang T, Du Z, Zhu F, Cao Z, An Y, Gao Y et al (2020) Comorbidities and multi-organ injuries in the treatment of COVID-19. Lancet 395(10228):e52. https://doi.org/10.1016/S0140-6736(20)30558-4
9. Yang J, Zheng Y, Gou X, Pu K, Chen Z, Guo Q et al (2020) Prevalence of comorbidities in the novel Wuhan coronavirus (COVID-19) infection: a systematic review and meta-analysis. Int J Infect Dis 94:91–95
10. Zhou F, Yu T, Du R, Fan G, Liu Y, Liu Z et al (2020) Clinical course and risk factors for mortality of adult inpatients with COVID-19 in Wuhan, China: a retrospective cohort study. Lancet 395(10229):1054–1062
11. Behrens EM, Koretzky GA (2017) Review: cytokine storm syndrome: looking toward the precision medicine era. Arthritis Rheum 69(6):1135–1143
12. Huang Q, Wu X, Zheng X, Luo S, Xu S, Weng J (2020) Targeting inflammation and cytokine storm in COVID-19. Pharmacol Res June 27;159:105051. https://doi.org/10.1016/j.phrs.2020.105051. Online ahead of print
13. Schmidt-Arras D, Rose-John S (2016) IL-6 pathway in the liver: from physiopathology to therapy. J Hepatol 64(6):1403–1415
14. Kaneko N, Kurata M, Yamamoto T, Morikawa S, Masumoto J (2019) The role of interleukin-1 in general pathology. Inflamm Regen 39:12. https://doi.org/10.1186/s41232-019-0101-5
15. Channappanavar R, Perlman S (2017) Pathogenic human coronavirus infections: causes and consequences of cytokine storm and immunopathology. Semin Immunopathol 39(5):529–539
16. Panigrahy D, Gilligan MM, Huang S, Gartung A, Cortes-Puch I, Sime PJ et al (2020) Inflammation resolution: a dual-pronged approach to averting cytokine storms in COVID-19? Cancer Metastasis Rev 39(2):337–340
17. Hammock BD, Wang W, Gilligan MM, Panigrahy D (2020) Eicosanoids: the overlooked storm in COVID-19? Am J Pathol July 7;S0002-9440(20)30332-1. https://doi.org/10.1016/j.ajpath.2020.06.010. Online ahead of print
18. Hoxha M (2020) What about COVID-19 and arachidonic acid pathway? Eur J Clin Pharmacol June 25;1–4. https://doi.org/10.1007/s00228-020-02941-w. Online ahead of print

19. Zuo Y, Yalavarthi S, Shi H, Gockman K, Zuo M, Madison JA et al (2020) Neutrophil extracellular traps in COVID-19. JCI Insight 5(11):e138999. https://doi.org/10.1172/jci.insight.138999
20. Gautier-Vargas G, Baldacini C, Benotmane I, Keller N, Perrin P, Moulin B et al (2020) Rapid resolution of cytokine release syndrome and favorable clinical course of severe COVID-19 in a kidney transplant recipient treated with tocilizumab. Kidney Int 98(2):508–509
21. Canna SW, Behrens EM (2012) Making sense of the cytokine storm: a conceptual framework for understanding, diagnosing, and treating hemophagocytic syndromes. Pediatr Clin N Am 59(2):329–344
22. Weaver LK, Behrens EM (2017) Weathering the storm: improving therapeutic interventions for cytokine storm syndromes by targeting disease pathogenesis. Curr Treatm Opt Rheumatol 3(1):33–48
23. Teoh N, Farrell G (2020) Statins as early therapy to mitigate COVID-19 (SARS-CoV-2)-associated ARDS and cytokine storm syndrome – time is of the essence. J Clin Transl Res 5(5):227–229
24. Hoiland RL, Stukas S, Cooper J, Thiara S, Chen LYC, Biggs CM et al (2020) Amelioration of COVID-19 related cytokine storm syndrome: parallels to chimeric antigen receptor-T cell cytokine release syndrome. Br J Haematol June 25;https://doi.org/10.1111/bjh.16961. Online ahead of print
25. Liu Y, Gayle AA, Wilder-Smith A, Rocklov J (2020) The reproductive number of COVID-19 is higher compared to SARS coronavirus. J Travel Med 27(2):taaa021. https://doi.org/10.1093/jtm/taaa021
26. Blanco-Melo D, Nilsson-Payant BE, Liu WC, Uhl S, Hoagland D, Moller R et al (2020) Imbalanced host response to SARS-CoV-2 drives development of COVID-19. Cell 181(5):1036–1045.e9
27. Hadjadj J, Yatim N, Barnabei L, Corneau A, Boussier J, Smith N et al (2020) Impaired type I interferon activity and inflammatory responses in severe COVID-19 patients. Science July 13;eabc6027. https://doi.org/10.1126/science.abc6027. Online ahead of print
28. McGonagle D, Sharif K, O'Regan A, Bridgewood C (2020) The role of cytokines including Interleukin-6 in COVID-19 induced pneumonia and macrophage activation syndrome-like disease. Autoimmun Rev 19(6):102537. https://doi.org/10.1016/j.autrev.2020.102537
29. Palanques-Pastor T, Lopez-Briz E, Poveda Andres JL (2020) Involvement of interleukin 6 in SARS-CoV-2 infection: siltuximab as a therapeutic option against COVID-19. Eur J Hosp Pharm June 4;ejhpharm-2020-002322. https://doi.org/10.1136/ejhpharm-2020-002322. Online ahead of print
30. Wang J, Jiang M, Chen X, Montaner LJ (2020) Cytokine storm and leukocyte changes in mild versus severe SARS-CoV-2 infection: review of 3939 COVID-19 patients in China and emerging pathogenesis and therapy concepts. J Leukoc Biol 108(1):17–41
31. Imazio M, Andreis A, De Ferrari GM (2020) COVID-19 and myocardial injury: is there a role for interleukin-1 inhibition? J Cardiovasc Med (Hagerstown) 21(7):465–466
32. Ni M, Tian FB, Xiang DD, Yu B (2020) Characteristics of inflammatory factors and lymphocyte subsets in patients with severe COVID-19. J Med Virol May 29;https://doi.org/10.1002/jmv.26070. Online ahead of print
33. Azkur AK, Akdis M, Azkur D, Sokolowska M, van de Veen W, Bruggen MC et al (2020) Immune response to SARS-CoV-2 and mechanisms of immunopathological changes in COVID-19. Allergy 75(7):1564–1581
34. Qin C, Zhou L, Hu Z, Zhang S, Yang S, Tao Y et al (2020) Dysregulation of immune response in patients with COVID-19 in Wuhan, China. Clin Infect Dis March 12;ciaa248. https://doi.org/10.1093/cid/ciaa248. Online ahead of print
35. Roncati L, Nasillo V, Lusenti B, Riva G (2020) Signals of Th2 immune response from COVID-19 patients requiring intensive care. Ann Hematol 99(6):1419–1420
36. Wei LL, Wang WJ, Chen DX, Xu B (2020) Dysregulation of the immune response affects the outcome of critical COVID-19 patients. J Med Virol June 16;https://doi.org/10.1002/jmv.26181. Online ahead of print
37. Mohamud AY, Griffith B, Rehman M, Miller D, Chebl A, Patel SC et al (2020) Intraluminal carotid artery thrombus in COVID-19: another danger of cytokine storm? AJNR Am J Neuroradiol July 2. https://doi.org/10.3174/ajnr.A6674. Online ahead of print
38. Benzakoun J, Hmeydia G, Delabarde T, Hamza L, Meder JF, Ludes B et al (2020) Excess out-of-hospital deaths during COVID-19 outbreak: evidence of pulmonary embolism as a main determinant. Eur J Heart Fail May 28;https://doi.org/10.1002/ejhf.1916. Online ahead of print
39. Garcia-Olive I, Sintes H, Radua J, Abad Capa J, Rosell A (2020) D-dimer in patients infected with COVID-19 and suspected pulmonary embolism. Respir Med May 13;169:106023. doi: https://doi.org/10.1016/j.rmed.2020.106023. Online ahead of print
40. Grifoni E, Valoriani A, Cei F, Lamanna R, Gelli AMG, Ciambotti B et al (2020) Interleukin-6 as prognosticator in patients with COVID-19. J Infect Jun 8;S0163-4453(20)30388-1. https://doi.org/10.1016/j.jinf.2020.06.008. Online ahead of print
41. Magro G (2020) SARS-CoV-2 and COVID-19: is interleukin-6 (IL-6) the 'culprit lesion' of ARDS onset? What is there besides Tocilizumab? SGP130Fc. Cytokine X May 14;2(2):100029. https://doi.org/10.1016/j.cytox.2020.100029. Online ahead of print
42. Smetana K Jr, Brabek J (2020) Role of Interleukin-6 in lung complications in patients with

COVID-19: therapeutic implications. In Vivo 34(3 Suppl):1589–1592
43. Chen D, Yang H, Cao Y, Cheng W, Duan T, Fan C et al (2020) Expert consensus for managing pregnant women and neonates born to mothers with suspected or confirmed novel coronavirus (COVID-19) infection. Int J Gynaecol Obstet 149(2):130–136
44. Cui Y, Tian M, Huang D, Wang X, Huang Y, Fan L et al (2020) A 55-day-old female infant infected with COVID 19: presenting with pneumonia, liver injury, and heart damage. J Infect Dis 221(11):1775–1781
45. Wu D, Yang XO (2020) TH17 responses in cytokine storm of COVID-19: An emerging target of JAK2 inhibitor Fedratinib. J Microbiol Immunol Infect 53(3):368–370
46. Pons S, Fodil S, Azoulay E, Zafrani L (2020) The vascular endothelium: the cornerstone of organ dysfunction in severe SARS-CoV-2 infection. Crit Care 24(1):353. https://doi.org/10.1186/s13054-020-03062-7
47. Casillo GM, Mansour AA, Raucci F, Saviano A, Mascolo N, Iqbal AJ et al (2020) Could IL-17 represent a new therapeutic target for the treatment and/or management of COVID-19-related respiratory syndrome? Pharmacol Res 156:104791. https://doi.org/10.1016/j.phrs.2020.104791
48. Mendoza VMM (2020) Interleukin-17: A potential therapeutic target in COVID-19. J Infect. June 3;S0163-4453(20)30350–9. https://doi.org/10.1016/j.jinf.2020.05.072. Online ahead of print
49. Wiche Salinas TR, Zheng B, Routy JP, Ancuta P (2020) Targeting the interleukin-17 pathway to prevent acute respiratory distress syndrome associated with SARS-CoV-2 infection. Respirology. June 17;https://doi.org/10.1111/resp.13875. Online ahead of print
50. De Biasi S, Meschiari M, Gibellini L, Bellinazzi C, Borella R, Fidanza L et al (2020) Marked T cell activation, senescence, exhaustion and skewing towards TH17 in patients with COVID-19 pneumonia. Nat Commun 11(1):3434. https://doi.org/10.1038/s41467-020-17292-4
51. Kempuraj D, Selvakumar GP, Ahmed ME, Raikwar SP, Thangavel R, Khan A et al (2020) COVID-19, mast cells, cytokine storm, psychological stress, and neuroinflammation. Neuroscientist July 18;1073858420941476. https://doi.org/10.1177/1073858420941476. Online ahead of print
52. Chen H, Guo J, Wang C, Luo F, Yu X, Zhang W et al (2020) Clinical characteristics and intrauterine vertical transmission potential of COVID-19 infection in nine pregnant women: a retrospective review of medical records. Lancet 395(10226):809–815
53. Mehta P, McAuley DF, Brown M, Sanchez E, Tattersall RS, Manson JJ et al (2020) COVID-19: consider cytokine storm syndromes and immunosuppression. Lancet 395(10229):1033–1034
54. Miao Y, Fan L, Li JY (2020) Potential treatments for COVID-19 related cytokine storm – beyond corticosteroids. Front Immunol 11:1445. https://doi.org/10.3389/fimmu.2020.01445
55. Mangalmurti N, Hunter CA (2020) Cytokine storms: understanding COVID-19. Immunity 53(1):19–25
56. Sinha P, Matthay MA, Calfee CS (2020) Is a "Cytokine Storm" relevant to COVID-19? JAMA Intern Med June 30. https://doi.org/10.1001/jamainternmed.2020.3313. Online ahead of print
57. Pearce L, Davidson SM, Yellon DM (2020) The cytokine storm of COVID-19: a spotlight on prevention and protection. Expert Opin Ther Targets June 27;1–8. https://doi.org/10.1080/14728222.2020.1783243. Online ahead of print
58. Das UN (2020) Can bioactive lipids inactivate coronavirus (COVID-19)? Arch Med Res 51(3):282–286
59. Liu J, Liu Y, Xiang P, Pu L, Xiong H, Li C et al (2020) Neutrophil-to-lymphocyte ratio predicts critical illness patients with 2019 coronavirus disease in the early stage. J Transl Med 18(1):206. https://doi.org/10.1186/s12967-020-02374-0
60. Middleton EA, He XY, Denorme F, Campbell RA, Ng D, Salvatore SP et al (2020) Neutrophil Extracellular Traps (NETs) contribute to immunothrombosis in COVID-19 acute respiratory distress syndrome. Blood June 29;blood.2020007008. https://doi.org/10.1182/blood.2020007008. Online ahead of print
61. Zuo Y, Yalavarthi S, Shi H, Gockman K, Zuo M, Madison JA et al (2020) Neutrophil extracellular traps (NETs) as markers of disease severity in COVID-19. medRxiv. https://doi.org/10.1101/2020.04.09.20059626
62. Zuo Y, Zuo M, Yalavarthi S, Gockman K, Madison JA, Shi H et al (2020) Neutrophil extracellular traps and thrombosis in COVID-19. medRxiv. https://doi.org/10.1101/2020.04.30.20086736
63. Tomar B, Anders HJ, Desai J, Mulay SR (2020) Neutrophils and neutrophil extracellular traps drive Necroinflammation in COVID-19. Cell 9(6). https://doi.org/10.3390/cells9061383
64. Mozzini C, Girelli D (2020) The role of neutrophil extracellular traps in Covid-19: only an hypothesis or a potential new field of research? Thromb Res 191:26–27
65. Barnes BJ, Adrover JM, Baxter-Stoltzfus A, Borczuk A, Cools-Lartigue J, Crawford JM et al (2020) Targeting potential drivers of COVID-19: neutrophil extracellular traps. J Exp Med 217(6). https://doi.org/10.1084/jem.20200652
66. Tabachnikova A, Chen ST (2020) Roles for eosinophils and basophils in COVID-19? Nat Rev Immunol Jun 22:1. https://doi.org/10.1038/s41577-020-0379-1. Online ahead of print
67. Rodriguez L, Pekkarinen P, Lakshmikanth T, Tan Z, Consiglio C, Pou C et al (2020) Systems-level immunomonitoring from acute to recovery phase of severe COVID-19. MedRxiv. https://doi.org/10.1101/2020.06.03.20121582

68. Sun DW, Zhang D, Tian RH, Li Y, Wang YS, Cao J et al (2020) The underlying changes and predicting role of peripheral blood inflammatory cells in severe COVID-19 patients: a sentinel? Clin Chim Acta 508:122–129
69. Luo Y, Mao L, Yuan X, Xue Y, Lin Q, Tang G et al (2020) Prediction model based on the combination of cytokines and lymphocyte subsets for prognosis of SARS-CoV-2 infection. J Clin Immunol July 13;1–10. https://doi.org/10.1007/s10875-020-00821-7. Online ahead of print
70. Quartuccio L, Sonaglia A, McGonagle D, Fabris M, Peghin M, Pecori D et al (2020) Profiling COVID-19 pneumonia progressing into the cytokine storm syndrome: results from a single Italian Centre study on tocilizumab versus standard of care. J Clin Virol May 15;129:104444. https://doi.org/10.1016/j.jcv.2020.104444. Online ahead of print
71. Chen R, Sang L, Jiang M, Yang Z, Jia N, Fu W et al (2020) Longitudinal hematologic and immunologic variations associated with the progression of COVID-19 patients in China. J Allergy Clin Immunol 146(1):89–100
72. Deng Z, Zhang M, Zhu T, Liu Z, Liu Z, Xiang R et al (2020) Dynamic changes of peripheral blood lymphocytes subsets in adult patients with COVID-19. Int J Infect Dis July 4:S1201-9712(20)30545-2. https://doi.org/10.1016/j.ijid.2020.07.003. Online ahead of print
73. Bellesi S, Metafuni E, Hohaus S, Maiolo E, Marchionni F, D'Innocenzo S et al (2020) Increased CD95 (Fas) and PD-1 expression in peripheral blood T lymphocytes in COVID-19 patients. Br J Haematol July 17. https://doi.org/10.1111/bjh.17034. Online ahead of print
74. Foldes D, Hinton R, Arami S, Bain BJ (2020) Plasmacytoid lymphocytes in SARS-CoV-2 infection (Covid-19). Am J Hematol 95(7):861–862
75. Weinberg SE, Behdad A, Ji P (2020) Atypical lymphocytes in peripheral blood of patients with COVID-19. Br J Haematol 190(1):36–39
76. Jin Y, Yang H, Ji W, Wu W, Chen S, Zhang W et al (2020) Virology, epidemiology, pathogenesis, and control of COVID-19. Viruses 12(4):372. https://doi.org/10.3390/v12040372
77. Kuba K, Imai Y, Rao S, Gao H, Guo F, Guan B et al (2005) A crucial role of angiotensin converting enzyme 2 (ACE2) in SARS coronavirus-induced lung injury. Nat Med 11(8):875–879
78. Richardson P, Griffin I, Tucker C, Smith D, Oechsle O, Phelan A et al (2020) Baricitinib as potential treatment for 2019-nCoV acute respiratory disease. Lancet 395(10223):e30–e31. https://doi.org/10.1016/S0140-6736(20)30304-4
79. Praveen D, Puvvada RC, VA M (2020) Janus kinase inhibitor baricitinib is not an ideal option for management of COVID-19. Int J Antimicrob Agents 55(5):105967. https://doi.org/10.1016/j.ijantimicag.2020.105967
80. Favalli EG, Biggioggero M, Maioli G, Caporali R (2020) Baricitinib for COVID-19: a suitable treatment? Lancet Infect Dis April 3:S1473-3099(20)30262-0. https://doi.org/10.1016/S1473-3099(20)30262-0. Online ahead of print
81. Haschke M, Schuster M, Poglitsch M, Loibner H, Salzberg M, Bruggisser M et al (2013) Pharmacokinetics and pharmacodynamics of recombinant human angiotensin-converting enzyme 2 in healthy human subjects. Clin Pharmacokinet 52(9):783–792
82. Khan A, Benthin C, Zeno B, Albertson TE, Boyd J, Christie JD et al (2017) A pilot clinical trial of recombinant human angiotensin-converting enzyme 2 in acute respiratory distress syndrome. Crit Care 21(1):234. https://doi.org/10.1186/s13054-017-1823-x
83. Monteil V, Kwon H, Prado P, Hagelkrüys A, Wimmer RA, Stahl M et al (2020) Inhibition of SARS-CoV-2 infections in engineered human tissues using clinical-grade soluble human ACE2. Cell 181(4):905–913.e7
84. Horby P, Lim WS, Emberson J, Mafham M, Bell J, Linsell L et al (2020) Effect of Dexamethasone in hospitalized patients with COVID-19: preliminary report. medRxiv. https://doi.org/10.1101/2020.06.22.20137273
85. Kotch C, Barrett D, Teachey DT (2019) Tocilizumab for the treatment of chimeric antigen receptor T cell-induced cytokine release syndrome. Expert Rev Clin Immunol 15(8):813–822
86. Xu X, Han M, Li T, Sun W, Wang D, Fu B et al (2020) Effective treatment of severe COVID-19 patients with tocilizumab. Proc Natl Acad Sci U S A 117(20):10970–10975
87. Luo P, Liu Y, Qiu L, Liu X, Liu D, Li J (2020) Tocilizumab treatment in COVID-19: a single center experience. J Med Virol 92(7):814–818
88. Capra R, De Rossi N, Mattioli F, Romanelli G, Scarpazza C, Sormani MP et al (2020) Impact of low dose tocilizumab on mortality rate in patients with COVID-19 related pneumonia. Eur J Intern Med 76:31–35
89. Morena V, Milazzo L, Oreni L, Bestetti G, Fossali T, Bassoli C et al (2020) Off-label use of tocilizumab for the treatment of SARS-CoV-2 pneumonia in Milan, Italy. Eur J Intern Med 76:36–42
90. Eloseily EM, Weiser P, Crayne CB, Haines H, Mannion ML, Stoll ML et al (2020) Benefit of Anakinra in treating Pediatric secondary Hemophagocytic Lymphohistiocytosis. Arthritis Rheum 72(2):326–334
91. Guilliams M, De Kleer I, Henri S, Post S, Vanhoutte L, De Prijck S et al (2013) Alveolar macrophages develop from fetal monocytes that differentiate into long-lived cells in the first week of life via GM-CSF. J Exp Med 210(10):1977–1992
92. Bosteels C, Maes B, Van Damme K, De Leeuw E, Declercq J, Delporte A et al (2020) Sargramostim to treat patients with acute hypoxic respiratory failure

due to COVID-19 (SARPAC): a structured summary of a study protocol for a randomised controlled trial. Trials 21(1):491. https://doi.org/10.1186/s13063-020-04451-7
93. Temesgen Z, Assi M, Vergidis P, Rizza SA, Bauer PR, Pickering BW et al (2020) First clinical use of Lenzilumab to neutralize GM-CSF in patients with severe COVID-19 pneumonia. medRxiv. https://doi.org/10.1101/2020.06.08.20125369
94. Subramaniam R, Hillberry Z, Chen H, Feng Y, Fletcher K, Neuenschwander P et al (2015) Delivery of GM-CSF to protect against influenza pneumonia. PLoS One 0(4):e0124593. https://doi.org/10.1371/journal.pone.0124593
95. Huang H, Li H, Zhou P, Ju D (2010) Protective effects of recombinant human granulocyte macrophage colony stimulating factor on H1N1 influenza virus-induced pneumonia in mice. Cytokine 51(2):151–157
96. Umstead TM, Hewage EK, Mathewson M, Beaudoin S, Chroneos ZC, Wang M et al (2020) Lower respiratory tract delivery, airway clearance, and preclinical efficacy of inhaled GM-CSF in a postinfluenza pneumococcal pneumonia model. Am J Phys Lung Cell Mol Phys 318(4):L571–L519
97. Tang W, Cao Z, Han M, Wang Z, Chen J, Sun W et al (2020) Hydroxychloroquine in patients with mainly mild to moderate coronavirus disease 2019: open label, randomised controlled trial. BMJ 369:m1849. https://doi.org/10.1136/bmj.m1849
98. Chen J, Liu D, Liu L, Liu P, Xu Q, Xia L et al (2020) A pilot study of hydroxychloroquine in treatment of patients with moderate COVID-19. Zhejiang Da Xue Xue Bao Yi Xue Ban 49(2):215–219
99. Geleris J, Sun Y, Platt J, Zucker J, Baldwin M, Hripcsak G et al (2020) Observational study of Hydroxychloroquine in hospitalized patients with Covid-19. N Engl J Med 382(25):2411–2418
100. Qian GQ, Yang NB, Ding F, Ma AHY, Wang ZY, Shen YF et al (2020) Epidemiologic and clinical characteristics of 91 hospitalized patients with COVID-19 in Zhejiang, China: a retrospective, multi-Centre case series. QJM 113(7):474–481
101. Qiao J (2020) What are the risks of COVID-19 infection in pregnant women? Lancet 395(10226):760–762
102. Deftereos SG, Siasos G, Giannopoulos G, Vrachatis DA, Angelidis C, Giotaki SG et al (2020) The Greek study in the effects of colchicine in COvid-19 complications prevention (GRECCO-19 study): rationale and study design. Hell J Cardiol 61(1):42–25
103. Stockmann H, Keller T, Büttner S, Jörres A, Kindgen-Milles D, Kunz JV et al (2020) CytoResc – "CytoSorb" rescue for critically ill patients undergoing the COVID-19 cytokine storm: a structured summary of a study protocol for a randomized controlled trial. Trials 21(1):577. https://doi.org/10.1186/s13063-020-04501-0
104. Corral P, Corral G, Diaz R (2020) Colchicine and COVID-19. J Clin Pharmacol 60(8):978. https://doi.org/10.1002/jcph.1684
105. Guo L, Ren L, Yang S, Xiao M, Chang D, Yang F et al (2020) Profiling early humoral response to diagnose novel coronavirus disease (COVID-19). Clin Infect Dis March 21;ciaa310. https://doi.org/10.1093/cid/ciaa310. Online ahead of print
106. Guo YR, Cao QD, Hong ZS, Tan YY, Chen SD, Jin HJ et al (2020) The origin, transmission and clinical therapies on coronavirus disease 2019 (COVID-19) outbreak – an update on the status. Mil Med Res 7(1):11. https://doi.org/10.1186/s40779-020-00240-0
107. Azzini AM, Dorizzi RM, Sette P, Vecchi M, Coledan I, Righi E et al (2020) A 2020 review on the role of procalcitonin in different clinical settings: an update conducted with the tools of the evidence based laboratory medicine. Ann Transl Med 8(9):610. https://doi.org/10.21037/atm-20-1855
108. Sun Q, Qiu H, Huang M, Yang Y (2020) Lower mortality of COVID-19 by early recognition and intervention: experience from Jiangsu Province. Ann Intensive Care 10(1):33. https://doi.org/10.1186/s13613-020-00650-2
109. Xu G, Yang Y, Du Y, Peng F, Hu P, Wang R et al (2020) Clinical pathway for early diagnosis of COVID-19: updates from experience to evidence-based practice. Clin Rev Allergy Immunol 59(1):89–100
110. Feng X, Li S, Sun Q, Zhu J, Chen B, Xiong M et al (2020) Immune-inflammatory parameters in COVID-19 cases: a systematic review and meta-analysis. Front Med (Lausanne) 7:301. https://doi.org/10.3389/fmed.2020.00301
111. Cheng FY, Joshi H, Tandon P, Freeman R, Reich DL, Mazumdar M et al (2020) Using machine learning to predict ICU transfer in hospitalized COVID-19 patients. J Clin Med 9(6). https://doi.org/10.3390/jcm9061668
112. Xing J, Loureiro J, Patel MT, Mikhailov D, Gusev AI (2020) Evaluation of a novel blood microsampling device for clinical trial sample collection and protein biomarker analysis. Bioanalysis July 20. https://doi.org/10.4155/bio-2020-0063. Online ahead of print

Does SARS-CoV-2 Threaten Male Fertility?

12

Amir Vahedian-Azimi, Leila Karimi, Somayeh Makvandi, Tannaz Jamialahmadi, and Amirhossein Sahebkar

Abstract

In the continuing COVID-19 pandemic, one of the most important concerns in reproductive health is the issue of male fertility of recovered patients. In this study, we discuss the potential mechanisms that justify the possible impact of COVID-19 on male fertility. The main point of entry of SARS-CoV-2 into the host cells appears to be through the viral spike protein which permits entry into cells via the angiotensin-converting enzyme 2 (ACE2 receptor). In human testes, ACE2 is enriched in Sertoli and Leydig cells and spermatogonia. Also, it seems that there is a mild or severe cytokine storm in patients with severe COVID-19, and such changes may affect fertility. It should also be mentioned that the orchitis caused by the SARS-CoV-2 virus may have an important impact on fertility. Prolonged and high fever may lead to changes in testicular temperature and destroy germ cells. In general, there is little

A. Vahedian-Azimi
Trauma research center, Nursing Faculty, Baqiyatallah University of Medical Sciences, Tehran, Iran

L. Karimi
Behavioral Sciences Research Center, Life style institute, Nursing Faculty, Baqiyatallah University of Medical Sciences, Tehran, Iran

S. Makvandi
Department of Midwifery, School of Nursing and Midwifery, Islamic Azad university Ahvaz branch, Ahvaz, Iran
e-mail: somayemakvandi@gmail.com

T. Jamialahmadi
Biotechnology Research Center, Pharmaceutical Technology Institute, Mashhad University of Medical Sciences, Mashhad, Iran

Department of Food Science and Technology, Quchan Branch, Islamic Azad University, Quchan, Iran

Department of Nutrition, Faculty of Medicine, Mashhad University of Medical Sciences, Mashhad, Iran

A. Sahebkar (✉)
Biotechnology Research Center, Pharmaceutical Technology Institute, Mashhad University of Medical Sciences, Mashhad, Iran

Halal Research Center of IRI, FDA, Tehran, Iran

Neurogenic Inflammation Research Center, Mashhad University of Medical Sciences, Mashhad, Iran

Polish Mother's Memorial Hospital Research Institute (PMMHRI), Lodz, Poland
e-mail: sahebkara@mums.ac.ir; amir_saheb2000@yahoo.com

P. C. Guest (ed.), *Clinical, Biological and Molecular Aspects of COVID-19*, Advances in Experimental Medicine and Biology 1321, https://doi.org/10.1007/978-3-030-59261-5_12

evidence for a definite conclusion, but there are facts that suggest that COVID-19 may affect male fertility. It is prudent for men of reproductive age who have recovered from COVID-19 to be evaluated for the presence of the virus in semen and fertility-related items. There is an urgent need to conduct quality studies on, in particular, the long-term effects of COVID-19 on the fertility of recovered males.

Keywords

ACE2 · Angiotensin-converting enzyme 2 · Male fertility · SARS-Cov-2 · Testis

12.1 Introduction

As of June 29, 2020, coronavirus disease 2019 (COVID-19) has been confirmed in 10,021,401 cases worldwide, with 499,913 deaths [1]. The SARS-CoV-2 virus that cause COVID-19 disease first appeared to affect the respiratory system, but later, it was found that it could affect other organs in the body, such as the cardiovascular and urogenital systems, gastrointestinal tract, and brain [2–6].

Recently, attention has been paid to the possible effects of the SARS-CoV-2 on the reproductive system. Scientific evidence suggests that in some viral diseases, there is a potential of spread to the semen in males. Twenty-seven viruses, including the Zika virus, have been identified to date with this capability [7]. The Zika virus was also found to remain in the semen of asymptomatic men for up to 1 year after recovery [8]. Factors such as the reproductive system immune response, alterations of the blood-testis barrier by inflammatory mediators, systemic immunosuppression, and viral stability can affect viral shedding into the semen [7]. However, the possibility that COVID-19 infection affects male fertility in this manner has not been adequately addressed. Also, viruses such as human immunodeficiency virus, mumps virus, hepatitis B and C viruses, Epstein-Barr virus, papillomavirus, and SARS-CoV have been shown to cause viral orchitis, which in turn may threaten male fertility [9, 10]. These findings suggest the need for andrological consultation and evaluation of gonadal function for male patients affected by COVID-19 disease [11].

In this study, we have reviewed the potential mechanisms that justify the possible impact of COVID-19 disease on male fertility and present the evidence that has led to these conclusions.

12.2 The Role of Angiotensin-Converting Enzyme 2

The SARS-CoV-2 virus shares 77% amino acid sequence identity with SARS-CoV, which caused a severe respiratory epidemic in 2002–2004. In addition, the spike proteins have similar three-dimensional structures in the receptor-binding domain. Like most viruses, coronaviruses that cause SARS and COVID-19 bind to host cell receptors, followed by endocytotic entry of the virus into the cells, genome proliferation, exocytosis, and budding. The main point of entry of SARS-Cov-2 into the cell appears to be through the viral spike protein attached to the human ACE2 receptor and the cellular transmembrane serine protease 2 (TMPRSS2) (Fig. 12.1) [12, 13]. Both SARS-CoV-2 and SARS-CoV bind to the ACE2 receptor although SARS-CoV-2 binds with approximately 20-fold greater affinity. This more efficient attachment could explain the more potent effects and interpersonal transmission of the SARS-CoV-2 virus [14, 15].

Both ACE2 and ACE1 are components of the renin-angiotensin system (RAS), a hormonal network that regulates blood pressure as well as fluid balance in the body [16, 17]. Renin is secreted by the juxtaglomerular apparatus in response to renal hypoperfusion and converts angiotensinogen (produced by the liver) into ACE1 which in turn is converted to ACE2 by the ACE enzyme [18]. ACE2 helps to maintain blood pressure due to its role in vasoconstriction [19]. ACE2 also stimulates the secretion of aldosterone from the cortical part of the adrenal gland. Aldosterone increases the absorption of sodium and water from the renal tubules. RAS overactivation leads to high blood pressure due to accumulation of water and sodium and the associated increased intravascular volume [20]. RAS

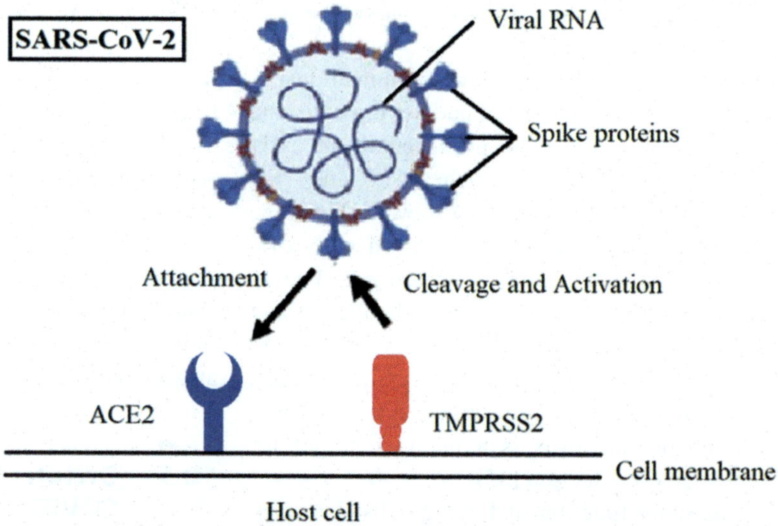

Fig. 12.1 Mechanism of SARS-CoV-2 entry into host cell

also has extra cardiovascular functions in organs such as the pancreas and reproductive system [21, 22]. In reproductive tissues, angiotensinogen has an autocrine/paracrine origin. There is evidence that the RAS plays a role in tubular contractility, spermatogenesis, sperm maturation, capacitation, acrosomal exocytosis, and fertilization [23].

ACE2 is a zinc-containing metalloprotease and a type 1 single-pass transmembrane protein, with its active domain located on the outer surface of endothelial cells, pneumocytes, and other cells. The extracellular domain of ACE2 is broken down by another enzyme called sheddase, and the resulting water-soluble protein is released into the bloodstream and excreted in the urine. The N-terminal peptidase domain of ACE2 is the target site for SARS-CoV [24, 25] and potentially SARS-CoV-2 [26] binding. Based on immunohistochemical analysis, Hikmet et al. showed that the extrapulmonary organs with high levels of ACE2 expression includes intestinal microvilli, renal proximal tubules, gallbladder epithelium, Sertoli and Leydig testicular cells, glandular seminal vesicle cells, and cardiomyocytes [27]. It does not appear to be highly expressed in some organs such as the spleen, thymus, lymph nodes, and bone marrow [28, 29].

ACE2 activity appears to counteract and moderate that of ACE1 [26]. ACE1 breaks down angiotensin 1 and converts it to angiotensin 2, which constricts the arteries. ACE2 removes the amino acid phenylalanine from angiotensin 2, converting it to angiotensin 1–7, which dilates blood vessels.

12.3 The Testis as a Site for the ACE2 Receptors

ACE2 is highly expressed in testicular cells [30]. According to the evidence, both seminiferous ducts and Leydig cells showed high levels of ACE2 expression. Therefore, testicular cells appear to be potential targets for the SARS-CoV-2 virus. The findings of some studies have shown that expression of ACE2 in the testis is limited to Leydig cells in mice and Leydig and Sertoli cells in humans. These studies also suggested the principal role for ACE2 in controlling testicular function, possibly via regulation of steroidogenesis, or some other Leydig cell function [16, 31, 32].

Similar to ACE2 expression pattern, the angiotensin 1–7 receptor Mas mRNA is located in the Leydig and Sertoli cells, with greater abundance in the Leydig cells [33]. In animal models, Mas plays an important role in regulating androgen metabolism in the male reproductive system, and the Mas deficiency affects the expression of enzymes involved in testosterone biosynthesis in Leydig cells [34].

Wang et al. showed that in human testes, ACE2 is enriched in Sertoli and Leydig cells and spermatogonia, the cells produced at an early stage in spermatozoa genesis. Gene Set Enrichment Analysis (GSEA) showed that Gene Ontology (GO) categories associated with viral reproduction and transmission are enriched in ACE2-positive spermatogonia. Cell-cell binding and immunity-related processes are higher in Leydig and Sertoli cells expressing ACE2. Together, these findings support the possibility that the human testis is a potential target for SARS-CoV-2 infection [35].

The main function of Leydig cells is to produce sex hormones, especially testosterone. The presence of Mas receptors may indicate a modification in testosterone secretion by angiotensin 1–7. The involvement of Sertoli and germ cells can also be justified by the presence of Mas receptors and angiotensin 1–7 in the seminiferous tubules [36].

12.4 Role of Cytokines in Male Fertility

Systemic inflammatory response syndrome is a condition that affects the whole body as a response against any infectious or noninfectious insult. Despite the name, both proinflammatory and anti-inflammatory responses are involved in the pathogenesis. One subtype of this response has been termed cytokine release syndrome, and some evidence suggests the occurrence of a mild or severe "cytokine storm" in patients with severe COVID-19 disease [37]. There is compelling evidence that interleukin 6 (IL6), as an important anti-inflammatory cytokine in the testicular immune system, can disrupt the integrity of the blood-testicular barrier in animal models (Fig. 12.2) [38, 39].

IL6 inhibits protein degradation or activates phosphorylated extracellular signal-regulated kinase in Sertoli cells. Through the Zfp637 transcription factor, it can directly affect spermatogonia and, therefore, interferes with the differentiation or destruction of germ cells [40]. Xu et al. found that in SARS-CoV infection, the inflammatory cells could interfere with the function of Leydig cells and inhibit testosterone production and destroy the seminiferous tubules. An autoimmune response and antibody formation in the seminiferous tubules could also be caused by the abnormal presence of cytokines [9]. Together, these findings suggest that alterations in the cytokine profile may be the underlying mechanism by which COVID-19 disease affects male fertility.

12.5 Orchitis Induced by COVID-19

One of the concerns of COVID-19 disease on male fertility is viral orchitis particularly in patients who are within the reproductive age range. Orchitis is defined as an inflammation of the testis. Increased incidence of orchitis, infertility, and testicular tumors were reported as complications in several studies of the 2002–2004 SARS outbreak [9].

12.6 Other Concerns

According to the evidence obtained so far in adults, the most common clinical symptom of COVID-19 is fever [41]. This is important as germ cell growth occurs ideally below 37 °C, and prolonged fever may lead to changes in testicular temperature and destroy germ cells. In line with this, previous studies have reported that fever leads to meiotic germ cell apoptosis [42]. Some studies of SARS-CoV have shown that fever may have an indirect effect on testicular function [9]. A study of prolonged fever in SARS infection reported mild fibrosis and testicular congestion without loss of germ cells or leukocyte infiltration. Besides, the treatment of COVID-19 with steroids can also affect spermatogenesis. Gao et al. found that corticosteroid exposure caused apoptosis in male rat testicular Leydig cells, potentially reducing testosterone levels [43]. Similarly, dexamethasone treatment

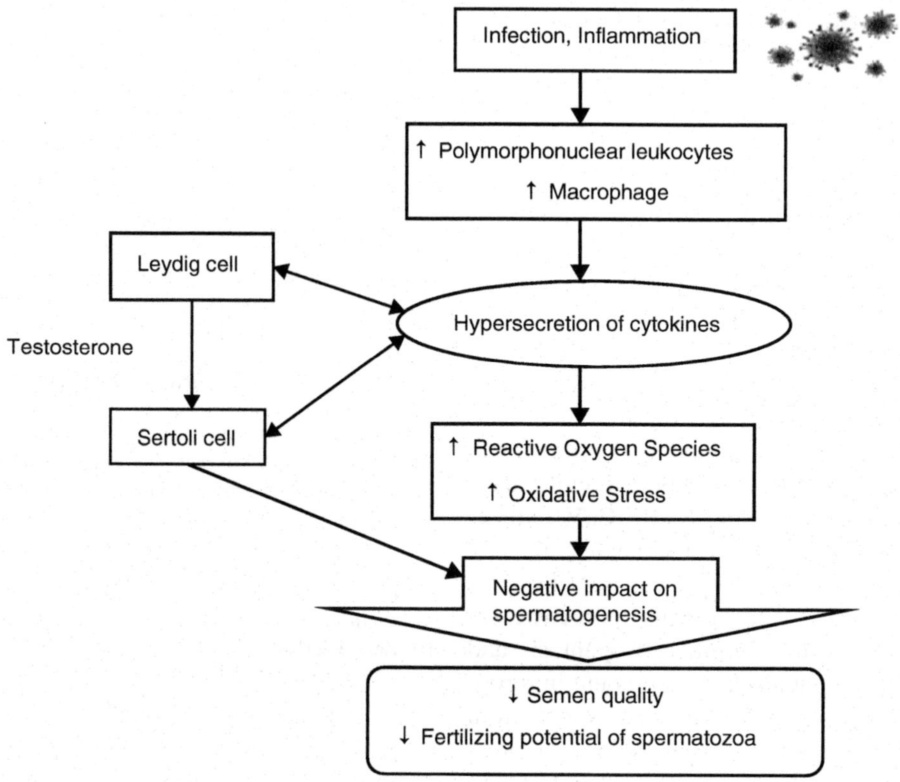

Fig. 12.2 Impact of inflammatory cytokines on male fertility

resulting in low testosterone levels has been reported [44].

12.7 Evidence So Far

To date, a few studies, some of which were not of high quality, have been performed on the presence of the SARS-CoV-2 in semen or biopsies taken from the male reproductive system in COVID-19 patients. Pan et al. believed that although their findings did not show the presence of the SARS-CoV-2 virus in semen, they could not definitively rule out the presence of the virus in the seminal fluid during the acute infection phase [45]. Li et al. reported that 6 of 38 patients had positive results for SARS-CoV-2 in semen samples, including 4 of 15 patients who were at the acute stage of infection and 2 of 23 patients who were in the recovery phase [46]. A postmortem examination of the testes from 12 COVID-19 male patients showed that in one case, the SARS-CoV-2 was detected in the semen [5]. However, due to fibrovascular tissue and few seminiferous tubules, the authors were not sure whether the virus was detected in the blood or the testes. Sertoli cells showed swelling, vacuolation, and cytoplasmic rarefaction, detachment from tubular basement membranes, and loss and sloughing into the luminal space of the intratubular cell mass. The mean number of Leydig cells in COVID-19 testes was significantly lower than in the control group. In the interstitium, there was edema and mild inflammatory infiltrates composed of T lymphocytes and histiocytes. Two studies did not detect the presence of SARS-CoV-2 RNA in semen samples of male patients recovering from COVID-19 [47, 48]. In the studies by Holtmann et al. [47] and Pan et al. [45], some of the patients had testicular discomfort at the time of COVID-19 confirmation, which raised the issue of orchitis. The findings of another study carried out in China showed that in male patients with COVID-19 disease, the level of serum luteinizing hormone was significantly increased, but the ratio of testosterone or follicle-stimulating hormone to luteinizing hormone was dramatically decreased [49].

12.8 Conclusions

The little evidence that is available suggests that COVID-19 disease may affect male fertility. The host receptor, ACE2, is present in human testis. The cytokine storm induced by severe COVID-19 infection cases may affect the function of testicular cells and the production of testosterone. In addition, reports of orchitis have been published following SARS-CoV-2 infection, which can also affect male fertility and high and persistent fever may disrupt the production of germ cells. Finally, corticosteroid therapies can cause Sertoli cell apoptosis and lower testosterone levels. It is prudent in reproductive-aged men who have recovered from COVID-19 disease to be evaluated for the presence of the virus in semen and other fertility-related body tissues. Of highest importance, there is an urgent need to conduct quality studies on the long-term effects of COVID-19 infection on the fertility of recovered males. This may require several years of follow-up studies to obtain a definitive answer.

Acknowledgments Thanks to guidance and advice from the Clinical Research Development Unit of Baqiyatallah Hospital.

Conflict of Interest No conflicts of interest are declared by the authors.

References

1. WHO Coronavirus disease 2019 (COVID-19) situation report – 161. June 29, 2020. https://www.who.int/docs/default-source/coronaviruse/situation-reports/20200629-covid-19-sitrep-161.pdf?sfvrsn=74fde64e_2
2. Zheng YY, Ma YT, Zhang JY, Xie X (2020) COVID-19 and the cardiovascular system. Nat Rev Cardiol 17(5):259–260
3. Cheng Y, Luo R, Wang K, Zhang M, Wang Z, Dong L et al (2020) Kidney disease is associated with in-hospital death of patients with COVID-19. Kidney Int 97(5):829–838
4. Gu J, Han B, Wang J (2020) COVID-19: gastrointestinal manifestations and potential fecal–oral transmission. Gastroenterology 158(6):1518–1519
5. Yang M, Chen S, Huang B, Zhong JM, Su H, Chen YJ et al (2020) Pathological findings in the testes of COVID-19 patients: clinical implications. Eur Urol Focus May 31;S2405-4569(20)30144-9. https://doi.org/10.1016/j.euf.2020.05.009. Online ahead of print
6. Wu Y, Xu X, Chen Z, Duan J, Hashimoto K, Yang L et al (2020) Nervous system involvement after infection with COVID-19 and other coronaviruses. Brain Behav Immun 87:18–22
7. Salam AP, Horby PW (2017) The breadth of viruses in human semen. Emerg Infect Dis 23(11):1922–1924
8. Kurscheidt FA, Mesquita CS, Damke GM, Damke E, Analine RA, Suehiro TT et al (2019) Persistence and clinical relevance of Zika virus in the male genital tract. Nat Rev Urol 16(4):211–230
9. Xu J, Qi L, Chi X, Yang J, Wei X, Gong E et al (2006) Orchitis: a complication of severe acute respiratory syndrome (SARS). Biol Reprod 74(2):410–416
10. Zhao J, Zhou G, Sun Y, Wang S, Yang J, Meng E et al (2003) Clinical pathology and pathogenesis of severe acute respiratory syndrome. Zhonghua Shi Yan He Ln Chuang Bing Du Xue Za Zhi 17(3):217–221
11. Corona G, Baldi E, Isidori A, Paoli D, Pallotti F, De Santis L et al (2020) SARS-CoV-2 infection, male fertility and sperm cryopreservation: a position statement of the Italian Society of Andrology and Sexual Medicine (SIAMS)(Società Italiana di Andrologia e Medicina della Sessualità). J Endocrinol Investg May 27;1–5. https://doi.org/10.1007/s40618-020-01290-w. Online ahead of print
12. Lukassen S, Chua RL, Trefzer T, Kahn NC, Schneider MA, Muley T et al (2020) SARS-CoV-2 receptor ACE 2 and TMPRSS 2 are primarily expressed in bronchial transient secretory cells. EMBO J 39(10):e105114. https://doi.org/10.15252/embj.20105114
13. Fehr AR, Perlman S (2015) Coronaviruses: an overview of their replication and pathogenesis. Methods Mol Biol 1282:1–23
14. Hoffmann M, Kleine-Weber H, Schroeder S, Krüger N, Herrler T, Erichsen S et al (2020) SARS-CoV-2 cell entry depends on ACE2 and TMPRSS2 and is blocked by a clinically proven protease inhibitor. Cell 181(2):271–280
15. Walls AC, Park YJ, Tortorici MA, Wall A, McGuire AT, Veesler D (2020) Structure, function, and antigenicity of the SARS-CoV-2 spike glycoprotein. Cell 181(2):281–292.e6
16. Donoghue M, Hsieh F, Baronas E, Godbout K, Gosselin M, Stagliano N et al (2000) A novel angiotensin-converting enzyme–related carboxypeptidase (ACE2) converts angiotensin I to angiotensin 1–9. Circ Res 87(5):e1–e9
17. Santos RAS, Sampaio WO, Alzamora AC, Motta-Santos D, Alenina N, Bader M et al (2018) The ACE2/

angiotensin-(1–7)/MAS axis of the renin-angiotensin system: focus on angiotensin-(1–7). Physiol Rev 98(1):505–553
18. Visniauskas B, Arita DY, Rosales CB, Feroz MA, Lefante J, Thethi TK et al (2019) The association between plasma soluble prorenin receptor and renin activity points toward renin angiotensin system activation and cardiovascular complications in women with type-2 diabetes. Hypertension 74(Suppl_1). https://doi.org/10.1161/hyp.74.suppl_1.073
19. Akishita M, Yamada H, Dzau VJ, Horiuchi M (1999) Increased vasoconstrictor response of the mouse lacking angiotensin II type 2 receptor. Biochem Biophys Res Commun 261(2):345–349
20. Ibrahim MM (2006) RAS inhibition in hypertension. J Hum Hypertens 20(2):101–108
21. Leung PS (2007) The physiology of a local renin–angiotensin system in the pancreas. J Physiol Pharmacol 580(1):31–37
22. Herr D, Bekes I, Wulff C (2013) Local renin-angiotensin system in the reproductive system. Front Endocrinol 4:150. https://doi.org/10.3389/fendo.2013.00150
23. Leung P, Sernia C (2003) The renin-angiotensin system and male reproduction: new functions for old hormones. J Mol Endocrinol 30(3):263–270
24. Li W, Zhang C, Sui J, Kuhn JH, Moore MJ, Luo S et al (2005) Receptor and viral determinants of SARS-coronavirus adaptation to human ACE2. EMBO J 24(8):1634–1643
25. Lambert DW, Yarski M, Warner FJ, Thornhill P, Parkin ET, Smith AI et al (2005) Tumor necrosis factor-α convertase (ADAM17) mediates regulated ectodomain shedding of the severe-acute respiratory syndrome-coronavirus (SARS-CoV) receptor, angiotensin-converting enzyme-2 (ACE2). J Biol Chem 280(34):30113–30119
26. Xiao L, Sakagami H, Miwa N (2020) ACE2: the key molecule for understanding the pathophysiology of severe and critical conditions of COVID-19: demon or angel? Viruses 12(5):491. https://doi.org/10.3390/v12050491
27. Hikmet F, Mear L, Uhlen M, Lindskog C (2020) The protein expression profile of ACE2 in human tissues. BioRxiv. https://doi.org/10.1101/2020.03.31.016048
28. Feng Y, Yue X, Xia H, Bindom SM, Hickman PJ, Filipeanu CM et al (2008) Angiotensin-converting enzyme 2 overexpression in the subfornical organ prevents the angiotensin II–mediated pressor and drinking responses and is associated with angiotensin II type 1 receptor downregulation. Circ Res 102(6):729–736
29. Crackower MA, Sarao R, Oudit GY, Yagil C, Kozieradzki I, Scanga SE et al (2002) Angiotensin-converting enzyme 2 is an essential regulator of heart function. Nature 417(6891):822–828
30. Fan C, Li K, Ding Y, Lu WL, Wang J (2020) ACE2 expression in kidney and testis may cause kidney and testis damage after 2019-nCoV infection. MedRxiv. https://doi.org/10.1101/2020.03.31.016048
31. Douglas GC, O'Bryan MK, Hedger MP, Lee DK, Yarski MA, Smith AI et al (2004) The novel angiotensin-converting enzyme (ACE) homolog, ACE2, is selectively expressed by adult Leydig cells of the testis. Endocrinology 145(10):4703–4711
32. Tipnis SR, Hooper NM, Hyde R, Karran E, Christie G, Turner AJ (2000) A human homolog of angiotensin-converting enzyme cloning and functional expression as a captopril-insensitive carboxypeptidase. J Biol Chem 275(43):33238–33243
33. Alenina N, Baranova T, Smirnow E, Bader M, Lippoldt A, Patkin E et al (2002) Cell type-specific expression of the Mas proto-oncogene in testis. J Histochem Cytochem 50(5):691–695
34. Xu P, Santos RA, Bader M, Alenina N (2007) Alterations in gene expression in the testis of angiotensin-(1–7)-receptor Mas-deficient mice. Regul Pept 138(2–3):51–55
35. Wang Z, Xu X (2020) scRNA-seq profiling of human testes reveals the presence of the ACE2 receptor, a target for SARS-CoV-2 infection in spermatogonia, Leydig and Sertoli cells. Cell 9(4):920. https://doi.org/10.3390/cells9040920
36. Reis AB, Araújo FC, Pereira VM, Dos Reis AM, Santos RA, Reis FM (2010) Angiotensin (1–7) and its receptor Mas are expressed in the human testis: implications for male infertility. J Mol Histol 41(1):75–80
37. Zhang C, Wu Z, Li JW, Zhao H, Wang GQ (2020) The cytokine release syndrome (CRS) of severe COVID-19 and Interleukin-6 receptor (IL-6R) antagonist Tocilizumab may be the key to reduce the mortality. Int J Antimicrob Agents 55(5):105954. https://doi.org/10.1016/j.ijantimicag.2020.105954
38. Pérez CV, Theas MS, Jacobo PV, Jarazo-Dietrich S, Guazzone VA, Lustig L (2013) Dual role of immune cells in the testis: protective or pathogenic for germ cells? Spermatogenesis 3(1):e23870. https://doi.org/10.4161/spmg.23870
39. Pérez CV, Sobarzo CM, Jacobo PV, Pellizzari EH, Cigorraga SB, Denduchis B et al (2012) Loss of occludin expression and impairment of blood-testis barrier permeability in rats with autoimmune orchitis: effect of interleukin 6 on Sertoli cell tight junctions. Biol Reprod 87(5):122. https://doi.org/10.1095/biolreprod
40. Loveland KL, Klein B, Pueschl D, Indumathy S, Bergmann M, Loveland BE et al (2017) Cytokines in male fertility and reproductive pathologies: immunoregulation and beyond. Front Endocrinol 8:307. https://doi.org/10.3389/fendo.2017.00307
41. Novel Coronavirus Pneumonia Emergency Response Epidemiology Team (2020) Vital surveillances: the epidemiological characteristics of an outbreak of 2019 novel coronavirus diseases (COVID-19)—China. Zhonghua Liu Xing Bing Xue Za Zhi 41(2):145–151

42. Xu J, Xu Z, Jiang Y, Qian X, Huang Y (2000) Cryptorchidism induces mouse testicular germ cell apoptosis and changes in bcl-2 and bax protein expression. J Environ Pathol Toxicol Oncol 19(1/2):25–34
43. Gao HB, Tong MH, Hu YQ, Guo QS, Ge R, Hardy MP (2002) Glucocorticoid induces apoptosis in rat leydig cells. Endocrinology 143(1):130–138
44. Drobnis EZ, Nangia AK (2017) Immunosuppressants and male reproduction. Impacts of medications on male fertility. Adv Exp Med Biol 1034:179–210
45. Pan F, Xiao X, Guo J, Song Y, Li H, Patel DP et al (2020) No evidence of severe acute respiratory syndrome–coronavirus 2 in semen of males recovering from coronavirus disease 2019. Fertil Steril 113(6):1135–1139
46. Li D, Jin M, Bao P, Zhao W, Zhang S (2020) Clinical characteristics and results of semen tests among men with coronavirus disease 2019. JAMA Netw Open 3(5):e208292. https://doi.org/10.1001/jamanetworkopen.2020.8292
47. Holtmann N, Edimiris P, Andree M, Doehmen C, Baston-Buest D, Adams O et al (2020) Assessment of SARS-CoV-2 in human semen-a cohort study. Fertil Steril May 29. https://doi.org/10.1016/j.fertnstert.2020.05.028:1-13. [Epub ahead of print]
48. Song C, Wang Y, Li W, Hu B, Chen G, Xia P et al (2020) Absence of 2019 novel coronavirus in semen and testes of COVID-19 patients. Biol Reprod 103(1):4–6
49. Ma L, Xie W, Li D, Shi L, Mao Y, Xiong Y et al (2020) Effect of SARS-CoV-2 infection upon male gonadal function: a single center-based study. MedRxiv. https://doi.org/10.1101/2020.03.21.20037267

COVID-19 and Vulnerable Populations in Sub-Saharan Africa

13

J. A. George, M. R. Maphayi, and T. Pillay

Abstract

The novel corona virus 2019 (COVID-19) outbreak which started in Hubei province in China has now spread to every corner of the earth. While the pandemic started later in Africa, it is now found in all African countries to varying degrees. It is thought that the prevalence and severity of disease is influenced by a number of non-communicable diseases (NCDs) which are all becoming increasingly prevalent in sub-Saharan Africa (SSA). In addition, SSA bears the major burden of human immunodeficiency virus (HIV) and tuberculosis (TB) infections. While data from Europe and the United States show that children are spared severe disease, it is uncertain if the same holds true in SSA where children suffer from sickle cell disease and malnutrition in addition to other infectious diseases. There is limited data from Africa on the effects of these conditions on COVID-19. In this review, we discuss the epidemiology of some of these conditions in Africa and the possible pathogenesis for the interactions of these with COVID-19.

J. A. George (✉) · M. R. Maphayi · T. Pillay
Department of Chemical Pathology, National Health Laboratory Services, University of the Witwatersrand, Johannesburg, South Africa

Keywords

COVID-19 · SARS-CoV-2 · Risk factors · Obesity · Diabetes · Hypertension · Kidney disease · HIV tuberculosis

13.1 Introduction

On the 11th of March 2020, the World Health Organization (WHO) declared COVID-19, which is caused by the novel severe acute respiratory syndrome (SARS) corona virus 2 (SARS-CoV-2), a pandemic [1]. The first case of COVID-19 in sub-Saharan Africa (SSA) was reported in Nigeria in January 2020 [2]. Initially, cases were imported from other countries; however, as with the rest of the world, local transmission has taken over with more than 844,542 infected people in Africa to date and 17,682 deaths reported [3]. Initial reports of ethnically homogenous populations from China suggested that key risk factors for severity include age, male sex, and non-communicable diseases (NCDs) such as cardiovascular disease, hypertension, and diabetes mellitus (DM) [4, 5]. More recent data from the United Kingdom and the United States show that black and other ethnic minorities are at increased risk for acquiring the infection and are also at increased risk of poor outcomes [6]. This has

P. C. Guest (ed.), *Clinical, Biological and Molecular Aspects of COVID-19*, Advances in Experimental Medicine and Biology 1321, https://doi.org/10.1007/978-3-030-59261-5_13

been attributed to increased prevalence of comorbidities as well as cultural and socioeconomic factors.

While the numbers of cases and mortality in Africa are relatively low, compared to the West, there are fears that African populations may be more vulnerable to severe disease. This is because of the burden of infectious diseases like tuberculosis (TB) and human immunodeficiency virus (HIV) with an increasing burden of NCDs.

Vulnerable people are those that are disproportionately exposed to risk, and these groups may differ in different regions of the world, depending on disease burden, age structure, as well as health systems and policy. Across the globe, older people were shown to be more vulnerable to more severe outcome from COVID-19 [4, 7]. Africa has a young age structure, with about 60% of the population below the age of 25 years [8]. While this may be a protective factor for severe disease and mortality, it is uncertain how infectious comorbidities and malnutrition will affect the disease course. At the same time, the number of people aged 60 and above is increasing at a faster rate than in the developed world [9].

The rapid but poorly organized growth of cities and movement of people from rural to urban areas expose the population to diseases of lifestyle such as overweight and obesity which are major determinants for hypertension, stroke, and DM [10]. The Global Burden of Diseases Study 2017 reported that a large amount of the NCD burden is due to cardiovascular disease and DM, and the prevalence of these differs by age and by region [11]. While publications on COVID-19 have exploded, there is limited data on the situation in Africa. Data on hospital admissions for 10,710 COVID-19 patients from the National Institute for Communicable Diseases (NICD) of South Africa shows that the median age of hospital admission for COVID-19 was 50 years. Fifty-four percent (5,778/10,700) of admissions were female. Among 8,245 (77%) patients with data on comorbid conditions, the most commonly reported were hypertension (3,419; 59%) and DM (2,813; 48%). There were 1,116 (19%) patients admitted with HIV, 240 (4%) with active TB, and 579 (10%) patients with previous history of TB. Obesity, while not consistently recorded for all reported COVID-19 admissions, was noted by clinicians as a risk factor in 297 (3%) patients [12].

In this review, we will discuss the epidemiology of obesity, DM mellitus, hypertension, HIV, and TB in Africa and the factors that predispose individuals with these comorbidities to severe COVID-19. We will further explore risk factors in African children that may predispose them to severe disease.

13.2 Obesity

A pooled analysis of population-based studies has shown that body mass index (BMI) has increased in African men and women over time with increases above the global average seen in Northern and Southern Africa [13–15]. The prevalence of obesity is higher in urban compared to rural populations, with marked differences noted across the continent [13, 16]. For example, in a study that compared BMI across four SSA countries, the prevalence of obesity and overweight in women from Soweto, South Africa, was 66% compared to 1.3% for women from Nanoro, Burkina Faso [17]. The same study also showed that the prevalence of obesity was greater in women than in men, ranging from 42.3–66.6% in women to 2.81–17.5% in men. This rise in obesity is at least partly responsible for the increasing burden of NCDs on the continent.

Several studies and meta-analysis have shown that increasing BMI is associated with more severe outcomes in COVID-19-infected patients [18–21]. For example, a retrospective analysis of 92 consecutively admitted patients with COVID-19-related pneumonia observed that more patients with overweight and obesity required mechanical ventilation and admission to intensive care units (ICUs). These patients were also younger than normal weight patients. This increased risk remained even after adjusting for age, sex, and comorbidities [22]. The increased severity of disease is seen in both overweight and in obese hospitalized patients [22–24]. It is thought that obesity may not only predispose to severe infection, but it may also predispose to acquiring COVID-19. Surveillance data from Mexico showed that proportionately more people who tested positive for COVID-19 were obese than those who tested negative [24]. Using New York City (NYC) data, El Chaar et al. sought to determine if the increased mortality seen in ethnic minorities was related to BMI. They looked at age adjusted mortality rates for the different NYC boroughs. Both the Bronx and Brooklyn were found to have the highest mortality rates (6.0%, 5.4%) and obesity rates at 32% and 27%, respectively. Hispanic and blacks had the highest obesity rates and were also found to have the highest age-adjusted mortality rates per 100,000 compared with the other ethnic groups [25].

The evidence for impaired immunity in obesity was seen with previous influenza outbreaks which showed worse outcomes with obesity [26]. Furthermore, obese individuals have high rates of vaccine failure [27]. The pathophysiology of COVID-19 infection involves an aggressive inflammatory response due to both the viral infection and the subsequent host response. Obesity itself induces a chronic inflammatory state which causes metabolic changes as well as immune dysfunction. The chronic inflammation of obesity is driven by adipose tissue hypertrophy and subsequent apoptosis with release of inflammatory cytokines such as interleukin (IL)-6, IL-8, and monocyte chemoattractant protein-1 (MCP-1) as well as adipokines like leptin [28]. These cytokines lead to an inflammatory cellular infiltrate [29]. Furthermore, these inflammatory cells themselves release cytokines which increase leptin, which is pro-inflammatory, and reduce adiponectin, an anti-inflammatory adipokine. Leptin also affects endothelial function, with impaired production of nitric oxide (NO) and increased circulating plasminogen activator inhibitor-1 (PAI-1), thus predisposing to thrombotic disease [30]. In addition to the alterations in the innate immune system, cellular immunity is affected with reduced numbers of naïve CD4 T cells and the more pro-inflammatory Th17 and Th22 cells present [31]. Obese individuals have a greater prevalence of comorbidities like hypertension, DM, and kidney disease. Other factors are thought to be responsible for the increased severity and mortality associated with obesity in COVID-19 patients including impaired respiratory mechanics, increased airway resistance, and impaired gas exchange, as well as other pathophysiological features of obesity, such as low respiratory muscle strength and lung volumes [32] (Fig. 13.1).

13.3 Hypertension

The global prevalence of hypertension is about 31.1% (CI, 30.0–32.2%) [33]. The prevalence is highest in Africa at 46%, with a relatively young group affected (mean age range: thirties to the forties). Less than 10% of those affected are on treatment [34]. A systematic review on the prevalence of underlying disease in COVID-19 hospitalized patients noted that hypertension was the most prevalent comorbidity found in 16.4% of cases [35]. However, it is difficult to draw any conclusions as there was significant heterogeneity across studies. The association between hypertension and COVID-19 may be a reflection of the community prevalence or it may be because hypertension is more common in older people and age tracks with other comorbidities like obesity and DM. However, a meta-analysis of 30 studies on COVID-19 and hypertension showed that hypertension was associated with disease severity, ICU admission, and mortality, and this was not influenced by age, DM, or chronic obstructive pulmonary disease [36].

Hypertension could predispose to COVID-19 via a number of mechanisms. It could be via effects on the immune system, treatment related, or co-incidental due to the increased prevalence in the older population. There is some evidence to suggest that alterations in innate and acquired immunity play a role in the pathogenesis of hypertension via the effect of inflammation on vascular remodeling and subsequent end organ damage [37, 38]. Another potential role is via the angiotensin-converting enzyme 2 (ACE-2) receptor which is used for viral entry. Theoretically,

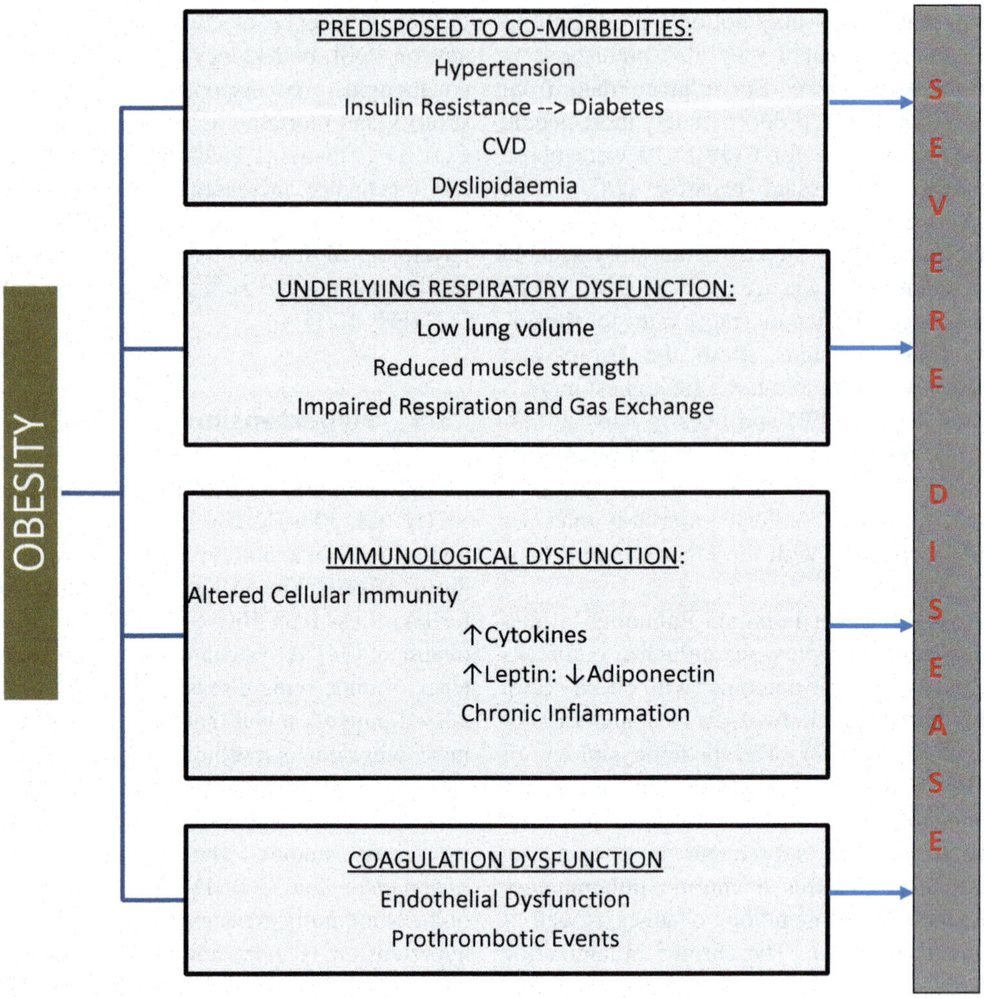

Fig. 13.1 Possible mechanisms by which obesity impacts COVID-19 severity

increased ACE-2 expression may be deleterious for COVID-19 infected patients. It is postulated that the use of renin angiotensin system (RAS) inhibitors may alter ACE-2 expression leading to increased virulence. ACE-2 displays considerable homology with angiotensin-converting enzyme (ACE). Unlike ACE, ACE-2 does not convert angiotensin I to angiotensin II, and ACE inhibitors (ACEi) do not block its activity. So the use of these is unlikely to affect virulence [39]. Angiotensin II receptor blockers (ARBs) on the other hand have been shown to upregulate ACE-2 in animal studies, but several observational clinical studies have not shown increased mortality in patients on ARBs [40, 41]. For example, a study from the Hubei Province in China that compared COVID-19 hypertensive patients on ACEi or ARBs to those on other hypertensives drugs showed that while blood pressure control was similar in both groups, those on ACEis and ARBs had lower concentrations of inflammatory markers [42]. A retrospective analysis from NYC of antihypertensive treatment and the likelihood of a COVID-19-positive test or disease severity showed that there was no association between class of medication used and likelihood of a positive test or disease severity [43]. Local data shows that only 27% of people with hypertension are aware of their status, on average only 18% are on treatment, and about 7%

have controlled blood pressure [44]. The reasons for this are multifactorial, including the alarming prevalence of obesity, absent or poor primary healthcare services, lack of medication, poor compliance, and poverty. Many of these factors also increase the risk of acquiring COVID-19 and of severe disease.

13.4 Diabetes Mellitus

SSA carries most of the DM burden compared to the rest of the continent with more than 19 million people currently estimated to have DM. The impact of DM and COVID-19 in this region is difficult to measure given that >60% are thought to be undiagnosed and many are unaware of their condition [45, 46].

Review of the data from past outbreaks like SARS coronavirus (SARS-CoV), Influenza A (H1N1), and Middle East Respiratory Syndrome coronavirus (MERS-CoV) suggests an increased morbidity and mortality associated with DM [47]. Following the H1N1 outbreak in 2009, a Canadian study of 239 patients with H1N1 influenza showed that patients with diabetes were more likely to require hospitalization and had increased odds for ICU admission [48].

Descriptive case series for diabetes and COVID-19 have reported varying prevalences of diabetes ranging from 5.3% to 58% of cases [47]. The prevalence varies according to national prevalence and age. It has been shown to be higher in older people. It is uncertain if diabetes predisposes to COVID-19 infection as the prevalence reported in some studies is similar to the national prevalence. For example, a meta-analysis of Chinese patients with COVID-19 reported a DM prevalence of 10.3% which was comparable to the 10.9% national prevalence [49]. Similarly, an Italian study found a prevalence of 8.9% in their COVID-19 population compared to a prevalence of 11% among similar aged individuals in the general population from the same region [49]. Data from an International Consensus reported by Caballero et al. suggests that diabetes predisposes to increased disease severity and mortality. In China, 5.3% of the total population were people living with diabetes (PLWD); however, the COVID-19 mortality was disproportionately higher (7.3%) compared to those without comorbidities (<1%) [45]. Increased risk of progression to severe disease and increased mortality have been reported across the globe [7, 50, 51].

In the case series of the first 100 patients seen at a designated COVID-19 hospital in Gauteng South Africa, 52 patients were found to have at least one coexisting medical condition. The most seen comorbidities were hypertension (n=31.3) and diabetes (n=18). Although they were unable to measure the BMI in all patients, they described 47% of patients as having an endomorphic phenotype [52].

The underlying reasons for the relationship between COVID-19 and DM are likely to be multifactorial and associated with many of the vulnerabilities of diabetes like older age, increased BMI, and metabolic dysfunction. The bidirectional relationship between COVID-19 and diabetes further complicates our understanding. While the presence of diabetes is considered to confer a greater risk of acquiring severe infection, COVID-19 is considered to have diabetogenic effects. It is implicated in either new onset DM or increasing risk of metabolic complications of DM [53]. Poorly controlled DM impairs the immune response to infection, which can predispose to the development of COVID-19 [54]. At the same time, SARS-CoV-2 may play a role in worsening glucose control and hyperglycemia, even in patients without underlying DM due to its action of the ACE-2-dependent damage of pancreatic islet cells [55].

The increased risk of developing poorer outcomes has been related to the chronic inflammatory state associated with hyperglycemia and insulin resistance [55]. The pro-inflammatory state is associated with increased glycosylation end products (AGEs), pro-inflammatory cytokines, oxidative stress, and adhesion molecules. This might facilitate the cytokine storm implicated in disease severity and mortality [55]. DM is also associated with a prothrombotic state and increased plasminogen which has been hypothesized to increase the virulence of SARS-CoV-2 [54].

Animal and human models have demonstrated increased expression of serum ACE-2 in DM, which is being touted as a potential predisposing factor for COVID-19 [56]. Evidence of ACE-2 expression in the islet cells of the pancreas suggests that SARS–CoV-2 enters islets using ACE-2 as its receptor and damages the cells causing acute DM [57, 56]. Furin, a membrane protease, has been shown to be increased in DM and has now consequently been implicated in the replication of SARS-CoV-2 [56].

13.5 Kidney Disease

Chronic kidney disease (CKD) is the end result of many infections and non-communicable diseases and is an independent risk factor for death from cardiovascular disease. Consequently, there is concern about the increase of prevalence of CKD, estimated to range between 8% and 16% globally [58]. Its prevalence in Africa was reported as 10.7% (CI 9.9–11.7), with regional differences noted [59]. The kidney is one of the main organs affected by COVID-19 infection. A cross-sectional analysis of primary healthcare data from the United Kingdom showed that CKD was independently associated with a positive test for SARS-CoV-2 [60]. South African surveillance data showed that among hospital admissions, reporting comorbid conditions, 7.5% had CKD [12].

Much less is known about acute kidney injury (AKI) in Africa. The reported prevalence varies from 1.9% of all admissions to 20% of medical admissions [61, 62]. The possible reason given is that renal function is not measured routinely in many hospitals. The main causes of AKI are sepsis, infections such as malaria and HIV, diarrheal diseases, use of nephrotoxic substances, and AKI in pregnancy. A meta-analysis of AKI in COVID-19 patients showed that the incidence rate was almost 10%, similar to the incidence rate in community acquired pneumonia [63]. The rate in critically ill patients may be much higher, and AKI in COVID-19 is an independent risk factor for mortality [64, 65]

One of the postulated mechanisms for kidney injury is direct injury due to the virus. The kidneys have higher ACE-2 expression than the lungs [66]. Postmortem examinations of kidney tissue from six patients showed severe acute tubular necrosis and lymphocyte infiltration as well as the presence of the nucleocapsid protein [67]. Other possible mechanisms are sepsis following the cytokine storm, AKI due to volume depletion, and rhabdomyolysis. A picture of collapsing focal segmental glomerulosclerosis was described in two African-Americans, with high-risk apolipoprotein L1 (*APOL1*) genotype. Africans homozygous for APOL1 are at increased risk for HIV-associated kidney disease and may also be at increased risk for kidney injury associated with COVID-19 disease [68, 69].

13.6 HIV and TB

Globally, the number of people living with HIV (PLWH) is estimated to be 37.9 million [70]. Furthermore, HIV is among the most significant contributors to disease burden in Africa with an estimated 3.9% of adults being affected. At present, the interaction of HIV with COVID-19 co-infection is uncertain and incompletely described but warrants further understanding in context of the high prevalence of HIV. There is concern that PLWH may be at increased risk for COVID-19 due to overall immunosuppression.

A number of studies suggest that PLWH who develop COVID-19 do not have more severe disease (Table 13.1). For example, an observational study aimed at describing COVID-19 in HIV-infected persons by Vizcarra et al. reviewed data of 2,873 PLWH in Madrid, Spain. Of 51 cases (35 laboratory confirmed and 16 suspected), they found no significant difference in age and CD4 count between COVID-19 and non-COVID-19 individuals. Their findings corroborated the higher risk conferred by comorbidities with 32 (63%) of COVID-19-affected individuals displaying at least one comorbidity (predominantly DM and hypertension). Clinical and radiological presentation was similar to that of the general pop-

Table 13.1 Studies that have reported on HIV and COVID-19

Study type/country	N(total)n(HIV with COVID-19)	Findings	References
Observational/Spain	2873/51	No difference in age or CD4 between HIV with COVID and HIV without COVID. No differences in clinical or radiological presentation	[71]
Retrospective/Italy	6000/47	Younger than HIV negative/COVID-19 positive. No difference in outcomes	[72]
Case series /Germany	33	No evidence for excess mortality among virologically suppressed PLWH	[74]
Retrospective/S Africa	100	79% Black HIV 11% No difference in HIV prevalence between symptomatic and asymptomatic	[52]
Case series/UK	18	17(94%) black Lower CD4 counts than HIV positive/COVID-19 negative	[80]
Case series/US	6	Four out of six died from ARDS or septic shock Virologically suppressed Inverse relationship between CD4 and death	[82]
Retrospective review of public sector patient data /S Africa	12522 COVID	HIV 18% in COVID-19 positive vs. 16% in COVID-19 negative Increased risk for death among HIV positive compared to HIV negative, after adjusting for comorbidities. Increased risk regardless of treatment with ARVs or of viral suppression	[12]

ulation [71]. An Italian study of 47 HIV-positive COVID-19 probable or proven cases noted that the HIV-positive group were about 10 years younger than the HIV-negative group, although they did not appear to have increased risk of severe disease or death. Most of the patients were virologically suppressed with acceptable immunological function [72]. Similar findings have been reported by other investigators [73, 74]. Some investigators attributed these findings to a possible protective effect of protease inhibitors [75]. In vitro studies have shown that protease inhibitors are involved in the activation of envelope glycoproteins used for viral entry [76]. However, based on the evidence available, a recent systematic review concluded that it is uncertain that anti-retroviral drugs prevent SARS-CoV2 infection or improve clinical outcomes [77]. It has also been suggested that PLWH might be at decreased risk for complications of COVID-19 due to their defective cellular immunity which reduces the possibility of the cytokine dysregulation (storm) that is associated with poorer outcomes and more severe cases of COVID-19 [78].

As these are largely case series of hospitalized patients, caution should be applied when interpreting the results. The numbers are small, and not all hospitalized patients were tested for HIV. As shown in Table 13.1, there are also studies that show increased morbidity and mortality. Patients with moderate or severe COVID-19 infection have reduced numbers of CD4 and CD8 T cells, and immune deficiency caused by both infections may result in poorer outcomes [79]. A UK study of hospitalized PLWH with confirmed COVID-19 reported a substantially increased risk of morbidity and mortality. The study group comprised predominantly black males, with longstanding HIV patients who were virologically suppressed. When compared to the overall HIV-positive population, those with COVID-19 were more likely to be black and have a lower CD4 count. In this study, ARVs did not provide protection against moderate or severe COVID-19 [80]. The major significance of this study was the reporting of ethnicity and associated risk which is particularly relevant in an African context. Similarly, a US study of nine hospitalized patients with HIV and COVID-19 showed a mortality rate of 94%. Seven out of nine patients were male, they had multiple comorbidities, and all were virally suppressed. The study reported an inverse relationship between CD4 count and mortality rate [80]. Furthermore, the cytokine storm associated with acute respiratory distress syndrome (ARDS) in COVID-19 infection is characterized by very high levels of IL-6, IL-2R, IL-10, and interferon (IFN)-γ [79]. Higher IL-6 levels are also seen in HIV patients who are older and have a higher BMI and lower nadir CD4 counts. It remains to be determined if preexisting high pro-inflammatory cytokine concentrations predispose people to more severe disease. This might not only infer a higher risk of co-infection with SARS-CoV-2 but also suggests the potential risk for other opportunistic infections in patients with HIV during co-infection [81]. Another mechanism for poorer outcome and more severe disease might be due to HIV-related lymphopenia. This produces an inadequate T-lymphocyte response to SARS-CoV-2, poor viral clearance, and disease progression [82]. Finally, PLWH have increased comorbidities, including an increased BMI, DM, hypertension, and cardiovascular disease [83, 84].

At present, there are only two reports from Africa (both from South Africa) that have investigated HIV status and COVID-19. A single center case series of the first 100 patients admitted to a large academic hospital in Gauteng province, which is currently the epicenter of the COVID-19 pandemic, found the prevalence of HIV in their study population to be comparable to that of the general population (11% and 14%, respectively). They also did not report any deaths among HIV-positive patients in this cohort [52]. In contrast to this, data presented at the 23rd International AIDS Conference from the Western Cape, which was the first to experience a surge of cases in South Africa, demonstrated that HIV, as well as a past or current history of TB, increased the risk of mortality from SARS-CoV-2 [12].

Drawing conclusions from the limited and seemingly conflicting data on SARS-CoV-2 and HIV co-infection should be done cautiously. Outcomes described in current case series have been inconsistent, not least because of the marked variability in the study designs and population groups from where the data come. Reports of false-negative PCR tests early in the course of disease with COVID-19 might present an additional challenge to diagnosis in the HIV population [85, 86]. The number of cases where this was noted is limited and incompletely described.

TB co-infection of PLWH is a major problem in SSA. In 2016, 2.5 million people fell ill with TB in the African region, accounting for a quarter of new TB cases worldwide. TB is a leading killer of HIV-positive people. In 2016, 40% of HIV deaths were due to TB [87]. Furthermore, global rates of latent TB are estimated to be as high as 25%.

There is a paucity of data on the impact of TB on COVID-19 infection. Tadolini et al. reported a case series of 49 patients from a European cohort with current or history of TB and SARS-CoV-2 infection [88]. They reported a mortality of 12.3% in the patients with dual infection which is much higher than in COVID-19 only disease. However, most patients that died had other risk factors such as age >60 years and more than one comorbidity. A preliminary analysis of 69 patients with TB and COVID-19 co-infection from 8 European countries reported a mortality rate of 11.6% [89]. Forty-three (62%) of the 69 patients were migrants. The migrant population was shown to be younger and to have fewer comorbidities and, interestingly, a lower mortality. They concluded that TB may not have a major effect on mortality. In contrast, data from the population cohort study in public health facilities from the Western Cape Province in South Africa reported an increased hazard of death from COVID-19 with both current and past TB [12]. The relationship between COVID-19 and latent TB appears to have implications both for development and progression of severe COVID-19 as well as the progression of TB. A recent study from China, which actively tested for latent TB, noted that the prevalence of latent TB was higher in COVID-19 patients than in the general population and that the percentage of patients with TB was also considerably higher than the percentage of patients with other comorbidities [90, 88]. Data from South Africa showed that patients with TB had a higher mortality from influenza [91]. Whether or not this is true of COVID-19 remains to be determined.

Factors that may exacerbate the development of SARS-CoV-2 in patients with TB are the dysfunctional immune responses as well as the upregulated expression of ACE-2 in epithelial cells of the respiratory tract [88]. The presence of TB with comorbidities like HIV and DM may also potentially influence mortality rate in COVID-19. In Africa, poverty and malnutrition may play an important role in increasing morbidity and mortality [92].

TB and COVID-19 have many similar clinical features such as fever, cough, and dyspnea [93]. The overlap of clinical features between TB and SARS-CoV-2 will pose a significant diagnostic challenge especially in a region with such a high prevalence of the former.

Both TB and SARS-CoV-2 have been proposed to cause injury and lung damage in the long term. However, the impact of co-infection on disease course and overall outcomes is uncertain and warrants further research and investigation [94]. TB causes chronic lung damage which might contribute to negative outcomes in patients who develop COVID-19 [81]. Based on evidence from chronic lung conditions like silicosis, it has been proposed that individuals with undiagnosed pulmonary TB, recently initiated TB therapy, or complex presentations like drug-resistant or disseminated forms might be susceptible to a more severe disease course if SARS-CoV-2 co-infection occurs [95].

13.7 COVID-19 in Children

Africa has a relatively younger population than the rest of the world. Children in the 0–14 age group make up 40.1%, while the elderly ≥65 age group make up only 3.5% of the population [92]. Studies from other parts of the world showed that most children with SARS-CoV-2 infection experience mild symptoms or are asymptomatic [96, 97]. However, children in SSA are exposed to different socioeconomic and health risk factors that could make them more vulnerable to more severe illness and death from COVID-19.

Current data from SSA, though limited, suggest that children are at lower risk of severe COVID-19 compared to adults. Data released by the South African NICD showed that 6% (3,025) of all patients (52,991) who tested positive by PCR for SARS-CoV-2 were children [12]. Also, children made up only 3.3% (230) of all COVID-19 (6,353) hospital admissions. Besides, those who died (three children) had underlying medical conditions, including dilated cardiomyopathy, leukemia, and hypertension. It is essential to note the number of COVID-19 infections in SA has since increased to over 400,000 with over 6,000 deaths [98]. It is not yet known how many of these are children. Another African study comprising 74 children admitted with COVID-19 infection showed that children experience mild to moderate symptoms [99]. Nonetheless, all children in this study had no malnutrition or other immunocompromising illnesses such as HIV except for one who had type 1 DM. More extensive studies will be needed to assess the real impact of COVID-19 in children with underlying conditions such as HIV and malnutrition in SSA.

Although most children do not develop severe illness, there are few reports of children presenting with inflammatory, multisystem syndrome following COVID-19 infection [100–103].

There are several reasons children rarely develop severe COVID-19 disease. The virus enters the target cell via ACE-2, which is found in the respiratory tract [104]. The spike protein on the surface of the virus is cleaved by the TMPRSSA protease which enables the virus to enter the host cell [104]. However, ACE-2 expression in the nasal epithelial increases with age and younger children with lower ACE-2 in the nasal epithelial may be protected from infection by a high dose of the virus [105]. Patients with severe COVID-19 disease have been found to have a higher viral load and shed the virus over a prolonged period [106]. Secondly, the immune response of children to COVID-19 infection seems to be different from that of adults. A study by Selva et al. analyzing serology features in different age groups found that although children have less exposure to coronavirus antigens, they have higher primary humoral immune response targeted at COVID-19 compared to adults, and this may enable faster elimination of the virus [107]. An increase in IgM is our first response to viral infections, and this immunoglobulin has wide-ranging reactivity and affinity. Younger children were found to have higher IgM, while adults had higher mature specific IgA and IgG [107]. It is thought that these mature IgG and IgA antibodies may cross-react with COVID-19 and cause an exaggerated immune response. Thirdly, most children do not have many of the comorbidities, such as DM and hypertension, which decreases their risk of infection and death to COVID-19.

Because the pandemic in SSA lags behind most parts of the world, there is currently a paucity of evidence to assess how the pandemic is affecting children in this region. Unlike European, North American, or Asian countries, children in SSA are being exposed to a variety of factors which may make them more vulnerable to severe COVID 19 illness. These include malnutrition, sickle cell disease, and infectious diseases such as HIV and TB.

13.8 Malnutrition

Africa is one of two regions that bear the highest share of all forms of malnutrition in the world [108]. Malnutrition refers to a clinical syndrome of anthropometric abnormalities, and it includes wasting and stunting in undernutrition and overweight and obesity in overnutrition. According to the United Nations Children's Fund (UNICEF)/WHO reports on malnutrition, about 39% of stunted, 28% of wasted, and 24% of overweight children under 5 years old live in Africa [108]. At the same time, malnutrition is one of the most common causes of mortality in children living in SSA. Both undernutrition and overnutrition are associated with chronic inflammation. In a study done in Zimbabwe, it was found that children who were stunted had higher inflammatory markers such as C-reactive protein, alpha-1 acid glycoprotein, and IL-6 compared to non-stunted children [109]. Children who are severely wasted have weak immune systems and are prone to infections [110].

There are several trials underway evaluating the safety and efficacy of different COVID-19 vaccines in humans. Vaccines are based on stimulating the immune system to form neutralizing antibodies against an inactive form of the virus to protect against subsequent infections. Response to vaccinations is inadequate in children with severe malnutrition [111, 112]. Therefore, similar trends may be observed in severely malnourished SSA children and will need to be considered when vaccines become available in SSA.

The total number of children who are overweight in Africa has increased significantly over the past years. About 24% of overweight children under 5 years old live in Africa and the majority of these in Southern Africa [108]. The increase in overweight among SSA children has been ascribed to the increase in middle-income families and the consumption of processed food and low physical activities. As in adults, obesity is associated with an increased risk of infection, severe illness, and death from COVID-19 [113, 114]. In a cohort of children admitted with multisystem inflammatory syndrome associated with COVID-19, obesity was the predominant underlying condition [100].

13.9 Sickle Cell Disease

The term sickle cell disease (SCD) refers to a heterogeneous group of disorders in which sickle hemoglobin (HbS) predominates. HbS is a structural variant of normal adult hemoglobin (HbA) caused by a single nucleotide polymorphism of the *HBB* gene. It is estimated that about 314,000 children are born with SCD worldwide every year, with the majority in SSA [115]. About half of all births come from just three countries: Nigeria, the Democratic Republic of Congo, and India [115]. Infections in patients with SCD can cause painful vaso-occlusive crises and acute chest syndrome and death. There is limited data on the impact of COVID-19 on SCD. Arlet et al. compared COVID-19 ICU admissions in those with SCD to all others and showed a lower rate of admission for those with SCD [116]. The authors cautioned that the study was not powered to detect statistical differences. Two small case series from the United Kingdom also suggested that SCD does not increase disease severity or mortality [117, 118]. In contrast, a larger US retrospective review of 179 patients with SCD suggests that those with SCD have more severe disease and have a high case fatality rate [119]. The increase in disease severity occurred in those who had mild-moderate and severe SCD genotypes [120]. At the time of this review, there is no data from African countries on SCD and COVID-19.

People with SCD have defects in innate immunity such as splenic dysfunction, impaired neutrophil chemotaxis and killing, reduced opsonization, and propensity to develop bacterial infections. They are also vulnerable to viral infections and infections with atypical bacteria which points to defects in cell-mediated immunity [121, 122]. Another potential mechanism for increased disease severity and mortality is by increased risk for thromboembolic disease. Both SCD and COVID-19 increase the risk of thrombosis, and the potential combination necessitates further studies [123].

13.10 Conclusions

Metabolic comorbidities do appear to predispose to severe COVID-19 disease, and these are common in SSA. There is limited data on the risk for acquiring COVID-19 in people who have infections like TB and HIV. This data will have to come from Africa. Finally, we still do not know what the risk of disease is for children in Africa, many of whom suffer from malnutrition and infections.

References

1. WHO (2020) Rolling updates on Corona Virus disease. https://www.who.int/emergencies/diseases/novel-coronavirus-2019/events-as-they-happen. Accessed 30 June 2020
2. Adepoju P (2020) Nigeria responds to COVID-19; first case detected in sub-Saharan Africa. Nat Med 26:444–448
3. Worldometer Coronavirus update. https://www.worldometers.info/coronavirus
4. Li X, Xu S, Yu M, Wang K, Tao Y, Zhou Y et al (2020) Risk factors for severity and mortality in adult COVID-19 inpatients in Wuhan. J Allergy Clin Immunol 146(1):110–118
5. Li B, Yang J, Zhao F, Zhi L, Wang X, Liu L et al (2020) Prevalence and impact of cardiovascular metabolic diseases on COVID-19 in China. Clin Res Cardiol 109(5):531–538
6. Cyrus E, Clarke R, Hadley D, Bursac Z, Trepka MJ, Dévieux JG et al (2020) The impact of COVID-19 on African American communities in the United States. medRxiv. https://doi.org/10.1101/2020.05.15.20096552
7. Docherty AB, Harrison EM, Green CA, Hardwick HE, Pius R, Norman L et al (2020) Features of 20133 UK patients in hospital with covid-19 using the ISARIC WHO clinical characterisation protocol: prospective observational cohort study. BMJ 369:m1985. https://doi.org/10.1136/bmj.m1985
8. UN (2019) World population prospects 2019: data booklet. Accessed 30 June 2020
9. Velkoff VA, Kowal PR (2006) Aging in Sub-Saharan Africa: the changing demography of the region. In: Cohen B, Menken J (eds) Aging in Sub-Saharan Africa. National Academies Press, Washigton DC. ISBN: 0-309-010281-2
10. Amuyunzu-Nyamongo M (2010) Need for a multi-factorial, multi-sectorial and multi-disciplinary approach to NCD prevention and control in Africa. Glob Health Promot 17(2 Suppl):31–32
11. Gouda HN, Charlson F, Sorsdahl K, Ahmadzada S, Ferrari AJ, Erskine H et al (2019) Burden of non-communicable diseases in sub-Saharan Africa, 1990–2017: results from the Global Burden of Disease Study 2017. Lancet Glob Health 7(10):e1375–e1387. https://doi.org/10.1016/S2214-109X(19)30374-2
12. NICD (2020) An update on the novel coronavirus disease 2019 outbreak, South Africa. Communicable Diseases Communiqué. April 2020, Vol. 19(4). NICD Communications Unit, Johannesburg
13. Kengne AP, Bentham J, Zhou B, Peer N, Matsha TE, Bixby H et al (2017) Trends in obesity and diabetes across Africa from 1980 to 2014: an analysis of pooled population-based studies. Int J Epidemiol 46:1421. https://doi.org/10.1093/ije/dyx078
14. Anderson AK (2017) Prevalence of anemia, overweight/obesity, and undiagnosed hypertension and

diabetes among residents of selected communities in Ghana. Int J Chronic Dis 2017:7836019. https://doi.org/10.1155/2017/7836019
15. Sartorius B, Veerman LJ, Manyema M, Chola L, Hofman K (2015) Determinants of obesity and associated population attributability, South Africa: empirical evidence from a national panel survey, 2008–2012. PLoS One 10:e0130218. https://doi.org/10.1371/journal.pone.0130218
16. Price AJ, Crampin AC, Amberbir A, Kayuni-Chihana N, Musicha C, Tafatatha T et al (2018) Prevalence of obesity, hypertension, and diabetes, and cascade of care in sub-Saharan Africa: a cross-sectional, population-based study in rural and urban Malawi. Lancet Diabetes Endocrinol 6(3):208–222
17. Ramsay M, Crowther NJ, Agongo G, Ali SA, Asiki G, Boua RP et al (2018) Regional and sex-specific variation in BMI distribution in four sub-Saharan African countries: the H3Africa AWI-gen study. Glob Health Action 11(sup2):1556561. https://doi.org/10.1080/16549716.2018.1556561
18. Simonnet A, Chetboun M, Poissy J, Raverdy V, Noulette J, Duhamel A et al (2020) High prevalence of obesity in severe acute respiratory syndrome coronavirus-2 (SARS-CoV-2) requiring invasive mechanical ventilation. Obesity (Silver Spring) 28(7):1195–1199
19. Palaiodimos L, Kokkinidis DG, Li W, Karamanis D, Ognibene J, Arora S et al (2020) Severe obesity, increasing age and male sex are independently associated with worse in-hospital outcomes, and higher in-hospital mortality, in a cohort of patients with COVID-19 in the Bronx, New York. Metab Clin Exp 108:154262. https://doi.org/10.1016/j.metabol.2020.154262
20. Lighter J, Phillips M, Hochman S, Sterling S, Johnson D, Francois F et al (2020) Obesity in patients younger than 60 years is a risk factor for Covid-19 hospital admission. Clin Infect Dis 71(15):896–897
21. Wu J, Li W, Shi X, Chen Z, Jiang B, Liu J et al (2020) Early antiviral treatment contributes to alleviate the severity and improve the prognosis of patients with novel coronavirus disease (COVID-19). J Intern Med 288(1):128–138
22. Cai Q, Chen F, Wang T, Luo F, Liu X, Wu Q et al (2020) Obesity and COVID-19 severity in a designated hospital in Shenzhen, China. Diabetes Care 43(7):1392–1398
23. Busetto L, Bettini S, Fabris R, Serra R, Dal Pra C, Maffei P et al (2020) Obesity and COVID-19: an Italian snapshot. Obesity (Silver Spring) 28:1600–1605. https://doi.org/10.1002/oby.22918. Online ahead of print
24. Denova-Gutiérrez E, Lopez-Gatell H, Alomia-Zegarra JL, López-Ridaura R, Zaragoza-Jimenez CA, Dyer-Leal DD et al (2020) The association between obesity, type 2 diabetes, and hypertension with severe COVID-19 on admission among Mexicans. Obesity (Silver Spring). https://doi.org/10.1002/oby.22946. Online ahead of print
25. El Chaar M, King K, Galvez Lima A (2020) Are black and Hispanic persons disproportionately affected by COVID-19 because of higher obesity rates? Surg Obes Relat Dis 16:1096. https://doi.org/10.1016/j.soard.2020.04.038. Online ahead of print
26. Korakas E, Ikonomidis I, Kousathana F, Balampanis K, Kountouri A, Raptis A et al (2020) Obesity and COVID-19: immune and metabolic derangement as a possible link to adverse clinical outcomes. Am J Physiol Endocrinol Metab 319(1):E105–E109. https://doi.org/10.1152/ajpendo.00198.2020
27. Green WD, Beck MA (2017) Obesity impairs the adaptive immune response to Influenza Virus. Ann Am Thorac Soc 14(Supplement_5):S406–S409
28. Maurizi G, Della Guardia L, Maurizi A, Poloni A (2018) Adipocytes properties and crosstalk with immune system in obesity-related inflammation. J Cell Physiol 233(1):88–97
29. Sartipy P, Loskutoff DJ (2003) Monocyte chemoattractant protein 1 in obesity and insulin resistance. Proc Natl Acad Sci USA 100(12):7265–7270
30. Pieterse C, Schutte R, Schutte AE (2015) Leptin links with plasminogen activator inhibitor-1 in human obesity: the SABPA study. Hypertens Res 38(7):507–512
31. Frydrych LM, Bian G, O'Lone DE, Ward PA, Delano MJ (2018) Obesity and type 2 diabetes mellitus drive immune dysfunction, infection development, and sepsis mortality. J Leukoc Biol 104(3):525–534
32. Dixon AE, Peters U (2018) The effect of obesity on lung function. Expert Rev Respir Med 12(9):755–767
33. Mills KT, Bundy JD, Kelly TN, Reed JE, Kearney PM, Reynolds K et al (2016) Global disparities of hypertension prevalence and control: A systematic analysis of population-based studies from 90 countries. Circulation 134(6):441–450
34. Nulu S, Aronow WS, Frishman WH (2016) Hypertension in Sub-Saharan Africa: a contextual view of patterns of disease, best management, and systems issues. Cardiol Rev 24(1):30–40
35. Emami A, Javanmardi F, Pirbonyeh N, Akbari A (2020) Prevalence of underlying diseases in hospitalized patients with COVID-19: a systematic review and meta-analysis. Arch Acad Emerg Med 8:e35
36. Pranata R, Lim MA, Huang I, Raharjo SB, Lukito AA (2020) Hypertension is associated with increased mortality and severity of disease in COVID-19 pneumonia: a systematic review, meta-analysis and meta-regression. J Renin-Angiotensin-Aldosterone Syst 21(2):1470320320926899. https://doi.org/10.1177/1470320320926899
37. Caillon A, Schiffrin EL (2016) Role of inflammation and immunity in hypertension: Recent epidemiological, laboratory, and clinical evidence. Curr Hypertens Rep 8(3):21. https://doi.org/10.1007/s11906-016-0628-7
38. Rodriguez-Iturbe B, Pons H, Johnson RJ (2017) Role of the immune system in hypertension. Physiol Rev 97(3):1127–1164

39. Danser AHJ, Epstein M, Batlle D (2020) Renin-Angiotensin system blockers and the COVID-19 Pandemic: at present there is no evidence to abandon Renin-Angiotensin system blockers. Hypertension 75(6):1382–1385
40. Khera R, Clark C, Lu Y, Guo Y, Ren S, Truax B et al (2020) Association of Angiotensin-converting enzyme inhibitors and angiotensin receptor blockers with the risk of hospitalization and death in hypertensive patients with Coronavirus Disease-19. medRxiv. https://doi.org/10.1101/2020.05.17.20104943
41. Mancia G, Rea F, Ludergnani M, Apolone G, Corrao G (2020) Renin-Angiotensin-aldosterone system blockers and the risk of Covid-19. N Engl J Med 382(25):2431–2440
42. Yang G, Tan Z, Zhou L, Yang M, Peng L, Liu J et al (2020) Effects of Angiotensin II receptor blockers and ACE (Angiotensin-Converting Enzyme) inhibitors on virus infection, inflammatory status, and clinical outcomes in patients with COVID-19 and hypertension: a single-center retrospective study. Hypertension 76(1):51–58
43. Reynolds HR, Adhikari S, Pulgarin C, Troxel AB, Iturrate E, Johnson SB et al (2020) Renin-Angiotensin-aldosterone system inhibitors and risk of Covid-19. N Engl J Med 382(25):2441–2448
44. Ataklte F, Erqou S, Kaptoge S, Taye B, Echouffo-Tcheugui JB, Kengne AP (2015) Burden of undiagnosed hypertension in Sub-Saharan Africa: a systematic review and meta-analysis. Hypertension 65(2):291–298
45. Caballero AE, Ceriello A, Misra A, Aschner P, McDonnell ME, Hassanein M et al (2020) COVID-19 in people living with diabetes: an international consensus. J Diabetes Complicat 34:107671. https://doi.org/10.1016/j.jdiacomp.2020.107671. Online ahead of print
46. Mbanya JC, Motala AA, Sobngwi E, Assah FK, Enoru ST (2010) Diabetes in sub-Saharan Africa. Lancet 375(9733):2254–2266
47. Singh AK, Gupta R, Ghosh A, Misra A (2020) Diabetes in COVID-19: prevalence, pathophysiology, prognosis and practical considerations. Diabetes Metab Syndr 14(4):303–310
48. Allard R, Leclerc P, Tremblay C, Tannenbaum TN (2010) Diabetes and the severity of pandemic influenza A (H1N1) infection. Diabetes Care 33(7):1491–1493
49. Hussain A, Bhowmik B, do Vale Moreira NC (2020) COVID-19 and diabetes: knowledge in progress. Diabetes Res Clin Pract 162:108142. https://doi.org/10.1016/j.diabres.2020.108142
50. Shang J, Wang Q, Zhang H, Wang X, Wan J, Yan Y et al (2020) The relationship between diabetes mellitus and COVID-19 prognosis: a retrospective cohort study in Wuhan, China. Am J Med:S0002-9343(20)30532-5 https://doi.org/10.1016/j.amjmed.2020.05.033. Online ahead of print
51. Pititto BA, Ferreira SRG (2020) Diabetes and covid-19: more than the sum of two morbidities. Rev Saude Publica 54:54. https://doi.org/10.11606/s1518-8787.2020054002577
52. Zamparini J, Venturas J, Shaddock E, Edgar J, Naidoo V, Mahomed A et al (2020) Clinical characteristics of the first 100 COVID-19 patients admitted to a tertiary hospital in Johannesburg, South Africa. Wits J Clin Med 2:105. https://doi.org/10.18772/26180197.2020
53. Rubino F, Amiel SA, Zimmet P, Alberti G, Bornstein S, Eckel RH et al (2020) New-onset diabetes in Covid-19. N Engl J Med 383:789. https://doi.org/10.1056/NEJMc2018688. Online ahead of print
54. Gupta R, Hussain A, Misra A (2020) Diabetes and COVID-19: evidence, current status and unanswered research questions. Eur J Clin Nutr 74(6):864–870
55. Maddaloni E, Buzzetti R (2020) Covid-19 and diabetes mellitus: unveiling the interaction of two pandemics. Diabetes Metab Res Rev: e33213321. https://doi.org/10.1002/dmrr.3321. Online ahead of print
56. Cristelo C, Azevedo C, Marques JM, Nunes R, Sarmento B (2020) SARS-CoV-2 and diabetes: new challenges for the disease. Diabetes Res Clin Pract 164:108228. https://doi.org/10.1016/j.diabres.2020.108228
57. Yang JK, Lin SS, Ji XJ, Guo LM (2010) Binding of SARS coronavirus to its receptor damages islets and causes acute diabetes. Acta Diabetol 47(3):193–199
58. Jha V, Garcia-Garcia G, Iseki K, Li Z, Naicker S, Plattner B et al (2013) Chronic kidney disease: global dimension and perspectives. Lancet 382(9888):260–272
59. George JA, Brandenburg J-T, Fabian J, Crowther NJ, Agongo G, Alberts M et al (2019) Kidney damage and associated risk factors in rural and urban sub-Saharan Africa (AWI-Gen): a cross-sectional population study. Lancet Glob Health 7(12):e1632–e1643
60. de Lusignan S, Dorward J, Correa A, Jones N, Akinyemi O, Amirthalingam G et al (2020) Risk factors for SARS-CoV-2 among patients in the Oxford Royal College of General Practitioners Research and Surveillance Centre primary care network: a cross-sectional study. Lancet Infect Dis 20:1934. https://doi.org/10.1016/S1473-3099(20)30371-6. Online ahead of print
61. Emem-Chioma PC, Alasia DD, Wokoma FS (2012) Clinical outcomes of dialysis-treated acute kidney injury patients at the university of port harcourt teaching hospital, Nigeria. ISRN Nephrol 2013:540526. https://doi.org/10.5402/2013/540526
62. Riley S, Diro E, Batchelor P, Abebe A, Amsalu A, Tadesse Y et al (2013) Renal impairment among acute hospital admissions in a rural Ethiopian hospital. Nephrology (Carlton) 18(2):92–96
63. Chen YT, Shao SC, Hsu CK, Wu IW, Hung MJ, Chen YC (2020) Incidence of acute kidney injury in COVID-19 infection: a systematic review and

meta-analysis. Crit Care 24(1):346. https://doi.org/10.1186/s13054-020-03009-y
64. Yang X, Yu Y, Xu J, Shu H, Ja X, Liu H et al (2020) Clinical course and outcomes of critically ill patients with SARS-CoV-2 pneumonia in Wuhan, China: a single-centered, retrospective, observational study. Lancet Respir Med 8(5):475–481
65. Cheng Y, Luo R, Wang K, Zhang M, Wang Z, Dong L et al (2020) Kidney disease is associated with in-hospital death of patients with COVID-19. Kidney Int 97(5):829–838
66. Du M, Cai G, Chen F, Christiani DC, Zhang Z, Wang M (2020) Multiomics evaluation of gastrointestinal and other clinical characteristics of COVID-19. Gastroenterology 158(8):2298–2301.e7
67. Diao B, Wang C, Wang R, Feng Z, Tan Y, Wang H et al (2020) Human kidney is a target for novel Severe Acute Respiratory Syndrome Coronavirus 2 (SARS-CoV-2) infection. medRxiv. https://doi.org/10.1101/2020.03.04.20031120
68. Larsen CP, Bourne TD, Wilson JD, Saqqa O, Sharshir MA (2020) Collapsing glomerulopathy in a patient with Coronavirus Disease 2019 (COVID-19). Kidney Int Rep 5(6):935–939
69. Kissling S, Rotman S, Gerber C, Halfon M, Lamoth F, Comte D et al (2020) Collapsing glomerulopathy in a COVID-19 patient. Kidney Int 98(1):228–231
70. UNAIDS (2019) UNAIDS DATA 2019. https://www.unaids.org/en/resources/documents/2019/2019-UNAIDS-data. Accessed 23 July 2020
71. Vizcarra P, Pérez-Elías MJ, Quereda C, Moreno A, Vivancos MJ, Dronda F et al (2020) Description of COVID-19 in HIV-infected individuals: a single-centre, prospective cohort. Lancet HIV 7:e554. https://doi.org/10.1016/S2352-3018(20)30164-8. Online ahead of print
72. Gervasoni C, Meraviglia P, Riva A, Giacomelli A, Oreni L, Minisci D et al (2020) Clinical features and outcomes of HIV patients with coronavirus disease 2019. Clin Infect Dis:ciaa579. https://doi.org/10.1093/cid/ciaa579. Online ahead of print
73. Ridgway JP, Farley B, Benoit JL, Frohne C, Hazra A, Pettit N et al (2020) A case series of five people living with HIV hospitalized with COVID-19 in Chicago, Illinois. AIDS Patient Care STDs 34:331. https://doi.org/10.1089/apc.2020.0103. Online ahead of print
74. Härter G, Spinner CD, Roider J, Bickel M, Krznaric I, Grunwald S et al (2020) COVID-19 in people living with human immunodeficiency virus: a case series of 33 patients. Infection: 1–6. https://doi.org/10.1007/s15010-020-01438-z. Online ahead of print
75. Blanco JL, Ambrosioni J, Garcia F, Martínez E, Soriano A, Mallolas J et al (2020) COVID-19 in patients with HIV: clinical case series. Lancet HIV 7(5):e314–e316
76. Zhou Y, Vedantham P, Lu K, Agudelo J, Carrion R Jr, Nunneley JW et al (2015) Protease inhibitors targeting coronavirus and filovirus entry. Antivir Res 116:76–84
77. Ford N, Vitoria M, Rangaraj A, Norris SL, Calmy A, Doherty M (2020) Systematic review of the efficacy and safety of antiretroviral drugs against SARS, MERS or COVID-19: initial assessment. J Int AIDS Soc 23(4):e25489. https://doi.org/10.1002/jia2.25489
78. Mascolo S, Romanelli A, Carleo MA, Esposito V (2020) Could HIV infection alter the clinical course of SARS-CoV-2 infection? When less is better. J Med Virol 92:1777. https://doi.org/10.1002/jmv.25881. Online ahead of print
79. Chen G, Wu D, Guo W, Cao Y, Huang D, Wang H et al (2020) Clinical and immunological features of severe and moderate coronavirus disease 2019. J Clin Invest 130(5):2620–2629
80. Childs K, Post FA, Norcross C, Ottaway Z, Hamlyn E, Quinn K et al (2020) Hospitalized patients with COVID-19 and HIV: a case series. Clin Infect Dis: ciaa657. doi: https://doi.org/10.1093/cid/ciaa657. Online ahead of print
81. Cainelli F, Dzudzor B, Lanzafame M, Goushchi A, Chhem S, Vento S (2020) HIV and SARS-Coronavirus-2 epidemics: possible interactions and need for studies, especially in Africa. Front Med (Lausanne) 7:216. https://doi.org/10.3389/fmed.2020.00216
82. Suwanwongse K, Shabarek N (2020) Clinical features and outcome of HIV/SARS-CoV-2 coinfected patients in The Bronx, New York city. J Med Virol. https://doi.org/10.1002/jmv.26077. Online ahead of print
83. Brennan AT, Jamieson L, Crowther NJ, Fox MP, George JA, Berry KM et al (2018) Prevalence, incidence, predictors, treatment, and control of hypertension among HIV-positive adults on antiretroviral treatment in public sector treatment programs in South Africa. PLoS One 13:e0204020. https://doi.org/10.1371/journal.pone.0204020
84. Brennan AT, Berry KM, Rosen S, Stokes A, Crowther NJ, George J et al (2019) Growth curve modelling to determine distinct BMI trajectory groups in HIV-positive adults on antiretroviral therapy in South Africa. AIDS 33(13):2049–2059
85. Fang Y, Zhang H, Xie J, Lin M, Ying L, Pang P et al (2020) Sensitivity of chest CT for COVID-19: comparison to RT-PCR. Radiology 296(2):E115–E117
86. Chen Z, Li Y, Wu B, Hou Y, Bao J, Deng X (2020) A patient with COVID-19 presenting a false-negative reverse transcriptase polymerase chain reaction result. Korean J Radiol 21(5):623–624
87. WHO (2018) Tuberculosis. www.afro.who.int. Accessed 02 July 2020
88. Tadolini M, Codecasa LR, García-García JM, Blanc FX, Borisov S, Alffenaar JW et al (2020) Active tuberculosis, sequelae and COVID-19 co-infection: first cohort of 49 cases. Eur Respir J 56(1):2001398. https://doi.org/10.1183/13993003.01398-2020

89. Motta I, Centis R, D'Ambrosio L, García-García JM, Goletti D, Gualano G et al (2020) Tuberculosis, COVID-19 and migrants: preliminary analysis of deaths occurring in 69 patients from two cohorts. Pulmonology 26(4):233–240
90. Yu C, Yaguo W, Joy F, Yanhong Y, Ye G, Chang L et al (2020) Active or latent tuberculosis increases susceptibility to COVID-19 and disease severity. Medrxiv. https://doi.org/10.1101/2020.03.10.20033795
91. Walaza S, Tempia S, Dawood H, Variava E, Wolter N, Dreyer A et al (2019) The impact of influenza and tuberculosis interaction on mortality among individuals aged ≥15 years hospitalized with severe respiratory illness in South Africa, 2010–2016. Open Forum Infect Dis 6(3):ofz020. https://doi.org/10.1093/ofid/ofz020
92. Africa UNECf (2016) The demographic profile of African countries. https://www.uneca.org. Accessed 25 July 2020
93. Amimo F, Lambert B, Magit A (2020) What does the COVID-19 pandemic mean for HIV, tuberculosis, and malaria control? Trop Med Health 48:32. https://doi.org/10.1186/s41182-020-00219-6
94. Togun T, Kampmann B, Stoker NG, Lipman M (2020) Anticipating the impact of the COVID-19 pandemic on TB patients and TB control programmes. Ann Clin Microbiol Antimicrob 19:21. https://doi.org/10.1186/s12941-020-00363-1
95. Boffa J, Mhlaba T, Sulis G, Moyo S, Sifumba Z, Pai M et al (2020) COVID-19 and tuberculosis in South Africa: a dangerous combination. S Afr Med J 110(5):341–342
96. Dong Y, Mo X, Hu Y, Qi X, Jiang F, Jiang Z et al (2020) Epidemiology of COVID-19 among children in China. Pediatrics 145(6):e20200702. https://doi.org/10.1542/peds.2020-0702
97. Götzinger F, Santiago-García B, Noguera-Julián A, Lanaspa M, Lancella L, Calò Carducci FI et al (2020) COVID-19 in children and adolescents in Europe: a multinational, multicentre cohort study. Lancet Child Adolesc Health 4:653. https://doi.org/10.1016/S2352-4642(20)30177-2. Online ahead of print
98. WHO Weekly bulletin on outbreak and other emergencies: week 30 World Health Organisation. Regional Office for Africa https://apps.who.in/iris/handle/10665/333517. Accessed 26 July 2020
99. Fakiri KE, Nassih HN, Sab IA, Draiss G, Bouskraoul M (2020) Epidemiology and clinical features of coronavirus disease 2019 in Moroccan children. Indian J Pediatr 57:808. S097475591600207. Online ahead of print
100. Dufort EM, Koumans EH, Chow EJ, Rosenthal EM, Muse A, Rowlands J et al (2020) Multisystem inflammatory syndrome in children in New York State. N Engl J Med 383(4):347–358
101. Latimer G, Corriveau C, DeBiasi RL, Jantausch B, Delaney M, Jacquot C et al (2020) Cardiac dysfunction and thrombocytopenia-associated multiple organ failure inflammation phenotype in a severe paediatric case of COVID-19. Lancet Child Adolesc Health 4(7):552–554
102. Whittaker E, Bamford A, Kenny J, Kaforou M, Jones CE, Shah P et al (2020) Clinical characteristics of 58 children with a pediatric inflammatory multisystem syndrome temporally associated with SARS-CoV-2. JAMA 324:e2010369. https://doi.org/10.1001/jama.2020.10369. Online ahead of print
103. Pouletty M, Borocco C, Ouldali N, Caseris M, Basmaci R, Lachaume N et al (2020) Paediatric multisystem inflammatory syndrome temporally associated with SARS-CoV-2 mimicking Kawasaki disease (Kawa-COVID-19): a multicentre cohort. Ann Rheum Dis 79(8):999–1006
104. Hoffmann M, Kleine-Weber H, Schroeder S, Kruger N, Herrler T, Erichsen S et al (2020) SARS-CoV-2 cell entry depends on ACE2 and TMPRSS2 and is blocked by a clinically proven protease inhibitor. Cell 181(2):271-280.e8
105. Bunyavanich S (2020) Nasal gene expression of angiotensin-converting enzyme 2 in children and adults. JAMA 323(23):2427–2429
106. Yang LM, Li-Meng Y, Lagen W, Tian-Xin X, Aiping L, Jia-Ming L et al (2020) Viral dynamics in mild and severe cases of COVID-19. Lancet Infect Dis 20(6):656–657
107. John K, van de Sandt C, Lemke M, Lee C, Shoffner S, Chua B et al (2020) Distinct systems serology features in children, elderly and COVID patients. medRxiv. https://doi.org/10.1101/2020.05.11.20098459
108. UNICEF/WHO/WB Joint child malnutrition estimates: levels and trends in child malnutrition. Key findings of the 2020 edition. https://apps.who.int/iris/bitstream/handle/10665/331621/9789240003576-eng.pdf. Accessed 26 July 2020
109. Prendergast AJ, Rukobo S, Chasekwa B, Mutasa K, Ntozini R, Mbuya MNN et al (2014) Stunting is characterized by chronic inflammation in Zimbabwean infants. PLoS One 9:e86928. https://doi.org/10.1371/journal.pone.0086928
110. Rytter MJH, Kolte L, Briend A, Friis H, Christensen VB (2014) The immune system in children with malnutrition—a systematic review. PLoS One 9:e105017. https://doi.org/10.1371/journal.pone.0105017
111. Haque R, Snider C, Liu Y, Ma JZ, Liu L, Nayak U et al (2014) Oral polio vaccine response in breast fed infants with malnutrition and diarrhea. Vaccine 32(4):478–482
112. Sinha DP, Bang FB (1976) Protein and calorie malnutrition, cell mediated immunity, and B.C.G vacination in children from rural West Bengal. Lancet 2(7985):531–534
113. Bello-Chavolla OY, Bahena-López JP, Antonio-Villa NE, Vargas-Vázquez A, González-Díaz A, Márquez-Salinas A et al (2020) Predicting mortality due to SARS-CoV-2: a mechanistic score relating obesity and diabetes to COVID-19 outcomes in Mexico. J Clin Endocrinol Metab 105(8):dgaa346. https://doi.org/10.1210/clinem/dgaa346

114. Garg S, Kim L, Whitaker M, O'Halloran A, Cummings C, Holstein R et al (2020) Hospitalization rates and characteristics of patients hospitalized with laboratory-confirmed Coronavirus Disease 2019 — COVID-NET, 14 States, March 1–30, 2020. MMWR Morb Mortal Wkly Rep 69:458. https://doi.org/10.15585/mmwr.mm6915e3
115. Williams TN (2016) Sickle cell disease in Sub-Saharan Africa. Hematol Oncol Clin North Am 30(2):343–358
116. Arlet JB, de Luna G, Khimoud D, Odièvre MH, de Montalembert M, Joseph L et al (2020) Prognosis of patients with sickle cell disease and COVID-19: a French experience. Lancet Haematol 7:e632. https://doi.org/10.1016/S2352-3026(20)30204-0. Online ahead of print
117. McCloskey KA, Meenan J, Hall R, Tsitsikas DA (2020) COVID-19 infection and sickle cell disease: a UK centre experience. Br J Haematol 190(2):e57–e58
118. Chakravorty S, Padmore-Payne G, Ike F, Tshibangu V, Graham C, Rees D et al (2020) COVID-19 in patients with sickle cell disease – a case series from a UK Tertiary Hospital. Haematologica:haematol.2020.254250. https://doi.org/10.3324/haematol.2020.254250. Online ahead of print
119. Panepinto JA, Brandow A, Mucalo L, Yusuf F, Singh A, Taylor B et al (2020) Coronavirus disease among persons with sickle cell disease, United States, March 20-May 21, 2020. Emerg Infect Dis 26(10):2473. https://doi.org/10.3201/eid2610.202792. Online ahead of print
120. Folsom AR, Tang W, Roetker NS, Kshirsagar AV, Derebail VK, Lutsey PL et al (2015) Prospective study of sickle cell trait and venous thromboembolism incidence. J Thromb Haemost 13(1):2–9
121. Cameron PU, Jones P, Gorniak M, Dunster K, Paul E, Lewin S et al (2011) Splenectomy associated changes in IgM memory B cells in an adult spleen registry cohort. PLoS One 6(8):e23164. https://doi.org/10.1371/journal.pone.0023164
122. Koffi KG, Sawadogo D, Meite M, Nanho DC, Tanoh ES, Attia AK et al (2003) Reduced levels of T-cell subsets CD4+ and CD8+ in homozygous sickle cell anaemia patients with splenic defects. Hematol J 4(5):363–365
123. Brousse V, Buffet P, Rees D (2014) The spleen and sickle cell disease: the sick(led) spleen. Br J Haematol 166(2):165–176

Overview of the Haematological Effects of COVID-19 Infection

14

T. M. Wiggill, E. S. Mayne, J. L. Vaughan, and S. Louw

Abstract

From its early origins, COVID-19 has spread extensively and was declared a global pandemic by the World Health Organization in March of 2020. Although initially thought to be predominantly a respiratory infection, more recent evidence points to a multisystem systemic disease which is associated with numerous haematological and immunological disturbances in addition to its other effects. Here we review the current knowledge on the haematological effects of COVID-19.

Keywords

COVID-19 · SARS-CoV-2 · Pandemic · Haematology · Immunology · Human immunodeficiency virus · Tuberculosis

T. M. Wiggill (✉) · J. L. Vaughan · S. Louw
Department of Molecular Medicine and Haematology, Faculty of Health Sciences, University of the Witwatersrand and National Health Laboratory Service, Johannesburg, South Africa
e-mail: tracey.wiggill@nhls.ac.za

E. S. Mayne
Department of Immunology, Faculty of Health Sciences, University of the Witwatersrand and National Health Laboratory Service, Johannesburg, South Africa

14.1 Introduction

The first cases of COVID-19 were diagnosed in China in November to December of 2019, with patients presenting with severe pneumonia and acute respiratory distress syndrome (ARDS) of uncertain origin [1, 2]. Next-generation sequencing and phylogenetic analysis identified the associated pathogen as a novel β-coronavirus strain which has been called severe acute respiratory syndrome coronavirus 2 (SARS-CoV-2) [1].

From its early origins, COVID-19 has spread extensively and was declared a global pandemic by the World Health Organization (WHO) on 11th March 2020. As of 15th November 2020, there have been >53,7 million cases reported worldwide and >1,3 million deaths [3]. Although the majority of people infected with COVID-19 are either asymptomatic or mildly symptomatic (>80%), approximately 19% of patients develop severe to critical illness [2] with a variably reported mortality rate of less than 1% to >16% dependent on the country of origin, the testing strategy adopted, and the manner of calculation of the rate [2, 4, 5].

COVID-19 was initially thought to be predominantly a respiratory infection; however, more recent evidence points to a multisystem, systemic disease which is associated with numerous haematological and immunological disturbances [6, 7]. Very little data on COVID-19 have

P. C. Guest (ed.), *Clinical, Biological and Molecular Aspects of COVID-19*, Advances in Experimental Medicine and Biology 1321, https://doi.org/10.1007/978-3-030-59261-5_14

been published from South Africa and the African continent as a whole, although South Africa had the fifth highest number of confirmed cases worldwide in July and still has the highest number of cases on the African continent (>751,000 reported cases as of 15th November 2020) [8]. This region is also known to have the highest number of people living with human immunodeficiency virus (PLWHIV) worldwide at 7,7 million, with a high burden of other infectious diseases (including tuberculosis) [9], which may also associate with haematological and immunological complications. The interaction between these epidemics is uncertain [9, 10].

14.2 Full Blood Count (FBC) Changes Associated with COVID-19

14.2.1 Red Cells

Anaemia may be present in COVID-19 but is frequently only mild to moderate in severity [11–14]. The presence of severe anaemia is thought to infer a poorer prognosis [15]. Anaemia may occur as an autoimmune complication of COVID-19 infection, and autoimmune haemolysis has been described in case studies and case series [16, 17].

Severe COVID-19 may be associated with dysregulation of iron metabolism [18, 19]. This is hypothesized to be due to viral mimicry, as a component of the COVID-19 spike glycoprotein cytoplasmic tail displays significant homology with the hepcidin protein [20]. Hepcidin is a major regulator of iron metabolism, and in conjunction with its target receptor ferroportin, it controls iron exit from cells such as macrophages. COVID-19 is thus associated with intracellular iron accumulation (increased ferritin) and with a corresponding decline in serum iron and haemoglobin levels (likely due to restricted iron bioavailability/ reticuloendothelial iron blockade) [19, 20]. Iron accumulation may also drive a pro-inflammatory phenotype within macrophages and exacerbate the cytokine storm typical of severe disease [20, 21], with a negative impact on prognosis.

14.2.2 White Cells

14.2.2.1 Leukocyte Counts

Total leukocyte counts are variable in patients with COVID-19, ranging from decreased to increased in different patient subgroups [11, 12].

14.2.2.2 Lymphopenia

The most significant change associated with COVID-19 is lymphopenia [6, 7, 22, 23]. This is described in the majority of admitted patients (ranging from 30% to more than 80% of patients) [12, 24–28]. The presence and severity of the lymphopenia, in conjunction with its persistence during disease progression, has a negative prognostic implication [12, 22, 25, 29]. It appears to predict patients who develop severe disease with a higher risk of ARDS and who may require intensive care unit (ICU) admission and ventilation, with a potentially fatal outcome [6, 12, 29].

14.2.2.3 Alterations in Lymphocyte Subsets

Abnormalities of lymphocyte subsets and immunological function are described within the setting of lymphopenia. This is typically due to a decrease in both CD4+ and CD8+ T cell subsets, often with decreases in natural killer and B cells [22, 23, 26, 30]. Associated with this immunological disturbance is the activation of the T cells, with CD4 T cells driving a T-helper cell 1 (TH1)-dependent monocyte and macrophage response [31, 32] and increased cytotoxicity within the CD8-positive subset [33, 34]. This may contribute to the presence of a cytokine storm [35], autoimmune complications [36, 37], and immunological lung injury [34]. The normalisation of these counts, with a decrease in naïve and an increase in certain memory and regulatory subsets, is associated with recovery [22, 34].

14.2.2.4 Neutrophils and Monocytes

Neutrophilia is common in severe COVID-19 disease [22, 23, 29] and reflects the immunological dysregulation. It may also indicate secondary bacterial infection. In contrast, peripheral blood monocyte numbers were reported to be decreased in one study among patients with COVID-19 [38], which is hypothesized to be due to monocyte trafficking into the lungs. Circulating monocytes also display severity-specific immunophenotypic changes, including downregulation of surface HLA-DR expression [39] and an enrichment of a pro-inflammatory CD14+CD16+ monocyte subset [32, 38, 40]. The latter produce interleukin-6 (IL-6) and are likely to be important contributors to the cytokine storm seen in severe COVID-19 infection. The neutrophil–lymphocyte ratio (NLR) [12, 23] and the monocyte–lymphocyte ratio (MLR) may both be increased, which are also thought to be predictive of poorer outcome [41].

14.2.2.5 Changes in Eosinophil and Basophil Counts

Eosinophils and basophils play a role in COVID-19 in driving the immune pulmonary hyper-reactivity which is a feature of more severe disease [42]. Decreased peripheral blood eosinophils and basophils are potential predictors of more severe disease [22, 23], with a normalisation of these counts documented in recovery from COVID-19 [42].

14.2.3 Platelets

Thrombocytopenia may be present, although the prognostic significance differs between studies with some showing no association with severe disease [6, 25, 29] and others suggesting that thrombocytopenia predicts a poorer outcome [12, 14, 24, 43, 44]. Immune thrombocytopenic purpura (ITP) and thrombotic thrombocytopenic purpura (TTP) have been associated as autoimmune complications of COVID-19 leading to decreased platelet counts [45–49]. Disseminated intravascular coagulopathy (DIC) may also contribute to thrombocytopenia [7, 50].

14.3 Peripheral Blood Smear Morphological Findings

Multi-lineage atypical peripheral smear morphological features have been described in COVID-19 patients [51, 52]. Dysplasia is noted within the granulocyte lineage in a large proportion of patients with severe disease, including acquired Pelger-Huët and monolobate neutrophils with left-shift and apoptotic cells [51–53]. Dysplastic platelet morphology has also been reported with large hyperchromatic platelets and pseudopodia formation [52].

Circulating large plasmacytoid lymphocytes [54], plasma cells [27, 53–58], and plasmablasts [27, 51, 57] have been documented morphologically and on printouts from haematology analysers, which is suggestive of COVID-19 infection [55, 56, 58]. However, these cells are not specific for COVID-19 as they have been described in other viral infections, such as HIV [59] and viral haemorrhagic fevers particularly dengue fever [54, 60]. These cells differ from the atypical lymphocytes which more typically associate with other viral infections (such as Epstein-Barr virus and cytomegalovirus) although these later atypical cells may also be seen with COVID-19 infection [27, 54, 56, 57]. Other reactive changes may also be seen, including the presence of atypically large and vacuolated monocytes [40].

14.4 Coagulation Abnormalities

A comprehensive overview of COVID-19-associated abnormalities of coagulation is beyond the scope of this review. In brief, coagulopathy has been described in severe COVID-19 infection, with increased D-dimers [7, 12, 25, 29, 61] and prolongation of the prothrombin time (PT)/international normalized ratio (INR) [29] suggesting a poorer outcome, with possible development of DIC [7, 50]. Arterial and venous thrombosis has been associated with COVID-19 and may contribute to multiorgan failure [62, 63].

14.5 Autoimmune Haematological Complications

The significantly deranged immunological function in patients, particularly those who develop severe disease [35], may manifest in autoimmune complications including haematological and non-haematological disorders. This is hypothesized to occur through viral molecular mimicry [36, 37, 64]. Viral proteins (such as the viral spike and nucleoproteins) can cross-react with human antigens and may lead to the production of autoantibodies by the host [37, 65]. There have been case reports and case series of autoimmune haematological complications, including autoimmune haemolytic anaemia (AIHA) [16, 17], cold agglutinin disease [16], ITP [45, 46, 48], TTP [47, 49, 66], antiphospholipid antibodies [67, 68], lupus anticoagulant [69], and development of antibodies directed against the endothelium. Formation of antibodies against the endothelium may occur particularly in situations when the vascular endothelium is already activated and under stress due to comorbidities such as diabetes, hypertension, and cardiovascular disease. These may contribute to the development of severe disease, including ARDS, multiorgan failure, arterial and venous thrombosis, and DIC [64].

14.6 Hemophagocytic Lymphohistiocytosis (HLH)

Hemophagocytic lymphohistiocytosis (HLH) is a severe life-threatening disease which may occur as an inherited disorder or as a secondary phenomenon in other conditions, including infections [70]. It is associated with fever, organomegaly, severe cytopenias, increased serum ferritin and triglyceride levels, and multiorgan failure. HLH and macrophage activation syndrome (MAS) have been described in COVID-19 and may reflect the severe hyperinflammatory state [39, 70, 71]. This may contribute to a poor outcome.

14.7 High-Risk Patients with Haematological Disorders

Recent evidence suggests that patients with certain haematological disorders should be considered as being at high risk, in particular, patients with haemoglobinopathies such as sickle cell disease (SSD) and red cell enzymopathies such as glucose-6-phosphate dehydrogenase (G6PD) deficiency [72–74].

COVID-19 infection may interact with SSD in a number of ways. COVID-19 is associated with the development of pneumonia and, in particular, may cause severe hypoxia. Hypoxia increases red cell sickling and haemolysis in patients with SSD, potentiating COVID-19 vascular disturbances and thrombosis, which may contribute to the onset of painful vaso-occlusive crises and acute chest syndrome [74, 75]. Many patients with SSD have chronic lung damage from recurrent acute chest syndrome which may predispose to COVID-19 pneumonia. SSD is associated with auto-splenectomy and immunosuppression which predisposes to severe infections, including bacterial infections which may present with COVID-like symptoms or complicate COVID-19 infection [72, 74]. It has been suggested that haemoglobinopathies, including SSD, should be considered a comorbidity which predisposes to severe COVID-19 disease [76].

G6PD deficiency is the most common enzyme deficiency worldwide and may lead to haemolysis due to infections, including COVID-19 infection [73, 77]. These patients may also be susceptible to haemolysis precipitated by potential drugs researched in the treatment of COVID-19 such as hydroxychloroquine [78].

There is a theoretically increased risk of severe COVID-19 infection in patients with haematological malignancies. However, the presence of immune suppression has been hypothesized to be protective against severe COVID-19 disease. This remains controversial with some studies showing no increased risk of COVID-19 in patients with haematological malignancy [79] and others showing a higher risk of more severe disease [79–81]. Some of the autoimmune haematological complications have been precipitated by COVID-19 in patients with underlying haematological malignancies including lymphoproliferative disorders (such as chronic lymphocytic leukaemia) and multiple myeloma [16].

14.8 COVID-19 and HIV

As COVID-19 is associated with lymphopenia and immunological dysfunction, concern is raised for its interaction with other infections which impair immunological function such as HIV and/or tuberculosis, especially in the setting of comorbid diabetes mellitus. This remains a poorly quantified risk [10, 82, 83], although case studies and small case series, predominantly in higher income countries, suggest that PLWHIV may not be at higher risk of developing COVID-19 or a more severe disease [84–87].

Of particular interest is the similarity noted between the haematological complications of HIV and SARS-CoV-2. Although HIV has a tendency to a more chronic disease course, it associates with many similar features, including lymphopenia, immunological disturbance, molecular mimicry, autoimmune complications (AIHA, ITP, TTP, anti-phospholipid syndrome), morphological dysplasia, coagulopathy, thrombosis, and endothelial activation [88–90]. This suggests that the potential interaction between the viruses may lead to more severe disease or a greater frequency of the above complications.

Data from areas where the majority of PLWHIV reside, such as South Africa, are limited. This is further impacted by socioeconomic issues and healthcare accessibility [10, 82]. Within these regions, not all patients are on effective antiretroviral therapy [9], and patients may not be virologically suppressed or show immunological recovery [82]. This may increase the risk and severity of lymphopenia in COVID-19 infection in PLWHIV with low CD4 T-cell counts. Preliminary data on a small group of PLWHIV in Johannesburg, South Africa, supported findings in other centres of a lack of increased risk in HIV-positive patients [13]. Despite this, early data originating from Cape Town, South Africa, suggests that HIV-positive patients and patients with current or previous tuberculosis are at approximately twofold higher risk of COVID-19-related mortality, particularly if not virologically suppressed and in the presence of lymphopenia with low CD4 T-cell counts [82].

14.9 Haematological Markers of COVID-19 Severity

Although many of the haematological indices may be significantly different when COVID-19 patients are compared to normal controls, the results may still be within the normal reference ranges and may thus not be useful in distinguishing patients with and without COVID-19 [11, 12]. In addition, the differences in counts between COVID-19 patients and patients with other viral causes of respiratory disease (such as influenza) may not be significantly different or useful in distinguishing causes of pneumonia [11].

The most useful parameters in suggesting the presence of COVID-19 and predicting its severity include the presence and severity of lymphopenia and neutrophilia, the NLR, coagulation disturbances, and increase in D-dimers [7, 11, 12, 22, 41, 50, 61].

Table 14.1 summarises the haematological parameters associated with a higher risk of severe COVID-19 infection, which may predict the development of ARDS and increase the risk of ICU admission and mechanical ventilation and potentially a fatal outcome.

14.10 Conclusions

Patients with COVID-19 may have changes in all haemopoietic cell lineages and show atypical peripheral smear morphological findings. Coagulation disturbances are marked, and autoimmune haematological complications are increasingly described. Haematological changes may be suggestive of COVID-19 disease, although they may not be conclusive in confirming the diagnosis. Haematological parameters may also be useful in predicting patients who have more severe disease and may require ICU admission and mechanical ventilation.

Table 14.1 Peripheral blood haematological results in admitted patients with severe COVID-19 disease which may predict ARDS and ICU-admission

Parameter	Comments/Prognostic implications	References
Lymphopenia	Severe lymphopenia, failure of lymphocyte recovery, and/or persistent decline in lymphocyte counts associated with progressive disease and demise	[7, 11, 12, 22, 23, 25, 28, 29, 42]
Lymphocyte subsets	Lower T-cell counts (CD4- and CD8-positive T cells), failure of T-cell subsets, and T-cell activation to normalise and disturbance of memory and regulatory subsets predicts poorer outcome	[22, 26, 30, 34, 42]
Neutrophils and monocytes	Neutrophilia associates with poorer outcome. Increased monocytes with an activated immunophenotype and decreased expression of molecules associated with antigen presentation (HLA-DR) are suggestive of progressive disease and macrophage activation syndrome	[12, 22, 23, 25, 29, 32, 38–40]
Increased NLR and MLR	Suggests more aggressive disease	[7, 11, 23, 41]
Eosinophils and basophils	Decreased circulating eosinophils and basophils have negative prognostic implications. These cells increase with recovery	[22, 23, 42]
Multi-lineage haemopoietic dysplasia, large activated monocytes, and circulating plasmacytoid lymphocytes and plasmablasts	Improvement in dysplasia associates with recovery	[31, 40, 52]
Anaemia and thrombocytopenia	Predictive value is contradictory in different studies. Some studies suggest this is more common in patients requiring ICU admission, although may only be mildly to moderately reduced. More severe anaemia and thrombocytopenia is associated with poorer outcome in some studies	[12, 14, 15, 24, 43, 44]
Iron metabolism	Disturbances in iron metabolism described in COVID-19. Low serum iron associates with a poorer outcome	[18, 19]
Coagulopathy	Raised D-Dimers and prolonged INR associates with poorer prognosis	[7, 12, 24, 25, 29, 50, 61]

ARDS Acute respiratory distress syndrome, *ICU* intensive care unit, *INR* international normalized ratio, *NLR* neutrophil:lymphocyte ratio, *MLR* monocyte: lymphocyte ratio

References

1. Ren LL, Wang YM, Wu ZQ, Xiang ZC, Guo L, Xu T et al (2020) Identification of a novel coronavirus causing severe pneumonia in human: a descriptive study. Chin Med J 133(9):1015–1024
2. Wu Z, McGoogan JM (2020) Characteristics of and important lessons from the Coronavirus Disease 2019 (COVID-19) outbreak in China: summary of a report of 72 314 cases from the Chinese Center for Disease Control and Prevention. JAMA 323(13):1239–1242
3. World Health Organization (2020) COVID-19 Weekly epidemiological update - 17 November 2020. https://www.who.int/publications/m/item/weekly-epidemiological-update---17-november-2020. Accessed 20 November 2020
4. Baud D, Qi X, Nielsen-Saines K, Musso D, Pomar L, Favre G (2020) Real estimates of mortality following COVID-19 infection. Lancet Infect Dis 20(7):773. https://doi.org/10.1016/s1473-3099(20)30195-x
5. Wilson N, Kvalsvig A, Barnard LT, Baker MG (2020) Case-fatality risk estimates for COVID-19 calculated by using a lag time for fatality. Emerg Infect Dis 26(6):1339–1441
6. Fan BE, Chong VCL, Chan SSW, Lim GH, Lim KGE, Tan GB et al (2020) Hematologic parameters in patients with COVID-19 infection. Am J Hematol 95(6):E131–E134
7. Terpos E, Ntanasis-Stathopoulos I, Elalamy I, Kastritis E, Sergentanis TN, Politou M et al (2020) Hematological findings and complications of COVID-19. Am J Hematol 95(7):834–847
8. COVID-19 South African coronavirus news and information portal (2020) https://sacoronavirus.co.za/2020/11/15/update-on-covid-19-15th-november-2020/. Accessed 20 November 2020

9. Feldman C (2020) Potential impact of SARS-CoV-2 infection in HIV-positive patients in South Africa. Wits J Clin Med 2(Special Issue 1):19–24
10. Bulled N, Singer M (2020) In the shadow of HIV & TB: a commentary on the COVID epidemic in South Africa. Glob Public Health 15:1231. https://doi.org/10.1080/17441692.2020.1775275
11. Peng J, Qi D, Yuan G, Deng X, Mei Y, Feng L et al (2020) Diagnostic value of peripheral hematologic markers for coronavirus disease 2019 (COVID-19): a multicenter, cross-sectional study. J Clin Lab Anal 00:e23475. https://doi.org/10.1002/jcla.23475
12. Zhou F, Yu T, Du R, Fan G, Liu Y, Liu Z et al (2020) Clinical course and risk factors for mortality of adult inpatients with COVID-19 in Wuhan, China: a retrospective cohort study. Lancet 395(10229):1054–1062
13. Zamparini J, Venturas J, Shaddock E, Edgar J, Naidoo V, Mahomed A et al (2020) Clinical characteristics of the first 100 COVID-19 patients admitted to a tertiary hospital in Johannesburg, South Africa. Wits J Clin Med 2(2):105–114
14. Lippi G, Plebani M (2020) Laboratory abnormalities in patients with COVID-2019 infection. Clin Chem Lab Med 58(7):1131–1134
15. Zhao J, Gao HY, Feng ZY, Wu QJ (2020) A retrospective analysis of the clinical and epidemiological characteristics of COVID-19 patients in Henan Provincial People's Hospital, Zhengzhou, China. Front Med (Lausanne) 7:286. https://doi.org/10.3389/fmed.2020.00286
16. Lazarian G, Quinquenel A, Bellal M, Siavellis J, Jacquy C, Re D et al (2020) Autoimmune haemolytic anaemia associated with COVID-19 infection. Br J Haematol 190(1):29–31
17. Lopez C, Kim J, Pandey A, Huang T, DeLoughery TG (2020) Simultaneous onset of COVID-19 and autoimmune haemolytic anaemia. Br J Haematol 190(1):31–32
18. Cavezzi A, Troiani E, Corrao S (2020) COVID-19: hemoglobin, iron, and hypoxia beyond inflammation. A narrative review. Clin Pract 10(2):1271. https://doi.org/10.4081/cp.2020.1271
19. Hippchen T, Altamura S, Muckenthaler MU, Merle U (2020) Hypoferremia predicts hospitalization and oxygen demand in COVID-19 patients. medRxiv. https://doi.org/10.1101/2020.06.26.20140525
20. Ehsani S (2020) Distant sequence similarity between hepcidin and the novel coronavirus spike glycoprotein: a potential hint at the possibility of local iron dysregulation in COVID-19. arXiv preprint arXiv:200312191. https://arxiv.org/ftp/arxiv/papers/2003/200312191.pdf
21. Vinchi F, Costa da Silva M, Ingoglia G, Petrillo S, Brinkman N, Zuercher A et al (2016) Hemopexin therapy reverts heme-induced proinflammatory phenotypic switching of macrophages in a mouse model of sickle cell disease. Blood 127(4):473–486
22. Sun D-w, Zhang D, Tian R-h, Li Y, Wang Y-s, Cao J et al (2020) The underlying changes and predicting role of peripheral blood inflammatory cells in severe COVID-19 patients: a sentinel? Clin Chim Acta 508:122–129
23. Qin C, Zhou L, Hu Z, Zhang S, Yang S, Tao Y et al (2020) Dysregulation of immune response in patients with Coronavirus 2019 (COVID-19) in Wuhan, China. Clin Infect Dis 71(15):762–768
24. Guan W-j, Ni Z-y, Hu Y, Liang W-h, Ou C-q, He J-x et al (2020) Clinical characteristics of Coronavirus Disease 2019 in China. N Engl J Med 382(18):1708–1720
25. Wang D, Hu B, Hu C, Zhu F, Liu X, Zhang J et al (2020) Clinical characteristics of 138 hospitalized patients with 2019 Novel Coronavirus–infected pneumonia in Wuhan, China. JAMA 323(11):1061–1069
26. Liu Z, Long W, Tu M, Chen S, Huang Y, Wang S et al (2020) Lymphocyte subset (CD4+, CD8+) counts reflect the severity of infection and predict the clinical outcomes in patients with COVID-19. J Infect 8(2):318–356
27. El Jamal SM, Salib C, Stock A, Uriarte-Haparnas NI, Glicksberg BS, Teruya-Feldstein J et al (2020) Atypical lymphocyte morphology in SARS-CoV-2 infection. Pathol Res Pract 216(9):153063. https://doi.org/10.1016/j.prp.2020.153063
28. Yang X, Yu Y, Xu J, Shu H, Ja X, Liu H et al (2020) Clinical course and outcomes of critically ill patients with SARS-CoV-2 pneumonia in Wuhan, China: a single-centered, retrospective, observational study. Lancet Respir Med 8(5):475–481
29. Wu C, Chen X, Cai Y, Ja X, Zhou X, Xu S et al (2020) Risk factors associated with acute respiratory distress syndrome and death in patients with Coronavirus Disease 2019 pneumonia in Wuhan, China. JAMA Intern Med 180(7):934–943. https://doi.org/10.1001/jamainternmed.2020.0994
30. Wang F, Nie J, Wang H, Zhao Q, Xiong Y, Deng L et al (2020) Characteristics of peripheral lymphocyte subset alteration in COVID-19 pneumonia. J Infect Dis 221(11):1762–1769
31. Tay MZ, Poh CM, Rénia L, MacAry PA, Ng LFP (2020) The trinity of COVID-19: immunity, inflammation and intervention. Nat Rev Immunol 20(6):363–374
32. Zhou Y, Fu B, Zheng X, Wang D, Zhao C, qi Y et al (2020) Aberrant pathogenic GM-CSF+ T cells and inflammatory CD14+CD16+ monocytes in severe pulmonary syndrome patients of a new coronavirus. bioRxiv. https://doi.org/10.1101/2020.02.12.945576
33. Xu Z, Shi L, Wang Y, Zhang J, Huang L, Zhang C et al (2020) Pathological findings of COVID-19 associated with acute respiratory distress syndrome. Lancet Respir Med 8(4):420–422
34. Song J-W, Zhang C, Fan X, Meng F-P, Xu Z, Xia P et al (2020) Immunological and inflammatory profiles in mild and severe cases of COVID-19. Nat Commun 11(1):3410. https://doi.org/10.1038/s41467-020-17240-2
35. Mehta P, McAuley DF, Brown M, Sanchez E, Tattersall RS, Manson JJ (2020) COVID-19: consider

cytokine storm syndromes and immunosuppression. Lancet 395(10229):1033–1034
36. Angileri F, Legare S, Marino Gammazza A, Conway de Macario E, Macario AJI, Cappello F (2020) Molecular mimicry may explain multi-organ damage in COVID-19. Autoimmun Rev 19(8):102591. https://doi.org/10.1016/j.autrev.2020.102591
37. Angileri F, Légaré S, Marino Gammazza A, Conway de Macario E, Macario AJL, Cappello F (2020) Is molecular mimicry the culprit in the autoimmune haemolytic anaemia affecting patients with COVID-19? Br J Haematol 190(2):e92–e93
38. Zhou Y, Fu B, Zheng X, Wang D, Zhao C, Qi Y et al (2020) Pathogenic T-cells and inflammatory monocytes incite inflammatory storms in severe COVID-19 patients. Natl Sci Rev 7(6):998–1002
39. Giamarellos-Bourboulis EJ, Netea MG, Rovina N, Akinosoglou K, Antoniadou A, Antonakos N et al (2020) Complex immune dysregulation in COVID-19 patients with severe respiratory failure. Cell Host Microbe 27(6):992-1000.e1003. https://doi.org/10.1016/j.chom.2020.04.009
40. Zhang D, Guo R, Lei L, Liu H, Wang Y, Wang Y et al (2020) COVID-19 infection induces readily detectable morphological and inflammation-related phenotypic changes in peripheral blood monocytes, the severity of which correlate with patient outcome. medRxiv. https://doi.org/10.1101/2020.03.24.20042655
41. Lagunas-Rangel FA (2020) Neutrophil-to-lymphocyte ratio and lymphocyte-to-C-reactive protein ratio in patients with severe coronavirus disease 2019 (COVID-19): a meta-analysis. J Med Virol 92:1733. https://doi.org/10.1002/jmv.25819. Online ahead of print
42. Rodriguez L, Pekkarinen P, Tadepally LK, Tan Z, Rosat Consiglio C, Pou C et al (2020) Systems-level immunomonitoring from acute to recovery phase of severe COVID-19. medRxiv. https://doi.org/10.1101/2020.06.03.20121582
43. Yang X, Yang Q, Wang Y, Wu Y, Xu J, Yu Y et al (2020) Thrombocytopenia and its association with mortality in patients with COVID-19. J Thromb Haemost 18(6):1469–1472
44. Lippi G, Plebani M, Henry BM (2020) Thrombocytopenia is associated with severe coronavirus disease 2019 (COVID-19) infections: a meta-analysis. Clin Chim Acta 506:145–148
45. Zulfiqar A-A, Lorenzo-Villalba N, Hassler P, Andrès E (2020) Immune thrombocytopenic purpura in a patient with COVID-19. N Engl J Med 382(18):e43. https://doi.org/10.1056/nejmc2010472
46. Bomhof G, Mutsaers PGNJ, Leebeek FWG, te Boekhorst PAW, Hofland J, Croles FN et al (2020) COVID-19-associated immune thrombocytopenia. Br J Haematol 190(2):e61–e64
47. Hindilerden F, Yonal-Hindilerden I, Akar E, Kart-Yasar K (2020) Covid-19 associated autoimmune thrombotic thrombocytopenic purpura: report of a case. Thromb Res 195:136–138
48. Lévesque V, Millaire É, Corsilli D, Rioux-Massé B, Carrier F-M (2020) Severe immune thrombocytopenic purpura in critical COVID-19. Int J Lab Hematol: 1–5. https://doi.org/10.1007/s12185-020-02931-9. Online ahead of print
49. Capecchi M, Mocellin C, Abbruzzese C, Mancini I, Prati D, Peyvandi F (2020) Dramatic presentation of acquired TTP associated with COVID-19. Haematologica. https://doi.org/10.3324/haematol.2020.262345. Online ahead of print
50. Tang N, Li D, Wang X, Sun Z (2020) Abnormal coagulation parameters are associated with poor prognosis in patients with novel coronavirus pneumonia. J Thromb Haemost 18(4):844–847
51. Lüke F, Orsó E, Kirsten J, Poeck H, Grube M, Wolff D et al (2020) Coronavirus disease 2019 induces multi-lineage, morphologic changes in peripheral blood cells. eJHaem 1:376. https://doi.org/10.1002/jha2.44
52. Zini G, Bellesi S, Ramundo F, d'Onofrio G (2020) Morphological anomalies of circulating blood cells in COVID-19. Am J Hematol 95(7):870–872
53. Nazarullah A, Liang C, Villarreal A, Higgins RA, Mais DD (2020) Peripheral blood examination findings in SARS-CoV-2 infection. Am J Clin Pathol 154:319. https://doi.org/10.1093/ajcp/aqaa108
54. Chong VCL, Lim KGE, Fan BE, Chan SSW, Ong KH, Kuperan P (2020) Reactive lymphocytes in patients with COVID-19. Br J Haematol 189(5):844–844
55. Foldes D, Hinton R, Arami S, Bain BJ (2020) Plasmacytoid lymphocytes in SARS-CoV-2 infection (Covid-19). Am J Hematol 95(7):861–862
56. Gérard D, Henry S, Thomas B (2020) SARS-CoV-2: a new aetiology for atypical lymphocytes. Br J Haematol 189(5):845. https://doi.org/10.1111/bjh.16730
57. Weinberg SE, Behdad A, Ji P (2020) Atypical lymphocytes in peripheral blood of patients with COVID-19. Br J Haematol 190(1):36–39
58. Osman J, Lambert J, Templé M, Devaux F, Favre R, Flaujac C et al (2020) Rapid screening of COVID-19 patients using white blood cell scattergrams, a study on 381 patients. Br J Haematol 190:718. https://doi.org/10.1111/bjh.16943. Online ahead of print
59. Wiggill TM, Mayne ES, Willem P (2013) Challenges in lymphoma diagnosis in HIV positive patients in the South African setting. Transfus Apher Sci 49(2):157–162
60. Gawoski JM, Ooi WW (2003) Dengue fever mimicking plasma cell leukemia. Arch Pathol Lab Med 127(8):1026–1027

61. Lippi G, Favaloro EJ (2020) D-dimer is associated with severity of Coronavirus Disease 2019: a pooled analysis. Thromb Haemost 120(5):876
62. Cui S, Chen S, Li X, Liu S, Wang F (2020) Prevalence of venous thromboembolism in patients with severe novel coronavirus pneumonia. J Thromb Haemost 18(6):1421–1424
63. Bavishi C, Bonow RO, Trivedi V, Abbott JD, Messerli FH, Bhatt DL (2020) Acute myocardial injury in patients hospitalized with COVID-19 infection: a review. Prog Cardiovasc Dis;S0033-0620(20)30123-7. https://doi.org/10.1016/j.pcad.2020.05.013. Online ahead of print
64. Cappello F (2020) Is COVID-19 a proteiform disease inducing also molecular mimicry phenomena? Cell Stress Chaperones 2(3):381–382
65. Vojdani A, Kharrazian D (2020) Potential antigenic cross-reactivity between SARS-CoV-2 and human tissue with a possible link to an increase in autoimmune diseases. Clin Immunol 217:108480. https://doi.org/10.1016/j.clim.2020.108480
66. Albiol N, Awol R, Martino R (2020) Autoimmune thrombotic thrombocytopenic purpura (TTP) associated with COVID-19. Ann Hematol 99(7):1673–1674
67. Zhang Y, Xiao M, Zhang S, Xia P, Cao W, Jiang W et al (2020) Coagulopathy and antiphospholipid antibodies in patients with Covid-19. N Engl J Med 382(17):e38. https://doi.org/10.1056/NEJMc2007575
68. Hossri S, Shadi M, Hamarsha Z, Schneider R, El-Sayegh D (2020) Clinically significant anticardiolipin antibodies associated with COVID-19. J Crit Care 59:32–34
69. Harzallah I, Debliquis A, Drénou B (2020) Lupus anticoagulant is frequent in patients with Covid-19. J Thromb Haemost n/a (n/a). https://doi.org/10.1111/jth.14867
70. Opoka-Winiarska V, Grywalska E, Roliński J (2020) Could hemophagocytic lymphohistiocytosis be the core issue of severe COVID-19 cases? BMC Med 18(1):214. https://doi.org/10.1186/s12916-020-01682-y
71. Dewaele K, Claeys R (2020) Hemophagocytic lymphohistiocytosis in SARS-CoV-2 infection. Blood 135(25):2323. https://doi.org/10.1182/blood.2020006505
72. De Luna G, Habibi A, Deux JF, Colard M, Pham Hung d'Alexandry d'Orengiani AL, Schlemmer F et al (2020) Rapid and severe Covid-19 pneumonia with severe acute chest syndrome in a sickle cell patient successfully treated with tocilizumab. Am J Hematol 95(7):876–878
73. Aydemir D, Ulusu NN (2020) Is glucose-6-phosphate dehydrogenase enzyme deficiency a factor in Coronavirus-19 (COVID-19) infections and deaths? Pathog Glob Health 114(3):109–110
74. Hussain FA, Njoku FU, Saraf SL, Molokie RE, Gordeuk VR, Han J (2020) COVID-19 infection in patients with sickle cell disease. Br J Haematol 189(5):851–852
75. Nur E, Gaartman AE, van Tuijn CFJ, Tang MW, Biemond BJ (2020) Vaso-occlusive crisis and acute chest syndrome in sickle cell disease due to 2019 novel coronavirus disease (COVID-19). Am J Hematol 95(6):725–726
76. de Sanctis V, Canatan D, Corrons JLV, Karimi M, Daar S, Kattamis C et al (2020) Preliminary data on COVID-19 in patients with hemoglobinopathies: a multicentre ICET-A study. Mediterr J Hematol Infect Dis 12(1):–e2020046. https://doi.org/10.4084/mjhid.2020.046
77. Jamerson BD, Haryadi TH, Bohannon A (2020) Glucose-6-phosphate dehydrogenase deficiency: an actionable risk factor for patients with COVID-19? Arch Med Res:S0188-4409(20)30948-6. https://doi.org/10.1016/j.arcmed.2020.06.006. Online ahead of print
78. Maillart E, Leemans S, Van Noten H, Vandergraesen T, Mahadeb B, Salaouatchi MT et al (2020) A case report of serious haemolysis in a glucose-6-phosphate dehydrogenase-deficient COVID-19 patient receiving hydroxychloroquine. Infect Dis (Lond) 52(9):659–661
79. He W, Chen L, Chen L, Yuan G, Fang Y, Chen W et al (2020) COVID-19 in persons with haematological cancers. Leukemia 34(6):1637–1645
80. El-Sharkawi D, Iyengar S (2020) Haematological cancers and the risk of severe COVID-19: exploration and critical evaluation of the evidence to date. Br J Haematol 190:336. https://doi.org/10.1111/bjh.16956. Online ahead of print
81. Sanchez-Pina JM, Rodríguez Rodriguez M, Castro Quismondo N, Gil Manso R, Colmenares R, Gil Alos D et al (2020) Clinical course and risk factors for mortality from COVID-19 in patients with haematological malignancies. Eur J Haematol:597–105. https://doi.org/10.1111/ejh.13493. Online ahead of print
82. Davies MA (2020) HIV and risk of COVID-19 death: a population cohort study from the Western Cape Province, South Africa. medRxiv. https://doi.org/10.1101/2020.07.02.20145185
83. Zhu F, Cao Y, Xu S, Zhou M (2020) Co-infection of SARS-CoV-2 and HIV in a patient in Wuhan city, China. J Med Virol 92(6):529–530
84. Blanco JL, Ambrosioni J, Garcia F, Martínez E, Soriano A, Mallolas J et al (2020) COVID-19 in patients with HIV: clinical case series. Lancet HIV 7(5):e314–e316
85. Byrd KM, Beckwith CG, Garland JM, Johnson JE, Aung S, Cu-Uvin S et al (2020) SARS-CoV-2 and HIV coinfection: clinical experience from Rhode

Island, United States. J Int AIDS Soc 23(7):e25573. https://doi.org/10.1002/jia2.25573
86. Toombs JM, Van den Abbeele K, Democratis J, Merricks R, Mandal AKJ, Missouris CG (2020) COVID-19 in three people living with HIV in the United Kingdom. J Med Virol. https://doi.org/10.1002/jmv.26178. Online ahead of print
87. Shalev N, Scherer M, LaSota ED, Antoniou P, Yin MT, Zucker J et al (2020) Clinical characteristics and outcomes in people living with human immunodeficiency virus hospitalized for Coronavirus Disease 2019. Clin Infect Dis:jiaa380. https://doi.org/10.1093/infdis/jiaa380. Online ahead of print
88. Opie J (2012) Haematological complications of HIV infection. S Afr Med J 102(6):465–468
89. Bhardwaj S, Almaeen A, Ahmed Wani F, Thirunavukkarasu A (2020) Hematologic derangements in HIV/AIDS patients and their relationship with the CD4 counts: a cross-sectional study. Int J Clin Exp Pathol 13(4):756–763
90. Volberding PA, Baker KR, Levine AM (2003) Human immunodeficiency virus hematology. Hematology 2003(1):294–313

The Novel Coronavirus and Haemostatic Abnormalities: Pathophysiology, Clinical Manifestations, and Treatment Recommendations

15

S. Louw, B. F. Jacobson, E. S. Mayne, and T. M. Wiggill

Abstract

The COVID-19 pandemic, caused by the SARS-COV-2 virus, was initially considered and managed in a similar manner to the previous SARS epidemic as they are both caused by coronaviruses. What has now become apparent is that a major cause of morbidity and mortality in COVID-19 is abnormal thrombosis. This thrombosis occurs on a macro- and microvascular level and is unique to this disease. The virus has been demonstrated in the endothelium of the pulmonary alveoli and as such is thought to contribute to the devastating respiratory complications encountered. D-dimer concentrations are frequently raised in COVID to levels not frequently seen previously. The optimal anticoagulation treatment in COVID remains to be determined, and the myriad of pathophysiologic effects caused by this virus in the human host have also yet to be fully elucidated.

Keywords

COVID-19 · SARS-CoV-2 · Coagulopathy · Thrombosis · Biomarkers · Treatment

S. Louw (✉) · B. F. Jacobson · T. M. Wiggill
Department of Molecular Medicine and Haematology, National Health Laboratory Service, Johannesburg, South Africa

Faculty of Health Sciences, University of the Witwatersrand,
Johannesburg, South Africa
e-mail: Susan.louw@nls.ac.za

E. S. Mayne
Faculty of Health Sciences, University of the Witwatersrand,
Johannesburg, South Africa

Department of Immunology, National Health Laboratory Service, Johannesburg, South Africa

15.1 Background

In December 2019, a disease (COVID-19), caused by the novel severe acute respiratory syndrome coronavirus (SARS-COV-2), was documented in the city of Wuhan in the Hubei province in China, which rapidly spread to the rest of the world. To date, according to the World Health Organisation (WHO), there have been more than 17 million COVID-19 infections and 677 thousand deaths in 216 countries, with an estimated mortality rate of approximately 3.9% (as of July 31, 2020). As the world struggles with the health, social, and economic impact of this pandemic, the medical fraternity is assessing various aspects of this viral disease with scientific publications

P. C. Guest (ed.), *Clinical, Biological and Molecular Aspects of COVID-19*, Advances in Experimental Medicine and Biology 1321, https://doi.org/10.1007/978-3-030-59261-5_15

detailing the involvement of multiple organ systems in the human host [1, 2].

15.2 Coagulation System

The coagulation system is integral to the innate immune response to severe infection and patients with severe COVID-19 disease commonly present with systemic coagulation abnormalities such as disseminated intravascular coagulation (DIC) and other thrombotic microangiopathies [3]. As such, the International Society on Thrombosis and Haemostasis (ISTH) recommends measuring D-dimers, prothrombin time (PT), activated partial thromboplastin time (aPTT), and platelet count in all hospitalised patients with COVID-19 [4]. Patients with the disease are also predisposed to both venous and arterial thromboembolism, related to a hypercoagulable state [5–7].

15.3 COVID-19 Coagulopathy

The SARS-COV-2 virus gains access to human cells by binding to the angiotensin-converting enzyme (ACE) 2 proteins on cell membranes of various human tissues including the heart, lungs, and brain with an estimated 20-fold increased affinity compared to SARS-COV [8, 9]. Viral DNA has consequently been detected in multiple organs including alveolar type II epithelial cells and monocytes [9]. Cell invasion by SARS-COV-2 also depends on the availability of the protease transmembrane protease serine 2 (TMPRSS-2) [10]. SARS-COV-2 viral infection consequently results in significant inflammation because of wide tissue distribution and release of pro-inflammatory cytokines culminating in a systemic inflammatory response syndrome (SIRS) with multiorgan dysfunction (MODS) related to accelerated cell death in many organs [11, 12]. SIRS results in endothelial cell (EC) damage secondary to complement activation, direct viral infection of ECs, and the cytolytic action of cytotoxic T cells. Damage to ECs with activation of the coagulation system manifests as an immunothrombotic syndrome with generalised small vessel vasculitis and microthrombosis resulting in a consumptive coagulopathy [11–13].

Approximately 71% of patients (in some studies) who died from COVID-19 met the ISTH scoring criteria for DIC (Table 15.1), and it is therefore recommended that laboratory tests in patients admitted with COVID-19 should include the markers of haemostasis included in this scoring system [14, 15].

Repeated measurements of these markers may be indicated to identify a worsening coagulopathy which may warrant an increase in the level of care, blood product support, and possibly therapies including anticoagulation and immunosuppression therapy [3, 13].

D-dimer levels on admission serves as a prognostic marker of the severity of COVID-19 disease with levels greater than 2.0 μg/mL (i.e., a fourfold increase) predicting in-hospital mortality [17–19]. The PT in COVID-19 non-survivors at admission was also mildly prolonged at a mean

Table 15.1 International Society on Thrombosis and Haemostasis (ISTH) scoring criteria for disseminated intravascular coagulation (DIC) [3, 16]

Criteria:	Assigned score:	Relevance in COVID-19
Platelet count (x 10^9/L)	≥100 = 0 50–99 = 1 <50 = 2	Thrombocytopaenia may be prognostic, but this is inconsistent [17]
Fibrinogen (mg/dL)	≥100 = 0 <100 = 1	Decreased in patients progressing to overt DIC [3]
Prothrombin time (PT) prolongation above upper limit of normal (ULN) (seconds)	<3 = 0 3–6 = 1 >6 = 2	May be subtly prolonged (15.5 s for survivors vs 13.6 s for non-survivors) [3]
D-dimer increase above upper limit of normal (ULN) (ng/mL)	<2 = 0 2–4 = 2 >4 = 3	Levels >2.0 ug/mL predict mortality [17–19]
Total score	<5 = No overt DIC ≥5 = Overt DIC	

level of 15.5 s versus 13.6 s in survivors. This subtle prolongation in PT may not be appreciated if the PT is converted to an international normalised ratio (INR) [3]. Low platelet count (thrombocytopenia) is an indicator of sepsis-related mortality in critically ill patients [20], but this is not consistently the case in COVID-19 patients. Studies have documented platelet counts of less than 100×10^9/L in as few as 5% of hospitalised COVID-19 patients on admission [21], but lower platelet counts correlate with mortality [22]. Thrombocytopenia at the time of admission may therefore be a prognostic marker but is inconsistent [17]. Measuring fibrinogen and antithrombin levels may also be of value as decreased levels are indicative of progression to an overt DIC [3].

15.4 Thrombotic Manifestations

There is an increased prevalence of venous thromboembolic disease (VTED) in patients with COVID-19 [23], and this is associated with poor prognosis. It is unclear whether the risk of VTED is greater for COVID-19-related sepsis than for other septic processes [24]. Hypercoagulability relates to cytokine release [including interleukin (IL)-1, IL-6 and tumour necrosis factor alpha (TNF-α)], hypoxia, immobilisation, and dehydration. Interestingly, DIC also appears to contribute to the increased risk of VTED [5, 25–29]. Increasing age and presence of a coagulopathy denoted by prolongation of the PT and aPTT, although counterintuitive, were independent predictors of thrombotic complications in a study on 184 Dutch patients admitted to the intensive care unit (ICU) with COVID-19 pneumonia [25]. Deep vein thromboses (DVTs) and pulmonary emboli (PE) have been documented in up to 27% of patients with COVID-19 admitted to ICU with PE constituting 80% of events [25]. VTED can occur despite prophylactic anticoagulation therapy [5, 25, 30]. Pulmonary thrombi in COVID-19 may not be embolic but rather related to in situ pulmonary thrombosis secondary to endotheliitis [24]. Thromboses also develop in extracorporeal circuits for continuous veno-venous hemofiltration (CVVH) and central venous catheters and can also manifest as extensive thrombophlebitis [5].

The clinical diagnosis of VTED can be challenging in ICU patients with COVID-19 because of the clinical overlap with the COVID-19 pneumonia and difficulties with clinical examination [5, 31]. Radiologic confirmation of VTED may be required consisting of Duplex Doppler investigation for DVT and CT pulmonary angiography (CTPA) for PE, which is preferred over ventilation perfusion scan (VQ scan) because of the COVID-19-related lung changes and the risk of aerosolised spread of the virus [24, 32].

Laboratory evidence of the hypercoagulable state in patients admitted with COVID-19 includes elevated fibrinogen and D-dimer levels as well as short clot formation time (CFT) and higher maximum clot firmness (MCF) on thromboelastometry [30]. Both D-dimer and fibrinogen levels are elevated, and D-dimer levels three–fourfold higher than the upper limit of normal are an indication for admission irrespective of the clinical presentation. Serial monitoring of D-dimers may be useful as a sudden elevation may denote the development of VTED (although a background raised D-dimer may be seen in most patients reducing the predictive value) [29]. Decreases in the natural anticoagulants, protein C, and antithrombin, and increases of plasminogen activator inhibitor-1 (PAI-1) levels as well as presence of antiphospholipid antibodies have also been documented in COVID-19 patients and contribute to the hypercoagulability [26].

15.5 Arterial Thromboses

15.5.1 Cardiac Events

Elevation of cardiac biomarkers, such as high-sensitivity cardiac troponin (hs-troponin) and creatinine kinase-MB (CKMB), is common in patients with COVID-19 with an overall prevalence of acute myocardial injury of up to 38% and acute myocardial infarction of 20% [7]. There is an apparent linear relationship between the magnitude of elevation of cardiac troponin and COVID-19 severity and overall prognosis [31, 33, 34]. Additional cardiac manifestations of COVID-19 include arrhythmias, worsening of existing or new onset of heart failure which can be exacerbated by electrolyte disturbances, hypoxia, and myocardial ischaemia [31, 35].

Although the pathogenesis of myocardial injury in COVID-19 is not completely understood, the following mechanisms are postulated: (1) respiratory failure with hypoxic damage to cardiac muscle [36]; (2) inflammatory cytokine storm resultant myocarditis [37]; (3) direct endothelial injury by SARS-CoV-2 viral infection and/or host inflammatory response [38]; (4) downregulation ofACE-2 receptor expression with loss of subsequent protective anti-inflammatory, antioxidative, and vasodilatory signalling pathways in cardiac myocytes [37, 39]; (5) general hypercoagulability with coronary microvascular thrombosis [40]; and (6) inflammation and cardiac strain with coronary plaque rupture and myocardial ischemia/infarction (MI) [7]. COVID-19 viral RNA has been detected in autopsied human heart specimens of patients who died from this viral infection [41].

Since acute myocardial injury in COVID-19 can be asymptomatic and indicated only by underlying elevation of cardiac troponins, various guidelines have been issued with respect to laboratory testing in patients with COVID-19. The Chinese Clinical Guidance for COVID-19 Pneumonia, the WHO document on the management of severe COVID-19, the Asian Critical Care Clinical Trials Group, and the British Medical Journal Best Practice guidelines all advocate for testing of troponins on admission with repeat testing when clinically indicated, but the American College of Cardiology supports testing only if clinically indicated [7]. Evaluation of pro-brain natriuretic peptide (BNP) should be considered if clinical evidence of cardiomyopathy is present [42]. Special investigations that may be indicated include echocardiography, right and left cardiac catheterization with placement of a pulmonary artery catheter for continuous hemodynamic monitoring, and cardiac magnetic resonance imaging (MRI) [7].

15.5.2 Cerebrovascular Accidents

The precise incidence of cerebrovascular accidents (CVAs), i.e., strokes, in patients with COVID-19 is not known, but it is apparent that this is an important complication and has been reported in up to 5.7% of patients with severe disease [43]. COVID-19-related CVAs are predominantly ischaemic, involving large cerebral vessels although haemorrhagic strokes have also been described [43]. These events seem to correlate with underlying severity of systemic disease, the presence of comorbidities such diabetes and hypertension and inflammation in older patients, and the degree of hypercoagulability in younger patients [44].

Human brain endothelial cells display ACE II receptors, and it is postulated that viral interaction with these receptors and the downstream effects are pathophysiological in COVID-related ischaemic CVAs [37]. Antiphospholipid antibodies have also been demonstrated in patients with COVID-19, and together with endothelial dysfunction and a general hypercoagulable state could be contributory [6]. The possibility also exists that strokes in patients with COVID may relate to thromboses in the pulmonary veins which migrate to the left atrium or paradoxical emboli via a patent foramen ovale created by the high pulmonary pressures.

15.6 Microvascular Thrombosis and Endothelitis

Thromboses in the microvascular compartment of the circulation play an important role in the pathogenesis of organ dysfunction in COVID-19. The microvascular thrombosis in COVID-19 and in related disorders is referred to as thromboinflammation since inflammation, together with hypercoagulability and loss of physiological inhibitory control of the coagulation and the innate immune system are pivotal in its development [45, 46]. Healthy microvascular endothelium has powerful antithrombotic properties including a heparin sulphate-containing endothelial glycocalyx [45]. It is not possible to detect thromboses in the microvasculature with common imaging techniques, and the diagnosis of thrombotic microvascular disease usually relies on clinical, laboratory, and histological observations [47].

Histopathological studies have highlighted the crucial role of ECs in the vascular dysfunction, inflammation, and immunothrombosis in COVID-19. Direct viral infection of endothelial cells and diffuse, inflammation-related endotheliitis with mononuclear cell infiltrate have been demonstrated as well as the presence of microvascular thrombosis in the various organs including the kidney, heart, lung, and liver [15, 38, 45]. ACE 2 and TMPRSS-2, present on the surface of arterial and venous ECs, mediate direct viral entry and damage. Pro-inflammatory cytokines in COVID-19 suppress natural antithrombotic and anti-inflammatory functions of ECs with coagulation system, complement and platelet activation, and leukocyte influx into the microvasculature [48]. Complement deposition with associated microvascular injury and thrombosis has been described in the lungs of COVID-19 patients [49]. Surrogate markers of EC dysfunction, von Willebrand factor and PAI-1 levels, are also increased [15]. However, the COVID-19 microangiopathic thrombosis does not present with the laboratory features of thrombotic thrombocytopenic purpura (TTP). The platelet counts in severe COVID-19 are usually relatively preserved, and red cell fragments (schistocytes) are not a prominent feature. There are only limited case studies in the literature describing patients with COVID-19 and features of TTP [50–52].

15.7 Treatment Recommendations for COVID-19-Related Coagulation Abnormalities

Complete guidance on the treatment recommendations of COVID-19-related haemostatic abnormalities is outside the scope of this review, but prophylactic treatment anticoagulation is recommended particularly in admitted patients.

Prophylactic dose low molecular weight heparin (LMWH) should be considered in patients with COVID-19 coagulopathy as inhibiting thrombin generation may improve outcomes in critically ill coagulopathic patients [53]. LMWH has anti-inflammatory properties and also protects against VTED. The contraindications for LMWH in COVID-19 patients with a coagulopathy include active bleeding and platelet counts $<25 \times 10^9$/L but not prolonged time-to-clot assays, such as PT and aPTT. LMWH activity monitoring with an anti-Factor Xa (a-FXa) assay is advised in all patients as the standard dose of anticoagulation may not protect against VTED in these patients and higher doses (enoxaparin at 0.5mg/kg twice daily) may be required. Bleeding is rare in patients with COVID-19, but the ISTH guidelines with respect to management can be followed if it does occur [54].

VTED prophylaxis is indicated in all patients with severe COVID-19 and for at least 7–14 days after discharge. Evidence is however emerging for all hospitalised patients with mild to moderate disease to also receive VTED prophylaxis in the absence of contraindications. Mechanical intermittent pneumatic compression is advocated in individuals with a contraindication to anticoagulation therapy [24].

Therapeutic dose anticoagulation is indicated in patients with confirmed VTED and in patients with a clinical suspicion of VTED and the presence of high-risk parameters if radiologic imaging is not feasible. Anti-FXa LMWH activity monitoring is indicated to ensure adequate anticoagulation [29]. Although possibly a reasonable approach, given the hypercoagulable state, the use of therapeutic doses of heparin in all hospitalised patients with severe COVID-19 is currently not supported by scientific evidence outside of treatment of VTED or as bridging therapy in patients on vitamin K antagonists (VKA) and cannot be recommended. Trials in this regard are ongoing [5, 24, 55]. Fibrinolytic therapy in patients with ST-elevation MI (STEMI) should be considered in addition to cardiac interventional therapies. Elevated troponins in patients with COVID-19 and a low pre-test probability of an acute coronary event could be a marker of systemic critical illness and should be reviewed in conjunction with other markers of inflammation [56]. The use of ACE inhibitors (ACEIs) and angiotensin receptor blockers (ARBs) in patients with COVID-19 infection is currently an area of active research [44]. For patients with acute COVID-19-related stroke, treatment with fibrinolytic therapy such as tissue plasminogen activator (tPA) may be considered, but it is currently not clear if other anticoagulants such as LMWH or full dose heparin have therapeutic benefit [6]. Novel therapies that impact endothelial dysfunction in COVID-19 patients, including complement inhibitors, are under evaluation [48].

15.8 Future Areas of Research

As our knowledge of the COVID-19 haematologic abnormalities and coagulopathy expand, it is apparent that additional scientific evidence and ongoing studies are required. The optimal dose and duration of heparin treatment as well as the subpopulation of COVID-19 patients with the best benefit-harm ratio must be clearly defined. The role of nonstandard anticoagulants such as thrombomodulin must be established. The effect of SARS-CoV-2 on platelet activation and the antiviral role of platelets need to be further investigated. The utility of non-routine coagulation assays such as thromboelastography and point-of-care haematology and coagulation assays must be clearly established and treatment and investigation algorithms compiled. The pathophysiological role of endotheliitis and complement activation in COVID-19 also warrants further research. There are currently over 450 registered clinical trials worldwide which will expand our knowledge of COVID-19 with the aim of improved patient outcomes [57].

References

1. Gupta A, Madhavan MV, Sehgal K, Nair N, Mahajan S, Sehrawat TS et al (2020) Extrapulmonary manifestations of COVID-19. Nat Med 26(7):1017–1032
2. Wu D, Wu T, Liu Q, Yang Z (2020) The SARS-CoV-2 outbreak: what we know. Int J Infect Dis 94:44–48
3. Tang N, Li D, Wang X, Sun Z (2020) Abnormal coagulation parameters are associated with poor prognosis in patients with novel coronavirus pneumonia. J Thromb Haemost 18(4):844–847
4. Akima S, McLintock C, Hunt BJ (2020) RE: ISTH interim guidance to recognition and management of coagulopathy in COVID-19. J Thromb Haemost 18:2057. https://doi.org/10.1111/jth.14853. Online ahead of print
5. Middeldorp S, Coppens M, van Haaps TF, Foppen M, Vlaar AP, Muller MCA et al (2020) Incidence of venous thromboembolism in hospitalized patients with COVID-19. J Thromb Haemost 19:1995. https://doi.org/10.1111/jth.14888. Online ahead of print
6. Hess DC, Eldahshan W, Rutkowski E (2020) COVID-19-related stroke. Transl Stroke Res 11(3):322–325
7. Bavishi C, Bonow RO, Trivedi V, Abbott JD, Messerli FH, Bhatt DL (2020) Acute myocardial injury in patients hospitalized with COVID-19 infection: a review. Prog Cardiovasc Dis:S0033-0620(20)30123-7. https://doi.org/10.1016/j.pcad.2020.05.013. Online ahead of print
8. Kannan S, Shaik Syed Ali P, Sheeza A, Hemalatha K (2020) COVID-19 (Novel Coronavirus 2019) – recent trends. Eur Rev Med Pharmacol Sci 24(4):2006–2011
9. Chen Y, Guo Y, Pan Y, Zhao ZJ (2020) Structure analysis of the receptor binding of 2019-nCoV. Biochem Biophys Res Commun 525(1):135–140
10. Hoffmann M, Kleine-Weber H, Schroeder S, Kruger N, Herrler T, Erichsen S et al (2020) SARS-CoV-2 cell entry depends on ACE2 and TMPRSS2 and is blocked by a clinically proven protease inhibitor. Cell 181(2):271-280.e8
11. Chousterman BG, Swirski FK, Weber GF (2017) Cytokine storm and sepsis disease pathogenesis. Semin Immunopathol 39(5):517–528
12. Ding YQ, Bian XW (2020) Analysis of coronavirus disease-19 (COVID-19) based on SARS autopsy. Zhonghua Bing Li Xue Za Zhi 49(4):291–293
13. Pons S, Fodil S, Azoulay E, Zafrani L (2020) The vascular endothelium: the cornerstone of organ dysfunction in severe SARS-CoV-2 infection. Crit Care 24(1):353. https://doi.org/10.1186/s13054-020-03062-7
14. Marietta M, Ageno W, Artoni A, De Candia E, Gresele P, Marchetti M et al (2020) COVID-19 and haemostasis: a position paper from Italian Society on Thrombosis and Haemostasis (SISET). Blood Transfus 18(3):167–169
15. Helms J, Tacquard C, Severac F, Leonard-Lorant I, Ohana M, Delabranche X et al (2020) High risk of thrombosis in patients with severe SARS-CoV-2 infection: a multicenter prospective cohort study. Intensive Care Med 46(6):1089–1098
16. Gando S, Levi M, Toh CH (2016) Disseminated intravascular coagulation. Nat Rev Dis Primers 2:16037. https://doi.org/10.1038/nrdp.2016.37
17. Thachil J, Tang N, Gando S, Falanga A, Cattaneo M, Levi M et al (2020) ISTH interim guidance on

recognition and management of coagulopathy in COVID-19. J Thromb Haemost 18(5):1023–1026
18. Zhang L, Yan X, Fan Q, Liu H, Liu X, Liu Z et al (2020) D-dimer levels on admission to predict in-hospital mortality in patients with Covid-19. J Thromb Haemost 18(6):1324–1329
19. Guan WJ, Ni ZY, Hu Y, Liang WH, Ou CQ, He JX et al (2020) Clinical characteristics of Coronavirus Disease 2019 in China. N Engl J Med 382(18):1708–1720
20. Williamson DR, Lesur O, Tetrault JP, Nault V, Pilon D (2013) Thrombocytopenia in the critically ill: prevalence, incidence, risk factors, and clinical outcomes. Can J Anaesth 60(7):641–651
21. Huang C, Wang Y, Li X, Ren L, Zhao J, Hu Y et al (2020) Clinical features of patients infected with 2019 novel coronavirus in Wuhan, China. Lancet 395(10223):497–506
22. Lippi G, Plebani M, Henry BM (2020) Thrombocytopenia is associated with severe coronavirus disease 2019 (COVID-19) infections: a meta-analysis. Clin Chim Acta 506:145–148
23. Cui S, Chen S, Li X, Liu S, Wang F (2020) Prevalence of venous thromboembolism in patients with severe novel coronavirus pneumonia. J Thromb Haemost 18(6):1421–1424
24. Kollias A, Kyriakoulis KG, Dimakakos E, Poulakou G, Stergiou GS, Syrigos K (2020) Thromboembolic risk and anticoagulant therapy in COVID-19 patients: emerging evidence and call for action. Br J Haematol 189(5):846–847
25. Klok FA, Kruip M, van der Meer NJM, Arbous MS, Gommers D, Kant KM et al (2020) Incidence of thrombotic complications in critically ill ICU patients with COVID-19. Thromb Res 191:145–147
26. Zhai Z, Li C, Chen Y, Gerotziafas G, Zhang Z, Wan J et al (2020) Prevention and treatment of venous thromboembolism associated with Coronavirus Disease 2019 infection: a consensus statement before guidelines. Thromb Haemost 120(6):937–948
27. Helms J, Severac F, Merdji H, Angles-Cano E, Meziani F (2020) Prothrombotic phenotype in COVID-19 severe patients. Intensive Care Med 46(7):1502–1503
28. Lodigiani C, Iapichino G, Carenzo L, Cecconi M, Ferrazzi P, Sebastian T et al (2020) Venous and arterial thromboembolic complications in COVID-19 patients admitted to an academic hospital in Milan, Italy. Thromb Res 191:9–14
29. Aryal MR, Gosain R, Donato A, Pathak R, Bhatt VR, Katel A et al (2020) Venous thromboembolism in COVID-19: towards an ideal approach to thromboprophylaxis, screening, and treatment. Curr Cardiol Rep 22(7):52. https://doi.org/10.1007/s11886-020-01327-9
30. Spiezia L, Boscolo A, Poletto F, Cerruti L, Tiberio I, Campello E et al (2020) COVID-19-related severe hypercoagulability in patients admitted to intensive care unit for acute respiratory failure. Thromb Haemost 120(6):998–1000
31. Zhou F, Yu T, Du R, Fan G, Liu Y, Liu Z et al (2020) Clinical course and risk factors for mortality of adult inpatients with COVID-19 in Wuhan, China: a retrospective cohort study. Lancet 395(10229):1054–1062
32. Zuckier LS, Moadel RM, Haramati LB, Freeman LM (2020) Diagnostic evaluation of pulmonary embolism during the COVID-19 pandemic. J Nucl Med 61(5):630–631
33. Li D, Chen Y, Jia Y, Tong L, Tong J, Wang W et al (2020) SARS-CoV-2-induced immune dysregulation and myocardial injury risk in China: insights from the ERS-COVID-19 study. Circ Res 27(3):397–399
34. Shi S, Qin M, Cai Y, Liu T, Shen B, Yang F et al (2020) Characteristics and clinical significance of myocardial injury in patients with severe coronavirus disease 2019. Eur Heart J 41(22):2070–2079
35. Goyal P, Choi JJ, Pinheiro LC, Schenck EJ, Chen R, Jabri A et al (2020) Clinical characteristics of Covid-19 in New York City. N Engl J Med 382(24):2372–2374
36. Kubasiak LA, Hernandez OM, Bishopric NH, Webster KA (2002) Hypoxia and acidosis activate cardiac myocyte death through the Bcl-2 family protein BNIP3. Proc Natl Acad Sci USA 99(20):12825–12830
37. Zhou R (2020) Does SARS-CoV-2 cause viral myocarditis in COVID-19 patients? Eur Heart J 41(22):2123. https://doi.org/10.1093/eurheartj/ehaa392
38. Varga Z, Flammer AJ, Steiger P, Haberecker M, Andermatt R, Zinkernagel AS et al (2020) Endothelial cell infection and endotheliitis in COVID-19. Lancet 395(10234):1417–1418
39. Gheblawi M, Wang K, Viveiros A, Nguyen Q, Zhong JC, Turner AJ et al (2020) Angiotensin-converting enzyme 2: SARS-CoV-2 receptor and regulator of the renin-angiotensin system: celebrating the 20th Anniversary of the discovery of ACE2. Circ Res 126(10):1456–1474
40. Han H, Yang L, Liu R, Liu F, Wu KL, Li J et al (2020) Prominent changes in blood coagulation of patients with SARS-CoV-2 infection. Clin Chem Lab Med 58(7):1116–1120
41. Tavazzi G, Pellegrini C, Maurelli M, Belliato M, Sciutti F, Bottazzi A et al (2020) Myocardial localization of coronavirus in COVID-19 cardiogenic shock. Eur J Heart Fail 22(5):911–915
42. Hu H, Ma F, Wei X, Fang Y (2020) Coronavirus fulminant myocarditis saved with glucocorticoid and human immunoglobulin. Eur Heart J:ehaa190. https://doi.org/10.1093/eurheartj/ehaa190. Online ahead of print
43. Mao L, Jin H, Wang M, Hu Y, Chen S, He Q et al (2020) Neurologic manifestations of hospitalized patients with Coronavirus Disease 2019 in Wuhan, China. JAMA Neurol 77(6):1–9
44. Koralnik IJ, Tyler KL (2020) COVID-19: a global threat to the nervous system. Ann Neurol 88(1):1–11
45. Bray MA, Sartain SE, Gollamudi J, Rumbaut RE (2020) Microvascular thrombosis: experimental and clinical implications. Transl Res:S1931-5244(20)30108-0. https://doi.org/10.1016/j.trsl.2020.05.006. Online ahead of print

46. Delvaeye M, Conway EM (2009) Coagulation and innate immune responses: can we view them separately? Blood 114(12):2367–2374
47. Go RS, Winters JL, Leung N, Murray DL, Willrich MA, Abraham RS et al (2016) Thrombotic microangiopathy care pathway: a consensus statement for the Mayo Clinic Complement Alternative Pathway-Thrombotic Microangiopathy (CAP-TMA) disease-oriented group. Mayo Clin Proc 91(9):1189–1211
48. Pons S, Arnaud M, Loiselle M, Arrii E, Azoulay E, Zafrani L (2020) Immune consequences of endothelial cells' activation and dysfunction during sepsis. Crit Care Clin 36(2):401–413
49. Magro C, Mulvey JJ, Berlin D, Nuovo G, Salvatore S, Harp J et al (2020) Complement associated microvascular injury and thrombosis in the pathogenesis of severe COVID-19 infection: a report of five cases. Transl Res 220:1–13
50. Capecchi M, Mocellin C, Abbruzzese C, Mancini I, Prati D, Peyvandi F (2020) Dramatic presentation of acquired TTP associated with COVID-19. Haematologica: haematol.2020.262345. https://doi.org/10.3324/haematol.2020.262345. Online ahead of print
51. Albiol N, Awol R, Martino R (2020) Autoimmune thrombotic thrombocytopenic purpura (TTP) associated with COVID-19. Ann Hematol 99(7):1673–1674
52. Hindilerden F, Yonal-Hindilerden I, Akar E, Kart-Yasar K (2020) Covid-19 associated autoimmune thrombotic thrombocytopenic purpura: report of a case. Thromb Res 195:136–138
53. Iba T, Levy JH, Warkentin TE, Thachil J, van der Poll T, Levi M et al (2019) Diagnosis and management of sepsis-induced coagulopathy and disseminated intravascular coagulation. J Thromb Haemost 17(11):1989–1994
54. Wada H, Thachil J, Di Nisio M, Mathew P, Kurosawa S, Gando S et al (2013) Guidance for diagnosis and treatment of DIC from harmonization of the recommendations from three guidelines. J Thromb Haemost. https://doi.org/10.1111/jth.12155. Online ahead of print
55. Al-Ani F, Chehade S, Lazo-Langner A (2020) Thrombosis risk associated with COVID-19 infection. A scoping review. Thromb Res 192:152–160
56. Szerlip M, Anwaruddin S, Aronow HD, Cohen MG, Daniels MJ, Dehghani P et al (2020) Considerations for cardiac catheterization laboratory procedures during the COVID-19 pandemic perspectives from the Society for Cardiovascular Angiography and Interventions Emerging Leader Mentorship (SCAI ELM) members and graduates. Catheter Cardiovasc Interv 96:586. https://doi.org/10.1002/ccd.28887. Online ahead of print
57. Harenberg J, Favaloro E (2020) COVID-19: progression of disease and intravascular coagulation – present status and future perspectives. Clin Chem Lab Med 58(7):1029–1036

Part II

Clinical Studies

COVID-19-Related Laboratory Analyte Changes and the Relationship Between SARS-CoV-2 and HIV, TB, and HbA1c in South Africa

16

R. Hesse, D. J. van der Westhuizen, and J. A. George

Abstract

We conducted a retrospective analysis on data of all adults tested for SARS-CoV-2 across our laboratory network in South Africa over a 4-month period. Out of 842,197 tests, 11.7% were positive and 88.3% negative. The prevalence of HIV was 6.25 and 6.31% in the SARS-CoV-2-positive and SARS-CoV-2-negative cohort, respectively ($p = 0.444$). However, the prevalence of HIV-positive individuals in the critical cohort (9.15%) was higher than in the noncritical group (6.24%) ($p = 0.011$). Active tuberculosis infection was approximately 50% less in SARS-CoV-2-positive than in SARS-CoV-2-negative individuals. The prevalence of uncontrolled diabetes was 3.4 times higher in SARS-CoV-2-positive cases but was not higher in the critical vs. noncritical cases ($p = 0.612$). The neutrophil-to-lymphocyte ratio, coagulation markers, urea, and cardiac- and liver-related analytes were significantly elevated in the critical compared to noncritical cases. Platelet count and creatinine concentration did not differ significantly between the two groups. These findings do not support increased prevalence of HIV or tuberculosis in individuals with SARS-CoV-2 infection but do suggest an association of increased disease severity with HIV-positive status. Uncontrolled diabetes was positively associated with a significantly higher prevalence of SARS-CoV-2, and our investigation into analyte changes associated with SARS-CoV-2 disease severity supported previous findings of raised inflammatory markers, coagulation markers, liver- and cardiac-related analytes, and urea but not for creatinine and platelet count.

Keywords

COVID-19 · SARS-CoV-2 · Tuberculosis · HIV · Diabetes mellitus · Coagulation · Inflammatory markers · Neutrophil-to-lymphocyte ratio

Authors R. Hesse, D. J. van der Westhuizen have equally contributed to this chapter.

R. Hesse · J. A. George (✉)
Department of Chemical Pathology, University of the Witwatersrand; National Health Laboratory Service, Johannesburg, South Africa

D. J. van der Westhuizen
Department of Chemical Pathology, University of Cape Town; National Health Laboratory Service, Cape Town, South Africa

P. C. Guest (ed.), *Clinical, Biological and Molecular Aspects of COVID-19*, Advances in Experimental Medicine and Biology 1321, https://doi.org/10.1007/978-3-030-59261-5_16

16.1 Introduction

The severe acute respiratory syndrome novel coronavirus 2 (SARS-CoV-2) that causes the coronavirus disease of 2019 (COVID-19) has resulted in a worldwide pandemic and continues to affect many countries. Although SARS-CoV-2 is part of the well-known genus of *Betacoronaviruses*, like that of the severe acute respiratory syndrome (SARS-CoV) and Middle Eastern respiratory syndrome (MERS), current research indicates that SARS-CoV-2 is less fatal but has much wider community spread due to its high reproductive number [1]. COVID-19 also differs from the seasonal influenza viral strains due to the atypical disease presentation and higher resultant mortality rate [2].

Although predominantly affecting the respiratory system, COVID-19 is now known to have a wide spectrum of clinical manifestations ranging from being totally asymptomatic to multiorgan and systemic dysfunction [3]. Organ systems affected other than the respiratory system include the cardiovascular system, the renal system, the gastrointestinal system, the immune system, the nervous system, and the coagulation cascade [3, 4]. Due to the wide range of potential systems affected, multiple analytes are influenced by COVID-19. The role of laboratory medicine in COVID-19 has evolved from a predominant diagnostic testing role in SARS-CoV-2 reverse transcriptase polymerase chain reaction (PCR) testing to include testing of analytes that assist with staging, prognostication, therapeutic monitoring, and epidemiological surveillance [5].

Several serum- and plasma-based tests available in routine laboratories have been shown to be of value in assessing COVID-19 disease severity (both for staging and prognostication), progression, and therapeutic monitoring that aid in management of the patient [5, 6]. These analytes span across different domains of the clinical laboratory and include biochemistry, hematology, and immunochemistry. The International Federation of Clinical Chemistry's (IFCC) task force evaluation of the latest evidence categorizes these markers into the following broad groups: (i) proinflammatory response markers (consistent with cytokine storm disease process) and (ii) progression to multiorgan damage/failure markers (that include system-specific markers for the hepatic system, the cardiac system, the coagulation system, and the renal system) [6].

Additionally, newer findings suggest a bidirectional relationship between COVID-19 and diabetes mellitus (DM). It has been previously demonstrated that DM is a major risk factor for a severe COVID-19 disease course and mortality, but now, there are also reports that describe patients with COVID-19 and new-onset diabetes presenting in severe metabolic acidosis with ketosis which suggests a COVID-19-mediated mechanism of glucose dysregulation [7, 8]. Another study demonstrated the value of a baseline serum cortisol concentration in the prediction of survival rate in patients admitted with COVID-19 [9]. Studies such as these show the potential value of using biochemical markers beyond those included by the IFCC's task force, such as hemoglobin A1c (HbA1C), fasting glucose, and serum cortisol, for both the evaluation of other COVID-19-mediated disease processes as well as prognostication.

Despite emerging evidence from other countries on the protective factors of the widely used bacille Calmette-Guerin (BCG) vaccine, South Africa has not been spared by COVID-19, and the number of daily new infections continue to grow [10, 11]. At the time of writing, the country has already recorded more than 500,000 confirmed cases, accounting for more than half of all cases on the African continent and the fifth highest case number in the world [12, 13]. Despite the high number of infections, COVID-19-related deaths in the country are surprisingly low with 8153 deaths recorded at the time of writing, ranking 36th in the world in terms of deaths/million people [13]. This anomaly, which is also seen in other African countries, is speculated to be due to the younger age demographic, the aggressive government response, and the exposure to a wide range of existing infectious diseases and treatment regimens or a combination of all these factors [14].

South Africa is a country already heavily burdened by the concurrent human immunodeficiency virus (HIV) epidemic, that causes acquired immunodeficiency syndrome (AIDS), as well as mycobacterium tuberculosis (TB) epidemic [15]. Studies that evaluate the effect of COVID-19 on patients with HIV have not found significantly worse outcomes but are limited to case series and single-center retrospective analyses with limited numbers [16–19]. Studies done on SARS-CoV and MERS' interaction with HIV have found lower risk of infection and lower burden of disease in patients with HIV, speculated to be due to either antiretroviral therapy or distinct immune response [18, 20]. It is further unclear to what extent, if any, immunological function (CD4) and HIV viral load impact the susceptibility and outcomes in COVID-19. The current literature indicates that patients with TB are both more susceptible to COVID-19 and experience a more severe disease course; however, the case numbers reported are also relatively low at this stage [21, 22].

Although the infectious diseases burden in South Africa is of great concern, the country also faces significant morbidity and mortality from noncommunicable diseases such as DM, hypertension, cancer, and chronic kidney disease (CKD) [23, 24]. Several studies have demonstrated the higher rate of susceptibility, as well as the poorer outcomes, in patients with concomitant COVID-19 and these comorbidities [25].

South Africa therefore faces a unique challenge of both a high communicable and noncommunicable disease burden in the face of a worsening COVID-19 pandemic and a limited health-care infrastructure. To our knowledge, no large studies have been done to evaluate the changes in laboratory analytes and the impact of this unique combination of factors on COVID-19.

The National Health Laboratory Service (NHLS) is the national state laboratory of South Africa and provides diagnostic services to more than 80% of the population [26]. All laboratory tests from the NHLS are stored in a single central data warehouse (CDW) of the NHLS.

The aim of this study was to describe the biochemical and hematological analyte changes seen in COVID-19 patients using data from the NHLS and to determine the effect of HIV, TB, and DM on the risk for acquiring SARS-CoV-2 and the outcomes as measured by intensive and high care admission.

16.2 Methods

16.2.1 Ethical Clearance

We obtained ethics clearance from the Human Research and Ethics Committee of the University of the Witwatersrand (clearance certificate number: M200424), and data acquisition permission was obtained from the NHLS' academic affairs and research office.

16.2.2 Database Extraction and Cleaning

We extracted data for all individuals who had at least one SARS-CoV-2 PCR test done via the NHLS between the periods of 1 March 2020 to 7 July 2020. Patient data was anonymized to prevent traceability. A 6-month period of demographic, biochemical, hematological, and infectious disease data was extracted for the individuals that had undergone SARS-CoV-2 testing.

Test data reflecting chronic diseases such as HIV, TB, and DM was taken from a 6-month period preceding the SARS-CoV-2 test, but the rest of the analyte data was restricted to a time window of 7 days prior to and 14 days after SARS-COV2 testing to ensure relevancy.

Stata and Excel were used for dataset handling, cleaning, and statistical analysis. Data cleaning was done by removing patients younger than 18 years of age and those with SARS-CoV-2 PCR results of indeterminate significance. Where more than one SARS-CoV-2 result was available for an individual, the individual was considered "positive" if at least one of the results were positive. In such cases, the analyte data accompanying the prior negative SARS-CoV-2 test was excluded from further analysis to ensure that only those analytes relevant to the final SARS-CoV-2 status was used.

16.2.3 Comorbid Disease Classification

HIV-positive diagnosis was determined first by HIV enzyme-linked immunosorbent assay (ELISA) test result, and if that test was not available, the presence of an HIV viral load or a CD4 count was used for HIV-positive classification. This approach was taken as national guidelines only indicate viral load testing for persons that tested HIV positive and started treatment. Similarly, HIV is the only disease for which the CD4 count is routinely requested and only indicated as a monitoring tool in patients already diagnosed with HIV [27].

Active TB infection was determined by ascertaining if a patient had a positive TB test result (by GeneXpert®) in the 6-month period prior to SARS-CoV-2 testing.

HbA1c data were used as proxy to classify patients with an HbA1c <6.5% as "nondiabetic or controlled diabetic" and those with an HbA1c result of ≥6.5% as "uncontrolled diabetic."

16.2.4 Descriptive Statistics

Descriptive statistics were performed on the whole cohort as well as the "positive" and "negative" subsets to calculate the overall prevalence of SARS-CoV-2 infection. We reported age as the mean and standard deviation and categorical data as proportions.

16.2.5 Categorization by Ward as Proxy for Disease Severity

Outcome data were not available, but intensive care unit (ICU), high-care, and critical-care admissions were ascertained by using the ward description data to identify if a patient was admitted to one of these wards by filtering for wards that contained the word "ICU," "intensive care," "high care," and "critical care." Patients admitted to any one of these wards with SARS-CoV-2-positive results were classified as "critical COVID-19." Patients without admission to any of these types of wards were classified as "noncritical COVID-19."

16.2.6 SARS-CoV-2-Associated Analyte Changes

We evaluated differences in commonly requested analytes, as well as those that have been shown to reflect SARS-CoV-2-associated changes in literature, between the "critical SARS-CoV-2" and "noncritical SARS-CoV-2" cohorts. We calculated the median and interquartile range (IQR) for all analytes. Analytes were grouped by physiological system as follows: inflammatory [C-reactive protein (CRP), IL-6, procalcitonin (PCT), ferritin, erythrocyte sedimentation rate (ESR)], coagulation (D-dimer, INR, fibrinogen), full blood count [white cell count (WCC) total, red cell count, hemoglobin, hematocrit, mean corpuscular volume, mean corpuscular hemoglobin, mean corpuscular hemoglobin concentration, red cell distribution width, platelet count], WCC differential [absolute count, neutrophil, lymphocyte, monocyte, eosinophil, basophil, as well as the neutrophil-to-lymphocyte ratio (NLR)], liver related [aspartate aminotransferase (AST), alanine aminotransferase (ALT), gamma-glutamyl transferase (GGT), lactate dehydrogenase (LDH), total bilirubin, albumin)], cardiac related [troponin T, troponin I, N-terminal pro b-type natriuretic peptide (NT-proBNP)], endocrine related (HbA1c), and renal function related [urea, creatinine, estimated glomerular filtration rate (eGFR)].

16.2.7 DM, HIV, and SARS-COV2 Infection

Individuals that had undergone at least one HbA1c measurement within the 6-month period preceding SARS-CoV-2 testing were further analyzed by stratifying their HbA1c measurement into one of four categories: (1) optimal control and normal (<5.7%), (2) controlled or prediabetic (5.7–6.49%), (3) uncontrolled diabetic (6.5–10%), and (4) poorly controlled diabetic (>10%).

Where more than one result was available for a patient, the mean of the HbA1c results was used. The prevalence of a SARS-CoV-2-positive test was then calculated for each of these categories to assess the effect of glycemic control on the probability of infection.

A similar approach was used to stratify individuals that had at least one CD4 count done in the 6-month period prior to SARS-CoV-2 testing. In the few instances where there was more than one CD4 count available per individual, the mean CD4 count was used. CD4 counts were categorized into one of eleven bins, with increments of 100 cells/uL between bins, ranging from 0–99 cells/uL (bin 1) to ≥1000 cells/uL (bin 11). SARS-CoV-2 prevalence was calculated for each of these bins to compare the influence of immunosuppression on the risk of infection.

To assess the effect of HIV coinfection on the severity of SARS-CoV-2, the HIV-positive prevalence was calculated for the severe SARS-CoV-2 and non-severe SARS-CoV-2 groups, as well as the mean CD4 count for these two groups.

16.2.8 Statistical Analysis

The statistical significance between groups for all results was calculated by Wilcoxon rank-sum test. For the difference in age, the Student's t-test was used. Pearson's chi-square test was used for evaluating the difference in sex and the prevalence of TB, HIV, and DM. A p-value of <0.05 was regarded as significant, regardless of the statistical method used. We used Stata SE 14.1 (StataCorp, Texas, USA) for all data analyses.

16.3 Results

16.3.1 Demographics and Prevalence of HIV, TB, and Uncontrolled DM

We retrieved 909,818 individuals' SARS-CoV-2 PCR results, and after excluding all indeterminate PCR results and results for individuals less than 18 years of age, the cohort consisted of a total of 842,197 unique individuals (Fig. 16.1). As seen in Table 16.1, 11.7% (98,335) of these individuals had at least one positive SARS-CoV-2 PCR test, and 88.3% (743,862) tested negative. The mean age for the positive group was 42.3 ± 15.0 years vs. 42.6 ± 14.7 years in the negative group, and female prevalence was 61.6% (60,545) vs. 56.3% (419,011) ($p < 0.001$ for both), respectively.

The overall prevalence of HIV was 6.3% and did not differ significantly between the groups. Active TB was found in 6256 of the 842,197 SARS-CoV-2-tested individuals, resulting in a prevalence of 743 cases per 100,000. The TB prevalence was significantly different between the two groups in our study, with the SARS-CoV-2-negative individuals displaying almost twice the number of active TB cases than the SARS-CoV-2-positive group (0.79% and 0.40%, respectively, $p < 0.001$).

SARS-CoV-2-positive patients had a significantly increased prevalence of uncontrolled DM compared to those that tested negative, with prevalence values of 4.61% and 1.36%, respectively ($p < 0.001$).

16.3.2 Laboratory Analyte Findings

Of the 98,335 individuals that tested positive, 12,270 had at least one additional test within the defined time period around their SARS-CoV-2 test. Of these individuals, only 142 (1.16%) were admitted to an ICU or high-care ward, meeting our criteria to be classified as "critical SARS-CoV-2."

The critical group displayed significantly elevated levels of CRP, PCT, and WCC, but not IL-6, ferritin, or ESR. In addition to the raised total WCC, the comparison of the differential count also revealed higher neutrophil and lower lymphocyte count in the severe group, resulting in a significantly increased NLR of 9.18. The eosinophil count did not differ significantly between the two groups (Table 16.2).

The D-dimer and INR values were significantly raised in the critical group. Despite the dif-

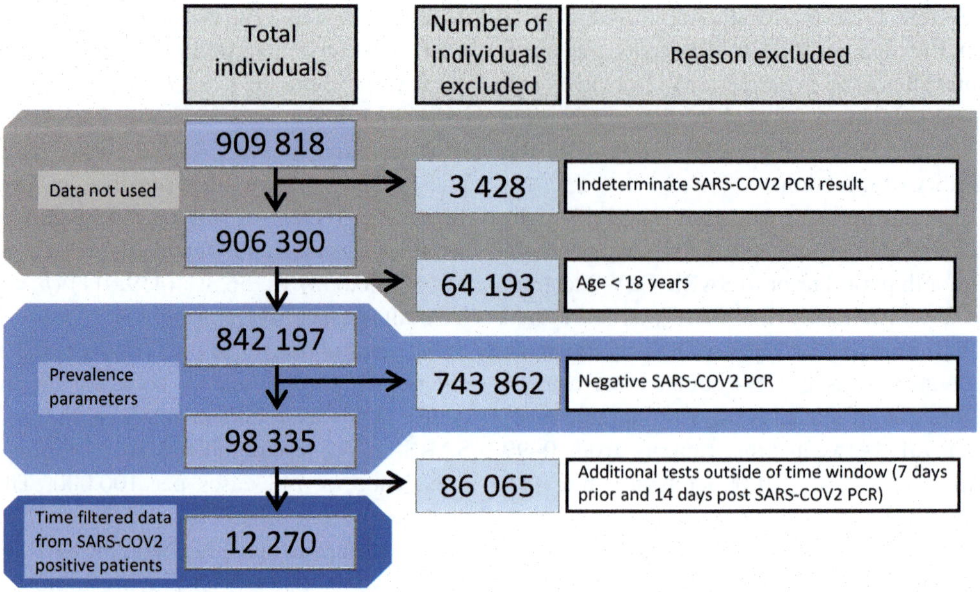

Fig. 16.1 Data exclusion process

Table 16.1 Descriptive statistics of the negative and positive cohorts of SARS-COV-2-tested individuals

	SARS-COV-2 negative	SARS-COV-2 positive	p-value
Number of individuals (n; %)	743,862 (88.3)	98,335 (11.7)	
Age (mean; ±SD)	42.6 (±14.7)	42.3 (±15.0)	<0.001
Female sex (n; %)	419,011 (56.3)	60,545 (61.6)	<0.001
HIV+ (n; %)	46,961 (6.31)	6146 (6.25)	0.444
Active TB (n; %)	5860 (0.79)	396 (0.40)	<0.001
Uncontrolled DM (n; %)	10,107 (1.36)	4534 (4.61)	<0.001

%, fraction expressed in percentage of total number of individuals in the negative or positive cohorts; *SD* standard deviation, *n* sample size, *TB* mycobacterium tuberculosis infection, *DM* diabetes mellitus

ference in the former two coagulation markers, fibrinogen was not significantly different between the groups.

Among the other hematological analytes, the red cell count (RCC), hemoglobin, hematocrit, and mean corpuscular hemoglobin concentration (MCHC) were all decreased in the critical group, while the red cell distribution width (RDW) was increased. No significant difference was found in the platelet count between the two groups.

The transaminases and GGT were higher in the critical group. Total bilirubin was not statistically different between the groups and was well within the normal range. LDH displayed increased values in the critical group, whereas the albumin concentration was decreased. Both troponin T and NT-proBNP were markedly raised in the critical group.

HbA1c did not differ significantly. The renal function markers indicated a heterogeneous result with the urea significantly raised in the critical patients, but both the creatinine and eGFR were failing to show a difference between the groups.

16.3.3 Impact of HIV Viral Load and CD4 Count on Prevalence and Severity

From Table 16.3, we can see that patients with critical SARS-CoV-2 had a higher prevalence of

Table 16.2 Comparison of analyte changes between non-severe and severe SARS-COV-2 infection

Measurement				Noncritical SARS-COV-2	Critical SARS-COV-2	
Category	Parameter	Count (n)	Unit			*p*-value
Demographics	Total individuals	12,270 (100%)	n (%)	12,128 (98.84)	142 (1.16)	
	Female sex	7055 (57.5%)	n (%)	6977 (98.89)	78 (1.11)	0.571
	Male sex	5144 (41.9%)	n (%)	5081 (98.59)	63 (1.22)	0.571
	Age	12,226	Years (SD)	52.21 (± 15.7)	48.81 (±14.0)	**0.012**
Inflammatory	CRP	8721	mg/L	115 (56–199)	136 (87–216)	**0.009**
	IL-6	32	pg/ml	44.4 (9–96)	503.8 (20.5–987)	0.350
	PCT	1387	ug/L	0.25 (0.09–1.07)	2.1 (0.6–11.1)	**<0.001**
	Ferritin	2073	ug/L	574 (245–1194)	592 (276–1429)	0.727
	ESR	881	mm/hr	56 (30–85)	50 (33–80)	0.936
Coagulation	D-dimer	4445	mg/L	0.61 (0.33–1.62)	1.37 (0.61–2.74)	**<0.001**
	INR	2288	–	1.11 (1.02–1.26)	1.17 (1.09–1.3)	**0.011**
	Fibrinogen	195	g/L	4.7 (3.4–6.4)	4.35 (3.05–6.7)	0.791
Full blood count	WCC (total)	5563	× 10^9/L	8.48 (6.2–11.7)	11.05 (7.7–16.3)	**<0.001**
	Red cell count	10,921	× 10^12/L	4.54 (4.07–4.99)	3.99 (3.5–4.8)	**<0.001**
	Hemoglobin	10,921	g/dL	13.1 (11.7–14.4)	11.5 (10.3–13.4)	**<0.001**
	Hematocrit	10,916	L/L	0.399 (0.359–0.436)	0.356 (0.320–0.42)	**<0.001**
	MCV	10,916	fL	88.1 (84.2–92)	88.8 (84.5–92.9)	0.287
	MCH	10,916	pg	28.9 (27.5–30.3)	28.7 (27.3–30.3)	0.390
	MCHC	10,916	g/dL	32.8 (31.8–33.6)	32.4 (31.4–33.3)	**0.002**
	RDW	10,912	%	14 (13.2–15)	14.7 (13.7–16.2)	**<0.001**
	Platelet count	10,900	×10^9/L	268 (207–343)	283 (207–344)	0.663
White blood cell differential	Neutrophil abs.	5672	×10^9/L	6.4 (4.3–9.5)	9.1 (6.3–14.2)	**<0.001**
	Lymphocyte abs.	5671	×10^9/L	1.28 (0.91–1.75)	0.98 (0.74–1.71)	**0.026**
	NLR	5671	–	5.12 (3.0–8.8)	9.18 (5.7–15.4)	**<0.001**
	Monocyte abs.	5670	×10^9/L	0.48 (0.33–0.68)	0.49 (0.31–0.89)	0.506
	Eosinophil abs.	5590	×10^9/L	0.023 (0.01–0.07)	0.02 (0.003–0.067)	0.442
	Basophil abs.	5573	×10^9/L	0.03 (0.01–0.06)	0.02 (0.01–0.05)	0.111

(continued)

Table 16.2 (continued)

Measurement				Noncritical SARS-COV-2	Critical SARS-COV-2	
Category	Parameter	Count (n)	Unit			*p*-value
Liver related	AST	4050	U/L	44 (30–71)	60 (35–107)	**0.002**
	ALT	6271	U/L	28 (18–46)	36.7 (18.8–60)	**0.023**
	GGT	4127	U/L	56 (32–106)	87(43–182)	**<0.001**
	LDH	3224	U/L	457 (337–657)	640 (421–831)	**<0.001**
	Total bilirubin	5482	umol/L	8 (6–13)	8 (5.4–15)	0.894
	Albumin	4483	g/L	34 (29–39)	29 (24–34)	**<0.001**
Cardiac related	Troponin T	977	ng/L	14 (7–33)	29 (11–68)	**0.025**
	Troponin I	954	ng/L	20 (7–79)	49 (22.4–139.5)	0.102
	NT-proBNP	723	pg/ml	260 (54–1253)	645 (407–1632)	**0.017**
Endocrine	HbA1c	3479	%	8.4 (6.5–11.8)	9.4 (6.8–12)	0.404
Renal function	Urea	11,170	mmol/L	5.65 (3.8–9.5)	9 (4.8–14.7)	**<0.001**
	Creatinine	11,700	umol/L	72.5 (54–89)	67.7 (52.8–96.4)	0.567
	eGFR (CKD-EPI)	11,658	ml/min/1.73 m^2	98.5 (70.1–112.4)	99.9 (59.9–116.8)	0.931

SD standard deviation; abs., absolute count; %, percentage

Table 16.3 Severity of SARS-COV-2 as stratified by HIV status

	SARS-COV-2 positive		
HIV metric	Noncritical (n; %)	Critical (n; %)	*p*-value
HIV- or unknown (n; %)	91,792 (93.8)	397 (90.8)	**0.011**
HIV+ (n; %)	6106 (6.2)	40 (9.2)	
CD4 count (median; IQR)	562 (291–660)	647 (412–661)	0.378

HIV than the noncritical SARS-CoV-2 group. The mean CD4 counts among the severe group were higher than that of the noncritical group, although not statistically significant ($p = 0.378$). There were only 10 patients in the critical group with a viral load result, thus limiting meaningful analysis on the effect of HIV viral load suppression on the severity of SARS-CoV-2. There were 26,315 individuals that had both a CD4 count and a SARS-CoV-2 PCR test. The effect of CD4 count on the prevalence of SARS-CoV-2 revealed a bimodal prevalence of SARS-CoV-2 among the different CD4 count stratified groups (Fig. 16.2). It was highest in those with CD4 counts of 300–399 cells/uL and 900–999 cells/uL at 17.6% and lowest in those with CD4 counts of 0–99 cells/uL (7.1%) and 600–699 cells/uL (8.6%) ($p < 0.001$).

16.3.4 Impact of DM on Severity

Across all the SARS-COV-2-positive individuals with HbA1c results, only 26.4% had an HbA1c of less than 6.5%. The prevalence of HbA1c >6.5% was higher in the critical group although this was not statistically significant (Fig. 16.3).

16.4 Discussion

We have reported on laboratory changes seen in patients admitted to public health facilities across South Africa. Our findings indicate a relatively young average age (42.3 ± 15.0 years). In addition, 1.16% of all cases were classified as critical, and we did not show a higher prevalence of TB with COVID-19-infected cases. Although HIV was more prevalent in our critical group, the CD4 count was not different between the critical and noncritical groups. The inflammatory-, coagulation-, cardiac-, and liver-related analytes were significantly elevated in the critical group compared to the noncritical group.

Age is a well-known risk factor for COVID-19 disease severity and mortality, with an average age for fatalities as high as 81 years reported in Italy [28, 29]. In this regard, the younger age structure of the South African population, also evident in SARS-CoV-2 testing results, differs and may confer a survival benefit. Although more females tested positive for SARS-CoV-2, males experienced statistically significantly more severe COVID-19. This pattern is not unique to South Africa but is seen across the world [30].

Our reported prevalence of critical cases was similar to the reported figures in South Africa [31]. Because we limited the analysis to individuals that had at least one other chemical or hematological test within the period around the SARS-CoV-2-positive test, the noncritical SARS-CoV-2 set was effectively a "moderate-severe" COVID-19 group as most of these individuals were admitted to hospital but not an ICU or high-care ward.

Although we did not show a higher prevalence of SARS-CoV-2 among patients with either HIV or TB, there were proportionately more HIV-positive cases who were critical than noncritical. The similar prevalence of HIV-positive individuals in both the SARS-CoV-2-positive and SARS-CoV-2-negative groups was in keeping with the findings of limited previous research indicating a similar or even lower prevalence of SARS-CoV-2, SARS-CoV, and MERS in HIV-positive individuals compared to the general population [18–20]. It has been suggested that this is due to the effect of antiretroviral medication and/or the unique immune reaction on the infectivity of novel coronaviruses [18, 19]. In contrast to our findings, some studies have suggested that CD4 counts are lower in patients with severe COVID-19; however, the findings are not consistent. For example, a recent meta-analysis reported significantly decreased CD4 levels in patients with severe COVID-19 and an increase in CD4

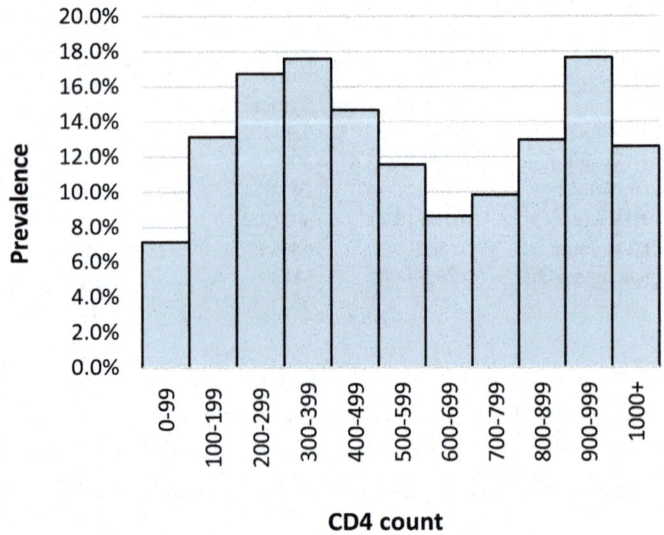

Fig. 16.2 Prevalence of SARS-COV-2 stratified by CD4 counts ($n = 26{,}315$; p-value <0.001)

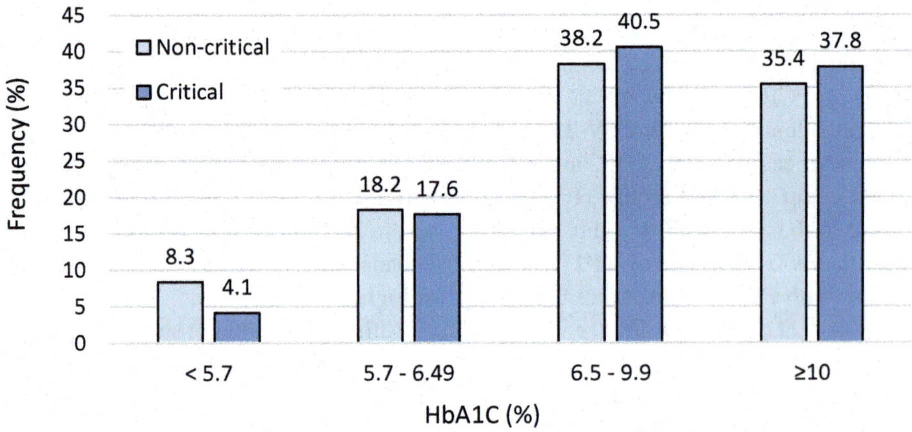

Fig. 16.3 Frequency of HbA1c control values in the critical and noncritical SARS-COV-2 groups ($n = 6158$; p-value = 0.612)

count when symptoms improved [32]. CD4 counts were shown to be lower in the 12 studies conducted in China but not in the one study outside of China. However, the meta-analysis did not indicate HIV prevalence [32]. The bimodal distribution, when assessing the relationship between the CD4 count and prevalence of SARS-CoV-2, was unexpected. We found no studies assessing the risk of contracting SARS-CoV-2 based on CD4 count, but because it is a marker of immunosuppression, we expected an increased prevalence among CD4-suppressed individuals [33]. We cannot explain this distribution, and although it was statistically significant, we cannot rule out the possible effect of confounders on the relationship. The finding of significantly lower active TB rates among SARS-CoV-2-positive individuals was also unexpected. Both active and previous TB have been reported as risk factors for increased COVID-19-related morbidity and mortality [34]. Unfortunately, our data on TB status of patients admitted to ICU and high-care wards were too limited to allow investigation of this in our study.

The increase in inflammatory markers, which was higher in critical patients, is in keeping with reports from other parts of the world [35]. Of note, many of these analytes were significantly raised even in the non-severe group. IL-6 has been widely reported as one of the early cytokines involved in the "cytokine storm" that causes the progression to severe COVID-19 [36, 37]. IL-6 is therefore also measured as a biochemical marker for disease severity and prognostic marker [37]. IL-6 measurement is not as widely offered in South African clinical laboratories. However, CRP, an acute-phase protein upregulated by IL-6, is offered widely and has also been shown to predict severity and the need for mechanical ventilation [38].

The WCC differential, especially the NLR, has also been proven to offer valuable information as a severity prediction marker [39]. Our analysis of the WCC differential is in keeping with the literature on the association between severe COVID-19 and lymphocytopenia as well as neutrophilia [35, 40]. This was also evident in the NLR, which was markedly raised in the critical group. The mechanism suggested for the increased neutrophil count is upregulation by proinflammatory cytokines as part of the "cytokine storm" process, while the lymphocytopenia may be multifactorial. The latter may be due to systemic inflammatory suppression of cellular immunity, resulting in a decrease in lymphocytes, as well as lysis and apoptosis. Lymphocytes express the ACE2 receptor, thus promoting viral entry and ultimate lysis, and cytokines like IL-6, IL-2, granulocyte-macrophage colony-stimulating factor (GM-CSF), and TNF-alpha promote apoptosis [40, 41]. In light of the promise shown by widely available CRP and the NLR as prognostic and severity markers, the utility of these markers should be explored further in Africa and other low-resource settings for the early identification of critically ill patients.

Our study also supported a lesser reported observation of severely depressed eosinophil counts in both the critical and noncritical groups. The reason for the dramatic decrease in eosinophils has not yet been elucidated, but it has been shown to be a good marker of prognosis and even diagnosis of COVID-19 (in combination with the NLR) and is one of the earliest analytes to return to baseline levels, signifying recovery [42, 43].

Abnormalities of the coagulation and fibrinolytic pathways are associated with disease severity and can be used to predict prognosis and outcomes [35, 36, 44]. The increased D-dimers that we report is clinically relevant in light of research that found that 75% of non-survivors in one study had a D-dimer of >1.3 mg/L [44]. Despite the markedly raised D-dimer levels, there was no statistically significant difference in the levels of fibrinogen between the two groups. This might be due to the use of prophylactic low molecular weight heparin administration to immobile hospitalized patients, particularly those admitted to ICU. Fibrinogen is also a positive acute-phase reactant, which could explain the increased levels in both groups.

Although not widely evaluated and reported, the erythrocyte concentration and red blood cell indices changes associated with SARS-CoV-2 and the severity of COVID-19 have been described in other studies [45, 46]. Our study also showed a significant decrease in hematocrit, hemoglobin, and RCC, as well as an increased RDW. The RDW is a numerical measure of size variability of circulating erythrocytes and has been reported to be a strong and independent indicator of adverse outcomes in the general population. It has also been shown to be associated with morbidity and mortality in people with previous myocardial infarction [47, 48]. The mechanism for anemia has not been clearly described, but theories on immune damage to the bone marrow, autoimmune hemolytic anemia, and anemia of chronic disease have been proposed [45, 49].

Our finding of relatively normal platelet counts with no significant difference between critical and noncritical patients is different to those reported in a meta-analysis on 1799 patients evaluating the relationship between COVID-19 severity and platelet count which showed that platelet counts decreased in severe cases [50, 35]. However, the studies included in the meta-analysis displayed high heterogeneity, with one study supporting increased platelet count, four studies supporting a decrease in platelet count, and four studies supporting no difference in platelet count between severity groups [50].

The widespread organ damage caused by COVID-19 is evidenced by the changes in cardiac-, liver-, and kidney-related analytes as well as the hematological manifestations of COVID-19 that we have described [51–53]. The effect of the cytokine storm on cardiac myocytes, direct myocyte injury by the virus (resulting in myocarditis), and hypoxia due to oxygen supply/demand mismatch have all been proposed as possible mechanisms for cardiac damage [52]. Regardless of the underlying mechanism, many previous studies evaluating the measurement of cardiac-related analytes such as troponin, NT-proBNP, and LDH clearly display an increase which is also related to disease severity and a strong indicator of prognosis [51, 52, 54, 55]. The reason for the discrepancy in the relationship between severity for troponin T and I in our study is unclear, but it was evident that the troponin I result had greater variation.

Similarly, we showed significant changes between the SARS-CoV-2 severity groups for all liver-related analytes, with the exception of total bilirubin in keeping with a number of other studies [53, 56]. The proposed mechanism behind the elevated liver-related enzymes is due to direct liver injury by the SARS-CoV-2 virus due to the abundance of ACE2 receptors in the liver [53]. It is uncertain how preexisting liver disease influences liver injury in patients with COVID-19, and this needs to be evaluated in our context where there is a high burden of liver disease from viral causes such as hepatitis A and B and HIV as well as from alcohol and nonalcoholic steatohepatitis.

LDH secretion is known to be triggered by cell membrane necrosis, which is a process that is highly upregulated in SARS-CoV-2 infection and increases with progression to severe COVID-19 [35, 57, 58]. Due to this strong relationship, also evident in our study, LDH has been proposed as a good biomarker for disease severity which also correlates with chest computed tomography scan data for pneumonia severity [35]. As albumin is a negative acute-phase reactant, its decrease has also been shown to strongly correlate with disease progression [35].

Despite evidence of CKD association with severe forms of COVID-19, as well as the reported propensity of acute kidney injury (AKI) to develop in severe cases, we did not find significantly raised creatinine or eGFR, and there was no significant difference between the cohorts [59]. Although unexpected, our study is not the first to report this finding [60, 61, 62]. The increased urea level associated with COVID-19 severity has also been described previously [54, 61]. The reported incidence of AKI among hospitalized patients ranges from 3 to 37%, and at least two other studies found urea to be more frequently elevated than creatinine in patients with COVID-19-related renal dysfunction [59, 62]. The proposed mechanisms for renal involvement in SARS-CoV-2 include thrombotic vascular processes, direct tubular cell injury, glomerulonephritis, fluid depletion, multiorgan failure, and rhabdomyolysis [63]. However, as we evaluated the analyte results in isolation, we did not attempt to classify patients according to the accepted AKI and CKD classification systems, which might have resulted in different findings. We also did not have the results of urine analytes, which have been proposed as a better marker for early renal impairment in COVID-19 than blood biochemistry in one preprint paper [62]. This finding warrants further studies to evaluate underlying mechanisms that could account for the raised urea and normal creatinine, such as the involvement of the renal tubules with sparing of the glomerulus, as has been suggested [64].

Our findings of more than threefold greater prevalence of uncontrolled DM among the SARS-CoV-2-positive patients are consistent with previous reports [65]. Data from Africa shows that most diabetics are unaware of their status, and less than 10% of those diagnosed are controlled [66]. However, the HbA1c level was not significantly elevated in the critical cohort compared with the noncritical cohort. At least two studies showed that when stratifying according to HbA1c levels, there was a trend to higher inflammatory markers evident in those with higher HbA1c [67, 68]. We could not find any studies comparing the median HbA1c values between critical and noncritical COVID-19 groups although studies measuring other aspects of diabetic control showed an association between worse control and poorer outcomes [69, 70].

Our study has limitations because it was a retrospective analysis of a laboratory database, and therefore, clinical information was lacking. Some clinical parameters, such as the severity (based on ward status) and chronic disease classification, could be deduced from the information available, but this approach is nonetheless less than ideal and may have resulted in some misclassification. In addition, the approach only allows for detection of associations, and therefore, a causative link between SARS-CoV-2 infection and the various effects on analytes and other parameters cannot be proven. Furthermore, we do not have data from the private sector which services patients who have medical insurance.

The strengths of this study include the large number of positive and negative SARS-CoV-2 individuals included and the time period over which the data were generated. There was also a large number of unique analytes compared between critical and noncritical COVID-19 cases. To our knowledge, this retrospective analysis of a large South African laboratory database comparing the effect of SARS-CoV-2 infection on multiple commonly tested analytes is the largest published analysis of its sort in terms of the number of individuals, different tests, and test results included.

In summary, in this large-scale retrospective analysis of laboratory data, we described the relationship between SARS-CoV-2 infection and HIV, TB, and DM as well as laboratory analyte changes associated with critical COVID-19 disease. Our findings did not support an increased prevalence of either HIV or TB in individuals with SARS-CoV-2 infection but did indicate an increase in disease severity with HIV-positive status. Our findings of clear differences in several commonly measured analytes between the critical and noncritical group suggest that these may be useful in our setting to triage patients.

Acknowledgments We thank the National Health Laboratory Service (NHLS) and their Academic Affairs and Research Office (AARMS) for permission to use the data and Mr. Thomas Papo for his assistance in data acquisition.

References

1. Petrosillo N, Viceconte G, Ergonul O, Ippolito G, Petersen E (2020) COVID-19, SARS and MERS: are they closely related? Clin Microbiol Infect 26:729e734. https://doi.org/10.1016/j.cmi.2020.03.026
2. Faust JS, Del Rio C (2020) Assessment of deaths from COVID-19 and from seasonal influenza. JAMA Intern. May 14. https://doi.org/10.1001/jamainternmed.2020.2306. Online ahead of print
3. Zaim S, Chong JH, Sankaranarayanan V, Harky A (2020) COVID-19 and multiorgan response. Curr Probl Cardiol 45(8):100618. https://doi.org/10.1016/j.cpcardiol.2020.100618
4. Tay MZ, Poh CM, Rénia L, MacAry PA, Ng LFP (2020) The trinity of COVID-19: immunity, inflammation and intervention. Nat Rev Immunol 20:363–374
5. IFCC Information Guide on COVID-19 Tuesday 14 Apr updates – IFCC. https://www.ifcc.org/ifcc-news/2020-03-26-ifcc-information-guide-on-covid-19/. Accessed 28 Jun 2020
6. Bohn MK, Lippi G, Horvath A, Sethi S, Koch D, Ferrari M et al (2020) Molecular, serological, and biochemical diagnosis and monitoring of COVID-19: IFCC taskforce evaluation of the latest evidence. Clin Chem Lab Med 58:1037–1052
7. Gupta R, Hussain A, Misra A (2020) Diabetes and COVID-19: evidence, current status and unanswered research questions. Eur J Clin Nutr 74:864–870
8. Rubino F, Amiel SA, Zimmet P, Alberti G, Bornstein S, Eckel RH et al (2020) New-onset diabetes in Covid-19. N Engl J Med. https://doi.org/10.1056/NEJMc2018688. Online ahead of print
9. Tan T, Khoo B, Mills EG, Phylactou M, Patel B, Eng PC et al (2020) Association between high serum total cortisol concentrations and mortality from COVID-19. Lancet Diabetes Endocrinol 8(8):659–660
10. Gallagher J, Watson C, Ledwidge M (2020) Association of Bacille Calmette-Guérin (BCG), adult pneumococcal and adult seasonal influenza vaccines with Covid-19 adjusted mortality rates in level 4 European countries. medRxiv. https://doi.org/10.1101/2020.06.03.20121624
11. Writer S (2020) Businesstech. 131,800 confirmed coronavirus cases in South Africa as deaths climb to 2,413. https://businesstech.co.za/news/trending/411383/131800-confirmed-coronavirus-cases-in-south-africa-as-deaths-climb-to-2413/. Accessed 28 Jun 2020
12. Coronavirus (COVID-19) | WHO | Regional Office for Africa. https://www.afro.who.int/health-topics/coronavirus-covid-19. Accessed 27 July 2020

13. Coronavirus Update (Live): 12,884,275 Cases and 568,561 Deaths from COVID-19 Virus Pandemic – Worldometer. https://www.worldometers.info/coronavirus/. Accessed 12 Jul 2020
14. Kamwendo EG (2020). Inter press service news agency. The curious case of Covid-19 in Africa. http://www.ipsnews.net/2020/06/curious-case-covid-19-africa/. Accessed 12 Jul 2020
15. Karim SSA, Churchyard GJ, Karim QA, Lawn SD (2009) HIV infection and tuberculosis in South Africa: an urgent need to escalate the public health response. Lancet 374:921–933
16. Riva A, Giacomelli A, Ridolfo A, Atzori C, Gervasoni C, Cattaneo D et al (2020) Clinical features and outcomes of HIV patients with coronavirus disease 2019. Clin Infect Dis ciaa579. https://doi.org/10.1093/cid/ciaa579. Online ahead of print
17. Vizcarra P, Pérez-Elías MJ, Quereda C, Moreno A, Vivancos MJ, Dronda F et al (2020) Description of COVID-19 in HIV-infected individuals: a single-Centre, prospective cohort. Lancet HIV S2352. https://doi.org/10.1016/S2352-3018(20)30164-8. Online ahead of print
18. Dauby N (2020) Potential impact of COVID-19 in people living with HIV: experience from previous 21st century coronaviruses epidemics. AIDS 34:1255–1256
19. Feldman C (2020) Potential impact of SARS-CoV-2 infection in HIV-positive patients in South Africa. Wits J Clin Med 2:19. https://doi.org/10.18772/26180197.2020.v2nsia3
20. Chen XP, Li GH, Tang XP, Xiong Y, Xie J, Cao Y (2020) Lack of severe acute respiratory syndrome in 18 AIDS patients hospitalized together. J Acquir Immune Defic Syndr. https://journals.lww.com/jaids/Citation/2003/10010/Letters_to_the_Editor__Lack_of_Severe_Acute.16.aspx
21. Liu Y, Bi L, Chen Y, Wang Y, Fleming J, Yu Y et al (2020) Active or latent tuberculosis increases susceptibility to COVID-19 and disease severity. medRxiv. https://doi.org/10.1101/2020.03.10.20033795
22. Crisan-Dabija R, Grigorescu C, Pavel CA, Artene B, Popa V, Cernomaz A et al (2020) Tuberculosis and COVID-19 in 2020: lessons from the past viral outbreaks and possible future outcomes. medRxiv. https://doi.org/10.1101/2020.04.28.20082917
23. Gouda HN, Charlson F, Sorsdahl K, Ahmadzada S, Ferrari AJ, Erskine H et al (2019) Burden of non-communicable diseases in sub-Saharan Africa, 1990–2017: results from the global burden of disease study 2017. Lancet Glob Heal 7:e1375–e1387
24. George JA, Brandenburg JT, Fabian J, Crowther NJ, Agongo G, Alberts M et al (2019) Kidney damage and associated risk factors in rural and urban sub-Saharan Africa (AWI-gen): a cross-sectional population study. Lancet Glob Heal 7:e1632–e1643
25. Pal R, Bhadada SK (2020) COVID-19 and non-communicable diseases. Postgrad Med J 96:429–430
26. National Health Laboratory Service (NHLS) – Overview. https://nationalgovernment.co.za/units/view/251/national-health-laboratory-service-nhls. Accessed 10 Jul 2020
27. SANDoH (2015) National consolidated guidelines for the prevention of mother-to-child transmission of HIV (PMTCT) and the management of HIV in children, adolescents and adults. https://www.health.gov.za/index.php/2014-03-17-09-09-38/policies-and-guidelines/category/230-2015p?download=937:national-art-guidelines-2015final. Accessed 28 July 2020
28. Petersen E, Koopmans M, Go U, Hamer DH, Petrosillo N, Castelli F et al (2020) Comparing SARS-CoV-2 with SARS-CoV and influenza pandemics. Lancet Infect Dis. https://doi.org/10.1016/S1473-3099(20)30484-9. Online ahead of print
29. De Giorgio A (2020) COVID-19 is not just a flu. Learn from Italy and act now. Travel Med Infect Dis 35:101655. https://doi.org/10.1016/j.tmaid.2020.101655
30. Scully EP, Haverfield J, Ursin RL, Tannenbaum C, Klein SL (2020) Considering how biological sex impacts immune responses and COVID-19 outcomes. Nat Rev Immunol 20:442–447
31. South Africa COVID-19 Corona Tracker. https://www.coronatracker.com/country/south-africa/. Accessed 25 Jul 2020
32. Zhang H, Wu T (2020) CD4+T, CD8+T counts and severe COVID-19: a meta-analysis J Infect S0163-4453(20)30417-5. https://doi.org/10.1016/j.jinf.2020.06.036. Online ahead of print
33. Ford N, Meintjes G, Vitoria M, Greene G, Chiller T (2017) The evolving role of CD4 cell counts in HIV care. Curr Opin HIV AIDS 12:123–128
34. Tadolini M, Codecasa LR, García-García JM, Blanc FX, Borisov S, Alffenaar JW et al (2020) Active tuberculosis, sequelae and COVID-19 co-infection: first cohort of 49 cases. Eur Respir J 56:2001398. https://doi.org/10.1183/13993003.01398-2020
35. Kermali M, Khalsa RK, Pillai K, Ismail Z, Harky A (2020) The role of biomarkers in diagnosis of COVID-19 – a systematic review. Life Sci 254:117788. https://doi.org/10.1016/j.lfs.2020.117788
36. Ponti G, Maccaferri M, Ruini C, Tomasi A, Ozben T (2020) Biomarkers associated with COVID-19 disease progression. Crit Rev Clin Lab Sci:1–11. https://doi.org/10.1080/10408363.2020.1770685. Online ahead of print
37. Magro G (2020) SARS-CoV-2 and COVID-19: is interleukin-6 (IL-6) the 'culprit lesion' of ARDS onset? What is there besides tocilizumab? SGP130Fc. Cytokine X 2(2):100029. https://doi.org/10.1016/j.cytox.2020.100029. Online ahead of print
38. Herold T, Jurinovic V, Arnreich C, Lipworth BJ, Hellmuth JC, von Bergwelt-Baildon M et al (2020) Elevated levels of IL-6 and CRP predict the need for mechanical ventilation in COVID-19. J allergy Clin Immunol 146:128-136.e4
39. Liu J, Liu Y, Xiang P, Pu L, Xiong H, Li C et al (2020) Neutrophil-to-lymphocyte ratio predicts critical illness patients with 2019 coronavirus disease

in the early stage. J Transl Med 18:206. https://doi.org/10.1186/s12967-020-02374-0
40. Yang AP, Liu J, Tao W, Li H (2020) The diagnostic and predictive role of NLR, d-NLR and PLR in COVID-19 patients. Int Immunopharmacol 84:106504. https://doi.org/10.1016/j.intimp.2020.106504
41. Xu H, Zhong L, Deng J, Peng J, Dan H, Zeng X et al (2020) High expression of ACE2 receptor of 2019-nCoV on the epithelial cells of oral mucosa. Int J Oral Sci 12:1–5
42. Xie G, Ding F, Han L, Yin D, Lu H, Zhang M (2020) The role of peripheral blood eosinophil counts in COVID-19 patients. Allergy. https://doi.org/10.1111/all.14465. Online ahead of print
43. Lu G, Wang J (2020) Dynamic changes in routine blood parameters of a severe COVID-19 case. Clin Chim Acta 508:98–102
44. Kavsak PA, de Wit K, Worster A (2020) Emerging key laboratory tests for patients with COVID-19. Clin Biochem 81:13–14
45. Wang C, Deng R, Gou L, Fu Z, Zhang X, Shao F et al (2020) Preliminary study to identify severe from moderate cases of COVID-19 using combined hematology parameters. Ann Transl Med 8:593–593
46. Yuan X, Huang W, Ye B, Chen C, Huang R, Wu F et al (2020) Changes of hematological and immunological parameters in COVID-19 patients. Int J Hematol 1:3. https://doi.org/10.1007/s12185-020-02930-w
47. Patel KV, Ferrucci L, Ershler WB, Longo DL, Gurainik JM (2009) Red blood cell distribution width and the risk of death in middle-aged and older adults. Arch Intern Med 169:515–523
48. Perlstein TS, Weuve J, Pfeffer MA, Beckman JA (2009) Red blood cell distribution width and mortality risk in a community-based prospective cohort. Arch Intern Med 169:588–594
49. Lazarian G, Quinquenel A, Bellal M, Siavellis J, Jacquy C, Re D et al (2020) Autoimmune haemolytic anaemia associated with COVID-19 infection. Br J Haematol 190:29–31
50. Lippi G, Plebani M, Henry BM (2020) Thrombocytopenia is associated with severe coronavirus disease 2019 (COVID-19) infections: a meta-analysis. Clin Chim Acta 506:145–148
51. Basu-Ray I, Almaddah NK, Adeboye A (2020) Cardiac Manifestations Of Coronavirus (COVID-19) [Updated 2020 Apr 23]. In: StatPearls [internet]. StatPearls Publishing, Treasure Island. Available from: https://www.ncbi.nlm.nih.gov/books/NBK556152/
52. Aikawa T, Takagi H, Ishikawa K, Kuno T (2020) Myocardial injury characterized by elevated cardiac troponin and in-hospital mortality of COVID-19: an insight from a meta-analysis, J Med Virol. https://doi.org/10.1002/jmv.26108. Online ahead of print
53. Ali N (2020) Relationship between COVID-19 infection and liver injury: a review of recent data. Front Med 7:458. https://doi.org/10.3389/fmed.2020.00458
54. Kermali M, Kaur Khalsa R, Pillai K, Ismail Z, Harky A (2020) The role of biomarkers in diagnosis of COVID-19-a systematic review. Life Sci 254:117788. https://doi.org/10.1016/j.lfs.2020.117788
55. Gao L, Jiang D, Wen XS, Cheng XC, Sun M, He B et al (2020) Prognostic value of NT-proBNP in patients with severe COVID-19. Respir Res 21:83. https://doi.org/10.1186/s12931-020-01352-w
56. Xu L, Liu J, Lu M, Yang D, Zheng X (2020) Liver injury during highly pathogenic human coronavirus infections. Liver Int 40:998–1004
57. Han Y, Zhang H, Mu S, Wei W, Jin C, Xue Y et al (2020) Lactate dehydrogenase, a risk factor of severe COVID-19 patients. medRxiv. https://doi.org/10.1101/2020.03.24.20040162
58. Zhang C, Shi L, Wang FS (2020) Liver injury in COVID-19: management and challenges. Lancet Gastroenterol Hepatol 5:428–430
59. Benedetti C, Waldman M, Zaza G, Riella LV, Cravedi P (2020) COVID-19 and the kidneys: an update. Front Med 7:423. https://doi.org/10.3389/fmed.2020.00423
60. Wang L, Li X, Chen H, Yan S, Li D, Li Y et al (2020) Coronavirus disease 19 infection does not result in acute kidney injury: an analysis of 116 hospitalized patients from Wuhan, China. Am J Nephrol 51:343–348
61. Hirsch JS, Ng JH, Ross DW, Sharma P, Shah HH, Barnett RL et al (2020) Acute kidney injury in patients hospitalized with COVID-19. Kidney Int 98:209–218
62. Zhou H, Zhang Z, Fan H, Li J, Li M, Dong Y et al (2020) Urinalysis, but not blood biochemistry, detects the early renal-impairment in patients with COVID-19. medRxiv. https://doi.org/10.2139/ssrn.3569863
63. Selby NM, Forni LG, Laing CM, Horne KL, Evans RDR, Lucas BJ et al (2020) Covid-19 and acute kidney injury in hospital: summary of NICE guidelines. BMJ 369:m1963. https://doi.org/10.1136/bmj.m1963
64. Wang S, Zhou X, Zhang T, Wang Z (2020) The need for urogenital tract monitoring in COVID-19. Nat Rev Urol 17:314–315
65. Erener S (2020) Diabetes, infection risk and Covid-19. Mol Metab:101044. https://doi.org/10.1016/j.molmet.2020.101044.. Online ahead of print
66. Mbanya JCN, Motala AA, Sobngwi E, Assah FK, Enoru ST (2010) Diabetes in sub-Saharan Africa. Lancet 375:2254–2266
67. Klein SJ, Fries D, Kaser S, Mathis S, Thomé C, Joannidis M (2020) Unrecognized diabetes in critically ill COVID-19 patients. Crit Care 24:406. https://doi.org/10.1186/s13054-020-03139-3
68. Wang Z, Du Z, Zhu F (2020) Glycosylated hemoglobin is associated with systemic inflammation, hypercoagulability, and prognosis of COVID-19 patients. Diabetes Res Clin Pract 164:108214. https://doi.org/10.1016/j.diabres.2020.108214
69. Roncon L, Zuin M, Rigatelli G, Zuliani G (2020) Diabetic patients with COVID-19 infection are at higher risk of ICU admission and poor short-term outcome. J Clin Virol 127:104354. https://doi.org/10.1016/j.jcv.2020.104354
70. Zhu L, She ZG, Cheng X, Qin JJ, Zhang XJ, Cai J et al (2020) Association of Blood Glucose Control and Outcomes in patients with COVID-19 and pre-existing type 2 diabetes. Cell Metab 31:1068–1077.e3

17 The Primary Outcomes and Epidemiological and Clinical Features of Coronavirus Disease 2019 (COVID-19) in Iran

Amir Vahedian-Azimi, Sara Ashtari, Gholamhosein Alishiri, Alireza Shahriary, Mohsen Saberi, Farshid Rahimi-Bashar, Yunes Panahi, Hossein Samadinia, Mojtaba Ferdosi, Iman Jafari, Ensieh Vahedi, Ali Ghazvini, Mohamad Nikpouraghdam, Keivan Goharimoghadam, Morteza Izadi, Nematollah Jonaidi-Jafari, Sajad Shojaee, Hamid Asadzadeh Aghdaei, Mohamad Amin Pourhoseingholi, and Amirhossein Sahebkar

Abstract

Aim

We aimed to describe the epidemiological and clinical characteristics of Iranian patients with COVID-19.

Methods

In this single-center and retrospective study, patients with confirmed COVID-19 infections were enrolled. Univariate and multivariate logistic regression methods were used to explore the risk factors associated with outcomes.

A. Vahedian-Azimi · I. Jafari
Trauma Research Center, Nursing Faculty, Baqiyatallah University of Medical Sciences, Tehran, Iran

S. Ashtari · H. A. Aghdaei
Basic and Molecular Epidemiology of Gastrointestinal Disorders Research Center, Research Institute for Gastroenterology and Liver Diseases, Shahid Beheshti University of Medical Sciences, Tehran, Iran

G. Alishiri · A. Shahriary · E. Vahedi · A. Ghazvini
Chemical Injuries Research Center, Systems Biology and Poisonings Institute, Baqiyatallah University of Medical sciences, Tehran, Iran

M. Saberi
Medicine, Quran and Hadith Research Center & Department of Community Medicine, Faculty of Medicine, Baqiyatallah University of Medical Sciences, Tehran, Iran

F. Rahimi-Bashar
Anesthesia and Critical Care Department, Hamadan University of Medical Sciences, Hamadan, Iran

Y. Panahi
Pharmacotherapy Department, Faculty of Pharmacy, Baqiyatallah University of Medical Sciences, Tehran, Iran

P. C. Guest (ed.), *Clinical, Biological and Molecular Aspects of COVID-19*, Advances in Experimental Medicine and Biology 1321, https://doi.org/10.1007/978-3-030-59261-5_17

Results
Of 179 patients with confirmed COVID-19 infection, 12 remained hospitalized at the end of the study and 167 were included in the final analysis. Of these, 153 (91.6%) were discharged and 14 (8.38%) died in hospital. Approximately half (50.9%) of patients suffered from a comorbidity, with diabetes or coronary heart disease being the most common in 20 patients. The most common symptoms on admission were fever, dyspnea, and cough. The mean durations from first symptoms to hospital admission was 8.64 ± 4.14 days, whereas the mean hospitalization time to discharge or death was 5.19 ± 2.42 and 4.35 ± 2.70 days, respectively. There was a significantly higher age in non-survivor patients compared with survivor patients. Multivariate regression showed increasing odds ratio (OR) of in-hospital death associated with respiratory rates >20 breaths/min (OR: 5.14, 95% CI: 1.19–22.15, p = 0.028) and blood urea nitrogen (BUN) >19 mg/dL (OR: 4.54, 95% CI: 1.30–15.85, *p* = 0.017) on admission. In addition, higher respiratory rate was associated with continuous fever (OR: 4.08, 95% CI: 1.18–14.08, *p* = 0.026) and other clinical symptoms (OR: 3.52, 95% CI: 1.05–11.87, *p* = 0.04).
Conclusion
The potential risk factors including high respiratory rate and BUN levels could help to identify COVID-19 patients with poor prognosis at an early stage in the Iranian population.

Keywords

COVID-19 · SARS · MERS · Iran · Respiratory rate · Blood urea nitrogen

17.1 Introduction

Novel coronavirus disease (COVID-19) was first reported in Wuhan, China, in December 2019, not long before the lead up to the Lunar New Year when China undertakes the world's largest mass travel event [1]. The COVID-19 outbreak has spread rapidly throughout the world as well as in Iran. Iran was the first Middle East country to report a death due to this coronavirus, which was officially announced on February 20 in Qom [2].

The clinical spectrum of COVID-19 infection appears to be wide, encompassing asymptomatic infection, mild upper respiratory tract illness, and severe viral pneumonia with respiratory failure and even death, with many patients being hospitalized with pneumonia [3, 4]. Previous studies in China have shown acute symptoms of severe respiratory infection in the early stages of this virus, with some patients rapidly developing acute respiratory distress syndrome (ARDS), acute respiratory failure, and other serious complications that can lead to death [5–7].

H. Samadinia
Nanobiotechnology Research Center, Baqiyatallah University of Medical Sciences, Tehran, Iran

M. Ferdosi
Students' Research Committee, Nursing Faculty, Baqiyatallah University of Medical Sciences, Tehran, Iran

M. Nikpouraghdam
Nephrology and Urology Research Center, Baqiyatallah University of Medical sciences, Tehran, Iran

K. Goharimoghadam
Internal Medicine, Shariati hospital, Tehran University of Medical Sciences, Tehran, Iran

M. Izadi · N. Jonaidi-Jafari
Health Research Center, Baqiyatallah University of Medical Sciences, Tehran, Iran

S. Shojaee · M. A. Pourhoseingholi (✉)
Gastroenterology and Liver Diseases Research Center, Research Institute for Gastroenterology and Liver Diseases, ShahidBeheshti University of Medical Sciences, Tehran, Iran
e-mail: aminphg@gmail.com

A. Sahebkar (✉)
Biotechnology Research Center, Pharmaceutical Technology Institute, Mashhad University of Medical Sciences, Mashhad, Iran

Neurogenic Inflammation Research Center, Mashhad University of Medical Sciences, Mashhad, Iran

Halal Research Center of IRI, FDA, Tehran, Iran

Polish Mother's Memorial Hospital Research Institute (PMMHRI), Lodz, Poland
e-mail: sahebkara@mums.ac.ir; amir_saheb2000@yahoo.com

As COVID-19 is a newly identified pathogen, the information on the clinical characteristics of affected patients is limited. A few studies on epidemiological and clinical characteristics of cases with novel coronavirus pneumonia have been conducted in China [8–10]. Despite the increasing number of confirmed cases in Iran, no clinical study of Iranian patients has yet been published. Additionally, details of the clinical and course of illness have not yet been well described. A better understanding of the clinical features of COVID-19 can be helpful in preventing and controlling the epidemic and understanding future developments and the potential effect of various interventions. In this paper, we have described the clinical characteristics, laboratory findings, and risk factors of in-hospital death for infected patients with COVID-19 during hospitalization to provide potential insights into the prevention and treatment.

17.2 Methods

17.2.1 Study Design and Participants

This single-center and retrospective study by focusing on the clinical characteristics of confirmed cases of COVID-19 was conducted on 179 adult patients, who referred to the Baqiyatallah Hospital in Tehran, Iran, between February 26 and March 15, 2020. All patients with COVID-19 enrolled in this study were diagnosed according to the World Health Organization interim guidance [11]. The patients were suspected to have COVID-19 infection according to the symptoms like fever, dry cough, shortness of breath, and aches. Upon admission, patients underwent chest computed tomography (CT) scans plus swab-based PCR tests. Since the scan results were readily available (compared to swab tests which took at least 24 h for the results), diagnosis was made based on the CT results. Moreover, the PCR results were dependent on methods of sampling, storage, handling, and transfer of specimens, which may cause a significant rate of false-negative results. All patients were monitored up to March 15, 2020, as the final date of follow-up. The study was approved by the Research Ethics Committee of the Baqiyatallah University of Medical Sciences, and written informed consent was obtained from patients before enrolment and data were collected retrospectively.

17.2.2 Data Collection

All demographic characteristics and clinical data for this retrospective study were collected from medical records of patients with COVID-19. Data recorded include demographics (age, gender, and occupation), smoking history, exposure history (details regarding infection), family history, comorbidities (coronary heart disease, hypertension, diabetes, lung disease, and malignancy), symptoms (fever, fatigue, dry cough, sore throat, headache, myalgia, dyspnea, chest pain, rhinorrhea, nausea, and vomiting), clinical features (respiratory rate, heart rate, blood pressure rate, and body temperature), laboratory findings [white blood cell (WBC) count, lymphocytopenia (lymphocyte count), platelet count, erythrocyte sedimentation rate (ESR), C-reactive protein (CRP), prothrombin time (PT), partial thromboplastin time (PTT), creatinine (Cr), alanine aminotransferase (ALT), aspartate aminotransferase (AST), blood urea nitrogen (BUN), lactic acid dehydrogenase (LDH)], blood gas analysis (saturated pressure of oxygen (SPO2), partial pressure of carbon dioxide ($PaCO_2$) and bicarbonate (HCO_3), clinical outcomes (respiratory failure, heart failure, ARDS, acute cardiac injury, and acute kidney injury), and treatment measures (antiviral therapy, respiratory support, and mechanical ventilation). In addition, the durations from first symptoms to hospital admission, hospitalization days, intensive care unit (ICU) admission, and patient status (death or recovery) at the end of study were recorded.

17.2.3 Definitions

The date of disease onset was defined as the day when the symptom was noticed. Acute respiratory distress syndrome (ARDS) was defined according to the Berlin definition [12]. Acute kidney injury was identified according to the Kidney Disease Improving Global Outcomes (KDIGO) definition [13]. Acute cardiac injury was diagnosed if serum levels of cardiac biomarkers [high sensitivity cardiac troponin I (cTnI)] were above the 99th percentile upper reference limit or if new abnormalities were shown in electrocardiography or echocardiography [14]. Fever was defined as axillary temperature of at least 37.3 °C.

17.2.4 Statistical Analysis

Categorical variables were described as frequency rates and percentages, and continuous variables were described using mean, median, and interquartile range (IQR) values. Means for continuous variables were compared using independent group t tests. In case of limited data or non-normality, the Mann-Whitney test was used. Chi-square test or Fisher exact test (in case of low sample numbers) were used to compare the distribution of categorical data. To explore the risk factors associated with in-hospital death, as well as the COVID-19 swab PCR test and symptom status at the end of hospitalization, univariate and multivariate logistic regression models were used. To avoid over-fitting in the multivariate model, only those factors which resulted in a p-value less than 0.2 in univariate analysis were selected for the multivariate model. The final model was selected according to forward conditional. All statistical analyses were performed using R version 3.6.1 software. A two-sided α of less than 0.05 was considered statistically significant.

17.3 Results

17.3.1 Demographic and Clinical Characteristics

All 179 hospitalized patients with confirmed COVID-19 were monitored up to March 15, 2020, and 12 patients were excluded from the study because they were still hospitalized up to the final date of follow-up. Therefore, we included 167 patients in the final analysis. The mean age of the 167 patients was 55.26 ± 13.01 years (range: 22–89 years), and 133 (79.5%) were men. Of the 167 patients, 153 (91.6%) were discharged, and 14 (8.38%) died in hospital. In 85 patients (50.9%), comorbidities were present which included diabetes (12%), coronary heart disease (12%), diabetes with hypertension (9.6%), and hypertension (8.4%). In addition, 26 (15.6%) patients had history of lung disease that included chronic obstructive pulmonary disease (COPD) (30.7%), history of chemical ventilators (30.7%), pneumonia (23.2%), and asthma (15.4%).

The most common symptoms on admission were fever (90.4%), dyspnea (85.6%), cough (83.8%), myalgia (68.9%), fatigue (50.9%), and headache (37.7%). Less common symptoms were sputum, chest pain, rhinorrhea, sore throat, nausea, and vomiting (Table 17.1). Compared to patients discharged from hospital (n = 153), patients who died in hospital (n = 14) were significantly older (64.78 ± 9.36 vs. 54.39 ± 12.96, p = 0.004). In terms of gender, family history, symptoms, and comorbidities, no statistically significant differences were found between survivor and non-survivor groups (p > 0.05) (Table 17.1).

According to chest CT scans, 58 (34.7%) patients showed consolidation, 52 (34.1%) showed bilateral pulmonary infiltration, and 52 (31.1%) had ground-glass opacity. Nasal swab PCR analysis for COVID-19 was positive in 108

Table 17.1 Baseline demographic and clinical characteristics of patients infected with COVID-19 on admission in survivor and non-survivor patients

Variables	Total (n = 167)	Survivor (n = 153)	Non-survivor (n = 14)	*P*-value
Age				
Mean ± SD (range)	55.26 ± 13.01 (22–89)	54.39 ± 12.96 (22–89)	64.78 ± 9.36 (53–87)	0.004[a]
Gender no (%)				
Male	133 (79.5)	121 (79.1)	12 (85.7)	0.737
Female	34 (20.4)	32 (20.9)	2 (14.3)	
Occupation no (%)				
Employee	52 (31.1)	51 (33.3)	1 (7.1)	0.265
Retired	58 (34.7)	51 (33.3)	7 (50)	
Housewife	29 (17.4)	27 (17.6)	2 (14.3)	
Unemployed	24 (14.4)	20 (13.1)	4 (28.6)	
Soldier	2 (1.2)	2 (1.3)	0	
Doctor	2 (1.2)	2 (1.3)	0	
Exposure history no (%)				
Community	138 (82.6)	124 (81)	14 (100)	0.360
Contact with confirmed cases	13 (7.8)	13 (8.5)	0	
Travel	14 (8.4)	14 (9.2)	0	
Hospital	2 (1.2)	2 (1.3)	0	
Family history no (%)				
Yes	40 (24.0)	39 (25.5)	1 (7.1)	0.191
Symptoms no (%)				
Fever	151 (90.4)	138 (90.2)	13 (92.9)	1.000
Cough	140 (83.8)	128 (83.7)	12 (85.7)	1.000
Sputum	43 (25.7)	41 (26.8)	2 (14.3)	0.523
Dyspnea	143 (85.6)	130 (85)	13 (92.9)	0.695
Myalgia	115 (68.9)	105 (68.6)	10 (71.4)	1.000
Headache	63 (37.7)	60 (39.2)	3 (21.4)	0.254
Fatigue	85 (50.9)	79 (51.6)	6 (42.9)	0.586
Chest pain	23 (13.8)	23 (15)	0	0.221
Rhinorrhea	23 (13.8)	22 (14.4)	1 (7.1)	0.695
Sore throat	23 (13.8)	22 (14.4)	1 (7.1)	0.695
Nausea/vomiting	33 (19.8)	30 (19.6)	3 (21.4)	1.000
Smoking history no (%)				
Yes	14 (8.4)	14 (9.2)	0	0.610
Duration onset of clinical symptoms to hospital admission, days				
Mean ± SD (range)	8.64 ± 4.14 (1–22)	8.69 ± 4.19 (1–22)	8.14 ± 3.63 (2–16)	0.636
Hospitalization, days				
Mean ± SD (range)	5.12 ± 2.45 (1–12)	5.19 ± 2.42 (1–12)	4.35 ± 2.70 (1–8)	0.222
Comorbidities no (%)				
Yes	85 (50.9)	77 (50.3)	8 (57.1)	0.625
The type of comorbidities no (%)				
Kidney disease	2 (1.2)	2 (1.3)	0	
Diabetes	20 (12.0)	17 (11.1)	3 (21.4)	
Hypertension	14 (8.4)	13 (8.5)	1 (7.1)	0.699
Coronary heart disease	20 (12.0)	18 (11.8)	2 (14.3)	
Diabetes and hypertension	16 (9.6)	16 (10.5)	0	
Others	13 (7.8)	11 (7.2)	2(14.3)	
Underlying lung disease no (%)				

(continued)

Table 17.1 (continued)

Variables	Total (n = 167)	Survivor (n = 153)	Non-survivor (n = 14)	*P*-value
Yes	26 (15.6)	23 (15)	3 (21.4)	0.460
The type of lung disease no (%)				
COPD	8 (4.8)	7 (4.6)	1 (7.1)	
Asthma	4 (2.4)	4 (2.6)	0	0.402
Pneumonia	6 (3.6)	6 (3.9)	0	
History of mechanical ventilators	8 (4.8)	6 (3.9)	2 (14.3)	
COVID-19 swab nose no (%)				0.363
Positive	108 (64.7)	97 (63.4)	11 (78.6)	
Negative	37 (22.2)	36 (23.5)	1 (7.1)	
Suspicious	22 (13.2)	20 (13.1)	2 (14.3)	
CT scan findings no (%)				
Consolidation	58 (34.7)	50 (32.7)	8 (57.1)	0.080
Ground-glass opacity	52 (31.1)	51 (33.3)	1 (7.1)	
Bilateral pulmonary infiltration	57 (34.1)	52 (34)	5 (35.7)	
Admitted situation no (%)				
Isolation wards	157 (94)	148 (96.7)	5 (35.7)	<0.001[a]
ICU	13 (7.8)	5 (3.3)	9 (64.3)	<0.001[a]
Antiviral therapy no (%)				
Monotherapy[a]	1 (0.6)	1 (0.7)	0	
Triple therapy[a]	68 (40.7)	64 (41.8)	4 (28.6)	0.586
Fourth therapy[a]	98 (58.7)	88 (57.5)	10 (71.4)	
Treatment no (%)				
Supplemental oxygen	139 (83.2)	125 (81.7)	14 (100)	0.130
NIV	14 (8.4)	6 (3.9)	8 (57.1)	<0.001[a]
IMV	11 (6.6)	1 (0.7)	10 (71.4)	<0.001[a]

[a]Monotherapy: oseltamivir + hydroxychloroquine; triple therapy: oseltamivir + hydroxychloroquine + lopinavir/ritonavir; fourth therapy: oseltamivir + hydroxychloroquine + lopinavir/ritonavir + ribavirin. Abbreviation: *NIV* noninvasive ventilation, *IMV* invasive mechanical ventilation

(64.4%) patients, negative in 37 (22.2%), and suspicious in 22 (13.2%) patients on the day of hospital admission. From 167 patients with COVID-19, 157 (94%) were admitted to isolation wards, and 13 (7.8%) were admitted and transferred to the ICU. The mean durations from first symptoms to hospital admission was 8.64 ± 4.14 (range: 1–22) days, whereas the mean hospitalization time to discharge and death were 5.19 ± 2.42 (range: 1–12) and 4.35 ± 2.70 (range: 1–8) days, respectively (Table 17.1).

17.3.2 Vital Signs, Laboratory Parameters, and Blood Gas Analysis

Heart rate, blood pressure, and body temperature on the day of hospital admission did not differ between survivor and non-survivor patients ($p > 0.05$). However, the median respiratory rate of non-survivor patients was significantly higher than survivor patients (18 vs. 20, $p = 0.031$). Baseline lymphocytopenia occurred in 123 patients (73.7%), with no significant difference between the two groups ($p = 0.732$). In addition, analyses of WBC and platelet count, ESR, CRP, LDH, BUN, Cr, ALT, AST, PTT, PT, INR, and PH showed no significant differences between survivor and non-survivor groups ($P > 0.05$) (Table 17.2). The median of SpO_2 was 90 mm Hg (IQR, 87–93), and the median of $PaCO_2$was 46.6 mm Hg (IQR, 34.6–59.9). The ratio of SpO2 was significantly lower in non-survivor patients than survivor cases (90 vs. 74 mm Hg, $p < 0.001$). However, the median of $PaCO_2$ was not significantly different between survivor and non-survivor groups ($p > 0.05$).

Table 17.2 Vital signs, laboratory parameters, and blood gas analysis of patients with COVID-19 on admission in survivor and non-survivor patients

Variables	Normal range	Total	Survivor	Non-survivor	*P*-value
Respiratory rate	12–20 min	18 (17–20)	18 (17–20)	20 (17–22)	0.031[a]
Heart rate	60–100 BPM	92 (82–106)	92 (84–106)	92.5 (80–102.5)	0.910
Blood pressure	120/80 mmHg	120/80 (110/70–135-80)	120/80 (110/70–135/80)	115/75 (110/67–136/57)	0.398
Temperature	36.1–37.2 C	38 (37–38.4)	38 (37–38.4)	38 (37.1–38.2)	0.712
SpO_2	90–92%	90 (87–93)	90 (88–93)	74 (55–85)	<0.001[a]
WBC	4-10 × 10^3 /L	6.2 (4.8–7.7)	6 (4.7–7.6)	7.4 (5.9–11.8)	0.282
Lymphocyte	1–3 × 10^3 /L	1.1 (0.89–1.5)	1.2 (0.86–1.5)	1.1 (0.98–1.6)	0.732
Platelet	145–45 × 10^3/L	169 (132–217)	169 (133–214)	162 (110–275)	0.395
ESR	Up to 15 mm/hr	39 (24–57)	40 (24–58)	35 (26.7–58)	0.952
CRP	Up to 5 mg/L	60 (24.2–94.7)	57.9 (23–94.7)	67.2 (49–94.7)	0.355
LDH	207–414 U/L	660 (252–824)	657 (508–819)	676 (557–852)	0.978
BUN	7–19 mg/dL	14 (11–18)	13 (11–17)	17.5 (11.7–23.5)	0.064
Cr	0.9–1.3 mg/dL	1.1 (0.9–1.3)	1.1 (0.9–1.3)	1.3 (0.9–1.4)	0.618
ALT	< 45 U/L	46 (29–77)	37 (23–73.5)	48 (24.2–95)	0.744
AST	<35 U/L	37 (23–76)	45 (29–76)	63 (40.5–89)	0.350
PTT	25–35 s	56 (35–76)	55 (35–76)	60 (35.5–76.5)	0.720
PT	≤ 13.5 s	14.2 (13–16.5)	13.8 (12.9–16.5)	15.1 (12.9–19.8)	0.073
INR	≤ 1	1.2 (1–1.3)	1.2 (1–1.4)	1.2 (1–1.3)	0.837
PH	7.38–7.42	7.39 (7.21–7.43)	7.39 (7.18–7.44)	7.4 (7.22–7.43)	0.814
$PaCO_2$	35–45 mm hg	46.6 (34.6–59.9)	47.3 (35.9–60.3)	37.9 (30.3–59.8)	0.440
HCO3	22–26 mEq/L	24.6 (21.3–27)	24.8 (21.3–27.5)	22 (15.7–24.1)	0.091

[a]Data are expressed as the median (IQR), *WBC* white blood cell, *SPO2* saturated pressure of oxygen, *CRP* C-reactive protein, *ESR* erythrocyte sedimentation rate, *LDH* lactic acid dehydrogenase, *BUN* blood urea nitrogen, *Cr* creatinine, *ALT* alanine aminotransferase, *AST* aspartate aminotransferase, *PTT* partial thromboplastin time, *PT* prothrombin time, *INR* international normalized ratio, *PH* pulmonary hypertension past history, *PaCO2* partial pressure of carbon dioxide, HCO_3 bicarbonate

17.3.3 Main Intervention and Complication

All patients in the current study received antiviral treatment, but the types of combination therapy varied among patients. Quadruple combination therapy (oseltamivir + hydroxychloroquine + lopinavir/ritonavir + ribavirin) was administered to 98 (58.7%) patients, 68 (40.7%) patients received triple combination therapy (oseltamivir + hydroxychloroquine + lopinavir/ritonavir), and only one patient received dual combination therapy (oseltamivir + hydroxychloroquine). Invasive mechanical ventilation (IMV) was required in 11 (6.6%) patients, 14 (8.4%) of patients received noninvasive ventilation (NIV), and the majority of patients, 139 (83.2%), received extracorporeal membrane oxygenation as rescue therapy. Compared with survivors, non-survivors were more likely to receive mechanical ventilation, either invasively or noninvasively (p < 0.001) (Table 17.1). In terms of complications, all patients developed ARDS, but no other complications such as acute cardiac injury and acute kidney injury were found in these patients.

17.3.4 Univariate and Multivariate Analysis

In univariate analysis, odds of in-hospital death were higher for older patients (OR = 1.07, 95%

Table 17.3 Risk factors associated with in-hospital death and COVID-19 swab test status

	In-hospital death (survive vs. non-survive)				COVID-19 swab test status (positive vs negative)			
Variable	Univariate OR (95%CI)	*P* value	Multivariate OR (95%CI)	*P* value	Univariate OR (95%CI)	*P* value	Multivariate OR (95%CI)	*P* value
Age, years	1.07 (1.02–1.12)	0.006			0.97 (0.95–1.00)	0.051	0.97 (0.95–1.00)	0.051
Female (vs. male)	0.63 (0.13–2.96)	0.56			0.99 (0.44–2.23)	0.98		
Smoker (vs. non-smoker)	–	–			0.44 (0.14–1.33)	0.15		
Underlying disease	1.31 (0.43–3.97)	0.63			0.43 (0.22–0.84)	0.14		
Chronic lung disease	1.54 (0.39–5.95)	0.53			1.72 (0.65–4.56)	0.27		
Respiratory rate (>20 vs. <20 per min)	5.72 (1.52–21.51)	0.01	5.14 (1.19–22.15)	0.028	1.21 (0.36–4.06)	0.75		
Temperature in admission (Celsius)	0.87 (0.45–1.70)	0.71			1.57 (1.03–2.38)	0.03		
SpO2 (<90 vs. >90)	5.47 (1.18–25.28)	0.03	4.33 (0.90–22.79)	0.067	0.97 (0.51–1.87)	0.93		
Lymphocytopenia (<1500 vs. >1500)	1.13 (0.33–3.81)	0.84			1.03 (0.49–2.16)	0.93		
Blood urea nitrogen (>19 vs. <19	5.62 (1.75–18.07)	0.004	4.54 (1.30–15.84)	0.017	0.62 (0.25–1.51)	0.29		
AST (>35 vs. <35)	3.76 (0.81–17.42)	0.09			0.63 (0.31–1.27)	0.20		
PTA (>13.5 vs. <13.5)	1.48 (0.47–4.61)	0.50			1.01 (0.52–1.93)	0.98		
INR (>1 vs. <1)	1.94 (0.58–6.48)	0.27			1.25 (0.65–2.41)	0.49		

CI: 1.02–1.12) (Table 17.3). For those patients with respiratory rates>20/min, the odds of mortality were 5.72 times more than for others (95% CI: 1.52–21.51). Low SpO_2 and high BUN were also found to be associated with a higher risk for in-hospital death according to univariate analysis, whereas multivariate analysis only revealed respiratory rate and BUN as risk factors of in-hospital mortality. For the results of COVID-19 swab PCR tests at the end of hospitalization, univariate analysis revealed that increased temperature on admission (1 °C) could increase the risk of having a positive COVID-19 swab test.

Table 17.4 presents the associated factors with fever and other symptoms at the end of hospitalization. Age still showed a significant factor in univariate analysis, but no significance was observed in multivariate analysis. Patients with higher respiratory rates had a higher chance of continuing fever (or other symptoms) in both univariate and multivariate models. SpO_2 also indicated a potential association with fever in the univariate model, and for those with SpO_2 less than 90 at admission, the odds of continuing fever were 2.52 times higher (95% CI: 1.001–6.39).

BUN was a risk factor for other clinical symptoms in univariate analysis (Table 17.4). However, the results for fever were only borderline ($p = 0.054$). Moreover, INR increased the chance

Table 17.4 Risk factors associated with fever and other clinical symptoms status at the end of hospitalization

	Fever (finish vs. continue)				Other clinical symptoms (finish vs. continue)			
Variables	Univariate OR (95%CI)	*P* value	Multivariate OR (95%CI)	*P* value	Univariate OR (95%CI)	*P* value	Multivariate OR (95%CI)	*P* value
Age, year	1.04 (1.003–1.07)	0.03			1.02 (0.98–1.05)	0.31		
Female (vs. male)					0.49 (0.13–1.73)	0.27		
Smoker (vs. nonsmoker)	0.39 (0.05–3.14)	0.38			0.94 (0.19–4.48)	0.94		
Underlying disease	0.95 (0.41–2.21)	0.92			0.72 (0.31–1.69)	0.45		
Chronic lung disease	0.98 (0.31–3.13)	0.97			0.71 (0.19–2.55)	0.59		
Respiratory rate (>20 vs. <20 per min)	3.49 (1.06–11.43)	0.04	4.08 (1.18–14.08)	0.026	3.69 (1.12–12.14)	0.03	3.52 (1.05–11.87)	0.04
Temperature in admission (Celsius)	0.88 (0.53–1.46)	0.63			0.89 (0.54–1.47)	0.65		
SpO2 (<90 vs. >90)	2.52 (1.001–6.39)	0.049			2.36 (0.92–6.007)	0.07		
Lymphocytopenia (<1500 vs. >1500)	1.60 (0.65–3.92)	0.30			1.10 (0.42–2.85)	0.84		
Blood urea nitrogen (<19 vs. >19)	2.68 (0.98–7.33)	0.054			2.85 (1.04–7.84)	0.04	2.74 (0.98–7.69)	0.055
AST (>35 vs. <35)	1.10 (0.46–2.65)	0.83			2.01 (0.75–5.32)	0.16		
PTA (>13.5 vs. <13.5)	1.98 (0.81–4.85)	0.13			1.84 (0.75–4.55)	0.18		
INR (>1 vs. <1)	2.85 (1.08–7.52)	0.03	3.12 (1.15–8.46)	0.025	2.11 (0.83–5.36)	0.12		

of continuous fever according to both univariate and multivariate models.

17.4 Discussion

This single-center retrospective study focused on the clinical characteristics of 167 confirmed cases with COVID-19, out of which 153 were discharged and 14 died in hospital. Diabetes and coronary heart disease were the most common comorbidities in these patients, and fever, dyspnea, and cough were the most common symptoms on admission. Our results showed significantly higher age in non-survivor patients compared with survivors, whereas gender, family history, symptoms, or comorbidities did not significantly alter survival. These findings contrast with more recent studies which found that male gender and the presence of comorbidities such as diabetes, hypertension, and heart disease are associated with poorer survival outcomes, including higher death rates [15, 16]. However, this report identified several risk factors for in-hospital death, continuous or completion of fever, and other clinical symptoms at the end of hospitalization up to the final date of follow-up. In particular, high respiratory rates of more than 20 per min and BUN levels greater than 19 mg/dL on admission were associated with a higher risk of death in-hospital. Additionally, a higher respiratory rate of more than 20 breaths per min on admission was associated with continuous fever and other clinical symptoms at the end of hospitalization. This is most likely linked with increased demand for oxygen due to the pneumonia-like symptoms.

The SARS-CoV-2 virus which causes COVID-19 disease is similar to the severe acute respiratory syndrome (SARS-CoV) and Middle Eastern respiratory syndrome (MERS-CoV) coronaviruses which resulted in 8096 and 2519 cases worldwide, respectively [17]. However, SARS-CoV-2 has had a much higher rate of infectivity with an estimated 11.9 million cases worldwide, as of July 7, 2020. Overall, global mortality rate in patients with COVID-19 is lower than that previously seen in patients with SARS and MERS [17–22]. According to the results of this study, 138 (82.6%) of patients were infected with the virus through contacts in the community, 14 (8.4%) by travelling, and 13 (7.3%) by contact with confirmed cases, and 2 (1.2%) patients were infected in hospital. Contrary to our study, the rate of hospital transmission was higher in China, which may be explained by low knowledge and experience in dealing with the virus in the early stages. A study by Wang et al. of 138 hospitalized patients with COVID-19 showed that nearly half of patients, 57 (41.3%), were infected in hospital [8]. However, 1 month after the outbreak of COVID-19, China gained increased control over the spread of the virus by adopting measures such as isolation of confirmed and suspected cases and lockdown and quarantine of Wuhan and surrounding areas. This kept the mortality rate less at less than 1% outside of the Hubei province [23, 24].

Our univariate analysis showed that age was the only demographic factor associated with increased death outcomes of COVID-19 infections although this was just outside of significance in the multivariate model. Previous studies showed that older age was an important independent predictor of mortality in SARS-CoV and MERS-CoV [14, 18], and it has been confirmed that increasing age is also associated with increased risk of death in patients with SARS-CoV-2 infections [4, 8, 15, 16]. In a study similar to our study in China by Zhou et al. [4], older age was associated with the high risk of mortality (OR:1.10, 95% CI: 1.03–1.17, $p < 0.001$).

In the current study, all patients developed ARDS, but we did not find other outcomes. In contrast, a previous study from China showed that sepsis was a common complication (59%) as well as other outcomes such as respiratory failure (54%), ARDS (31%), heart failure (23%), septic shock (20%), acute cardiac injury (17%), acute kidney injury (15%), and secondary infection (15%) [4]. In terms of treatment, all patients in the current study received antiviral treatments, but the types of combination therapies used varied between patients. Given the small numbers of patients studied and the lack of a control group, it is impossible to determine whether or not these treatments led to improved outcomes and increased survival. Additionally, the majority of patients (83.2%) received supplemental oxygen as a rescue therapy. Invasive mechanical ventilation (IMV) was required in 11 patients, and 14 patients received noninvasive ventilation (NIV). Compared with survivors, non-survivors were more likely to receive mechanical ventilation, either invasively or noninvasively.

Our study has some limitations. First, due to the retrospective study design, not all laboratory tests were done in all patients, including measurements of D-dimer, IL-6, and serum ferritin. Therefore, their role might be underestimated in predicting in-hospital death. Second, due to the lack of degree of organ dysfunction in the patients, we were not able to calculate the sequential organ failure assessments (SOFA) score. Third, the interpretation of our findings might be limited by the small sample size.

17.5 Conclusions

We found that a high respiratory rate more than 20 breaths per min and BUN levels greater than 19 (mg/dL) on admission were associated with a higher risk of death in hospital. Additionally, higher respiratory rate more than 20 per min on admission was associated with continuous fever and other clinical symptoms at the end of hospitalization. Further studies are needed to increase our understanding of this virus and to aid in the control of future outbreaks. As it stands now, it is still not certain when this pandemic will diminish to negligible levels as the infection rate is still on the rise in some countries such as Iran, the United States, Brazil, Mexico, and India [21, 22]. Therefore, a more complete understanding of COVID-19 which could be used to inform world policies and help prevent future outbreaks might not be achievable for several years to come.

Declarations Ethics Approval and Consent to Participate

The study was approved by the Research Ethics Committee of the Baqiyatallah University of Medical Sciences, and written informed consent was obtained from patients involved before enrolment when data were collected retrospectively. Thanks to guidance and advice from the Clinical Research Development Unit of Baqiyatallah Hospital.

Funding/Support This work was supported by the Baqiyatallah University of Medical Sciences, Tehran, Iran.

Availability of Data and Materials Data associated with this study is available from the corresponding authors on a reasonable request.

References

1. Phelan AL, Katz R, Gostin LO (2020) The novel coronavirus originating in Wuhan, China: challenges for Global Health governance. JAMA. https://doi.org/10.1001/jama.2020.1097. Online ahead of print
2. Stancu C, Sima A (2001) Statins: mechanism of action and effects. J Cell Mol Med 5(4):378–387
3. Young BE, Ong SWX, Kalimuddin S, Low JG, Tan SY, Loh J et al (2020) Epidemiologic features and clinical course of patients infected with SARS-CoV-2 in Singapore. JAMA 323(15):1488–1494
4. Zhou F, Yu T, Du R, Fan G, Liu Y, Zh L et al (2020) Clinical course and risk factors for mortality of adult inpatients with COVID-19 in Wuhan, China: a retrospective cohort study. Lancet 395(10229):1054–1062
5. Albarello F, Pianura E, Di Stefano F, Cristofaro M, Petrone A, Marchioni L et al (2020) 2019-novel coronavirus severe adult respiratory distress syndrome in two cases in Italy: an uncommon radiological presentation. Int J Infect Dis 93:192–197
6. Shi H, Han X, Jiang N, Cao Y, Alwalid O, Gu J et al (2020) Radiological findings from 81 patients with COVID-19 pneumonia in Wuhan, China: a descriptive study. Lancet Infect Dis 20(4):425–434
7. Liu L, Yang Y, Gao Z, Li M, Mu X, Ma X et al (2018) Practice of diagnosis and management of acute respiratory distress syndrome in mainland China: a cross-sectional study. J Thorac Dis 10(9):5394–5404
8. Wang D, Hu B, Hu C, Zhu F, Liu X, Zhang J et al (2020) Clinical characteristics of 138 hospitalized patients with 2019 novel coronavirus-infected pneumonia in Wuhan, China. JAMA 323(11):1061–1069
9. Xu XW, Wu XX, Jiang XG, Xu KJ, Ying LJ, Ma CL et al (2020) Clinical findings in a group of patients infected with the 2019 novel coronavirus (SARS-Cov-2) outside of Wuhan: retrospective case series. 368:m792. https://doi.org/10.1136/bmj.m792. Chen N, Zhou M, Dong X, Qu J, Gong F, Han Y et al (2020) Epidemiological and clinical characteristics of 99 cases of 2019 novel coronavirus pneumonia in Wuhan, China: a descriptive study. Lancet 395(10223):507–513
10. World Health Organization (2020) Clinical management of severe acute respiratory infection when novel coronavirus (nCoV) infection is suspected: interim guidance. Published 28 Jan 2020. Accessed 31 Jan 2020. https://www.who.int/publications-detail/clinical-managementof-severe-acute-respiratory-infection-when-novelcoronavirus-(ncov)-infection-is-suspected
11. Ranieri VM, Rubenfeld GD, Thompson BT, Ferguson ND, Caldwell E, Fan E et al (2012) Acute respiratory distress syndrome: the Berlin definition. JAMA 307(23):2526–2533
12. Kellum JA, Lameire N, Aspelin P, Barsoum RS, Burdmann EA, Goldstein SL et al (2012) Kidney disease: improving global outcomes (KDIGO) acute kidney InjuryWork group. KDIGO clinical practice guideline for acute kidney injury. Kidney Int Suppl 2(1):1–138
13. Huang C, Wang Y, Li X, Ren L, Zhao J, Hu Y et al (2020) Clinical features of patients infected with 2019 novel coronavirus in Wuhan, China. Lancet 395(10223):497–506
14. Parry J (2020) Wuhan: Britons to be evacuated as scientists estimate 44 000 cases of 2019-nCOV in the city. 368:m351. doi: https://doi.org/10.1136/bmj.m351
15. Ambrosino I, Barbagelata E, Ortona E, Ruggieri A, Massiah G, Giannico OV et al (2020) Gender differences in patients with COVID-19: a narrative review. Monaldi Arch Chest Dis 90:2. https://doi.org/10.4081/monaldi.2020.1389
16. Espinosa OA, Zanetti ADS, Antunes EF, Longhi FG, Matos TA, Battaglini PF (2020) Prevalence of comorbidities in patients and mortality cases affected by SARS-CoV2: a systematic review and meta-analysis.

Rev Inst Med Trop Sao Paulo 62:e43. https://doi.org/10.1590/S1678-9946202062043
17. Xiao J, Fang M, Chen Q, He B (2020) SARS, MERS and COVID-19 among healthcare workers: a narrative review. J Infect Public Health 13(6):843–848
18. Yang X, Yu Y, Xu J, Shu H, Xia J, Liu H et al (2020) Clinical course and outcomes of critically ill patients with SARS-CoV-2 pneumonia in Wuhan, China: a single-centered, retrospective, observational study. Lancet Respir Med 8(5):475–481
19. Rothe C, Schunk M, Sothmann P, Bretzel G, Froeschl G, Wallrauch C et al (2020) Transmission of 2019-nCoV infection from an asymptomatic contact in Germany. N Engl J Med 382(10):970–971
20. Gomersall CD, Joynt GM, Lam P, Li T, Yap F, Lam D et al (2004) Short-term outcome of critically ill patients with severe acute respiratory syndrome. Intensive Care Med 30(3):381–387
21. https://www.worldometers.info/coronavirus/#countries
22. https://coronavirus.jhu.edu/map.html
23. Chan KW, Wong VT, Tang SCW (2020) COVID-19: an update on the epidemiological, clinical, preventive and therapeutic evidence and guidelines of integrative Chinese-Western medicine for the management of 2019 novel coronavirus disease. Am J Chin Med 48(3):737–762
24. Wang Y, Wang Y, Chen Y, Qin Q (2020) Unique epidemiological and clinical features of the emerging 2019 novel coronavirus pneumonia (COVID-19) implicate special control measures. J Med Virol 92(6):568–576

Acute Respiratory Distress Syndrome and COVID-19: A Scoping Review and Meta-analysis

18

Mehdi Jafari-Oori, Fatemeh Ghasemifard, Abbas Ebadi, Leila Karimi, Farshid Rahimi-Bashar, Tannaz Jamialahmadi, Paul C. Guest, Amir Vahedian-Azimi, and Amirhossein Sahebkar

Abstract

Acute respiratory distress syndrome (ARDS) is a fatal complication of the new severe acute respiratory syndrome coronavirus (SARS-CoV-2), which causes COVID-19 disease. This scoping review was carried out with international, peer-reviewed research studies and gray literature published up to July 2020 in Persian and English languages. Using keywords derived from MESH, databases including Magiran, IranMedex, SID, Web of Sciences, PubMed, Embase via Ovid, Science Direct, and Google Scholar were searched. After screening titles and abstracts, the full texts of selected articles were evaluated, and those which passed the criteria were analyzed and synthesized with inductive thematic analysis.

M. Jafari-Oori · A. Vahedian-Azimi (✉)
Trauma Research Center, Nursing Faculty, Baqiyatallah University of Medical Sciences, Tehran, Iran

F. Ghasemifard
Department of Occupational Therapy, University of Social Welfare and Rehabilitation Sciences (USWR), Tehran, Iran

A. Ebadi · L. Karimi
Behavioral Sciences Research Center, Life style institute, Nursing Faculty, Baqiyatallah University of Medical Sciences, Tehran, Iran

F. Rahimi-Bashar
Anesthesia and Critical Care Department, Hamadan University of Medical Sciences, Hamadan, Iran

T. Jamialahmadi
Department of Food Science and Technology, Quchan Branch, Islamic Azad University, Quchan, Iran

Department of Nutrition, Faculty of Medicine, Mashhad University of Medical Sciences, Mashhad, Iran

P. C. Guest
Laboratory of Neuroproteomics, Department of Biochemistry and Tissue Biology, Institute of Biology, University of Campinas (UNICAMP), Campinas, Brazil

A. Sahebkar (✉)
Biotechnology Research Center, Pharmaceutical Technology Institute, Mashhad University of Medical Sciences, Mashhad, Iran

Neurogenic Inflammation Research Center, Mashhad University of Medical Sciences, Mashhad, Iran

School of Pharmacy, Mashhad University of Medical Sciences, Mashhad, Iran

Polish Mother's Memorial Hospital Research Institute (PMMHRI), Lodz, Poland
e-mail: sahebkara@mums.ac.ir

P. C. Guest (ed.), *Clinical, Biological and Molecular Aspects of COVID-19*, Advances in Experimental Medicine and Biology 1321, https://doi.org/10.1007/978-3-030-59261-5_18

Study quality was also evaluated using a standard tool. The overall prevalence of ARDS was estimated using a random-effects model. This led to identification of 23 primary studies involving 2880 COVID-19 patients. All articles were observational with a cross-sectional, retrospective, case report, and cohort design with moderate to strong quality. The main findings showed that COVID-19-related ARDS has a high prevalence and is different to ARDS due to other etiologies. Elderly and patients with comorbidities and organ failure should be closely surveyed for respiratory organ indications for several weeks after the onset of respiratory symptoms. There is currently no definitive treatment for ARDS in COVID-19 disease, and supportive therapies and their effects are somewhat controversial.

Keywords

Covid-19 · Acute respiratory distress syndrome (ARDS) · Scoping review · Meta-analysis

18.1 Introduction

In December 2019, the world was faced with the sudden outbreak of the SARS-CoV-2 virus which causes COVID-19 disease and was deemed a substantial threat to worldwide health by the World Health Organization (WHO) [1–3]. By March 6, 2020, more than 100,000 persons in 93 countries had been infected by SARS-CoV-2, and this rapidly progressed to over 12 million persons in 213 countries and territories by July 10, 2020 (Fig. 18.1) [4]. Clinical manifestations caused by the virus have varied from fever, cough, fatigue, sputum production, shortness of breath, sore throat, headache, gastrointestinal symptoms like diarrhea and vomiting, as well as acute respiratory distress syndrome (ARDS) and organ failure [5, 6]. Compared to other organs, the COVID-19 virus appears to most affect the respiratory system in the initial stages and in some patients develops to ARDS [7]. Although ARDS patients receive respiratory and therapeutic support in the intensive care unit (ICU), these patients still show a high mortality [8].

According to the Berlin definition, ARDS is a type of acute diffuse, inflammatory lung injury, which can lead to increased pulmonary vascular permeability, increased lung weight, and loss of aerated lung tissue [9]. The clinical manifestations are hypoxemia, bilateral opacities in computed tomography (CT) imaging, increased venous admixture, and physiological dead space. One way of measuring this is to determine the ratio of the arterial partial pressure of oxygen (PaO_2) to the fraction of inspired oxygen (FiO_2). Based on the degree of hypoxemia, ARDS can be classified as mild (PaO_2/FIO_2: >200 – ≤300 mm Hg), moderate (>100 – ≤200), or severe (<100) [9]. In the most severe cases, inspired oxygen cannot reach the blood circulation, leading to organ failure and death of the patient (Fig. 18.2).

Some previous review studies have estimated the prevalence of ARDS as 14–28% in COVID-19 patients [10, 11]. Given the critical impact of ARDS on patient outcomes and the current lack of effective treatment options, we have carried out a scoping review and meta-analysis to systematically assess its prevalence in COVID-19 patients. It was also of interest to identify clinical features that could be used to predict or monitor disease severity as a means of informing treatment options.

18.2 Methods

This scoping review study was conducted systematically according to the Preferred Reporting Items for Systematic Reviews and Meta-Analyses (PRISMA) guidelines [12–15]. A flowchart of the study selection procedure is shown in Fig. 18.3.

18.2.1 Search Strategy

In this review, any original studies conducted worldwide and published in the English or Persian languages in internal and external databases were evaluated up to July 2020. To search for ARDS and COVID-19 studies, we employed the national databases including Magiran, IranMedex, Iranian Archive for Scientific Documents Center (IASD),

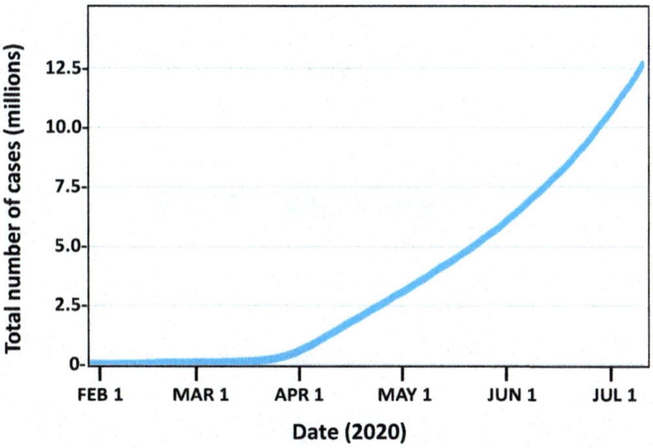

Fig. 18.1 Total number of COVID-19 cases globally from the end of January to July 10, 2020

and Iranian National Library (INL) and international databases such as MEDLINE (PubMed, Ovid), Scopus, Web of Science, Embase, ProQuest, Google, and Google Scholar. Gray literature and reference lists of the extracted primary articles were also reviewed to discover potential related studies. The keywords and subject headings used to search for these databases are shown in Table 18.1. These were used separately and in combination as a syntax using Boolean operators such as "AND," "OR," "NOT," and the "*" sign.

18.2.2 Study Selection Process

The inclusion and exclusion criteria for selecting primary articles are indicated in Table 18.2. Firstly all feasible studies were recognized. Secondly, the titles and abstracts of the identified articles were separately screened by two investigators, and the full texts of relevant articles were obtained and their eligibility was determined. In case of disagreement between the two researchers, the article in question was reviewed by an additional author who was an expert in review studies.

18.2.3 Quality Assessment and Data Extraction

To evaluate the quality of the studies, five-item tools were used, as described in previous studies [16–19]. The five items were related to the research design, sampling method, sample size, comparison group, and psychometric properties. Each item scaled from 0 to 3 and the overall score was from 0 to 15 [17]. Based on this approach, the studies were divided into three categories related to quality, which were defined as either weak (score 4 or below), medium (score 5 to 10), or strong (score above 10). The assessment was performed by two authors (MJO and FRB), and the disagreements were resolved by the senior author (AVA). A data extraction form was used to assemble the information in each article as follows: first author, year of study conduction, design and purpose of the study, setting, main findings and conclusions, limitations, and language. To guarantee accuracy, two other authors examined the extracted data for the final review.

18.2.4 Synthesis of Data and Analysis

Inductive thematic analysis was performed using the results of preliminary studies to find potentially important emerging themes [20]. The results of each of these studies were assessed and compared until the primary themes were outlined. To estimate the overall ARDS prevalence, each study was assessed using calculations of the binary distribution variance. Weighted averages were used to combine the prevalence values of the studies. The weight assigned to each study was an inverse of its vari-

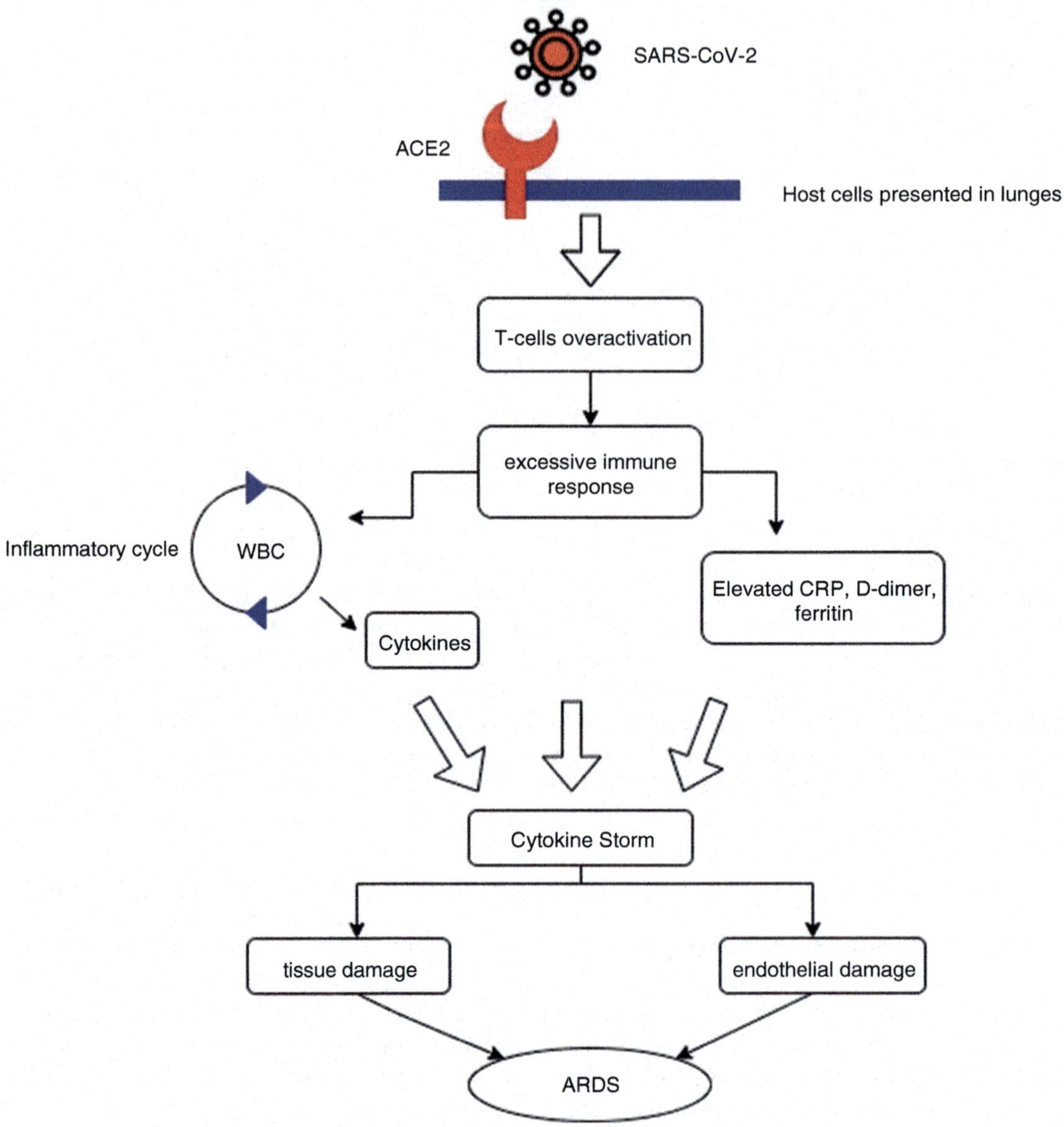

Fig. 18.2 The effects of ARDS on breathing in COVID-19 disease

ance. The heterogeneity of the studies was assessed by the I^2 index. Data heterogeneity was divided into three classes of less than 25% (low heterogeneity), 25 to 75% (medium heterogeneity), and more than 75% (high heterogeneity). Due to the resulting high heterogeneity of the data in this review, a random-effects model was applied. The analysis was conducted with STATA 12 software (StataCorp LLC; College Station, TX, USA).

18.3 Results

18.3.1 Literature Search

The flowchart of the selection process is revealed in Fig. 18.3. A total of 1100 potential study references were acquired after searching the database and search engines and the pertinent reference lists. After the first screening of titles and abstracts, 95 studies were selected to assess the

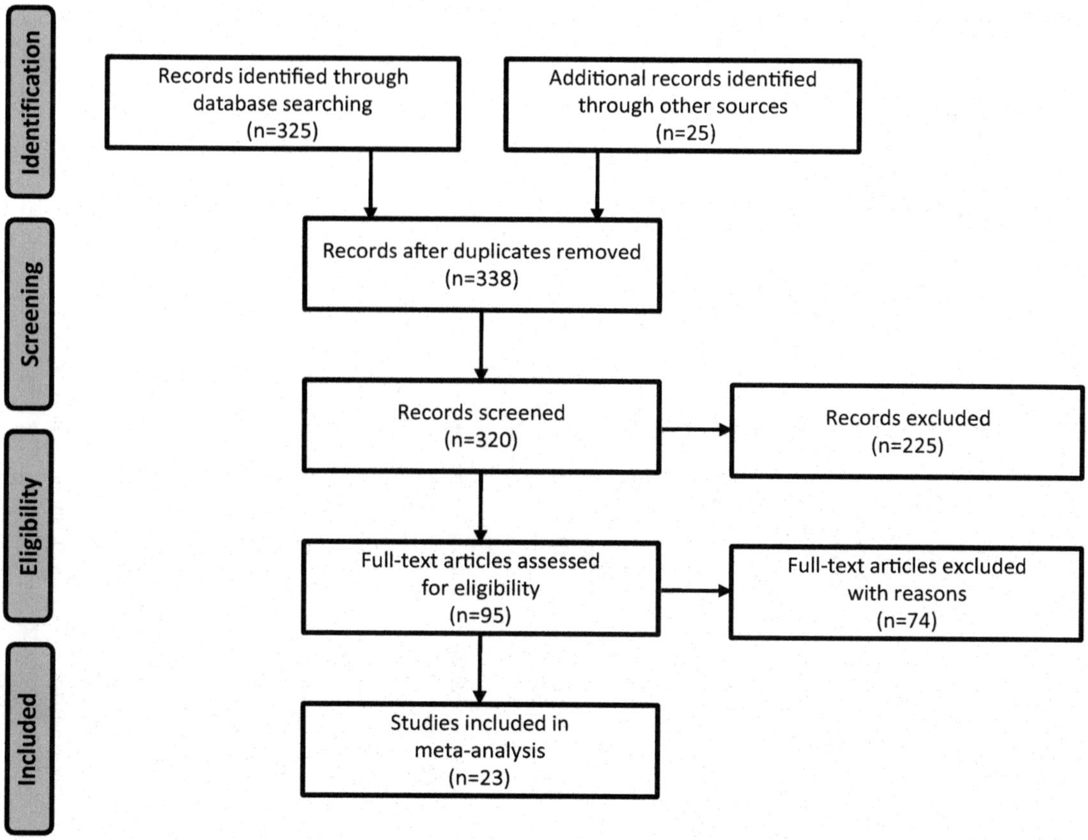

Fig. 18.3 Identification and selection of studies

Table 18.1 Keywords and subject headings used during the search

Search terms
"lung recruitment maneuver" OR "recruitment manoeuver" OR "lung volume recruitment" OR "lung recruitment" OR "recruit manoeuver" OR "recruit manoeuver" OR "recruit manoeuver" OR "recruit manoeuver" OR "PEEP setting" OR "recruitment ARDS" "titrated PEEP" OR "PEEP titration" OR "low PEEP" OR "protective lung strategy" OR "high PEEP" OR ards OR "acute respiratory distress" OR "acute chest syndrome" OR "respiratory distress syndrome" OR "shock lung" OR "adult respiratory distress syndrome" OR "respiratory distress syndrome, acute" AND "Novel coronavirus" OR "Novel coronavirus 2019" OR "2019 novel coronavirus" OR "2019 nCoV" OR "Wuhan coronavirus" OR "Wuhan pneumonia" OR Covid-19 OR "2019-nCoV" OR "SARS-CoV-2" OR "coronavirus 2019" OR "2019-nCoV"

Table 18.2 Inclusion and exclusion criteria for selected primary articles

Inclusion criteria	Exclusion criteria
Peer-reviewed, primary studies	Not peer-reviewed, primary research
Published in English or Persian	Not written in English or Persian
Pointing to the ARDS on patients with Covid-19	Not related to the empirical data (letters, editorials, news, etc.)
Published until July 2020	Low-quality studies Studies that did not include the keywords of "ARDS," "coronavirus 2019," or "Covid-19"

Table 18.3 Characteristics of the primary articles

No.	Author	Year	Design	Sample size	Country	Data type	Aim of the study	Main findings	Conclusion	Limitations	Quality score
1	Chu et al. [21]	2003	Cohort	133 patients with confirmed SARS	China	RT-PCR	N/A	Incidence of ARDS: 32 (24.1%). Risk factors: ARDS: age (p =0.001), comorbidity (p=0.002), initial LDH (p=0.034), and qPCR result of nasopharyngeal specimen (ps=0.025)	N/A	Limited to two centers	Strong
2	Albarello et al. [22]	2020	Case report	Two COVID-19 patients	Italy	X-ray and CT imaging	Present imaging findings of first two patients identified in Italy with COVID-19 infection traveling from Wuhan, China	Moderate to severe lung infiltrates, in bilateral and multi-segmental extension of lung opacities. Follow-up: tubular and enlarged appearance of pulmonary vessels with sudden caliber reduction and mediastinal lymphadenopathy on day 4 ARDS	Pulmonary vessel enlargement is early radiological sign of lung impairment	Low sample size and short time interval for follow-up CT	Medium
3	Beerkenset al. [23]	2020	Case report	21-year-old man with sickle cell disease	USA	Single-cell RNA sequencing in blood and urine	COVID-19 causing acute chest syndrome	On HD 5: atelectasis in the right lower lung. On HD 11: diffuse GGOs and reticular opacities	N/A	N/A	Medium
4	Chen et al. [24]	2020	RCT retrospective	44 patients with H7N9 in control and 17 in intervention group	China	Laboratory and imaging findings	Assessing effect of mesenchymal stem cell transplant on treatment of COVID-19-related ARDS	Intervention: transplantation of mesenchymal stem cells (MSCs) Results: mortality in intervention group < control group and without harmful effects	MSCs can be used in treatment of COVID-19-related ARDS	Low sample size	Strong
5	Yang et al. [25]	2020	Single-centered, retrospective, observational	52 critically ill adult patients with COVID-19 pneumonia	China	Demographic, symptoms, laboratory values, comorbidities, treatments, clinical outcomes	Comparing clinical manifestation of survivors and non-survivors	Survivors: ARDS in 9 patients Non-survivors: ARDS in 26 patients Mortality rate of patients with ARDS: 61.5%	Survival time of non-survivors likely to be within 1–2 weeks after ICU admission. ARDS: increased risk of death	Low sample size	Medium
6	Xu et al. [26]	2020	Retrospective case report	A 50-year-old with COVID-19	China	Laboratory test results, chest CT scans	Assessing pathogenesis of COVID-19	Symptoms: fever, chills, cough, fatigue, and shortness of breath.	N/A	N/A	Strong

								Chest X-ray: multiple patchy shadows in both lungs, progressive infiltrate and diffuse gridding shadow in both lungs. HD 13: oxygen saturation >95%.			
								HD14: oxygen saturation 60% and sudden cardiac arrest. Lung biopsy samples: bilateral diffuse alveolar damage with cellular fibromyxoid exudates. Right lung showed evident desquamation of pneumocystis and hyaline membrane formation, indicating ARDS. Left lung: pulmonary edema with hyaline membrane formation suggesting early-phase ARDS.			
								Treatment: corticosteroids, ventilator support, interferon alpha-2b, moxifloxacin			
7	Wu et al. [27]	2020	Retrospective cohort	201 patients with confirmed COVID-19 pneumonia	China	Epidemiology, demographics, clinical, laboratory, management, treatment, and outcome data	Describe clinical characteristics and outcomes in patients with COVID-19 pneumonia who developed ASRS or died	84 patients developed ARDS, and of these patients, 44 died. Patients with ARDS had more dyspnea, hypertension, and diabetes than patients without ARDS. Older age, neutrophilia, and organ and coagulation dysfunction were risk factors associated with development of ARDS and progression from ARDS to death. Among patients with ARDS, treatment with methylprednisolone decreased risk of death	Older age greater risk of ARDS and death. High fever associated with ARDS and better outcomes. Treatment with methylpre dnisolone may be beneficial for patients who develop ARDS	Only patients with severe COVID-19 pneumonia Single hospital with limited sample size, selection bias when identifying factors that influence clinical outcomes	Strong
8	Wang et al. [28]	2020	Cross-sectional	339 elderly patients with COVID-19	China	Demographics, clinical features, comorbidities, laboratory tests and complications	To investigate characteristics and prognostic factors in elderly patients with COVID-19	ARDS: 21% comorbidities including cardiovascular disease, and COPD and ARDS were strong predictors of death	N/A	Only one hospital	Strong

(continued)

Table 18.3 (continued)

No.	Author	Year	Design	Sample size	Country	Data type	Aim of the study	Main findings	Conclusion	Limitations	Quality score
9	Wang et al. [29]	2020	Case series	Three cases with COVID-19 and ARDS (75-year-old male, 59-year-old female, and 49-year-old)	China	N/A	Tissue plasminogen activator (tPA) treatment for COVID-19-associated ARDS	Case 1: hydroxychloroquine and azithromycin given for 5 days. tPA given intravenously over 2 h, followed by infusion over subsequent 22 h. 11 h into tPA infusion the P/F ratio improved to 408, a twofold improvement from pre-tPA. By HD 11: multiple organ failure with refractory hypotension secondary to arrhythmia and superimposed bacterial infection and died.	tPA: in all three cases, patients demonstrated initial improvement in their P/F ratio, with improvements ranging from a 38% improvement (case 3) to a ~100% improvement (Case 1). The improvements were transient in all three patients. Formal studies needed to determine whether this resulted from tPA therapy or due to unrelated /random effects	N/A	Medium
								Case 2: oxygen requirement progressed over 2 days from nasal cannula O2 to a 100% mask with a PaO2 of 137. On HD 4 patient required intubation. Vasopressor for hemodynamic support. IVtPA administered as intravenous bolus over 2 h, followed by infusion over subsequent 22 h.			
								Case 3: Same as case 2			

10	Tang et al. [30]	2020	Retrospective case control	Patients with either COVID-19 (n=73) or H1N1 (n=75) with ARDS	China	N/A	Explore different clinical presentations between COVID-19 and influenza A (H1N1) pneumonia in patients with ARDS	P/F ratio of 198.2 mmHg in COVID-19 patients significantly higher than the P/F ratio of 107.0 mmHg of H1N1 patients ($p<0.001$). GGOs more common in COVID-19 patients than in H1N1 patients ($p<0.001$)	Many differences between COVID-19 and H1N1-induced ARDS patients	Condition of H1N1 patients more severe than that of COVID-19 patients	Strong
11	Shi et al. [31]	2020	Cohort	416 hospitalized patients with COVID-19	China	Clinical laboratory, radiological, and treatment data	Explore association between cardiac injury and mortality in COVID-19 patients	97 patients had ARDS	N/A	Data from larger populations and multiple centers are needed to confirm findings	Strong
12	Shen et al. [32]	2020	Case series	Five critically ill patients with laboratory-confirmed COVID-19 and ARDS	China	Body temperature, sequential organ failure assessment P/F, viral load, serum antibody titer, blood biochemical index, ARDS, and ventilatory and ECMO supports before and after convalescent plasma transfusion	Test if convalescent plasma transfusion may be beneficial in treatment of critically ill COVID-19 patients with severe ARDS	ARDS resolved in four patients at 12 days after transfusion, and three patients were weaned from mechanical ventilation within 2 weeks of treatment. Treatment with: convalescent plasma, steroids, and antiviral (lopinavir/ritonavir; interferon alpha-1b; favipiravir, arbidol; darunavir)	All five patients improved in clinical status	Limited sample size and study design	Medium
13	Du et al. [33]	2020	Cross-sectional	85 fatal cases with COVID-19	China	Medical history, exposure history, comorbidities, symptoms, signs, laboratory findings, CT scans, and clinical management	Report clinical features of 85 fatal cases with COVID-19 in two hospitals in Wuhan	ARDS: 63	N/A	Only fatal cases of COVID-19 included	Medium
14	Chen et al. [34]	2020	Retrospective single-center	99 patients with COVID-19 pneumonia	China	Epidemiologic, demographic, clinical, and radiological features and laboratory data	Clarifying epidemiologicaland clinical characteristics of COVID-19 pneumonia	17 with ARDS	N/A	Low sample size One center	Medium

(continued)

Table 18.3 (continued)

No.	Author	Year	Design	Sample size	Country	Data type	Aim of the study	Main findings	Conclusion	Limitations	Quality score
15	Chen et al. [35]	2020	Retrospective case series	Cohort of 799 patients: 113 died and 161 recovered	China	Clinical characteristics and laboratory findings	Delineate clinical characteristics of patients with COVID-19 who died	ARDS: 113	N/A	N/A	Strong
16	Ferrey et al. [36]	2020	Case study	A 56-year-old male with ESRD and COVID-19	USA	Clinical characteristics and laboratory findings	Case analysis	CT showing new multifocal bilateral patchy GGOs with predominantly peripheral distribution. Chest X-ray sequence with interval increasing patchy opacities in both lungs consistent with evolving infectious process	N/A	N/A	Medium
17	Chung et al. [37]	2020	Retrospective case series	21 patients with symptoms and COVID-19	China	CT findings	Assessing imaging features COVID-19	Of 21 initial chest CT scans, three showed normal. Of 18 patients with GGOs, consolidation, or both, 12 had only GGOs, and no patient demonstrated consolidation without GGOs. Of the 21 patients, one had one affected lobe, two had two affected lobes, three had three affected lobes, four had four affected lobes, and eight had disease affecting all five lobes. 7/21 patients demonstrated GGOs and/or consolidative opacities with a rounded morphology, three demonstrated a predominantly linear abnormality, four had crazy-paving pattern, and seven patients (21%) had peripheral distribution of disease. No patients had cavitation in the lung, discrete pulmonary nodules, pleural effusion(s), lymphadenopathy, underlying pulmonary emphysema, or fibrosis	N/A	N/A	Medium

18	Huang et al. [1]	2020	Cross-sectional	41 patients with COVID-19	China	Epidemiologic, clinical, laboratory, and radiological characteristics and treatment /outcome data	Reporting epidemiology, clinical, laboratory, and radiological characteristics and treatment / clinical outcomes of COVID-19 patients	ARDS: 12	N/A	Low sample size One center	Medium
19	Beloncle et al. [38]	2020	Case series	25 patients with SARS-Cov-2-associated ARDS	France	Gas exchanges, respiratory system compliance, and hemodynamics assessed at two levels of PEEP (15 cmH2O and 5 cmH2O) within 36 h (day 1) and from 4 to 6 days (day 5) after intubation	Describe characteristics of respiratory mechanics of COVID-19-associated ARDS and, in particular, whether lungs are recruitable with high levels of PEEP	Systematic recruitment-to-inflation ratio assessment may help to guide initial PEEP titration to limit harmful effects of unnecessary high PEEP in COVID-19 crisis	N/A	Low sample size One center	Medium
20	Zhang et al. [39]	2020	Cross-sectional	17 patients with COVID-19 infection	China	Multi-detector CT scans	Assessing CT images of 17 patients with COVID-19 infection	12 patients had GGOs, and five had consolidation and GGOs. Distribution of abnormalities was in subpleural lung regions in 12, bilateral in 14, and unilateral in three patients. Both upper and lower lobes involved in 15 patients, upper lobe only was involved in two patients. Follow-up HRCT scan performed in five patients showed that three had markedly decreased consolidation, fibrotic changes developed, and the other two patients showed mild progression with increased extent and density of opacities	N/A	Low sample size One center	Medium

(continued)

Table 18.3 (continued)

No.	Author	Year	Design	Sample size	Country	Data type	Aim of the study	Main findings	Conclusion	Limitations	Quality score
21	Wu et al. [40]	2020	Retrospective cross-sectional	80 patients diagnosed with COVID-19	China	Multi-detector CT scans	Investigate chest CT scan findings in patients with confirmed COVID-19, and evaluate relationship with clinical features	GGOs, 73/80 cases; consolidation, 50 cases; and interlobular septal thickening, 47 cases.	The common chest CT findings of COVID-19 are multiple GGOs, consolidation,	Small number of patients with proven SARS-CoV-2 infection. No measure of histopathology changes	Medium
								Most lesions were multiple with an average of 12 ± 6 lung segments involved. The most common involved lung segments were the following: dorsal segment of right lower lobe, 69; posterior basal segment of the right lower lob,e 68; lateral basal segment of the right lower lobe, 64; dorsal segment of the left lower lobe, 61; and posterior basal segment of the left lower lobe, 65 patients. Correlation analysis showed pulmonary inflammation index was significantly correlated with lymphocyte and monocyte counts, CRP, procalcitonin, days from illness onset, and body temperature ($p<0.05$).	and interlobular septal thickening in both lungs, which are mostly distributed under the pleura		
								CT features – GGO, 73; consolidation, 50; interlobular septal thickening, 47; crazy-paving pattern, 23; spider web sign, 20; subpleural line, 16; bronchial wall thickening, 9; lymph node enlargement, 3; pericardial effusion, 4; pleural effusion, 5 patients			

22	Wang et al. [41]	2020	Retrospective single-center case series	138 patients with COVID-19	China	Epidemiologic, demographic, clinical, laboratory, radiological, and treatment data	Describe epidemiology and clinical characteristics COVID-19-infected pneumonia	ARDS: 22	N/A	N/A	Medium
23	Zhou et al. [42]	2020	Retrospective, multicenter cohort	191 patients with COVID-19 of whom 137 were discharged and 54 died in hospital	China	Epidemiologic, demographic, clinical, laboratory, radiological, and treatment data	N/A	Consolidation 112, GGO 136,bilateral pulmonary infiltration 143, ARDS 59	N/A	N/A	Medium

ARDS acute respiratory distress syndrome, *N/A* not available, *HD* hospital day, *GGO* ground-glass opacity, *SARS* severe acute respiratory syndrome, *RT-PCR* reverse transcription polymerase chain reaction, *LDH* lactate dehydrogenase, *qPCR* quantitative polymerase chain reaction, *COVID-19* coronavirus disease 19, *CT scan* computed tomography scan, *USA* the United States of America, *RNA* ribonucleic acid, *RCT* randomized controlled trial, *MSCs* mesenchymal stem cells, *ICU* intensive care unit, *COPD* chronic obstructive pulmonary disease, *tPA* tissue plasminogen activator, *P/F ratio* PaO2/FiO2 ratio, *ECMO* extracorporeal membrane oxygenation, *PEEP* positive end expiratory pressure, *HRCT* high-resolution computed tomography, *CRP C-reactive protein*

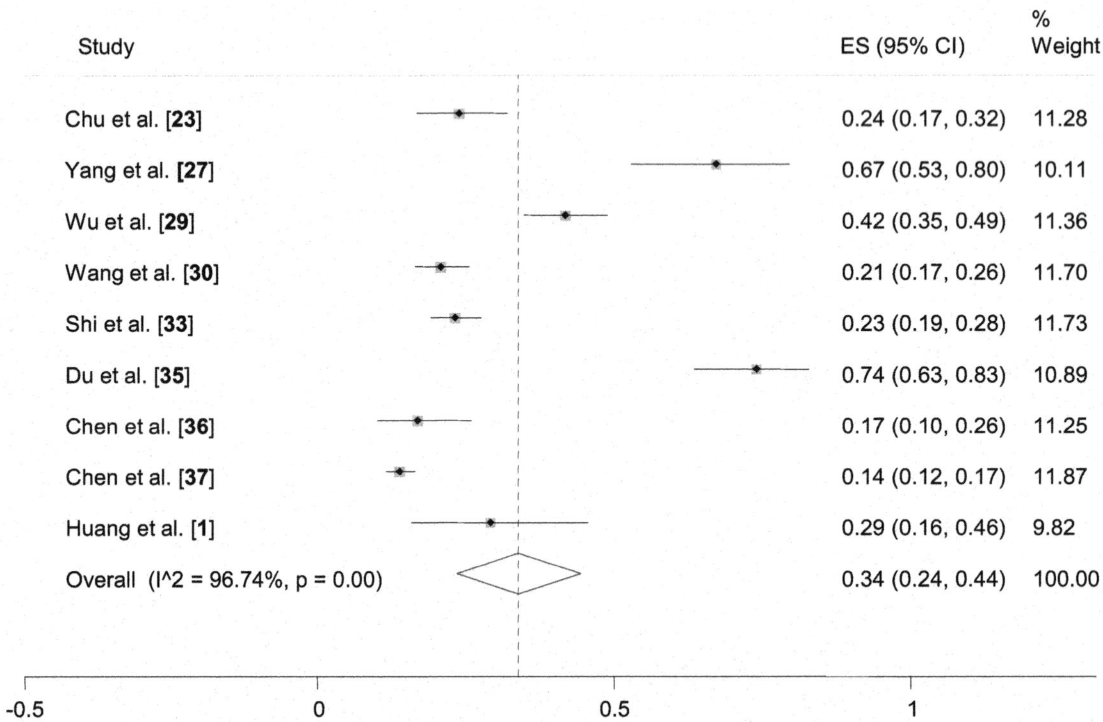

Fig. 18.4 The overall prevalence of COVID-19-induced ARDS determined from nine studies

full texts, while duplicate and unrelated articles were removed from the study. Lastly, 23 primary studies that met inclusion criteria were selected for appraisal and analysis [1, 21–42].

18.3.2 Study Descriptions

The study designs were diverse including cross-sectional, retrospective, case studies, or case series. All studies were published in the English language, most of them were conducted during the early phase of the COVID-19 pandemic in China [1, 24–35, 37, 39–42], two were carried out in the USA [23, 36], one was performed in Italy [22], one was in France [38], and one study was carried out during the SARS outbreak in China in 2003 [21]. The detailed characteristics of the 23 final primary articles are given in Table 18.3.

18.3.3 Methodological Quality Appraisal

The quality of all articles was evaluated using five-item mentioned tools, as shown in Table 18.3. Based on this scale, eight studies were ranked as having a strong quality, and 15 had a medium quality.

18.3.4 Narrative Summary and Brief of Themes

The data of the original articles were classified and discussed according to the following categories: (1) pathology of COVID-19, (2) diagnosis, (3) risk factors, (4) prevalence, and mortality, and (5) treatment.

18.3.4.1 Pathology of the COVID-19 Virus and ARDS

Biopsy samples taken from lung, liver, and heart tissue of a patient who died from COVID-19-related ARDS showed bilateral diffuse alveolar damage with cellular exudates, evident desquamation of pneumocytes, and hyaline membrane formation [26]. The left lung showed pulmonary edema with hyaline membrane formation. Both lungs showed interstitial mononuclear inflammatory infiltrates and multinucleated syncytial cells with atypical appearance in the intra-alveolar spaces. Blood analysis indicated an over-activation of T cells. An investigation using lung CT scans identified ground-glass opacities (GGOs) in 12 of 17 patients, and the remaining five patients showed a combination of GGOs and consolidation in the subpleural regions [39]. Follow-up scans performed on five patients showed development of fibrotic changes in three patients and mild progression with increased extent and density of GGOs in two patients. Another CT study of 80 COVID-19 patients found GGOs, consolidation, and interlobular thickening in the subpleural regions of both lungs in most cases [40]. Other studies showed that the medium time to development of ARDS from onset of symptoms in admitted patients ranged from 8 to 12 days [1, 41, 42].

Four studies of COVID-19 cases reported heterogeneous clinical manifestations, ranging from minimal symptoms such as fever, dry cough, or dyspnea to severe ARDS [23, 26, 30, 36]. Another study of two Chinese patients who developed symptoms while on holiday in Italy found GGOs with a crazy-paving pattern, slight unilateral pleural effusion, and mediastinal lymphadenopathies [22]. A study of an adult sickle cell patient with COVID-19 pneumonia using chest X-ray analysis found multifocal ill-defined opacities which worsened to GGOs and reticular opacities [23]. A retrospective case control study of patients with ARDS infected with either COVID-19 or H1N1 influenza found a higher frequency of coughing, fatigue, and gastrointestinal symptoms in the COVID-19 patients and higher sequential organ failure assessment scores in the H1N1 patients [30]. In addition, the COVID-19 patients had higher PaO_2/FiO_2 ratios, a higher proportion of GGOs, and lower mortality than the H1N1 patients. In addition, an end-stage renal disease patient in the USA who contracted COVID-19 disease while in hospital was found to have bilateral interstitial pneumonia, which worsened to ARDS [36].

18.3.4.2 Risk Factors

A one study of 133 patients with confirmed SARS (2003 outbreak of SARS-CoV) showed that advanced age, the presence of comorbidities, initial lactate dehydrogenase (LDH) levels, PCR test results of the nasopharyngeal specimen, and initial viral load were independently associated with ARDS [21]. Another study showed that advanced age and the presence of neutrophilia and organ and coagulation dysfunctions were contributing factors to development of ARDS and the progression from ARDS to death [27].

18.3.4.3 Diagnosis

Four studies showed that a decrease in the PaO_2/FiO_2 ratio, chest CT, and X-ray scan results could be used in the diagnosis of ARDS associated with COVID-19 infection [23, 26, 39, 40]. In addition, the severity of ARDS could be assessed as mild, medium, or severe using the PaO_2/FiO_2 ratio, as described earlier.

18.3.4.4 Prevalence and Mortality

The prevalence of COVID-19-related ARDS was reported by nine studies as follows: 32 out of 133 cases [21], 35 out of 52 critically ill cases [25], 84 out of 201 patients [27], 71 out of 339 cases [28], 97 out of 416 patients [31], 63 out of 85 cases) [33], 17 out of 99 patients [34], 113 out of 799 patients [35], and 12 out of 41 cases [1]. The overall prevalence of COVID-19-related ARDS was estimated as 34% using a random model (Fig. 18.4). The incidence of COVID-19-related ARDS in high-risk and critically ill patients was higher than for other cases [25, 33]. In addition, COVID-19-related ARDS resulted in a higher mortality rate. For example, the mortality rate was 60–61.5% in two studies of COVID-19 patients with ARDS **{Huang, 2020 #3}**[21, 25].

18.3.4.5 Treatment

Treatment and management of ARDS has been mostly supportive in nature. In the study by Chen et al. {Huang, 2020 #3}, transplantation of mesenchymal stem cells (MSCs) significantly lowered the mortality rate to 17.6% compared to 54.5% in the control group in patients with H7N9 influenza, demonstrating potential utility in COVID-19 ARDS cases [24]. In addition, this procedure had no adverse effects in the case of four patients who were followed over 5 years. A case series study by Wang et al. tested the effects of fibrinolytic therapy with tissue plasminogen activator in three COVID-19-related ARDS patients with transient improvements seen in all three cases [29]. Steroids such as methylprednisolone, antivirals (lopinavir, ritonavir, interferon alpha-1b, favipiravir, arbidol, darunavir), and antibacterial drugs have also been used for treatments, along with respiratory support [26, 27, 32]. However, the therapeutic benefits of these treatment strategies remain in question.

18.4 Discussion

Our findings indicate that approximately one-third of COVID-19 patients develop ARDS with common features of decreased PaO_2/FiO_2 ratios along with increased appearance of GGOs with consolidation by CT imaging. We found other risk factors for progression of COVID-19 patients to ARDS included advanced age, the presence of comorbidities, initial LDH levels, initial viral load, neutrophilia, as well as coagulation and organ dysfunction.

Increasing our understanding of the infection process of the SARS-CoV-2 virus is vital to improving patient outcomes during the current COVID-19 pandemic and future coronavirus and influenza outbreaks. It is now widely known that SARS-CoV-2 infection begins with entry of the virus via the upper airway followed by its migration down the respiratory tract to reach the gas exchange units of the lung. This leads to entry of the virus into the lung alveolar cells via ACE2 receptors. Once inside, the virus subverts the normal cellular processes enabling its replication and spreading to other organs and tissues of the body. During this process, an inflammatory response known as a "cytokine storm" can lead to pulmonary infiltration, cell death, and fibrosis, with development of ARDS [43].

The WHO interim guidelines recommend the use of extracorporeal membrane oxygenation (ECMO) to COVID-19 patients with ARDS [44]. Corticosteroids along with antiviral and antibacterial drugs have also been considered as potential treatments for ARDS because of their effects on reducing inflammation and fibrosis [3, 45]. However, none of these therapies are definitive cures for ARDS, and there have been some conflicting findings in this area. A recent trial in the UK found that the corticosteroid dexamethasone reduced mortality in COVID-19 patients receiving respiratory support [46], and early results from another trial conducted by Gilead showed that treatment with the antiviral drug remdesivir led to a nonsignificant reduction in the time to clinical improvement and duration of invasive mechanical ventilation compared to standard care [47]. However, further trials are required to confirm or refute the findings of both studies.

There are a number of limitations to the current study which makes definitive conclusions difficult. Most of these relate to the fact that we are still in the early phase of the COVID-19 pandemic, and a more complete picture is not likely to be available for several years to come. Therefore, we were only able to provide an estimate of the overall prevalence of ARDS in COVID-19 patients due to the low number of studies which have addressed this question thus far. For this reason, we also did not perform statistical methods of meta-analysis such as subgroup analysis or publication bias assessment and have only provided a summary of the results.

18.5 Conclusions

The current findings indicate that approximately one-third of patients with COVID-19 develop ARDS. Patients with ARDS are more likely to have a worse outcome and increased risk of death. For this reason, early detection of those patients who are at increased risk of ARDS is critical. Further studies should be performed to identify biomarkers which indicate progression to ARDS such as the changes in the PaO_2/FiO_2 ratio and CT and X-ray imaging signs described here, as well as molecular biomarkers which can be sampled in body fluids. Likewise, further studies on the pathophysiology and distinct ARDS phenotypes should be conducted to enable identification of new treatment possibilities for improved patient outcomes.

Acknowledgments We give thanks for guidance and advice from the Clinical Research Development Unit of Baqiyatallah Hospital.

Competing Interests None declared.

References

1. Huang C, Wang Y, Li X, Ren L, Zhao J, Hu Y et al (2020) Clinical features of patients infected with 2019 novel coronavirus in Wuhan, China. Lancet 395(10223):497–506
2. Zhu N, Zhang D, Wang W, Li X, Yang B, Song J et al (2019) A novel coronavirus from patients with pneumonia in China. N Engl J Med 382(8):727–733
3. Li X, Ma X (2020) Acute respiratory failure in COVID-19: is it "typical" ARDS? Crit Care 24(1):198. https://doi.org/10.1186/s13054-020-02911-9
4. Goh KJ, Choong MC, Cheong EH, Kalimuddin S, Duu Wen S, Phua GC et al (2020) Rapid progression to acute respiratory distress syndrome: review of current understanding of critical illness from COVID-19 infection. Ann Acad Med Singap 49(1):1–9
5. Guan W, Ni Z, Hu Y, Liang W, Ou C, He J et al (2020) Clinical characteristics of coronavirus disease 2019 in China. N Engl J Med 382(18):1708–1720
6. Phan LT, Nguyen TV, Luong QC, Nguyen TV, Nguyen HT, Le HQ et al (2020) Importation and human-to-human transmission of a novel coronavirus in Vietnam. N Engl J Med 382(9):872–874
7. Chan JFW, Yuan S, Kok KH, To KKW, Chu H, Yang J et al (2020) A familial cluster of pneumonia associated with the 2019 novel coronavirus indicating person-to-person transmission: a study of a family cluster. Lancet 395(10223):514–523
8. Burki TK (2020) Coronavirus in China. Lancet Respir Med 8(3):238. https://doi.org/10.1016/S2213-2600(20)30056-4
9. Ranieri VM, Rubenfeld GD, Thompson BT, Ferguson ND, Caldwell E, Fan E et al (2012) Acute respiratory distress syndrome: the Berlin definition. JAMA J Am Med Assoc 307(23):2526–2533
10. Sun P, Qie S, Liu Z, Ren J, Li K, Xi J (2020) Clinical characteristics of hospitalized patients with SARS-CoV-2 infection: a single arm meta-analysis. J Med Virol 92(6):612–617
11. Cao Y, Liu X, Xiong L, Cai K (2020) Imaging and clinical features of patients with 2019 novel coronavirus SARS-CoV-2: a systematic review and meta-analysis. J Med Virol. https://doi.org/10.1002/jmv.25822. Online ahead of print
12. Levac D, Colquhoun H, O'Brien KK (2012) Scoping studies: advancing the methodology. Implement Sci 5(1). https://doi.org/10.1186/1748-5908-5-69
13. Moher D, Shamseer L, Clarke M, Ghersi D, Liberati A, Petticrew M et al (2015) Preferred reporting items for systematic review and meta-analysis protocols (PRISMA-P) 2015 statement. Syst Rev 4(1):1. https://doi.org/10.1186/2046-4053-4-1
14. Chung C, McKenna L, Cooper SJ (2020) Patients' experiences of acute deterioration: a scoping review. Int J Nurs Stud 101:103404. https://doi.org/10.1016/j.ijnurstu.2019.103404
15. Liberati A, Altman DG, Tetzlaff J, Mulrow C, Gøtzsche PC, Ioannidis JPA et al (2009) The PRISMA statement for reporting systematic reviews and meta-analyses of studies that evaluate health care interventions: explanation and elaboration. PLoS Med 6(7):e1000100. https://doi.org/10.1371/journal.pmed.1000100
16. Hoodin F, Weber S (2003) A systematic review of psychosocial factors affecting survival after bone marrow transplantation. Psychosomatics 44(3):181–195
17. Silverman MK, Kopf AW, Grin CM, Bart RS, Levenstein MJ (1991) Recurrence rates of treated basal cell carcinomas. Part 2: curettage-electrodesiccation. J Dermatol Surg Oncol 17(9):720–726
18. Jafari Oori M, Mohammadi F, Norouzi-Tabrizi K, Fallahi-Khoshknab M, Ebadi A (2019) Prevalence of medication adherence in patients with hypertension in Iran: a systematic review and meta-analysis of studies published in 2000–2018. ARYA Atheroscler 15(2):82–92
19. Oori MJ, Mohammadi F, Norozi K, Fallahi-Khoshknab M, Ebadi A, Gheshlagh RG (2019) Prevalence of HTN in Iran: meta-analysis of published studies in 2004–2018. Curr Hypertens Rev 15(10):113–122
20. Braun V, Clarke V (2012) Thematic analysis. In: Cooper H, Camic PM, Long DL, Panter AT, Rindskopf D, Sher KJ (eds) APA handbook of research methods in psychology, Vol 2: research designs: quantitative, qualitative, neuropsychological, and biological. American Psychological Association, Washington, DC, pp 57–71. ISBN-13: 978-1-4338-1003-9
21. Chu CM, Poon LLM, Cheng VCC, Chan KS, Hung IFN, Wong MML et al (2004) Initial viral load and the outcomes of SARS. CMAJ 171(11):1349–1352
22. Albarello F, Pianura E, Di Stefano F, Cristofaro M, Petrone A, Marchioni L et al (2020) 2019-novel coronavirus severe adult respiratory distress syndrome in two cases in Italy: an uncommon radiological presentation. Int J Infect Dis 93:192–197
23. Beerkens F, John M, Puliafito B, Corbett V, Edwards C, Tremblay D (2020) COVID-19 pneumonia as a

cause of acute chest syndrome in an adult sickle cell patient. Am J Hematol 95(7):E154–E156
24. Chen J, Hu C, Chen L, Tang L, Zhu Y, Xu X et al (2020) Clinical study of mesenchymal stem cell treatment for acute respiratory distress syndrome induced by epidemic influenza a (H7N9) infection: a hint for COVID-19 treatment. Engineering (Beijing), Feb 28. https://doi.org/10.1016/j.eng.2020.02.006. Online ahead of print
25. Yang X, Yu Y, Xu J, Shu H, Xia J, Liu H et al (2020) Clinical course and outcomes of critically ill patients with SARS-CoV-2 pneumonia in Wuhan, China: a single-centered, retrospective, observational study. Lancet Respir Med 8(5):475–481
26. Xu Z, Shi L, Wang Y, Zhang J, Huang L, Zhang C et al (2020) Pathological findings of COVID-19 associated with acute respiratory distress syndrome. Lancet Respir Med 8(4):420–422
27. Wu C, Chen X, Cai Y, Xia J, Zhou X, Xu S et al (2020) Risk factors associated with acute respiratory distress syndrome and death in patients with coronavirus disease 2019 pneumonia in Wuhan, China. JAMA Intern Med 180(7):1–11
28. Wang L, He W, Yu X, Hu D, Bao M, Liu H et al (2020) Coronavirus disease 2019 in elderly patients: characteristics and prognostic factors based on 4-week follow-up. J Infect 80(6):639–645
29. Wang J, Hajizadeh N, Moore EE, McIntyre RC, Moore PK, Veress LA et al (2020) Tissue plasminogen activator (tPA) treatment for COVID-19 associated acute respiratory distress syndrome (ARDS): a case series. J Thromb Haemost. https://doi.org/10.1111/jth.14828. [Epub ahead of print]
30. Tang X, Du RH, Wang R, Cao TZ, Guan LL, Yang CQ et al (2020) Comparison of hospitalized patients with acute respiratory distress syndrome caused by COVID-19 and H1N1. Chest 158(1):195–205
31. Shi S, Qin M, Shen B, Cai Y, Liu T, Yang F et al (2020) Association of Cardiac Injury with Mortality in hospitalized patients with COVID-19 in Wuhan, China. JAMA Cardiol:e200950. https://doi.org/10.1001/jamacardio.2020.0950. Online ahead of print
32. Shen C, Wang Z, Zhao F, Yang Y, Li J, Yuan J et al (2020) Treatment of 5 critically ill patients with COVID-19 with convalescent plasma. JAMA 323(16):1582–1589
33. Du Y, Tu L, Zhu P, Mu M, Wang R, Yang P et al (2020) Clinical features of 85 fatal cases of COVID-19 from Wuhan. A retrospective observational study. Am J Respir Crit Care Med 201(11):1372–1379
34. Chen N, Zhou M, Dong X, Qu J, Gong F, Han Y et al (2020) Epidemiological and clinical characteristics of 99 cases of 2019 novel coronavirus pneumonia in Wuhan, China: a descriptive study. Lancet 395(10223):507–513
35. Chen T, Wu D, Chen H, Yan W, Yang D, Chen G et al (2020) Clinical characteristics of 113 deceased patients with coronavirus disease 2019: retrospective study. BMJ 368:m1091. https://doi.org/10.1136/bmj.m1091
36. Ferrey AJ, Choi G, Hanna RM, Chang Y, Tantisattamo E, Ivaturi K et al (2020) A case of novel coronavirus disease 19 in a chronic hemodialysis patient presenting with gastroenteritis and developing severe pulmonary disease. Am J Nephrol 51(5):337–342
37. Chung M, Bernheim A, Mei X, Zhang N, Huang M, Zeng X et al (2020) CT imaging features of 2019 novel coronavirus (2019-NCoV). Radiology 295(1):202–207
38. Beloncle FM, Pavlovsky B, Desprez C, Fage N, Olivier PY, Asfar P et al (2020) Recruitability and effect of PEEP in SARS-Cov-2-associated acute respiratory distress syndrome. Ann Intensive Care. https://doi.org/10.1186/s13613-020-00675-7
39. Zhang S, Li H, Huang S, You W, Sun H (2020) High-resolution computed tomography features of 17 cases of coronavirus disease 2019 in Sichuan province, China. Eur Respir J 55(4):2000334. https://doi.org/10.1183/13993003.00334-2020
40. Wu J, Wu X, Zeng W, Guo D, Fang Z, Chen L et al (2020) Chest CT findings in patients with coronavirus disease 2019 and its relationship with clinical features. Investig Radiol 55(5):257–261
41. Wang D, Hu B, Hu C, Zhu F, Liu X, Zhang J et al (2020) Clinical characteristics of 138 hospitalized patients with 2019 novel coronavirus-infected pneumonia in Wuhan, China. JAMA 323(11):1061–1069
42. Zhou F, Yu T, Du R, Fan G, Liu Y, Liu Z et al (2020) Clinical course and risk factors for mortality of adult inpatients with COVID-19 in Wuhan, China: a retrospective cohort study. Lancet 395(10229):1054–1062
43. Mason RJ (2020) Pathogenesis of COVID-19 from a cell biology perspective. Eur Respir J 55(4):2000607. https://doi.org/10.1183/13993003.00607-2020
44. Ramanathan K, Antognini D, Combes A, Paden M, Zakhary B, Ogino M et al (2020) Planning and provision of ECMO services for severe ARDS during the COVID-19 pandemic and other outbreaks of emerging infectious diseases. Lancet Respir Med 8(5):518–526
45. Meduri GU, Bridges L, Shih MC, Marik PE, Siemieniuk RAC, Kocak M (2016) Prolonged glucocorticoid treatment is associated with improved ARDS outcomes: analysis of individual patients' data from four randomized trials and trial-level meta-analysis of the updated literature. Intensive Care Med 42(5):829–840
46. https://www.ox.ac.uk/news/2020-06-16-dexamethasone-reduces-death-hospitalised-patients-severe-respiratory-complications
47. https://www.ox.ac.uk/news/2020-04-29-results-remdesivir-trial-released

Depression, Anxiety, and Stress Among Patients with COVID-19: A Cross-Sectional Study

19

Malihe Sadat Moayed, Amir Vahedian-Azimi, Golshan Mirmomeni, Farshid Rahimi-Bashar, Keivan Goharimoghadam, Mohamad Amin Pourhoseingholi, Mohsen Abbasi-Farajzadeh, Mostafa Hekmat, Thozhukat Sathyapalan, Paul C. Guest, and Amirhossein Sahebkar

Abstract

Aim
Patients with confirmed COVID-19 infection can develop several psychological consequences. Epidemiological data on mental health and psychological disorder inpatients infected with COVID-19 pneumonia are not available in Iranian patients. The purpose of this study was to evaluate the anxiety, stress, and depression of patients with COVID-19.

Material and Methods
This cross-sectional survey was conducted in 2020. All confirmed patients with COVID-19 were included in the study by census sampling. Assessment of depression, stress, and anxiety was performed using the DASS-21 questionnaire. All statistical analyses were performed using R version 3.5.1.

Results
The questionnaires were completed by 221 patients with COVID-19 infection (204 males,

Authors Malihe Sadat Moayed, Amir Vahedian-Azimi have equally contributed to this chapter.

M. S. Moayed · A. Vahedian-Azimi (✉)
Trauma Research Center, Nursing Faculty, Baqiyatallah University of Medical Sciences, Tehran, Iran
e-mail: amirvahedian63@gmail.com

G. Mirmomeni
Hearing Research Center, Ahvaz Jundishapur University of Medical Sciences, Ahvaz, Iran

F. Rahimi-Bashar
Anesthesia and Critical Care Department, Hamadan University of Medical Sciences, Hamadan, Iran

K. Goharimoghadam
Internal Medicine, Shariati hospital, Tehran University of Medical Sciences, Tehran, Iran

M. A. Pourhoseingholi
Gastroenterology and Liver Diseases Research Center, Research Institute for Gastroenterology and Liver Diseases, ShahidBeheshti University of Medical Sciences, Tehran, Iran

M. Abbasi-Farajzadeh
Marine Medicine Research Center, Baqiyatallah University of Medical Sciences, Tehran, Iran

P. C. Guest (ed.), *Clinical, Biological and Molecular Aspects of COVID-19*, Advances in Experimental Medicine and Biology 1321, https://doi.org/10.1007/978-3-030-59261-5_19

17 females). The mean age was 45.90 ± 7.73 years. Our results indicated that the mean scores of depression and anxiety were at "extremely severe" levels, while stress levels were "severe." The prevalence of "extremely severe" symptoms of depression and anxiety was 54.29% and 97.29%, respectively. The prevalence of severe stress was 46.61%.

Conclusion
In this study, patients infected with COVID-19 reported severe and extremely severe experience psychological distress. Further studies should focus on the combined use of psychological and molecular biomarker testing to increase accuracy. Overall, the findings demonstrate the necessity of special intervention programs for the confirmed patients with emerging infectious disease COVID-19 to promote mental health needs.

Keywords

COVID-19 · Coronavirus · Patient · Depression · Anxiety · Stress · Iran

19.1 Introduction

Coronavirus disease 2019 (COVID-19) is mainly a respiratory system infection with a newly diagnosed coronavirus thought to have originated as a zoonotic virus which has human pathogenicity [1]. The virus transmits from person to person through close contact or airborne probably as respiratory droplets [2]. At the time of writing (April 28, 2020), there have been approximately three million cases worldwide, with approximately 923,000 having recovered and 211,000 having died [3].

Therefore, virtually all patients with confirmed or suspected COVID-19 experience fear due to complications of the disease relating to the severe disability and potential death, in the absence of any definitive treatment or vaccine. In addition, symptoms such as fever, hypoxia, and cough, as well as adverse effects of treatment, could lead to worsening of anxiety and mental distress [4, 5]. In the early phases of the severe acute respiratory syndrome (SARS) and Middle East respiratory syndrome (MERS) outbreaks, a range of psychiatric morbidities, including persistent depression, anxiety, panic attack, psychomotor excitement, psychotic symptoms, delirium, and even suicidality, were reported [6, 7]. The persistence of depression in MERS survivors leads to prolonged chronic post-traumatic stress symptoms even 18 months after the infection [8]. A 14-day quarantine period and restrictions on social contact, which form part of the public health responses to the COVID-19 pandemic, could cause individuals to experience signs such as boredom, a higher level of depression, loneliness, and anger. It also could increase patients' guilt and anxiety about the effects of pathogenesis, quarantine, and stigma on their families and friends [4, 9]. Among the survivors of the SARS

M. Hekmat
Health Management Research Center, Baqiyatallah University of Medical Sciences, Tehran, Iran

T. Sathyapalan
Academic Diabetes, Endocrinology and Metabolism, Hull York Medical School, University of Hull, Hull, UK

P. C. Guest
Laboratory of Neuroproteomics, Department of Biochemistry and Tissue Biology, Institute of Biology, University of Campinas (UNICAMP), Campinas, Brazil

A. Sahebkar (✉)
Neurogenic Inflammation Research Center, Mashhad University of Medical Sciences, Mashhad, Iran

Biotechnology Research Center, Pharmaceutical Technology Institute, Mashhad University of Medical Sciences, Mashhad, Iran

Polish Mother's Memorial Hospital Research Institute (PMMHRI), Lodz, Poland

Halal Research Center of IRI, FDA, Tehran, Iran
e-mail: sahebkara@mums.ac.ir; amir_saheb2000@yahoo.com

outbreak, stress levels were persistently elevated 1 year later, compared to people who were not infected with SARS. SARS survivors had high levels of psychological distress even 1 year after the outbreak. This evidence suggests that the long-term psychological implications of infectious diseases should not be ignored and mental health services could play an important role in the rehabilitation of patients [6].

To date, epidemiological data on the mental health of patients with COVID-19 in Iran has not been established. The main purpose of this study was to measure the prevalence and severity of this psychological distress and gauge the current mental health burden on patients with a diagnosis of COVID-19 infection.

19.2 Material and Methods

19.2.1 Study Design

This cross-sectional survey was conducted in February and March, 2020.

19.2.2 Settings and Participants

Baqiyatallah Hospital is one of the main referral centers for the specialized diagnosis and treatment of patients with COVID-19 in Tehran, Iran. Participants were recruited mainly as patients referred to this hospital and patients from other centers who gave consent to participate in the study. Other centers also participated in this study, but since the number of participants from other centers was not high enough to obtain a separate ethics code, the Dean of each center approved data collection for their respective institutions. All patients over 18 years of age who were interested in participating in this study, who could read and write, with no preexisting physical disabilities or mental disorders, were included. All participants were confirmed COVID-19 positive. Census sampling was used.

19.2.3 Sample Size

Cochran's sample size estimation formula in the epidemiologic study was used [10]. The first and second type errors were considered five-hundredth and two-tenths, respectively. A 50% satisfaction probability was assumed to estimate the maximum sample size. The sample size was calculated at 87 patients. According to the nature of the study and the probability of sample size drop, a 20% dropout was considered and the final sample size was therefore calculated to be 110.

19.2.4 Research Tools

Demographic characteristics were self-reported by participants. These included sex, age, job, marital status, educational qualification, and history of chronic disease. The Depression Anxiety Stress Scale (DASS) was used to collect data. This questionnaire was designed and validated by Lovibond in 1995 [11] to measure psychological distress among the community with 21 items. This questionnaire included three subscales and each subscale had seven questions. In the translated version, the choices were never, little, moderate, and many for each question. The lowest score for each question was 0 and the highest score was 3. Validity and reliability of this questionnaire were previously established in Iran. According to the original questionnaire, the three subscales had high internal consistency with Cronbach's alpha values of 0.77, 0.79, and 0.78 for depression, anxiety, and stress, respectively [12]. Another study reported Cronbach's alpha values greater than 0.80 for all scales in an analysis of the effect of the 2008 Sichuan earthquake in China [13].

The questions 3, 5, 10, 13, 16, 17, and 21 were related to depression. Questions 1, 6, 8, 11, 12, 14, and 18 assessed stress. Q questions 2, 4, 7, 9, 15, 19, and 20 were related to anxiety. The cutoff points of the subscales were as follows:

19.2.5 Depression Scores

0–4 = normal
5–6 = mild
7–10 = average
11–13 = severe
> 14 = very severe

19.2.6 Stress Scores

0–7 = normal
8–9 = mild
10–12 = average
13–16 = severe
>17 = very severe

19.2.7 Anxiety Scores

0–3 = normal
4–5 = mild
6–7 = average
8–9 = severe
>10 = very severe

19.2.8 Ethical Consideration

The study was approved by the Ethics Committee of Baqiyatallah University of Medical Sciences with code IR.BMSU.REC.1398.438. The objectives of the study were explained to all patients, and informed consent was obtained from the patients to participate in the study. They were also assured of confidentiality.

19.2.9 Statistical Analysis

All analyses were performed using R statistical software (version 3.5.1). The variables included gender (male, female), age (>30, 41–50, <40 years), marital status (married, unmarried), job (governmental, nongovernmental, unemployed, student, and housewife), background disease (yes, no), and qualification (diploma or lower, associate, bachelor's degree, master's degree, or higher). The values for each variable were presented as the mean and standard deviation. Independent samples *t*-test was applied as a parametric test to compare psychological symptom (stress, depression, and anxiety) scores by gender and marital status. One-way analysis of variance (ANOVA) test was applied to compare the mean differences in psychological factors in terms of age group, job, and educational qualification. A p-value of <0.05 was considered significant.

19.3 Results

The DASS-21 questionnaires were completed by 221 patients with COVID-19 (204 males, 17 females). The demographic characteristics are summarized in Table 19.1. The mean age was 45.9 ± 7.73 years. A high proportion of patients reported a nongovernmental (37.1%) job. Most of the participants were unmarried (55.21%). In addition, 31.22% and 29.41% of the patients had "Bachelor" or "Associate" education level, respectively (Table 19.1).

The study results demonstrated that the comparison of mean scores for stress, depression, and anxiety subscales was not statistically significant in terms of "age," "gender," "job," "marital status," "background disease," and "qualification" variables (Table 19.1).

Table 19.2 shows the prevalence and score severity ratings of psychological symptoms among patients with COVID-19 infection. Our results indicated that the mean scores of symptoms of depression and anxiety were "extremely severe," while stress was at "severe" levels. The prevalence of "extremely severe" symptoms of depression and anxiety was 54.29% and 97.29%, respectively. The prevalence of "severe" symptom of stress was 46.61%.

19.4 Discussion

The purpose of this study was to evaluate the anxiety, stress, and depression in hospitalized Iranian patients with confirmed COVID-19 infec-

Table 19.1 The mean (±SD) scores for psychological symptoms in terms of age group, gender, marital status, job, education level, and history of background disease (n = 221)

Variable		Frequency (%)	Anxiety score	Stress score	Depression score
Age	<40 years	70 (31.67)	27.83 ± 4.82	28.97 ± 5.21	27.86 ± 4.70
	41–50 years	103 (46.61)	27.51 ± 5.17	28.54 ± 5.34	28.25 ± 5.02
	>50 years	48 (21.72)	27.58 ± 5.55	28.17 ± 4.88	28.00 ± 5.75
p-value			0.923	0.705	0.876
Gender	Male	204 (92.31)	27.68 ± 5.07	28.47 ± 5.01	28.11 ± 4.84
	Female	17 (7.69)	27.06 ± 5.88	30.12 ± 6.98	27.65 ± 7.46
p-value			0.634	0.354	0.805
Marital status	Married	99 (44.79)	27.33 ± 5.05	28.42 ± 5.03	28.48 ± 5.47
	Unmarried	122 (55.21)	27.87 ± 5.19	28.74 ± 5.33	27.74 ± 4.71
p-value			0.441	0.656	0.277
Job	Governmental	34 (15.39)	28.00 ± 5.72	28.00 ± 4.90	27.59 ± 4.29
	Nongovernmental	82 (37.10)	27.73 ± 5.16	28.66 ± 5.38	28.44 ± 5.71
	Unemployed	54 (24.43)	27.07 ± 4.77	28.74 ± 5.43	28.67 ± 4.88
	Student	39 (17.65)	27.74 ± 5.30	28.31 ± 5.14	27.28 ± 4.91
	Housewife	12 (5.43)	28.00 ± 4.75	30.17 ± 3.95	26.83 ± 3.56
p-value			0.922	0.790	0.536
Qualification	Diploma or lower	30 (13.57)	26.73 ± 4.25	28.93 ± 4.63	28.27 ± 4.75
	Associate	65 (29.41)	28.21 ± 5.11	28.31 ± 5.79	27.97 ± 5.45
	Bachelor's degree	69 (31.22)	27.77 ± 5.29	28.14 ± 5.24	28.43 ± 4.63
	Master's degree or higher	57 (25.73)	27.26 ± 5.39	29.30 ± 4.70	27.65 ± 5.37
p-value			0.553	0.598	0.846
Background disease	None	144(65.16)	28.14 ± 5.08	28.81 ± 5.35	28.20 ± 5.13
	Cardiovascular	16 (7.24)	27.13 ± 6.41	29.13 ± 3.72	26.75 ± 5.05
	Diabetic	21(9.33)	26.95 ± 5.24	29.33 ± 5.03	28.00 ± 5.10
	Hypertension	16 (7.24)	25.88 ± 4.98	26.50 ± 4.10	27.75 ± 4.31
	Allergy	12 (5.44)	26.50 ± 4.10	25.17 ± 5.15	28.00 ± 5.66
	Chronic kidney	7 (3.33)	27.14 ± 5.27	31.14 ± 6.62	30.28 ± 5.71
	Chronic liver	5 (2.26)	26.40 ± 4.56	29.20 ± 2.28	26.80 ± 4.60
p-value			0.581	0.105	0.823

Table 19.2 Prevalence and score severity ratings of depression, anxiety, and stress among patients with coronavirus infection (n = 221)

Psychological variable		Frequency	%
Depression	Moderate	16	7.25
	Severe	85	38.46
	Extremely severe	120	54.29
	MEAN ± SD	28.07 ± 5.07	
Anxiety	Severe	6	2.71
	Extremely severe	215	97.29
	MEAN ± SD	27.62 ± 5.13	
Stress	Mild	1	0.45
	Moderate	94	42.53
	Severe	103	46.61
	Extremely severe	23	10.41
	MEAN ± SD	28.59 ± 5.19	

tion. This can serve as important evidence to manage the promotion of mental health among patients with COVID-19. The results of the current study indicated that the patients with confirmed COVID-19 revealed a high prevalence of symptoms associated with mental disorders. All of the patients reported varying levels of depression, anxiety, and stress. Extreme anxiety was found in 97.27% of patients and severe depression signs were reported by 54.29% of patients.

All patients who participated in the study had signs of stress. The prevalence of severe symptoms of stress was 46.61%. Similar results were found in 90% of SARS-infected patients. There was an increased prevalence of general stress and negative psychological effects in patients infected with SARS [14]. Another study reported that patients with MERS infection had an increased incidence of clinically relevant depressive and post-traumatic stress disorder (PTSD) symptoms [8]. Mac et al. reported that 44% and 47.8% of survivors of SARS suffered from respective depressive symptoms and PTSD after infection [15].

None of the demographic variables appeared to contribute to the mean scores of depression, anxiety, and stress. We did not observe any effect of age in our sample of patients. However, Yang et al. reported that older adults in crisis conditions could experience significantly more distress. Therefore, the older population may need relative more mental health intervention [16]. Stress levels were not related to educational qualifications indicating that patients with all levels of educational qualifications who are positive for COVID-19 were adversely affected.

There is growing evidence that in confirmed or suspected COVID-19 infection, patients will need more advanced mental healthcare [17]. Despite the mental health problems reported among patients with COVID-19, few of the healthcare workers on the frontlines had received training in providing mental healthcare [8, 18, 19]. For individuals with suspected or confirmed COVID-19 infection who are under treatment (quarantine or at home), health service personnel should provide medical care and mental healthcare [20]. Special attention needs to be paid for the behavioral and mood changes of these patients. Insomnia, anxiety, anger, rumination, decreased concentration, low mood, and loss of energy are listed as warning symptoms that should be evaluated and managed by mental healthcare professionals [17]. In view of this, there is an urgent need to develop and recommend online and on-site mental health interventions such as psychotherapy [21]. This should involve multidisciplinary mental health teams with expertise in specialized psychiatric treatments to provide appropriate mental health services during and after this COVID-19 epidemic with specific treatment plans, progress reports, and health status updates. There is a need to secure services to provide psychological counseling in this group of patients. There should be more use of electronic devices and applications for affected patients as well as their families and members of the public. There is also a need to establish safe communication channels between patients and families. To date, mental health interventions are only provided for those presenting with more severe mental health problems [5]. Since personnel such as clinical psychiatrists, psychologists, and mental health social workers are unable to enter isolation wards for patients with COVID-19, frontline healthcare workers should be trained to provide psychological interventions for patients with COVID-19 in hospitals [20].

One of the limitations of this study was the single-center cross-sectional survey design which limits the generalizability of the findings. In addition, this study was conducted at the onset of the COVID-19 outbreak, and, therefore, there might be further changes in the mental health status of these patients. Finally, we did not assess the risk factors that may have affected depression, anxiety, and stress in patients with COVID-19 infection, and the prior absence of mental disease in these patients was self-reported. Further studies should focus on the incorporation of molecular biomarkers into such tests to increase accuracy. Such biomarkers should be easily accessible for ease of sampling and to minimize patient discomfort. For example, this could include analysis of stress-related biomarkers such as cortisol and alpha-amylase A in saliva [22, 23] and proinflammatory cytokines such as tumor necrosis factor alpha and interleukin (IL)-1 and IL-6 in blood serum or plasma [24, 25].

19.5 Conclusions

In this study, patients infected with COVID-19 reported severe and extremely severe psychological distress. There is a need for introduction of the mental health interventions in this patient group. As the situation has now progressed to a pandemic, the effects on mental health could be even more profound. This is also complicated by the risk of further outbreaks, the lack of treatments and a vaccine, as well as the effects on the economy at a global level. Given these challenges, it will be important to incorporate the use of molecular biomarkers to increase the accuracy of assessing the dynamic changes in mental health during the evolution of the pandemic.

Acknowledgments Thanks to guidance and advice from the Clinical Research Development Unit of Baqiyatallah Hospital.

Conflict of Interest The author(s) declare no conflicts of interest with respect to the authorship and publication of this article.

References

1. Fedson DS (2015) A practical treatment for patients with Ebola virus disease. J Infect Dis 211(4):661–662
2. Li Q, Guan X, Wu P, Wang X, Zhou L, Tong Y et al (2020) Early transmission dynamics in Wuhan, China, of novel coronavirus–infected pneumonia. N Engl J Med 382(13):1199–1207
3. Worldometer. https://www.worldometers.info/coronavirus/#countries. Accessed 28 Apr 2020)
4. Xiang YT, Yang Y, Li W, Zhang L, Zhang Q, Cheung T et al (2020) Timely mental health care for the 2019 novel coronavirus outbreak is urgently needed. Lancet Psychiatry 7(3):228–229
5. Park SC, Park YC (2020) Mental health care measures in response to the 2019 novel coronavirus outbreak in Korea. Psychiatry Investig 17(2):85–86
6. Lee AM, Wong JG, McAlonan GM, Cheung V, Cheung C, Sham PC et al (2007) Stress and psychological distress among SARS survivors 1 year after the outbreak. Can J Psychiatr 52(4):233–240
7. Moldofsky H, Patcai J (2011) Chronic widespread musculoskeletal pain, fatigue, depression and disordered sleep in chronic post-SARS syndrome; a case-controlled study. BMC Neurol 11:37. https://doi.org/10.1186/1471-2377-11-37
8. Lee SH, Shin HS, Park HY, Kim JL, Lee JJ, Lee H et al (2019) Depression as a mediator of chronic fatigue and post-traumatic stress symptoms in middle east respiratory syndrome survivors. Psychiatry Investig 16(1):59–64
9. Purssell E, Gould D, Chudleigh J (2020) Impact of isolation on hospitalised patients who are infectious: systematic review with meta-analysis. BMJ Open 10(2):e030371. https://doi.org/10.1136/bmjopen-2019-030371

10. Kasiulevičius V, Šapoka V, Filipavičiūtė R (2006) Sample size calculation in epidemiological studies. Geron 7(4):225–231
11. Lovibond PF (1998) Long-term stability of depression, anxiety, and stress syndromes. J Abnorm Psychol 107(3):520–516
12. Sahebi A, Asghari MJ, Salari R (2005) Validation of depression anxiety and stress scale (DASS-21) for an Iranian population. J Dev Psychol. http://jip.azad.ac.ir/article_512443_en.html
13. Chan RC, Xu T, Huang J, Wang Y, Zhao Q, Shum DH et al (2012) Extending the utility of the depression anxiety stress scale by examining its psychometric properties in Chinese settings. Psychiatry Res 200(2–3):879–883
14. Wu P, Fang Y, Guan Z, Fan B, Kong J, Yao Z et al (2009) The psychological impact of the SARS epidemic on hospital employees in China: exposure, risk perception, and altruistic acceptance of risk. Can J Psychiatr 54(5):302–311
15. Mak IWC, Chu CM, Pan PC, Yiu MGC, Chan VL (2009) Long-term psychiatric morbidities among SARS survivors. Gen Hosp Psychiatry 31(4):318–326
16. Yang Y, Li W, Zhang Q, Zhang L, Cheung T, Xiang YT (2020) Mental health services for older adults in China during the COVID-19 outbreak. Lancet Psychiatry 7(4):e19. https://doi.org/10.1016/S2215-0366(20)30079-1
17. Troyer EA, Kohn JN, Hong S (2020) Are we facing a crashing wave of neuropsychiatric sequelae of COVID-19? Neuropsychiatric symptoms and potential immunologic mechanisms. Brain Behav Immun 13. pii: S0889-1591(20)30489-X. https://doi.org/10.1016/j.bbi.2020.04.027. [Epub ahead of print]
18. Roberts T, Esponda GM, Krupchanka D, Shidhaye R, Patel V, Rathod S (2018) Factors associated with health service utilisation for common mental disorders: a systematic review. BMC Psychiatry 18(1):262. https://doi.org/10.1186/s12888-018-1837-1
19. Chen Q, Liang M, Li Y, Guo J, Fei D, Wang L et al (2020) Mental health care for medical staff in China during the COVID-19 outbreak. Lancet Psychiatry 7(4):e15–e16. https://doi.org/10.1016/S2215-0366(20)30078-X
20. Duan L, Zhu G (2020) Psychological interventions for people affected by the COVID-19 epidemic. Lancet Psychiatry 7(4):300–302
21. Jiang X, Deng L, Zhu Y, Ji H, Tao L, Liu L et al (2020) Psychological crisis intervention during the outbreak period of new coronavirus pneumonia from experience in Shanghai. Psychiatry Res Feb 28(286):112903. https://doi.org/10.1016/j.psychres.2020.112903. [Epub ahead of print]
22. Hunter MR, Gillespie BW, Chen SY (2019) Urban nature experiences reduce stress in the context of daily life based on salivary biomarkers. Front Psychol 10:722. https://doi.org/10.3389/fpsyg.2019.00722
23. Engeland CG, Bosch JA, Rohleder N (2019) Salivary biomarkers in psychoneuroimmunology. Curr Opin Behav Sci 28:58–65
24. Young JJ, Bruno D, Pomara N (2014) A review of the relationship between proinflammatory cytokines and major depressive disorder. J Affect Disord 169:15–20
25. Quagliato LA, Nardi AE (2018) Cytokine alterations in panic disorder: a systematic review. J Affect Disord 228:91–96

Survey of Immediate Psychological Distress Levels Among Healthcare Workers in the COVID-19 Epidemic: A Cross-Sectional Study

20

Malihe Sadat Moayed, Amir Vahedian-Azimi, Golshan Mirmomeni, Farshid Rahimi-Bashar, Keivan Goharimoghadam, Mohamad Amin Pourhoseingholi, Mohsen Abbasi-Farajzadeh, Mostafa Hekmat, Thozhukat Sathyapalan, Paul C. Guest, and Amirhossein Sahebkar

Abstract

Aim

The outbreak of COVID-19 has laid unprecedented psychological stress on healthcare workers (HCWs). We aimed to assess the immediate psychological impact of COVID-19 epidemic on the HCWs at Baqiyatallah Hospital in Tehran, Iran.

Material and Methods

We conducted a cross-sectional survey of HCWs using questionnaires in February and March 2020 in Baqiyatallah Hospital, Tehran. We evaluated depression, stress, and anxiety levels using the DASS-21 questionnaire. Participants were selected by using census sampling. All statistical analyses were performed using R version 3.5.1.

Authors Malihe Sadat Moayed, Amir Vahedian-Azimi have equally contributed to this chapter.

M. S. Moayed · A. Vahedian-Azimi (✉)
Trauma Research Center, Nursing Faculty, Baqiyatallah University of Medical Sciences, Tehran, Iran
e-mail: amirvahedian63@gmail.com

G. Mirmomeni
Hearing Research Center, Ahvaz Jundishapur University of Medical Sciences, Ahvaz, Iran

F. Rahimi-Bashar
Anesthesia and Critical Care Department, Hamadan University of Medical Sciences, Hamadan, Iran

K. Goharimoghadam
Internal Medicine, Shariati Hospital, Tehran University of Medical Sciences, Tehran, Iran

M. A. Pourhoseingholi
Gastroenterology and Liver Diseases Research Center, Research Institute for Gastroenterology and Liver Diseases, ShahidBeheshti University of Medical Sciences, Tehran, Iran

M. Abbasi-Farajzadeh
Marine Medicine Research Center, Baqiyatallah University of Medical Sciences, Tehran, Iran

M. Hekmat
Health Management Research Center, Baqiyatallah University of Medical Sciences, Tehran, Iran

T. Sathyapalan
Academic Diabetes, Endocrinology and Metabolism, Hull York Medical School, University of Hull, Hull, UK

P. C. Guest (ed.), *Clinical, Biological and Molecular Aspects of COVID-19*, Advances in Experimental Medicine and Biology 1321, https://doi.org/10.1007/978-3-030-59261-5_20

Results
The study population included 217 HCWs (111 male, 116 female) and the mean age of the study group was 39.6 years old. Approximately two-thirds of the HCWs stayed in the hospital for 2–3 weeks. The mean scores of depression and stress were at a "severe" level, while anxiety scores were at an "extremely severe" level. The prevalence of severe scores was 38.71%, 2.30%, and 48.97% for depression, anxiety, and stress, and the prevalence of extremely severe scores was 46.54%, 97.24%, and 4.98% depression, anxiety, and stress, respectively. In stress subscale, moderate stress was 47.46%. Female HCWs reported higher levels of depression compared with males.

Conclusion
In this study, HCWs reported experiencing severe and extremely severe psychological burdens. Timely interventions to promote mental health in HCWs exposed to patients with COVID-19 need to be immediately implemented, with female nurses requiring particular attention. This process could be facilitated via tests for molecular biomarkers in accessible body fluids, such as saliva, plasma, and serum.

Keywords

COVID-19 · Coronavirus · Healthcare worker · Depression · Anxiety · Stress · Iran

20.1 Introduction

In late December 2019, an outbreak of the 2019 novel coronavirus (COVID-19) caused a substantial public health crisis in China, and then this spread around the world [1]. On 19 February 2020, the first patient was diagnosed with COVID-19 in Iran. Healthcare workers (HCWs) on the frontline are directly involved in the diagnosis, treatment, and care of patients with COVID-19 [2]. In the early stages of the COVID-19 epidemic, it was reported that infected HCWs accounted for 29 percent of all hospitalized COVID-19 patients [3]. HCWs are vulnerable not only to increased risk of infection but are also at increased risk for mental health disorders. HCWs develop a psychological burden from psychological distress and other mental health symptoms [2]. The global number of COVID-19 cases has already exceeded that of most previous epidemics, and the death toll has also exceeded that of the severe acute respiratory syndrome (SARS) viral outbreak [4].

It is likely that HCWs experience a fear of spreading and transmitting COVID-19 to their colleagues, friends, or families [5]. HCWs who worked in high-risk clinical settings had significantly more post-traumatic stress symptoms than those who were not exposed to these settings. Even years after the SARS outbreak, relatively high levels of post-traumatic stress disorder (PTSD) symptoms were experienced by hospital employees who had been at a high risk of contracting the virus [6]. It was reported that the mental stress of HCWs increased significantly during the SARS epidemic since they had to wear thick isolation clothes, care for a large number of anxious patients, and work in a relatively con-

P. C. Guest
Laboratory of Neuroproteomics, Department of Biochemistry and Tissue Biology, Institute of Biology, University of Campinas (UNICAMP), Campinas, Brazil

A. Sahebkar (✉)
Neurogenic Inflammation Research Center, Mashhad University of Medical Sciences, Mashhad, Iran

Biotechnology Research Center, Pharmaceutical Technology Institute, Mashhad University of Medical Sciences, Mashhad, Iran

Polish Mother's Memorial Hospital Research Institute (PMMHRI), Lodz, Poland

Halal Research Center of IRI, FDA, Tehran, Iran
e-mail: sahebkara@mums.ac.ir; amir_saheb2000@yahoo.com

fined space. The high-intensity work further negatively affected the mental health of the medical staff, especially in the cases of those who came from all over the country to support HCWs working in the region [7]. HCWs who worked during the SARS outbreak reported depression, fear, anxiety, and disappointment 1 year after the outbreak. Also, HCW SARS survivors had higher levels of stress and psychological distress than other SARS survivors [8].

There is a paucity of evidence on the psychological disorders of HCWs in the past epidemics. Acute mental healthcare interventions for health professionals and mental health interventions targeting frontline healthcare workers are relatively rare [2]. The aim of the current study was to evaluate mental health outcomes among healthcare workers treating patients with COVID-19 by quantifying the magnitude of symptoms of depression, anxiety, and stress. The aim of this study was to provide an assessment of the mental health of Iranian healthcare workers in Baqiyatallah Hospital, Tehran.

20.2 Material and Methods

20.2.1 Study Design

A cross -sectional survey-based study was conducted in February and March 2020 at Baqiyatallah Hospital, one of the main and referral centers for specialized diagnosis and treatment of patients with COVID-19 in Tehran.

20.2.2 Participants

All HCWs including nurses, physicians, and other medical personnel who had at least 1-week involvement in providing direct services to confirmed or suspected COVID-19 patients were enrolled after informed consent. The samples were analyzed by the census method. People with pre-existing physical and mental disorders were excluded. The degree of symptoms of depression, anxiety, and stress was assessed by the Iranian version of DASS-21.

20.2.3 Sample Size

The target sample size calculated the first-type error of five-hundredth and the second-type error was considered two-tenths, and a 50% satisfaction was assumed to estimate the maximum sample size, using Cochran's sample size estimation formula [9]. The sample size was estimated to be 87 people. According to the nature of the study and the probability of participant dropouts, a 20% dropout rate was considered and the final sample size was estimated to be 110 participants.

20.2.4 Outcomes, Covariates, and Research Tools

The symptoms of depression, anxiety, and stress were assessed in all participating HCWs. The modified DASS-21 was used to collect data. In 1995, Lovibond designed and validated this questionnaire [10] to measure psychological distress among people with 21 items. These items included three subscales of depression, anxiety, and stress. Each subscale included seven questions. Each item was given a rating of "never," "little," "moderate," and "many." The lowest score was 0 and the highest score was 3.

In this questionnaire, questions 3, 5, 10, 13, 16, 17, and 21 are related to depression; questions 2, 4, 7, 9, 15, 19, and 20 are related to anxiety; and questions 1, 6, 8, 11, 12, 14, and 18 assessed stress. The cutoff points are as follows for each score range: depression, 0–4 = normal, 5–6 = mild, 7–10 = average, 11–13 = severe, and ≥ 14 = very severe; stress, 0–7 = normal, 8–9 = mild, 10–12 = average, 13–16 = severe, and ≥ 17 = very severe; and anxiety, 0–3 = normal, 4–5 = mild, 6–7 = average, 8–9 = severe, and ≥ 10 = very severe.

Validity and reliability of this questionnaire in Iran was performed by Sahebi et al. [11]. The validation found that the three subscales of questionnaire had high internal consistency with Cronbach's alpha values of 0.77, 0.79, and 0.78 for depression, anxiety, and stress, respectively [11]. A previous study carried out in China showed similar Cronbach's alpha values of 0.94 for depression, 0.92 for anxiety 0.82 for stress [12].

Demographic characteristics were self-reported by participants, including their sex, age, working experience, marital status, and hospital stay time.

20.2.5 Ethical Considerations

The study was approved by the Research Ethics Committee and vice-chancellor in Research and Technology of Baqiyatallah University of Medical Sciences (code IR.BMSU.REC.1398.442). The objectives of the study were stated to the HCWs and informed consent was obtained from each one for their participation in the study. They were also assured of confidentiality.

20.2.6 Statistical Analysis

All statistical analysis was conducted using R Statistical Software (version 3.5.1; R Foundation for Statistical Computing). The Shapiro-Wilk test was conducted to assess normality of data distribution. The qualitative variables were presented by frequency and percentage. The outcome numeric variables were presented by means and standard deviation (SD). Independent sample *t*-tests were used to test the differences in means of depression, anxiety, and stress symptom scores with respect to gender and marital status, and one-way ANOVA tests were conducted to determine statistical differences in scores with respect to age and hospital stay time. The level of statistical significance was set at $P < 0.05$.

20.3 Results

The DASS questionnaires were completed by 217 medical staff (111 males, 106 females). Their demographic characteristics are summarized in Table 20.1. Participants ranged in age from 28 to 62 years old and were 39.57 years old on average. Approximately 72% of subjects (n = 158) were married. The majority of staff stayed in the hospital for 2-week (31.79%) or 3-week (32.72%) durations (Table 20.1). According to "working experience in ward," a similar number of staff had been there less than and more than 10 years.

The results of our study showed that the comparison of mean scores for all the three DASS subscales were not statistically significant in terms of "age group," "marital status," "working experience in ward," and "hospital stay time" variables (Table 20.1). Our findings indicated that the depression scores for females were higher than males ($p = 0.011$), but these differences were not significant for the stress and anxiety components.

The mean scores of symptoms of depression, anxiety, and stress were found to be "severe" (Table 20.2). The prevalence of these severe symptoms of depression, anxiety, and stress was 46.54%, 97.24%, and 38.71%, respectively.

20.4 Discussion

The aim of this cross-sectional survey was to provide an assessment of the mental health of Iranian healthcare workers in the Baqiyatallah Hospital, in Tehran, Iran, which can serve as guidance in the promotion of mental well-being among healthcare workers. The results of the current study suggested that there was a high prevalence of mental disorder symptoms among HCWs treating patients with COVID-19. All HCWs reported different levels of depression, anxiety, and stress, including severe anxiety in 97.24% and severe depression in 46.54% of HCWs. In this study, 86.11% of the HCWs reported severe and extremely severe stress signs. A study by Jianbo et al. showed that a significant proportion of HCWs in China experienced anxiety, depression, and insomnia symptoms, and psychological distress was reported by more than 70% [2]. In one study reporting the immediate psychological impact on HCWs in Wuhan, the levels of stress, depres-

Table 20.1 The mean (±SD) scores for psychological symptoms in terms of age group, gender, marital status, working experience in ward, and hospital stay time (n = 217)

Variable		Frequency (%)	Anxiety score	Stress score	Depression score
Age	<46 years	107 (44.40)	26.21 ± 4.829	27.51 ± 4.183	26.45 ± 4.616
	46–55 years	98 (40.66)	25.76 ± 4.084	27.35 ± 4.566	26.14 ± 4.114
	>55 years	36 (14.94)	26.33 ± 4.840	26.83 ± 4.494	24.94 ± 4.394
p-value			0.713	0.724	0.206
Gender	Male	111 (51.15)	26.50 ± 4.16	26.25 ± 5.55	25.33 ± 5.38
	Female	106 (48.85)	25.79 ± 4.32	26.21 ± 5.72	27.08 ± 4.64
p-value			0.217	0.953	0.011[a]
Marital status	Married	158 (72.81)	25.86 ± 4.15	25.91 ± 5.52	26.57 ± 4.98
	Unmarried	59 (27.19)	26.95 ± 4.42	27.08 ± 5.84	25.15 ± 5.31
p-value			0.093	0.172	0.078
Working experience in ward	≤10 years	109 (50.2)	26.31 ± 4.24	25.95 ± 5.29	26.04 ± 4.82
	>10 years	108 (49.8)	26.00 ± 4.25	26.52 ± 5.94	26.33 ± 5.38
p-value			0.589	0.454	0.669
Hospital stay time	1-week	17 (7.84)	26.47 ± 4.33	25.06 ± 5.57	26.16 ± 5.39
	2-week	69 (31.79)	25.83 ± 4.27	25.83 ± 5.35	25.06 ± 5.57
	3-week	71 (32.72)	25.89 ± 3.81	26.65 ± 5.69	25.83 ± 5.17
	4-week	60 (27.65)	26.77 ± 4.69	26.53 ± 5.90	26.96 ± 4.29
p-value			0.567	0.648	0.412

[a]Statistically significant

Table 20.2 Prevalence and score severity ratings of depression, anxiety, and stress among medical staff (n = 217)

Psychological variable		Frequency	%
Depression	Moderate	32	14.75
	Severe	101	46.54
	Extremely severe	84	38.71
	MEAN ± SD	26.18 ± 5.09	
Anxiety	Moderate	1	0.46
	Severe	211	97.24
	Extremely severe	5	2.30
	MEAN ± SD	26.16 ± 4.24	
Stress	Mild	19	8.76
	Moderate	103	47.46
	Severe	84	38.71
	Extremely severe	11	5.07
	MEAN ± SD	26.23 ± 5.62	

sion, and anxiety were 29.8%, 13.5%, and 24.1%, respectively [13]. The above finding that Chinese HCWs reported lower levels of stress, anxiety, and depression compared to Iranian HCWs [13] may be due to the use of different questionnaires to assess psychological symptoms. We used a modified DASS-21 questionnaire that measures symptoms of all three psychological domains at the same time. A study of the SARS outbreak in Taiwan showed that 77.4% of HCWs reported anxiety and worry [14].

The majority of staff stayed in the hospital for a 2-week or 3-week duration, although no difference is reported between the mean scores of psychological distress and the duration of staying in the hospital. Female HCWs reported signs of depression more than males, and this was statistically significant. In the Wuhan study, the females reported higher levels of depression [15], and other studies have also confirmed that females have a higher risk of depression, anxiety, and psychological stress [2, 16]. This could be attributed to the multiple accepted roles in the lives of females, and they generally carry more responsibility toward families and children. Therefore, specific attention is warranted regarding the mental health well-being of female HCWs treating patients with COVID-19.

Another variable in our study was the experience of working in the ward. We found no significant difference between >10 years and <10 years experience of working in the ward. However, another study demonstrated that anxiety, depression, and acute stress increased with years of working [15]. The reason for the discrepancy in these two studies is not clear but warrants further study.

Well-timed mental health needs for HCWs should be addressed urgently. We need a comprehensive assessment of risk factors leading to various psychological disorders to design appropriate psychological interventions. For example, in the COVID-19 outbreak, the Chinese government implemented rapid and comprehensive emergency interventions. A series of guidelines have been established since the first confirmed case emerged in dealing with COVID-19 which includes (a) education and training of staff, (b) mental healthcare, (c) monitoring of the physical condition, (d) management of protective equipment, and (e) reassignment of medical resources [7].

The limitations of this study include the point that this was a single-center cross-sectional survey which limits the generalization of our findings. In addition, the study was conducted at the onset of the COVID-19 outbreak and may therefore not account for changes in mental health symptoms that are likely to change as the epidemic progresses. Also, we could not assess various risk factors that affected the symptoms of depression, anxiety, and stress in HCWs. We recommend construction of an algorithm that allows simultaneous assessment of HCWs at both the psychological and molecular levels for increased accuracy of testing over the changing course of the virus. For example, studies of physicians experiencing burnout syndrome have found elevated levels of both salivary and serum cortisol levels, as well as increased serum levels of glycated hemoglobin (HbA1C) [17]. Another study which assessed the effect of the working environment on nurses found that stress symptoms were negatively correlated with serum levels of dehydroepiandrosterone sulfate and positively correlated with serum interleukin-6 levels [18].

20.5 Conclusions

In this study, the healthcare workers reported severe or extremely severe symptoms of depression, anxiety, and stress. Timely focused interventions to promote mental health in HCWs exposed to COVID-19 should be performed in a timely manner, especially in the case of female nurses. In conclusion, the COVID-19 outbreak is a significant threat to international health, especially for the psychological well-being of healthcare workers. For the protection of medical staff, further research needs to be done to assess the risk factors for mental health in these critically important individuals. Such research should take into account the changing status of the COVID-19 outbreak and the effects on societies and financial institutions across the globe. Finally, molecular biomarkers could be combined with psychological assessments to increase accuracy of testing.

Acknowledgments Thanks to guidance and advice from the Clinical Research Development Unit of Baqiyatallah Hospital.

Conflict of Interest The author(s) declare no conflicts of interest with respect to the authorship and publication of this article.

References

1. Huang C, Wang Y, Li X, Ren L, Zhao J, Hu Y et al (2020) Clinical features of patients infected with 2019 novel coronavirus in Wuhan, China. Lancet 395(10223):497–506
2. Lai J, Ma S, Wang Y, Cai Z, Hu J, Wei N et al (2020) Factors associated with mental health outcomes among

health care workers exposed to coronavirus disease 2019. JAMA Netw Open 3(3):e203976–e203976
3. Wang D, Hu B, Hu C, Zhu F, Liu X, Zhang J et al (2020) Clinical characteristics of 138 hospitalized patients with 2019 novel coronavirus–infected pneumonia in Wuhan, China. JAMA 7. https://doi.org/10.1001/jama.2020.1585. [Epub ahead of print]
4. Coronavirus COVID-19 Global Cases by Johns Hopkins CSSE. https://www.iafc.org/topics-and-tools/resources/resource/coronavirus-covid-19-global-cases-(johns-hopkins-university-csse)
5. Maunder R, Hunter J, Vincent L, Bennett J, Peladeau N, Leszcz M et al (2003) The immediate psychological and occupational impact of the 2003 SARS outbreak in a teaching hospital. CMAJ 168(10):1245–1251
6. Wu P, Fang Y, Guan Z, Fan B, Kong J, Yao Z et al (2009) The psychological impact of the SARS epidemic on hospital employees in China: exposure, risk perception, and altruistic acceptance of risk. Can J Psychiatr 54(5):302–311
7. Wang H, Wang S (2020 Mar 18) Yu K (2020) COVID-19 infection epidemic: the medical management strategies in Heilongjiang Province. China Crit Care 24(1):107. https://doi.org/10.1186/s13054-020-2832-8
8. Lee AM, Wong JG, McAlonan GM, Cheung V, Cheung C, Sham PC et al (2007) Stress and psychological distress among SARS survivors 1 year after the outbreak. Can J Psychiatry 52(4):233–240
9. Kasiulevičius V, Šapoka V, Filipavičiūtė R (2006) Sample size calculation in epidemiological studies. Geron 7(4):225–231
10. Lovibond PF (1998) Long-term stability of depression, anxiety, and stress syndromes. J Abnorm Psychol 107(3):520–526
11. Sahebi A, Asghari MJ, Salari R (2005) Validation of depression anxiety and stress scale (DASS-21) for an Iranian population. J Dev Psychol. http://jip.azad.ac.ir/article_512443_en.html
12. Chan RC, Xu T, Huang J, Wang Y, Zhao Q, Shum DH et al (2012) Extending the utility of the depression anxiety stress scale by examining its psychometric properties in Chinese settings. Psychiatry Res 200(2–3):879–883
13. Roberts T, Esponda GM, Krupchanka D, Shidhaye R, Patel V, Rathod S (2018) Factors associated with health service utilisation for common mental disorders: a systematic review. BMC Psychiatry 18(1):262. https://doi.org/10.1186/s12888-018-1837-1
14. Koh D, Lim MK, Chia SE, Ko SM, Qian F, Ng V et al (2005) Risk perception and impact of severe acute respiratory syndrome (SARS) on work and personal lives of healthcare Workers in Singapore What can we Learn? Med Care 43(7):676–682
15. Zhu Z, Xu S, Wang H, Liu Z, Wu J, Li G et al (2020) COVID-19 in Wuhan: immediate psychological impact on 5062 health workers. medRxiv. https://doi.org/10.1101/2020.02.20.20025338
16. Audet MC (2019) Stress-induced disturbances along the gut microbiota-immune-brain axis and implications for mental health: does sex matter? Front Neuroendocrinol 54:100772. https://doi.org/10.1016/j.yfrne.2019.100772
17. Deneva T, Ianakiev Y, Keskinova D (2019) Burnout syndrome in physicians-psychological assessment and biomarker research. Medicina (Kaunas) 55(5). pii: E209. https://doi.org/10.3390/medicina55050209
18. Arnetz J, Sudan S, Goetz C, Counts S, Arnetz B (2019) Nurse work environment and stress biomarkers: possible implications for patient outcomes. J Occup Environ Med 61(8):676–681

Coronavirus (COVID-19)-Associated Psychological Distress Among Medical Students in Iran

21

Malihe Sadat Moayed, Amir Vahedian-Azimi, Golshan Mirmomeni, Farshid Rahimi-Bashar, Keivan Goharimoghadam, Mohamad Amin Pourhoseingholi, Mohsen Abbasi-Farajzadeh, Azam Khatibzadeh, Thozhukat Sathyapalan, Paul C. Guest, and Amirhossein Sahebkar

Abstract

Aim
The COVID-19 was declared a pandemic in early 2020 and is associated with high public anxiety all over the world. The healthcare community is at the highest risk of infection and thereby prone to most distress. The aim of this study was to explore and evaluate the degree of depression, anxiety, and stress levels among medical college students during the COVID-19 epidemic in Iran.

Methods
A cross-sectional study was conducted in February and March 2020, 3 weeks after the first reported COVID-19 infection was identified in Iran. All medical college students who entered clinical courses were eligible for the study. Depression, stress, and anxiety were evaluated in these students using the DASS-21 questionnaire. Participants were selected by using availability sampling. All statistical analyses were performed using R version 3.5.1.

Malihe Sadat Moayed and Amir Vahedian-Azimi have been equally contributed for this chapter.

M. S. Moayed · A. Vahedian-Azimi (✉)
Trauma Research Center, Nursing Faculty, Baqiyatallah University of Medical Sciences, Tehran, Iran
e-mail: amirvahedian63@gmail.com

G. Mirmomeni
Hearing Research Center, Ahvaz Jundishapur University of Medical Sciences, Ahvaz, Iran

F. Rahimi-Bashar
Anesthesia and Critical Care Department, Hamadan University of Medical Sciences, Hamadan, Iran

K. Goharimoghadam
Internal Medicine, Shariati Hospital, Tehran University of Medical Sciences, Tehran, Iran

M. A. Pourhoseingholi
Gastroenterology and Liver Diseases Research Center, Research Institute for Gastroenterology and Liver Diseases, Shahid Beheshti University of Medical Sciences, Tehran, Iran

M. Abbasi-Farajzadeh
Marine Medicine Research Center, Baqiyatallah University of Medical Sciences, Tehran, Iran

P. C. Guest (ed.), *Clinical, Biological and Molecular Aspects of COVID-19*, Advances in Experimental Medicine and Biology 1321, https://doi.org/10.1007/978-3-030-59261-5_21

Results
The total number of participants was 207, with 143 males and 64 females. More than half of the participants (57.97%) were married. The mean duration of working experience among students with COVID-19 infection and experience in a medical ward was 3.00 ± 1.27 days and 17.40 ± 7.26 months, respectively. The majority of students had 2 or 3 days working experience with COVID-19 infection. The mean anxiety score of participants was 28.56 ± 4.68, the depression score was 29.36 ± 4.42, and the stress score was 28.99 ± 4.53. Our findings indicated that the mean scores of depression were at an "extremely severe" level, while stress and anxiety were at "severe" levels. The prevalence of "severe" symptoms of depression, stress, and anxiety was 69.57%, 60.87%, and 99.04%, respectively.

Conclusions
There is a high prevalence of anxiety and depression among medical students who were exposed to COVID-19-infected patients. Our results highlight the need to establish psychological support programs, training, and self-care for medical college students in relation to mental health. We recommend incorporation of molecular biomarker tests into an algorithm to aid in assessments and consideration of the appropriate therapeutic responses.

Keywords

COVID-19 · Coronavirus · Medical student · Depression · Anxiety · Stress · Iran

A. Khatibzadeh
Health Management Research Center, Baqiyatallah University of Medical Sciences, Tehran, Iran

T. Sathyapalan
Academic Diabetes, Endocrinology and Metabolism, Hull York Medical School, University of Hull, Hull, UK

P. C. Guest
Laboratory of Neuroproteomics, Department of Biochemistry and Tissue Biology, Institute of Biology, University of Campinas (UNICAMP), Campinas, Brazil

21.1 Introduction

The global outbreak coronavirus disease 2019 (COVID-19) started with the infection of a large number of people in China in December 2019 [1]. On 19 February 2020, the first patient was diagnosed with COVID-19 in Iran. In addition to healthcare workers, most of the hospitals were supported by a number of medical students in the last years of their education who were providing voluntarily aid in managing patients with COVID-19 infections.

The absence of a comprehensive and definitive treatment or vaccination program to manage this disease has caused fear and anxiety in people [2, 3]. In the early stages of the COVID-19 outbreak, it was reported that around 29 percent of COVID-19 patients consisted of infected healthcare workers [4]. The outbreak has therefore imposed a significant psychological stress that could lead to undesirable effects on the overall psychological health of medical students attending patients with COVID-19 [5]. In line with this, Al-Rabiaah et al. reported that medical students had different levels of anxiety with most reporting minimal anxiety levels during the SARS outbreak [6]. In another study, junior medical students expressed a significantly greater degree of anxiety compared to the more senior medical students [7].

Many other studies reported psychological disorders in the aftermath of an epidemic which may not reflect the actual stress subjects felt during the actual event. Also, no studies have investigated the psychological distress of university medical students during the COVID-19 outbreak, although some studies have addressed similar

A. Sahebkar (✉)
Biotechnology Research Center, Pharmaceutical Technology Institute, Mashhad University of Medical Sciences, Mashhad, Iran

Neurogenic Inflammation Research Center, Mashhad University of Medical Sciences, Mashhad, Iran

Polish Mother's Memorial Hospital Research Institute (PMMHRI), Lodz, Poland

Halal Research Center of IRI, FDA, Tehran, Iran
e-mail: sahebkara@mums.ac.ir; amir_saheb2000@yahoo.com

issues among hospital workers and residents [3, 8, 9].

The aim of this study was to explore and evaluate the degree of stress, anxiety, and depression among medical students during the COVID-19 epidemic in Iran.

21.2 Material and Methods

21.2.1 Study Design

A cross-sectional survey was conducted in February and March 2020. The study was carried out in Tehran, the capital city of Iran 3 weeks after the first patient of COVID-19 was identified in Iran.

21.2.2 Participants

We conducted this survey on medical students exposed to patients with corona in Baqiyatallah Hospital. Baqiyatallah Hospital is one of the main referral centers for the diagnosis and treatment of patients with COVID-19 in Tehran. The medical students were mainly recruited from this hospital. Other centers also participated in this study, but since the number of medical students from other centers was not high enough for obtaining a separate ethics code, the dean of each center approved data collection for their respective centers. None of the participants had any physical disabilities or mental health disorders.

21.2.3 Sample Size

Cochran's sample size estimation formula was applied [10]. We used the following parameters: $\alpha = 0.05$; $\beta = 0.2$; and an assumed 50% satisfaction probability to estimate the maximum sample size. The sample size was calculated to be 87 people. According to the nature of the study and the probability of sample size dropouts, a 20% dropout rate was considered and the final sample size was 110 participants.

21.2.4 Research Tools

Demographic characteristics were collected using a questionnaire, and the parameters included age, gender, marital status, working experience in the ward, and working experience with COVID-19 patients. We focused on symptoms of depression, anxiety, and stress using this self-report questionnaire. The Depression Anxiety Stress Scale (DASS-21) was used to collect the data. This questionnaire was designed and validated by Lovibond in 1995 [11] to measure psychological distress among the community using 21 items. The items included three subscales and each subscale included seven questions.

In the translated version, each item had choices of never, little, moderate, and many. The lowest score linked to each question was 0 and the highest score was 3. In this questionnaire, the depression subscale included questions 3, 5, 10, 13, 16, 17, and 21; and questions 1, 6, 8, 11, 12, 14, and 18 assessed stress; and questions 2, 4, 7, 9, 15, 19, and 20 related to the anxiety subscale. The sum of all individual scores related to each section was multiplied by two. Validity and reliability studies of this questionnaire were carried out in Iran. The Persian version of the scale with appropriate psychometric properties was used in our study. For the total score of DASS-21, the Cronbach alpha was 0.94. The Cronbach alpha for depression, anxiety, and stress scales was 0.85, 0.85, and 0.87, respectively [12].

21.2.5 Ethical Considerations

This study was approved by the Ethics Committee of Baqiyatallah University of Medical Sciences (code IR.BMSU.REC.1398.439). In addition, the objectives of the study were explained, and informed consent was obtained from each student regarding their participation in the study. All students were assured that all of the resulting data would be treated with confidentiality.

21.2.6 Statistical Analysis

All statistical analyses were performed using R software, version 3.5.1. Categorical data (age group, gender, marital status, working experience in the ward, and working experience with COVID-19) were described with numbers and percentages, and continuous data (depression, anxiety, and stress scores) were given as means and standard deviation (SD). Differences in the mean scores of the DASS-21 subscales among groups with different sociodemographic and clinical characteristics were calculated using an independent sample t-test or one-way analysis of variance (ANOVA). All tests were two-tailed, with a statistical level set at $p < 0.05$.

21.3 Results

The total number of participants was 207, with 143 males (69.08%) and 64 females (30.82%). More than half of the participants (57.97%) were married. The mean student working experience with COVID-19 and experience in the ward was 3.00 ± 1.27 days and 17.40 ± 7.26 months, respectively. The majority of students had 2 or 3 weeks working experience with COVID-19 or in the ward (Table 21.1).

According to the results of t-tests and one-way ANOVA, the comparison of mean scores for stress, depression, and anxiety DASS-21 subscales was not statistically significant among the variables of "age," "gender," "marital status," and "working experience in ward" (Table 21.1). Furthermore, depression and stress scores were not significantly different across "working experience with corona" categories. However, the anxiety level was significantly different in terms of "working experience with COVID-19" (p-value = 0.016).

The anxiety score of participants was 28.56 ± 4.68, the depression score was 29.36 ± 4.42, and the stress score was 28.99 ± 4.53 (Table 21.2). Our findings indicated that the mean scores of depression were at the "extremely severe" level, while stress and anxiety levels were "severe." The prevalence of "severe" symptoms of depression, stress, and anxiety was 69.57%, 60.87%, and 99.04%, respectively.

21.4 Discussion

The COVID-19 outbreak in Iran and virtually all other countries and territories of the world is one of the most significant threats to national and international public health that has occurred in more than a century since the 1918 Spanish flu. The main purpose of this study was to measure the current prevalence and severity of psychological distress in Iranian medical college students during the early phase of the COVID-19 outbreak. The results confirmed that the amount of depression, anxiety, and stress was severe and extremely severe during the first days of the outbreak in medical students exposed to COVID-19 patients. The prevalence of extremely severe symptoms of depression was 69.57%. The prevalence of severe symptoms of stress and anxiety was 60.87% and 99.04%, respectively.

Anxiety is a worry about future events, and fear is a reaction to the current events. The increasing number of COVID-19 cases and its geographical expansion, the infectious cause, epidemiological characteristics, rapid transmission pattern, and insufficient preparedness has raised significant fear and anxiety about the outbreak [13]. In another study, the mean of perceived stress score in healthcare students was 18.4 and in non-healthcare students this was 19.6, which are both more than the mean community scores [14]. Nursing students had a higher level of perceived stress in comparison to other students. It has been reported that direct care of patients with SARS probably resulted in the high levels of stress in the nursing students. The causes of higher stress in the non-healthcare students were perceived as a lower opportunity for treatment and a higher risk of death from SARS. In our study, the mean scores for stress, depression, and anxiety were not significantly different due to age group, gender, marital status, or working experience in the ward. Similar to the results of our study, Wong et al. reported no dif-

Table 21.1 The mean (±SD) scores for psychological symptoms in terms of age group, gender, marital status, working experience with COVID-19, and working experience in ward (n = 207)

Variable		Frequency (%)	Anxiety score	Stress score	Depression score
Age	≤26 years	88 (42.5)	28.25 ± 4.94	28.61 ± 4.61	29.59 ± 4.16
	>26 years	119 (57.5)	28.79 ± 4.48	29.28 ± 4.47	29.19 ± 4.62
p-value			0.413	0.299	0.524
Gender	Male	143 (69.08)	28.63 ± 4.21	28.71 ± 4.44	29.45 ± 4.43
	Female	64 (30.82)	28.41 ± 5.63	29.63 ± 4.71	29.16 ± 4.43
p-value			0.777	0.182	0.655
Marital status	Married	120 (57.97)	28.62 ± 4.52	29.20 ± 4.32	29.02 ± 4.30
	Unmarried	87 (42.03)	28.48 ± 4.92	28.71 ± 4.82	29.84 ± 4.57
p-value			0.840	0.446	0.187
Working experience with corona	1-day	23 (11.12)	27.22 ± 5.14	28.70 ± 5.06	29.30 ± 3.84
	2-day	57 (27.55)	27.02 ± 3.984	28.28 ± 4.57	29.12 ± 4.24
	3-day	60 (28.98)	29.40 ± 4.89	29.60 ± 4.37	28.87 ± 4.74
	4-day	37 (17.87)	29.46 ± 4.89	29.68 ± 4.18	29.89 ± 4.67
	5-day	24 (11.59)	30.00 ± 3.91	27.92 ± 4.88	30.17 ± 3.38
	6-day	6 (2.89)	28.67 ± 4.84	31.00 ± 3.74	30.33 ± 7.42
p-value			0.016[a]	0.312	0.775
Working experience in ward	6-month	21 (10.14)	27.81 ± 4.85	29.81 ± 3.74	28.67 ± 4.68
	12-month	67 (32.36)	29.64 ± 3.98	28.87 ± 4.93	29.04 ± 3.86
	18-month	57 (27.55)	28.74 ± 5.34	29.89 ± 4.56	29.86 ± 4.45
	24-month	40 (19.32)	28.01 ± 4.18	27.75 ± 4.30	29.70 ± 4.58
	30-month	17 (8.21)	27.18 ± 5.48	28.94 ± 3.82	28.71 ± 4.76
	36-month	5 (2.42)	24.40 ± 1.67	27.20 ± 4.82	30.40 ± 4.36
p-value			0.237	0.067	0.785

[a]Statistically significant

Table 21.2 Prevalence and score severity ratings of depression, anxiety, and stress among students (n = 207)

Psychological Variable		Frequency	%
Depression	Moderate	3	1.45
	Severe	144	69.57
	Extremely severe	60	28.98
	MEAN ± SD	29.36 ± 4.42	
Anxiety	Severe	205	99.04
	Extremely severe	2	0.96
	MEAN ± SD	28.56 ± 4.68	
Stress	Mild	1	0.48
	Moderate	63	30.43
	Severe	126	60.87
	Extremely severe	17	8.22
	MEAN ± SD	28.99 ± 4.53	

ference between perceived stress levels between genders or other groups [14]. Another study reported that female students had significantly higher mean stress levels than male students [6]. In our study, there was no difference between junior and senior medical students and the number of days they had been in the COVID-19 ward. This finding contrasts with studies of the SARS epidemic, which showed that the stress levels in medical students were higher in junior compared to senior medical students [6]. The current study showed that the anxiety score was significantly different for students with different durations of working experience with COVID-

19. This showed that anxiety scores were higher among the medical students who had worked with COVID-19 patients for higher numbers of days.

Medical students, by having more experience in hospital setting, are thought to have a realistic assessment of the infectious diseases and should be able to control their stress levels more effectively. The results of one study have shown that increased knowledge about the SARS virus led to a reduction in their stress levels about this infection [15]. Knowledge toward prevention and control of the disease is necessary among the health students as they are at a higher risk of newly emerging diseases due to increased exposure via contact with patients. Another study confirmed that medical students would benefit in this way from learning about emerging infectious diseases [5].

A possible limitation of this study is that we recruited participants by convenience sampling. Therefore, the results cannot be interpreted and generalized to all medical students. In addition, this study was conducted at the onset of the COVID-19 outbreak. Therefore it is possible that with the continuing epidemic the mental health symptoms of medical students may also change. Finally, risk factors for depression, anxiety, and stress were not assessed in this cohort of medical students. To improve the accuracy of testing, we recommend combined assessment of molecular biomarkers in readily accessible body fluids. For example, studies have shown that cortisol levels in saliva are associated with symptoms of anxiety, depression, and post-traumatic avoidance [16], and salivary amylase levels could be useful for assessment of individuals working in stressful and isolated environments [17]. Furthermore, a systematic review found that serum levels of inflammatory biomarkers such as interleukin (IL)-1β, IL-5, and IL-6 could be used to identify individuals with panic disorder [18]. These studies help to illustrate the emerging point that mental illnesses can be characterized by biomarkers in body fluids through mind-body feed-forward and feedback systems.

21.5 Conclusions

In this study, medical college students reported severe and extremely severe psychological distress during the COVID-19 epidemic in Iran and through their experiences with patients. As it is still certain whether or not there will be a second wave of this virus, it will be critical to establish mental health intervention programs to help medical students and other healthcare works adapt to this challenging unprecedented situation.

Acknowledgments Thanks to guidance and advice from the Clinical Research Development Unit of Baqiyatallah Hospital.

Conflict of Interest The author(s) declare no conflicts of interest with respect to the authorship and publication of this article.

References

1. National Health Commission of the People's Republic of China (2020) Update on epidemic situation of novel coronavirus-infected pneumonia by 24:00 on Feb 4, 2020
2. Stein MB (2020) EDITORIAL: COVID-19 and anxiety and depression in 2020. Depress Anxiety 37(4):302. https://doi.org/10.1002/da.23014
3. Dai Y, Hu G, Xiong H, Qiu H, Yuan X (2020) Psychological impact of the coronavirus disease 2019 (COVID-19) outbreak on healthcare workers in China. medRxiv. https://doi.org/10.1101/2020.03.03.20030874
4. Wang D, Hu B, Hu C, Zhu F, Liu X, Zhang J et al (2020) Clinical characteristics of 138 hospitalized patients with 2019 novel coronavirus–infected pneumonia in Wuhan, China. JAMA. https://doi.org/10.1001/jama.2020.1585. [Epub ahead of print]
5. Alqahtani AS (2017) Knowledge and attitude toward Middle East respiratory syndrome coronavirus among heath colleges' students in Najran, Saudi Arabia. Int J Community Med Public Health 4:2641–2647
6. Al-Rabiaah A, Temsah M-H, Al-Eyadhy AA, Hasan GM, Al-Zamil F, Al-Subaie S et al (2020) Middle East Respiratory Syndrome-Corona Virus (MERS-CoV) associated stress among medical students at a university teaching hospital in Saudi Arabia. J Infect Public Health. pii: S1876-0341(20)30005-8. https://doi.org/10.1016/j.jiph.2020.01.005. [Epub ahead of print]

7. Loh LC, Ali AM, Ang TH, Chelliah A (2006) Impact of a spreading epidemic on medical students. Malays J Med Sci 13(2):30–36
8. Jiang X, Deng L, Zhu Y, Ji H, Tao L, Liu L et al (2020) Psychological crisis intervention during the outbreak period of new coronavirus pneumonia from experience in Shanghai. Psychiatry Res 286:112903. https://doi.org/10.1016/j.psychres.2020.112903
9. Chen Q, Liang M, Li Y, Guo J, Fei D, Wang L et al (2020) Mental health care for medical staff in China during the COVID-19 outbreak. Lancet Psychiatry 7(4):e15–e16
10. Kasiulevičius V, Šapoka V, Filipavičiūtė R (2006) Sample size calculation in epidemiological studies. Gerontologija 7(4):225–231
11. Lovibond PF (1998) Long-term stability of depression, anxiety, and stress syndromes. J Abnorm Psychol 107(3):520–526
12. Asghari A, Saed F, Dibajnia P (2008) Psychometric properties of the depression anxiety stress Scales-21 (DASS-21) in a non-clinical Iranian sample. Int J Psychol 2(2):82–102
13. Xiang YT, Yang Y, Li W, Zhang L, Zhang Q, Cheung T et al (2020) Timely mental health care for the 2019 novel coronavirus outbreak is urgently needed. Lancet Psychiatry 7(3):228–229
14. Wong JG, Cheung EP, Cheung V, Cheung C, Chan MT, Chua SE et al (2004) Psychological responses to the SARS outbreak in healthcare students in Hong Kong. Med Teach 26(7):657–659
15. Koh D, Lim MK, Chia SE, Ko SM, Qian F, Ng V et al (2005) Risk perception and impact of severe acute respiratory syndrome (SARS) on work and personal lives of healthcare Workers in Singapore What can we Learn? Med Care 43(7):676–682
16. Aardal-Eriksson E, Eriksson TE, Holm AC, Lundin T (1999) Salivary cortisol and serum prolactin in relation to stress rating scales in a group of rescue workers. Biol Psychiatry 46(6):850–855
17. Rai B, Kaur J, Foing BH (2012) Salivary amylase and stress during stressful environment: three Mars analog mission crews study. Neurosci Lett 518(1):23–26
18. Quagliato LA, Nardi AE (2018) Cytokine alterations in panic disorder: a systematic review. J Affect Disord 228:91–96

22 A Survey of Psychological Distress Among the Community in the COVID-19 Epidemic: A Cross-Sectional Study

Malihe Sadat Moayed, Amir Vahedian-Azimi, Golshan Mirmomeni, Farshid Rahimi-Bashar, Keivan Goharimoghadam, Mohamad Amin Pourhoseingholi, Mohsen Abbasi-Farajzadeh, Mansour Babaei, Thozhukat Sathyapalan, Paul C. Guest, and Amirhossein Sahebkar

Abstract

Aim
The ongoing COVID-19 outbreak has not only had an impact on physical health but also on psychological health. The aim of this study was to measure the prevalence and severity of psychological distress in the community due to the COVID-19 pandemic.

Methods
This cross-sectional survey was conducted in February and March 2020 in Tehran, Iran. We analyzed demographic characteristics and assessed depression, anxiety, and stress levels in 241 people using convenience sampling and the DASS-21 questionnaire. All statistical analyses were performed using R.

Malihe Sadat Moayed and Amir Vahedian-Azimi contributed equally with all other contributors.

M. S. Moayed · A. Vahedian-Azimi (✉)
Trauma Research Center, Nursing Faculty, Baqiyatallah University of Medical Sciences, Tehran, Iran
e-mail: amirvahedian63@gmail.com

G. Mirmomeni
Hearing Research Center, Ahvaz Jundishapur University of Medical Sciences, Ahvaz, Iran

F. Rahimi-Bashar
Anesthesia and Critical Care Department, Hamadan University of Medical Sciences, Hamadan, Iran

K. Goharimoghadam
Internal Medicine, Shariati hospital, Tehran University of Medical Sciences, Tehran, Iran

M. A. Pourhoseingholi
Gastroenterology and Liver Diseases Research Center, Research Institute for Gastroenterology and Liver Diseases, Shahid Beheshti University of Medical Sciences, Tehran, Iran

M. Abbasi-Farajzadeh
Marine Medicine Research Center, Baqiyatallah University of Medical Sciences, Tehran, Iran

M. Babaei
Health Management Research Center, Baqiyatallah University of Medical Sciences, Tehran, Iran

T. Sathyapalan
Academic Diabetes, Endocrinology and Metabolism, Hull York Medical School, University of Hull, Hull, UK

P. C. Guest (ed.), *Clinical, Biological and Molecular Aspects of COVID-19*, Advances in Experimental Medicine and Biology 1321, https://doi.org/10.1007/978-3-030-59261-5_22

Results
The study population included 241 community-dwelling participants, of whom 145 were women and 96 were males. The mean age was 49.16 ± 8.01 years. Approximately two-thirds of participants ($n = 158$) reported no history of comorbid illness. The mean scores of depression and stress were at a "severe" level, while anxiety levels were at an "extremely severe" level. The prevalence of severe and extremely severe depression readings was 51.45 and 38.17%, respectively. In the anxiety subscale, the prevalence of severe and extremely severe depression was 95.90 and 4.1%, and in the stress subscale the prevalence was 48.97 and 4.98%, respectively.

Conclusion
In this study, people reported experiencing severe and extremely severe psychological distress. Therefore, there is an urgent need to implement mental health intervention policies to cope with this ongoing challenge. We suggest that the incorporation of molecular biomarker tests into the algorithm could aid in assessment of patients and guide the most appropriate therapeutic response.

Keywords

Anxiety · Coronavirus · COVID-19 · Depression · Iran · Psychological distress · Stress

22.1 Introduction

On 19 February 2020, the first patient with COVID-19 infection was identified in Iran. The increasing number of coronavirus cases and its geographical expansion has raised significant concerns around the world. The mental health of the community is also at risk due to the highly infective nature of the disease, the epidemiological characteristics, the lack of preparedness of the health authorities and healthcare systems, and an insufficient supply of protective equipment [1]. In addition, the absence of a comprehensive and definitive treatment protocol or vaccination program against this disease led to the introduction of home quarantine to limit transmission of the virus on the basis of recommendations from health organizations [2]. This resulted in the closure of all schools, universities, and recreation centers and restrictions on commuting were also imposed. These conditions can lead to various negative psychological impacts, such as post-traumatic syndrome disorder (PTSD), confusion, and anger in society. Quarantine, fear of infection, frustration, boredom, lack of information, loss of property, and stigma are known stressors that can affect psychological health [3].

The fear of the unknown effects of the novel 2019 coronavirus raised anxiety levels in healthy persons as well those with preexisting mental health conditions [4]. One study has shown the persistence of these mental disorders 4–6 months after the Middle East respiratory syndrome

P. C. Guest
Laboratory of Neuroproteomics,
Department of Biochemistry and Tissue Biology,
Institute of Biology, University of Campinas
(UNICAMP), Campinas, Brazil

A. Sahebkar (✉)
Biotechnology Research Center, Pharmaceutical Technology Institute, Mashhad University of Medical Sciences, Mashhad, Iran

Neurogenic Inflammation Research Center, Mashhad University of Medical Sciences, Mashhad, Iran

Polish Mother's Memorial Hospital Research Institute (PMMHRI), Lodz, Poland

Halal Research Center of IRI, FDA, Tehran, Iran
e-mail: sahebkara@mums.ac.ir

(MERS) epidemic [5]. The underlying causes for the continued anxiety and stress included worries about symptoms, inadequate equipment, absence of social networking, and a history of psychiatric illness. It has been suggested that these individuals need psychological help and long-term follow-up. Anxiety and stigma were also reported as the most important psychological issues in the severe acute respiratory syndrome (SARS) epidemic [6]. In addition, pharmaco-epidemiological studies have confirmed an increase in the rate of prescription and use of antidepressant drugs after various disasters and natural events, which reflect increased anxiety and depression among the population [7]. In addition to the above effects of the MERS and SARS outbreaks, the COVID-19 pandemic has led to reduced face-to-face communication, associated with a range of mental disorders such as panic, stress, and depression. For the first time, we are facing a widespread epidemic in the Iranian community. Therefore, we need to provide a concrete basis for tailoring and implementing relevant mental health intervention policies to cope with this challenge efficiently and effectively.

So far, there is no epidemiological data on mental health and psychological outcomes of COVID-19 infection. The main aim of this study was to measure the prevalence and severity of psychological distress to compute the current mental health burden of COVID-19 pandemic on Iranian society.

22.2 Material and Methods

22.2.1 Study Design

This cross-sectional survey was conducted in February and March 2020.

22.2.2 Setting

Since Tehran is the capital city in Iran and people from all over Iran live there, it was selected for sampling.

22.2.3 Participants

All adults over the age of 18 who were interested in participating in the study and who could read and write with no known physical disability or mental disorder were selected using available sampling.

22.2.4 Sample Size

We used a first-type error of five-hundredths, a second-type error of two-tenths, and an assumed 50% satisfaction probability to estimate maximum sample size. The sample size was calculated to have 87 people. According to the nature of the study and the probability of dropouts, we allowed for a 20% increase of the calculated size, which resulted in 110 individuals being selected. Cochran's sample size estimation formula in the epidemiologic study was used [8].

22.2.5 Outcomes, Covariates, and Research Tools

We focused on symptoms of depression, anxiety, and stress for all participants, using the Iranian version of validated measurement tools. The Depression Anxiety Stress Scale (DASS-21) was used to collect data. This questionnaire was designed and validated in 1995 to measure psychological distress among the community with 21 items [9]. The scale includes three subscales, and each subscale includes seven questions. In the translated version, each item has choices of never, little, moderate, and many. The lowest score is equivalent to 0 and the highest score is 3. In this questionnaire, questions 2, 4, 7, 9, 15, 19 and 20 are related to anxiety; questions 3, 5, 10, 13, 16, 17, and 21 concern depression; and questions 1, 6, 8, 11, 12, 14, and 18 are for assessment of stress.

(a) Stress: scores from 0 to 7 are considered normal, 8–9 mild, 10–12 average, 13–16 severe, and higher than 17 is very severe.

(b) Anxiety: 0–3 is normal, 4–5 mild, 6–7 average, 8–9 severe, and higher than 10 is very severe.
(c) Depression: a score from 0 to 4 is considered normal, 5–6 mild, 7–10 average, 11–13 severe, and higher than 14 is very severe.

The validity and reliability of this questionnaire have already been established in Iran. For instance, in a study done on 970 students and armies, the authors reported that the translated questionnaire was comparable with the original, with high internal correlations of 0.77, 0.79, and 0.78 for depression, anxiety, and stress, respectively [10]. This was comparable with a study carried out in China, which reported Cronbach's alpha values greater than 0.80 for all scales in an analysis of the effect of the 2008 Sichuan earthquake [11].

Demographic characteristics were self-reported on questionnaire by participants and include sex, age, job, marital status, and educational qualifications.

22.2.6 Ethical Considerations

The study was approved by the Ethics Committee of Baqiyatallah University of Medical Sciences with the code IR.BMSU.REC.1398.441. The objectives of the study were explained, and informed consent was obtained from the participants in the study, and they were assured of confidentiality.

22.2.7 Statistical Analysis

All statistical analyses were performed using R version 3.5.1. The inferential statistical analyses were conducted using parametric tests since the data were found to be normally distributed with homogeneous variances, as shown by screening the data using the Kolmogorov-Smirnov and Levene's tests, respectively. Independent sample *t*-tests were carried out to test the differences in the mean values of the psychological factors (depression, anxiety, and stress) by gender and marital status, and one-way ANOVA tests were used to determine the mean differences in psychological factors relative to age, job, and educational qualification. The level of statistical significance was set at $P < 0.05$.

22.3 Results

The study population included 241 community-dwelling participants in Tehran. Of these, 145 were women and 96 were men. The participants were between 37 and 74 years of age, and the mean age of the study group was 49.16 ± 8.01 years. In addition, 151 patients were married, 82 had a nongovernmental job, 158 reported no background disease, and the majority had a Bachelor's or higher education degree ($n = 156$). There were no significant differences in "age," "marital status," "history of disease," and "job" variables across the different DASS subscales (Table 22.1). Although female participants showed higher depression scores than males (independent sample t-test; $p = 0.02$), the mean stress and anxiety scores between males and females were not significantly different. However, the mean stress scores were significantly different for the "educational qualification" variable, such that individuals with a higher educational degree (e.g., a PhD or Master's degree) experienced higher levels of stress (Table 22.1).

The mean scores of depression, anxiety, and stress were at a "severe" level. The prevalence of "severe" symptoms of depression, anxiety, and stress w 51.45, 95.90, and 48.97%, respectively (Table 22.2).

22.4 Discussion

The main purpose of this study is to measure the prevalence and severity of psychological distress and to compute the current mental health burden on society during the COVID-19 outbreak in Iran. The results confirmed that the amount of psychological distress in the community ranged from severe to extremely severe. More than

Table 22.1 Mean (±SD) scores for psychological symptoms in terms of age group, gender, marital status, job, education level, and history of background disease

Variable		Frequency (%)	Anxiety score	Stress score	Depression score
Age	<45 years	107 (44.40)	26.21 ± 4.829	27.51 ± 4.183	26.45 ± 4.616
	46–55 years	98 (40.66)	25.76 ± 4.084	27.35 ± 4.566	26.14 ± 4.114
	>55 years	36 (14.94)	26.33 ± 4.840	26.83 ± 4.494	24.94 ± 4.394
p-value			0.713	0.724	0.206
Gender	Male	96 (39.83)	26.00 ± 4.561	27.21 ± 4.574	25.29 ± 4.372
	Female	145 (60.17)	26.07 ± 4.524	27.43 ± 4.254	26.63 ± 4.344
p-value			0.908	0.695	0.020[a]
Marital status	Married	151 (62.65)	25.83 ± 4.636	27.21 ± 4.473	26.26 ± 4.090
	Unmarried	90 (37.35)	26.40 ± 4.347	27.58 ± 4.224	25.82 ± 4.877
p-value			0.344	0.524	0.451
Job	Governmental	31 (12.86)	26.39 ± 4.688	27.35 ± 4.176	26.97 ± 4.191
	Nongovernmental	82 (34.03)	25.88 ± 4.831	27.44 ± 4.600	26.15 ± 4.691
	Unemployed	64 (26.55)	26.84 ± 4.137	27.16 ± 4.752	26.50 ± 4.335
	Student	49 (20.33)	25.27 ± 4.177	27.55 ± 4.026	24.98 ± 4.023
	Housewife	15 (6.23)	25.33 ± 5.164	26.93 ± 3.283	26.00 ± 4.408
p-value			0.395	0.983	0.295
Qualification	Diploma or lower	26 (10.78)	24.23 ± 4.320	24.85 ± 3.885	25.77 ± 3.892
	Associate	59 (24.48)	26.61 ± 4.222	27.56 ± 4.427	26.61 ± 4.874
	Bachelor	78 (32.37)	26.49 ± 4.339	27.21 ± 4.145	25.13 ± 4.123
	Master's degree or higher	78 (32.37)	25.77 ± 4.899	28.15 ± 4.475	26.79 ± 4.329
p-value			0.105	0.009[a]	0.081
History of background diseases	None	158 (65.56)	25.82 ± 4.391	27.23 ± 4.530	26.08 ± 4.382
	Cardiovascular	10 (4.15)	25.00 ± 4.447	27.60 ± 4.195	25.60 ± 3.748
	Diabetic	16 (6.64)	25.88 ± 4.815	27.63 ± 5.018	25.75 ± 4.553
	Hypertension	21 (8.72)	26.29 ± 3.481	27.81 ± 4.600	26.57 ± 4.106
	Allergy	20 (8.30)	27.30 ± 4.911	27.80 ± 3.302	25.90 ± 4.564
	Chronic kidney	9 (3.73)	27.33 ± 6.164	26.22 ± 2.728	26.89 ± 5.110
	Chronic liver	7 (2.90)	26.86 ± 7.010	27.71 ± 4.536	26.29 ± 6.157
p-value			0.751	0.972	0.992

[a]Statistically significant

Table 22.2 Prevalence and score severity ratings of depression, anxiety, and stress among community population ($n = 241$)

Psychological variable		Frequency	%
Depression	Moderate	25	10.38
	Severe	124	51.45
	Extremely severe	92	38.17
	Mean ± SD	26.09 ± 4.39	
Anxiety	Severe	231	95.9
	Extremely severe	10	4.1
	Mean ± SD	26.04 ± 4.53	
Stress	Mild	5	2.08
	Moderate	106	43.98
	Severe	118	48.97
	Extremely severe	12	4.97
	Mean ± SD	27.34 ± 4.37	

95.9% of the respondents experienced severe anxiety, and around 90% of participants reported depressive symptoms. Furthermore, more than half of the participants had severe or extremely severe stress.

In comparison with the results from a recent study in China which showed that 35% of the respondents experienced psychological distress during the COVID-19 outbreak there, more than half of our participants reported severe or extremely severe psychological distress [12]. Another study reported that SARS survivors experienced similar psychological distress, with anxiety and depressive features occurring in 52.2 and 45.4% of the subjects, respectively [13]. Thus, the observed public fear and anxiety are an expected consequence of COVID-19 pandemic [14].

The results of the current study demonstrated that there were no significant differences between demographic characteristics and psychological distress apart from the level of education. Our findings suggest that the mean stress subscale was significantly different among "educational qualification" levels, such that individuals with a higher educational degree (e.g., PhD or Master's degree) experienced higher levels of stress. This is in concordance with the other two other studies which showed that people with a higher education experience more distress, potentially due to increased self-awareness of their own health as well as other impacts of the virus on the economy, healthcare systems, and society in general [12, 15].

In accordance with the results of other research studies, female respondents showed significantly higher psychological distress (depression) than their male counterparts [12]. This is similar to the results from previous research which concluded that women are more vulnerable to stress and are more likely to develop post-traumatic stress disorder [16].

In our study, there were no differences between age of participants and the psychological stress levels observed. However, another study showed that young adults older than 60 years had the highest distress scores [12, 17]. Also, we showed that there was no significant difference between the presence of various background diseases and psychological distress. To date, the evidence suggests that the two groups of the community who are at a higher risk of getting severe COVID-19 disease are older people (over 60 years old) and those with underlying chronic diseases (diabetes, chronic respiratory disease, cardiovascular disease, and cancer) [18].

The unpredictable nature of the COVID-19 epidemic has been stimulated by myths and inaccurate information, often driven by incorrect news reporting and misunderstanding of public well-being messages, causing anxiety in the community [19]. This suggests that timely mental healthcare needs to be developed specifically for this situation and to help prepare in case of a second wave of the virus. Public health interventions should be based on a comprehensive assessment of risk factors leading to psychological issues such as the elevation in depressive anxiety and stress-related symptoms seen in this study.

Various countries have implemented different programs, strategies, and protocols for overcoming COVID-19-related psychological distress [20]. In line with this, the Ministry of Health in Iran has instituted various applied management models for overcoming this crisis such as using the capabilities of social media and television for public education (e.g., providing educational clips with more attention to vulnerable groups such as the young, the elderly, women, and migrant workers), training in the use of personal protective equipment, reducing gatherings with the campaign "stay at home" to prevent spreading of the infection, offering training for access to medical resources and the public health service system, and providing governmental financial support for the vulnerable population. There is also a movement toward increased screening, referral, and targeted intervention for reducing psychological distress to prevent further mental health problems. Some of the recommendations to aid in this include ensuring that sources of information regarding the COVID-19 situation are reliable; maintaining contact with family, friends, and colleagues; and seeking help as needed.

In this study, we could not assess the various factors affecting the observed psychological distress, and factors such as history of mental disease were self-reported. It should also be acknowledged that the questionnaire used in this study was optimized for use in Iran and may therefore not be generalizable to other cultures. Finally, these assessments were carried out based on reported symptoms only. We suggest that accuracy could be increased through combined assessment of easily accessible molecular biomarkers. For example, a study in 1999 showed that evening salivary cortisol levels are associated with anxiety, depressiveness, and post-traumatic avoidance [21]. Another study showed that salivary amylase levels could be useful for assessment of individuals working in a stressful and isolated environment [22]. A study showed that increased circulating biomarkers of inflammation such as high-sensitivity C-reactive protein, pro-inflammatory cytokines, and decreased vitamin D levels are associated with post-stroke depression [23]. In line with this, a systematic review found that some circulating inflammatory biomarkers such as interleukin (IL)-6, IL-1β, and IL-5 may be useful for identification of individuals with panic disorder [24]. Finally a number of meta-analyses have confirmed that circulating levels of brain-derived neurotrophic factor (BDNF) are correlated with the course of schizophrenia and depressive disorders [25]. These studies illustrate the connection between the mind and body in the maintenance of physiological homeostasis and mental well-being.

22.5 Conclusions

The results of the study show that the community in Tehran is experiencing severe and extremely severe psychological burdens due to the COVID-19 outbreak. Given that the situation is still ongoing, new mental health intervention policies are urgently needed to help individuals cope. Just as it is important to test for the virus, we also recommend testing for detection of changes in psychological symptoms. This may lead to development of an algorithm which incorporates both symptoms and molecular biomarkers to aid in selection of the most appropriate therapeutic response.

Acknowledgments Thanks to guidance and advice from the Clinical Research Development Unit of Baqiyatallah Hospital.

Conflict of Interest The authors declare no conflict of interest with respect to the authorship and publication of this article.

References

1. Xiang YT, Yang Y, Li W, Zhang L, Zhang Q, Cheung T et al (2020) Timely mental health care for the 2019 novel coronavirus outbreak is urgently needed. Lancet Psychiatry 7(3):228–229
2. Fedson DS (2015) A practical treatment for patients with Ebola virus disease. J Infect Dis 211(4):661–662
3. Brooks SK, Webster RK, Smith LE, Woodland L, Wessely S, Greenberg N et al (2020) The psychological impact of quarantine and how to reduce it: rapid review of the evidence. Lancet 395(10227):912–920
4. Shigemura J, Ursano RJ, Morganstein JC, Kurosawa M, Benedek DM (2020) Public responses to the novel 2019 coronavirus (2019-nCoV) in Japan: mental health consequences and target populations. Psychiatry Clin Neurosci 74(4):281–282
5. Jeong H, Yim HW, Song YJ, Ki M, Min JA, Cho J et al (2016) Mental health status of people isolated due to Middle East Respiratory Syndrome. Epidemiol Health 38:e2016048. https://doi.org/10.4178/epih.e2016048
6. Chua SE, Cheung V, Cheung C, McAlonan GM, Wong JW, Cheung EP et al (2004) Psychological effects of the SARS outbreak in Hong Kong on high-risk health care workers. Can J Psychiatr 49(6):391–393
7. Han KM, Kim KH, Lee M, Lee SM, Ko YH, Paik JW (2017) Increase in the prescription rate of antidepressants after the Sewol Ferry disaster in Ansan, South Korea. J Affect Disord 219:31–36
8. Kasiulevičius V, Šapoka V, Filipavičiūtė R (2006) Sample size calculation in epidemiological studies. Geron 7(4):225–231
9. Lovibond PF (1998) Long-term stability of depression, anxiety, and stress syndromes. J Abnorm Psychol 107(3):520–526
10. Sahebi A, Asghari MJ, Salari R (2005) Validation of depression anxiety and stress scale (DASS-21) for an Iranian population. J Dev Psychol http://jip.azad.ac.ir/article_512443_en.html
11. Chan RC, Xu T, Huang J, Wang Y, Zhao Q, Shum DH et al (2012) Extending the utility of the depression anxiety stress scale by examining its psychometric properties in Chinese settings. Psychiatry Res 200(2–3):879–883
12. Qiu J, Shen B, Zhao M, Wang Z, Xie B, Xu Y (2020) A nationwide survey of psychological distress among Chinese people in the COVID-19 epidemic: implications and policy recommendations. Gen Psychiatry 33(2):e100213. https://doi.org/10.1136/gpsych-2020-100213
13. Cheng SK, Wong C, Tsang J, Wong K (2004) Psychological distress and negative appraisals in survivors of severe acute respiratory syndrome (SARS). Psychol Med 34(7):1187–1195
14. Rubin GJ, Wessely S (2020) The psychological effects of quarantining a city. BMJ 368:m313. https://doi.org/10.1136/bmj.m313
15. Roberts T, Esponda GM, Krupchanka D, Shidhaye R, Patel V, Rathod S (2018) Factors associated with health service utilisation for common mental disorders: a systematic review. BMC Psychiatry 18(1):262
16. Sareen J, Erickson J, Medved MI, Asmundson GJ, Enns MW, Stein M et al (2013) Risk factors for post-injury mental health problems. Depress Anxiety 30(4):321–327
17. Cheng C, Huang J, Baoyong L (2014) Psychological health diathesis assessment system: a nationwide survey of resilient trait scale for Chinese adults. Stud Psychol Behav 12:735–742
18. World Health Organization (2020) Coronavirus disease 2019 (COVID-19): situation report, 51. https://www.who.int/docs/default-source/coronaviruse/situation-reports/20200311-sitrep-51-covid-19.pdf?sfvrsn=1ba62e57_10
19. Bao Y, Sun Y, Meng S, Shi J, Lu L (2020) 2019-nCoV epidemic: address mental health care to empower society. Lancet 395(10224):e37–e38
20. Duan L, Zhu G (2020) Psychological interventions for people affected by the COVID-19 epidemic. Lancet Psychiatry 7(4):300–302
21. Aardal-Eriksson E, Eriksson TE, Holm AC, Lundin T (1999) Salivary cortisol and serum prolactin in relation to stress rating scales in a group of rescue workers. Biol Psychiatry 46(6):850–855
22. Rai B, Kaur J, Foing BH (2012) Salivary amylase and stress during stressful environment: three Mars analog mission crews study. Neurosci Lett 518(1):23–26
23. Levada OA, Troyan AS (2018) Poststroke depression biomarkers: a narrative review. Front Neurol 9:577. https://doi.org/10.3389/fneur.2018.00577
24. Quagliato LA, Nardi AE (2018) Cytokine alterations in panic disorder: a systematic review. J Affect Disord 228:91–96
25. Peng S, Li W, Lv L, Zhang Z, Zhan X (2018) BDNF as a biomarker in diagnosis and evaluation of treatment for schizophrenia and depression. Discov Med 26(143):127–136

Gender Susceptibility to COVID-19 Mortality: Androgens as the Usual Suspects?

23

Amir Vahedian-Azimi,
Mohamad Amin Pourhoseingholi, Mohsen Saberi,
Behzad Behnam, and Amirhossein Sahebkar

Abstract

Identification of the causal risk factors of COVID-19 would allow better risk stratification and designing effective therapies. Epidemiological data have shown a higher incidence and mortality of COVID-19 in males compared to females. Here, we have used logistic regression analysis modeling to determine the association between gender and COVID-19 mortality in the Iranian population. The records of 2293 patients with COVID-19 infection were analyzed. The odds of death due to COVID-19 were 1.7 times higher in males compared to females after adjustment for age and background diseases. The gender difference was mainly observed at higher ages, suggesting an adjusted 2.32-fold higher risk of mortality in males aged >59.5 years old compared to females within the same age group. This finding suggests the male gender is a potential predisposing factor for mortality due to COVID-19 infection. The potential role of male hormones, particularly testosterone, as therapeutic targets deserves further investigation.

A. Vahedian-Azimi
Trauma Research Center, Nursing Faculty, Baqiyatallah University of Medical Sciences, Tehran, Iran

M. A. Pourhoseingholi (✉)
Gastroenterology and Liver Diseases Research Center, Research Institute for Gastroenterology and Liver Diseases, Shahid Beheshti University of Medical Sciences, Tehran, Iran
e-mail: aminphg@gmail.com

M. Saberi
Medicine, Quran and Hadith Research Center & Department of Community Medicine, Faculty of Medicine, Baqiyatallah University of Medical Sciences, Tehran, Iran

B. Behnam
Pharmaceutics Research Center, Institute of Neuropharmacology, Kerman University of Medical Sciences, Kerman, Iran

Herbal and Traditional Medicines Research Center, Kerman University of Medical Sciences, Kerman, Iran

A. Sahebkar (✉)
Neurogenic Inflammation Research Center, Mashhad University of Medical Sciences, Mashhad, Iran

Biotechnology Research Center, Pharmaceutical Technology Institute, Mashhad University of Medical Sciences, Mashhad, Iran

Polish Mother's Memorial Hospital Research Institute (PMMHRI), Lodz, Poland

Halal Research Center of IRI, FDA, Tehran, Iran
e-mail: sahebkara@mums.ac.ir; amir_saheb2000@yahoo.com

P. C. Guest (ed.), *Clinical, Biological and Molecular Aspects of COVID-19*, Advances in Experimental Medicine and Biology 1321, https://doi.org/10.1007/978-3-030-59261-5_23

Keywords

Coronavirus · COVID-19 · Sex · Mortality · Testosterone · Estrogen · Iran

23.1 Introduction

The rapidly progressing outbreak of coronavirus 2019 (COVID-19) disease has called for attempts to better identify and control the causal risk factors. Consistent with the previous coronavirus outbreaks, severe acute respiratory syndrome (SARS) and Middle East respiratory syndrome (MERS) [1, 2], there is a male preponderance in COVID-19 cases. In addition, the case fatality rate (CFR) is significantly higher in males compared to females, according to reports from China [3] and unpublished reports from Italy and the USA. This suggests that males may contract a more severe form of the disease. Iran is also among the countries facing a high disease burden with COVID-19 cases, yet information on the gender difference in mortality rates has not been reported there. Here, we used logistic regression analysis modeling to estimate the crude and adjusted association between gender and mortality in the Iranian population.

23.2 Methods

The records of 2293 patients (mean age 53.7 ± 14.6 years), who were hospitalized in the Baqiyatallah University of Medical Sciences (Tehran, Iran) from February 17 to March 30, 2020, were entered in this analysis. All patients had a diagnosis based by polymerase chain reaction (PCR) or computerized tomography (CT) findings, and men constituted 71.1% of the population group. Out of the 2293 patients, 97 (4.2%) died and the rest of recovered or partially recovered up to March 30, 2020. The most frequent comorbid diseases were chronic heart disease (7.9%) followed by hypertension (6.9%), diabetes (6.4%), cancer (6.3%), chronic lung disease (5.5%), and other chronic diseases (3.1%). Out of the total patient group, 1463 (63.8%) reported no background diseases.

23.3 Results

The results of logistic regression analysis indicated that the odds of death due to COVID-19 infection was 1.7 times higher in males compared to females after adjustment for age and background diseases (Table 23.1). At a defined cutoff of 59.5 years old (selected from a AUC-ROC analysis to predict mortality by age with a sensitivity of 65% and specificity of 68% in the total population), the risk of death for males was 2.32 times higher than that of females at age >59.5 years old, while the between-gender difference was not significant at <59.5 years old. Selecting a cutoff value of 48.6 years old as the average menopause age in the Iranian women, postmenopausal women were found to have significantly higher odds of death compared with premenopausal women. However, applying the same age cutoff for men showed an even greater mortality rate (Table 23.1).

Table 23.1 Association between gender and mortality due to COVID-19 (male vs. female as the reference group)

	Crude model		Adjusted model[a]	
Male vs. female	OR (95% CI)	P-value	OR (95% CI)	P-value
All age groups	1.25 (0.78–1.99)	0.35	1.71 (1.04–2.81)	0.036[a]
Age <59.5 years	0.86 (0.41–1.82)	0.69	1.02 (0.46–2.26)	0.95[a]
Age >59.5 years	1.88 (1.02–3.47)	0.044	2.32 (1.22–4.40)	0.01[a]
All men vs. women <48.6 years	4.12 (1.01–16.95)	0.049	1.39 (0.31–6.12)	0.66[a]
All men vs. women >48.6 years	0.98 (0.61–1.61)	0.95	1.73 (1.03–2.91)	0.039[a]
Age >48.6 vs. <48 years (for women)	4.18 (0.97–17.96)	0.045	4.65 (1.01–21.44)	0.048[b]
Age >48.6 vs. <48 years (for men)	8.43 (3.38–21.03)	<0.001	8.47 (3.39–21.16)	<0.001[b]

[a]Adjusted for age and background diseases
[b]Adjusted for background diseases

23.4 Discussion

The reason behind the observed gender difference is yet to be discovered. However, we elaborate on some lines of evidence which are suggestive of a potential role for sex hormones. The angiotensin-converting enzyme 2 (ACE2) has a key role in the entry of coronavirus particles into the cells and thus in the pathogenesis of the disease. On the other hand, ACE2 reduces angiotensin 2 levels and has a protective effect on the cardiopulmonary system [4]. Therefore, careful evaluation of this double-edged sword among the sexes and age groups can be helpful. There is evidence indicating a higher rate of ACE2 expression in males [5], which can serve as a susceptibility factor for the viral infection. However, the sex-based pattern for ACE2 is still controversial. A recent study using single-cell RNA expression profiling showed that ACE2 is expressed in higher rates in Asian male populations [5]. On the other hand, another recent report (pre-print) showed no effect of gender or age on the ACE2 expression levels in the lung. Surprisingly, another other factor, cigarette smoking, might be more important as it can significantly increase the expression of ACE2 in the lungs [6]. The data for cigarette smoking was not available for our report, although a previous meta-analysis showed a six times higher rate of smoking among Iranian men compared with women [7].

A plausible explanation for the higher rate of ACE2 expression in males could be the action of testosterone [5]. Testosterone can also reduce the immune system responses, while estrogens exert enhancing actions [6, 7]. Estrogens have also been shown to protect against adverse outcomes in SARS [8]. This is consistent with the reports suggesting that the immune response to microbial and viral infections is more efficient in females. Moreover, women elicit a more efficient immune response to influenza vaccination compared with men. The higher testosterone levels in males are predictive of a lower response to vaccination [9]. The male hormone-based pattern of COVID-19 mortality can also explain the extremely lower death rate in children and adolescent groups [10, 11], who have naturally lower levels of testosterone. While epidemiological findings on the gender susceptibility of COVID-19 mortality still need to evolve, early findings might point to a plausible role for sex hormones, particularly testosterone, as a predisposing factor for adverse outcomes. Further investigations on the possible therapeutic impact of intervening androgen and androgen receptor pathways are encouraged (Fig. 23.1).

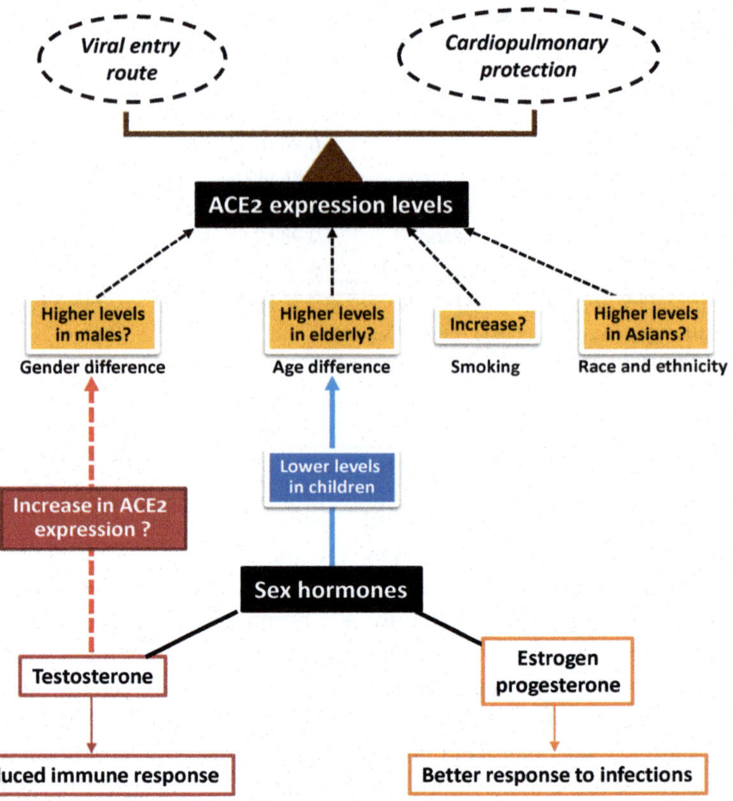

Fig. 23.1 Possible role of sex hormones in COVID-19 infection and several unanswered questions

Acknowledgments Thanks to guidance and advice from the Clinical Research Development Unit of Baqiyatallah Hospital.

References

1. Karlberg J, Chong D, Lai W (2004) Do men have a higher case fatality rate of severe acute respiratory syndrome than women do? Am J Epidemiol 159(3):229–231
2. Leung GM, Hedley AJ, Ho LM, Chau P, Wong IO, Thach TQ et al (2004) The epidemiology of severe acute respiratory syndrome in the 2003 Hong Kong epidemic: an analysis of all 1755 patients. Ann Intern Med 141(9):662–673
3. Epidemiology Working Group for NCIP Epidemic Response, Chinese Center for Disease Control and Prevention (2020) The epidemiological characteristics of an outbreak of 2019 novel coronavirus diseases (COVID-19) in China. Zhonghua Liu Xing Bing Xue Za Zhi 41(2):145–151
4. Shenoy V, Ferreira AJ, Qi Y, Fraga-Silva RA, Díez-Freire C, Dooies A et al (2010) The angiotensin-converting enzyme 2/angiogenesis-(1–7)/Mas axis confers cardiopulmonary protection against lung fibrosis and pulmonary hypertension. Am J Respir Crit Care Med 182(8):1065–1072
5. Dalpiaz P, Lamas A, Caliman I, Ribeiro R Jr, Abreu G, Moyses M et al (2015) Sex hormones promote opposite effects on ACE and ACE2 activity, hypertrophy and cardiac contractility in spontaneously hypertensive rats. PLoS One 0(5):e0127515. https://doi.org/10.1371/journal.pone.0127515
6. Channappanavar R, Fett C, Mack M, Ten Eyck PP, Meyerholz DK, Perlman S (2017) Sex-based differences in susceptibility to severe acute respiratory syndrome coronavirus infection. J Immunol 198(10):4046–4053
7. Oertelt-Prigione S (2012) The influence of sex and gender on the immune response. Autoimmun Rev 11(6–7):A479–A485
8. Klein SL, Flanagan KL (2016) Sex differences in immune responses. Nat Rev Immunol 16(10):626–638
9. Furman D, Hejblum BP, Simon N, Jojic V, Dekker CL, Thiebaut R et al (2014) Systems analysis of sex differences reveals an immunosuppressive role for testosterone in the response to influenza vaccination. Proc Natl Acad Sci U S A 111(2):869–874
10. Sinha IP, Harwood R, Semple MG, Hawcutt DB, Thursfield R, Narayan O et al (2020) COVID-19 infection in children. Lancet Respir Med 8(5):446–447
11. Livingston E, Bucher K (2020) Coronavirus disease 2019 (COVID-19) in Italy. JAMA. https://doi.org/10.1001/jama.2020.4344. Online ahead of print

24

Identification, Monitoring, and Prediction of Disease Severity in Patients with COVID-19 Pneumonia Based on Chest Computed Tomography Scans: A Retrospective Study

Ramezan Jafari, Sara Ashtari, Mohamad Amin Pourhoseingholi, Houshyar Maghsoudi, Fatemeh Cheraghalipoor, Nematollah Jonaidi Jafari, Hassan Saadat, Farshid Rahimi-Bashar, Amir Vahedian-Azimi, and Amirhossein Sahebkar

Abstract

Background and Aims
Non-contrast chest computed tomography (CT) scans can accurately evaluate the type and extent of lung lesions. The aim of this study was to investigate the chest CT features associated with critical and non-critical patients with coronavirus disease 2019 (COVID-19).

Methods
A total of 1078 patients with COVID-19 pneumonia who underwent chest CT scans, includ-

R. Jafari · H. Maghsoudi · F. Cheraghalipoor
Department of Radiology, Health Research Center, Life Style Institute, Baqiyatallah University of Medical Sciences, Tehran, Iran

S. Ashtari · M. A. Pourhoseingholi
Gastroenterology and Liver Diseases Research Center,Research Institute for Gastroenterology and Liver Diseases, Shahid Beheshti University of Medical Sciences, Tehran, Iran

N. J. Jafari
Health Research Center, Life Style Institute, Baqiyatallah University of Medical Sciences, Tehran, Iran

H. Saadat
Behavioral Sciences Research Center, Life Style Institute, Baqiyatallah University of Medical Sciences, Tehran, Iran

F. Rahimi-Bashar
Anesthesia and Critical Care Department, Hamadan University of Medical Sciences, Hamadan, Iran

A. Vahedian-Azimi (✉)
Trauma Research Center, Nursing Faculty, Baqiyatallah University of Medical Sciences, Tehran, Iran
e-mail: amirvahedian63@gmail.com

A. Sahebkar (✉)
Neurogenic Inflammation Research Center, Mashhad University of Medical Sciences, Mashhad, Iran

Biotechnology Research Center, Pharmaceutical Technology Institute, Mashhad University of Medical Sciences, Mashhad, Iran

Polish Mother's Memorial Hospital Research Institute (PMMHRI), Lodz, Poland

Halal Research Center of IRI, FDA, Tehran, Iran
e-mail: sahebkara@mums.ac.ir; amir_saheb2000@yahoo.com

P. C. Guest (ed.), *Clinical, Biological and Molecular Aspects of COVID-19*, Advances in Experimental Medicine and Biology 1321, https://doi.org/10.1007/978-3-030-59261-5_24

ing 169 critical cases and 909 non-critical cases, were enrolled in this retrospective study. The scans of all participants were reviewed and compared in two groups of study. In addition, the risk factors associated with disease in critical and non-critical patients were analyzed.

Results

Chest CT scans showed bilateral and multifocal involvement in most (86.4%) of the participants, with 97.6 and 84.3% reported in critical and non-critical patients, respectively. The incidences of pure consolidation (p = 0.019), mixed ground-glass opacities (GGOs) and consolidation ($p < 0.001$), pleural effusion ($p < 0.001$), and intralesional traction bronchiectasis ($p = 0.007$) were significantly higher in critical compared to non-critical patients. However, non-critical patients showed higher incidence of pure GGOs than the critical patients ($p < 0.001$). Finally, the total opacity scores of the critical patients were significantly higher than those of non-critical patients (13.71 ± 6.26 versus 4.86 ± 3.52, $p < 0.001$), with an area under the curve of 0.91 (0.88–0.94) for COVID-19 detection.

Conclusions

Our results revealed that the chest CT examination was an effective means of detecting pulmonary parenchymal abnormalities in the natural course of COVID-19. It can distinguish the critical patients from the non-critical patients (AUC = 0.91), which is helpful for the judgment of clinical condition and has important clinical value for the diagnosis and follow-up of COVID-19 pneumonia.

Keywords

COVID-2019 · Chest CT scan · Risk factor · Prediction · Prognosis · Critical · Non-critical

24.1 Introduction

The 2019 novel coronavirus disease (COVID-19) has spread rapidly throughout the world from its initial detection in China [1]. The disease is caused by the severe acute respiratory syndrome coronavirus 2 (SARS-CoV-2) [2]. The global pandemic of COVID-19 infection has emerged as a highly pathological and widespread virus in humans as a serious disease in the field of public health and global concern with high morbidity and mortality rates [3]. There were 2,830,289 confirmed cases, 197,263 deaths, and 801,917 recovered cases in 210 countries on April 24, 2020 [4]. The first COVID-19 patient was identified in Iran on February 19, 2020. Iran was the first Middle East country to report a death due to COVID-19. On April 24, 2020, Iran has the eighth-highest number of COVID-19 cases in the world with 88,194 confirmed cases, after the USA, Spain, Italy, France, Germany, the UK, and Turkey [4].

Accurate diagnostic tests are essential for rapid diagnosis, isolation, and treatment of infected patients with COVID-19. According to the Centers for Disease Control and Prevention (CDC), the COVID-19-related laboratory test kits mainly depend on the reverse transcription polymerase chain reaction (RT-PCR) technique, for detection of the viral RNA [5]. One of the disadvantages of this approach is the possibility of false-positive and false-negative results due to laboratory errors or indefinable viral material in the specimens [6, 7]. However, studies have demonstrated that X-ray and computed tomography (CT) imaging are helpful for suspected COVID-19 pneumonia cases with negative RT-PCR readings [8, 9].

Non-contrast chest CT has been shown to be an effective tool in detection, quantification, and follow-up of disease. According to recent studies, the lungs of patients with COVID-19 symptoms have certain visual hallmarks, such as ground-glass opacities (GGOs), and areas of increased lung density known as consolidations [10, 11]. These features became more frequent and more likely to spread across both lungs the longer a person is infected [12]. Bernheim et al. examined 121 chest CT studies from four centers in China that were obtained in the early, intermediate, and late stages of infection [13]. They also described GGOs as characteristic of the disease, particularly bilateral and peripheral GGOs and consolidative pulmonary opacities. They noted greater severity of disease with increasing time from onset of symptoms, and they described later signs of disease to include increased lung involvement, linear opacities, a "crazy-paving," pattern and "reverse halo" signs. In addition, this study showed that there was bilateral lung involvement in 28% of early patients, 76% of intermediate patients, and 88% of patients in the late stages of infection.

As of April 24, the publications on this topic were exclusively from China, and it remains unknown how COVID-19 pneumonia appears on chest radiographic and CT images of patients in other countries. Therefore, we conducted this retrospective study to investigate the association between chest CT findings in critical and non-critical patients with the COVID-19 pneumonia in Iran.

24.2 Material and Methods

24.2.1 Study Design and Participants

Patients with COVID-19 who underwent chest CT in Baqiyatallah Hospital, Tehran, Iran, from March to April 2020 were enrolled in the study. In all cases, the diagnosis of COVID-19 infection was confirmed by positive RT-PCR test of specimens collected on nasopharyngeal swabs [14]. A total of 1078 patients with confirmed COVID-19 infection were eligible to be enrolled in this study, and demographic (sex and age) characteristics and chest CT data were collected and evaluated. Based on clinical follow-up data, the participants were classified into either critical and non-critical groups. Critical patients (n = 169) were those who were admitted to the intensive care unit (ICU) (n = 65) or died (n = 104). Non-critical patients (n = 909) were those who were admitted to the routine hospital ward (n = 55) or discharged (n = 854). This retrospective study was approved by the Ethics Committee of Baqiyatallah University of Medical Sciences, Tehran, Iran (code:IR.BMSU.REC.1399.024), and patients were enrolled after giving written informed consent.

24.2.2 CT Examination and Image Analysis

Chest CT, especially high-resolution CT (HRCT), can detect small areas of GGOs [5] and is therefore a promising imaging tool for monitoring the disease. It is common practice for radiologists to evaluate the pneumonia severity qualitatively or semiquantitatively by visual scoring (6). All CT scan examinations were obtained using an Optima 16-slice detector CT scanner (General Electric Company; Boston, MA, USA). The detailed parameters for CT acquisition for COVID-19 evaluation using a low-dose thoracic CT scan protocol were as follows: tube voltage, 100 kVp, 120 mA; slice thickness, 2.5 mm; reconstruction interval, 1.25 mm; pitch, 1.75, speed, 35 mm/rot; detector configuration, 16*1.25; and CT dose index, 3.5 mGy.

Non-contrast chest CT scan images were obtained with the patient in the supine position at full inspiration. Two chest radiologists with 12 and 10 years of experience who were blinded to the clinical data evaluated the CT findings and came to a consensus conclusion. For all patients, the initial chest CT images were evaluated for the following characteristics based on the Fleischner Society Nomenclature recommendations and similar studies [15, 16]: GGO, consolidation, mixed GGO and consolidation, reversed halo, intralesional traction bronchiectasis, crazy-paving pattern, linear opacities, lymph node enlargement, pleural effusion, and pericardial effusion. To quantify the extent of lesions, a thin-section CT involvement score was assigned on the basis of all abnormal areas involved [17]. Each of the five lobes of the lungs was assigned a score according to the following: 0, 0% involvement; 1, <5% involvement; 2, 5–25% involvement; 3, 26–49% involvement; 4, 50–75% involvement; and 5, >75% involvement. Therefore, there were possible scores of 0–5 for each lobe, with a total possible score of 0–25 for the five combined lobes.

24.2.3 Statistical Analysis

The data obtained were analyzed using the Statistical Package for the Social Sciences (SPSS) (v 21.0; IBM SPSS Statistics; Chicago, IL, USA). Categorical variables were expressed as counts (%), and continuous variables as mean ± standard deviation (SD). The difference between demographic characteristics and chest CT findings in the two groups of study was assessed by t-test for continuous variables. The Chi-square test and Fisher's exact test (in case of low sample numbers) were used to compare the distribution of categorical data. Receiver operating characteristic area under the curve (ROC-AUC) analysis was used to determine the value of opacity scores for distinguishing the critical and non-critical types and identifying the corresponding cutoff value. In addition, to explore the risk factors associated with critical patients with COVID-19, univariate and multivariate logistic regression models were used. To avoid over-fitting in the multivariate model, only those factors which led to p-values less than 0.2 in univariate analysis were selected. The final model was selected according to forward conditional, and $p < 0.05$ (two-sided) was considered statistically significant.

24.3 Results

24.3.1 Demographic Characteristics and Chest CT Findings

A total of 1078 patients with COVID-19 who underwent chest CT, including 169 critical and 909 non-critical cases, were enrolled in this study. Out of all participants, 737 (68.4%) were male, 341 (31.6) were female, and the mean age was 53 ± 14.37 years (range, 14–92 years). The demographic characteristics and chest CT findings of the patients in two groups of study are shown in Table 24.1. The mean age in the critical group was significantly higher than the non-critical group (61.24 ± 13.48 vs. 51.47 ± 14.02, $p < 0.001$), but there was no statistical difference between the two groups in terms of gender ($p = 0.179$). Chest CT scans showed bilateral and multifocal involvement in most individuals (86.4%) of the participants, with 97.6% and 84.3% reported in critical and non-critical patients, respectively (Fig. 24.1). The common chest CT features in all participants were mixed GGOs and consolidations ($n = 489$, 45.4%), pure GGOs ($n = 414$, 38.4%), linear opacities ($n = 174$, 16.1%), mixed GGOs with crazy-paving pattern ($n = 133$, 12.3%), and pure consolidations ($n = 42$, 3.9%) (Figs. 24.2 and 24.3). The incidences of pure consolidations ($p = 0.019$), mixed GGOs and consolidations ($p < 0.001$), pleural effusions ($p < 0.001$), and intralesional traction bronchiectasis ($p = 0.007$) in critical patients were significantly higher than those of the non-critical patients. However, non-critical patients showed higher incidence of pure GGOs compared to the critical patients ($p < 0.001$). Among all participants, 181 (16.8%) had lung diffuse opacities of greater than 75% involvement in each lobe, and the presence of diffuse opacities was significantly higher in critical patients than in the non-critical ones (69.8% vs. 6.9%, $p < 0.001$) (Fig. 24.4). The number of lung lobes involved for the critical group was significantly higher than that for the non-critical group

Table 24.1 Demographic characteristics and chest CT findings in critical and non-critical patients with COVID-19

Variables	Total patients (*n* = 1078)	Critical patients (*n* = 169)	Non-critical patients (*n* = 909)	*p*-value
Age				
Mean ± SD (range)	53 ± 14.37 (14–92)	61.24 ± 13.48 (25–92)	51.47 ± 14.02 (14–91)	<0.001[a]
Gender (%)				
Male	737 (68.4)	123 (72.8)	614 (67.5)	0.179
Female	341 (31.6)	46 (27.2)	295 (32.5)	
Lesion distribution (%)				
Bilateral + multifocal	931 (86.4)	165 (97.6)	766 (84.3)	<0.001[a]
Others	147 (13.6)	4 (2.4)	143 (15.7)	
Lesion type (%)				
GGO [#]	414 (38.4)	13 (7.7)	401 (44.1)	<0.001[a]
GGO + crazy paving	133 (12.3)	19 (11.2)	114 (12.5)	0.637
Consolidation	42 (3.9)	12 (7.1)	30 (3.3)	0.019[a]
GGO + consolidation	489 (45.4)	125 (74)	364 (40)	<0.001[a]
Other findings (%)				
None	695 (64.5)	78 (46.2)	617 (67.9)	
Linear opacity	174 (16.1)	24 (14.2)	150 (16.5)	0.455
Reversed halo sign	49 (4.5)	6 (3.6)	43 (4.7)	0.499
Pleural effusion	55 (5.1)	34 (20.1)	21 (2.3)	<0.001[a]
Intralesional traction bronchiectasis	61 (5.7)	17 (10.1)	44 (4.8)	0.007[a]
Lymphadenopathy	44 (4.1)	10 (5.9)	34 (3.7)	0.189
Underlying diseases (%)				
None	1041 (96.6)	159 (94.1)	882 (97)	
Pulmonary	7 (0.6)	1 (0.6)	6 (0.7)	0.919
Cardiac	28 (2.6)	8 (4.7)	20 (2.2)	0.057
Kidney	2 (0.2)	1 (0.6)	1 (0.1)	0.289
Presence of diffuse opacity (%)				
Yes	181 (16.8)	118 (69.8)	63 (6.9)	<0.001[a]
No	897 (83.2)	51 (30.2)	846 (93.1)	
Number of involved lobe with diffuse opacities (%)				
1	6 (0.6)	1 (0.6)	5 (0.6)	1
2	49 (4.5)	33 (19.5)	10 (1.8)	<0.001[a]
3	50 (4.6)	35 (20.7)	15 (1.7)	<0.001[a]
4	43 (4)	30 (17.8)	13 (1.4)	<0.001[a]
5	33 (3.1)	19 (11.2)	14 (1.5)	<0.001[a]
Total opacity score				
Mean ± SD	6.24 ± 5.19	13.71 ± 6.26	4.86 ± 3.52	<0.001[a]

[#]*GGOs* ground-glass opacities
[a]$p < 0.05$ was statistically significant

($p < 0.001$). In addition, the total opacity scores of the critical patients were significantly higher than those of the non-critical patients (13.71 ± 6.26 vs. 4.86 ± 3.52, $p < 0.001$). We also observed reversed halo signs in a small number of patients (Fig. 24.5) in both groups and unilateral lung involvement in 13.6% of patients (Fig. 24.6).

24.3.2 Factors Associated with Critical Patients with COVID-19

Table 24.2 showed the results of univariate and multivariate logistic regression analyses in relation to critical patients with COVID-19 disease. In terms of demographic characteristics, older age was a risk factor for critical COVID-19 pneumonia according to both univariate (OR 1.052, 95% CI 1.038–1.056, $p < 0.001$) and multivariate (OR 1.031, 95% CI 1.015–1.047, $p < 0.001$) models. Imaging features associated with critical patients were pure consolidations (OR 0.379, 95% CI 0.182–0.792, $p = 0.01$); mixed GGOs with crazy-paving pattern (OR 0.193, 95% CI 0.094–0.394, $p < 0.001$); linear opacities (OR 3.65, 95% CI 1.43–9.29, $p = 0.007$); intralesional traction bronchiectasis (OR 3.55, 95% CI 1.78–15.87, $p < 0.001$); presence of diffuse opacities (score 5; OR 5.83, 95% CI 2.86–11.85, $p < 0.001$); and high opacity scores (OR 1.15, 95% CI 1.08–1.22, $p < 0.001$).

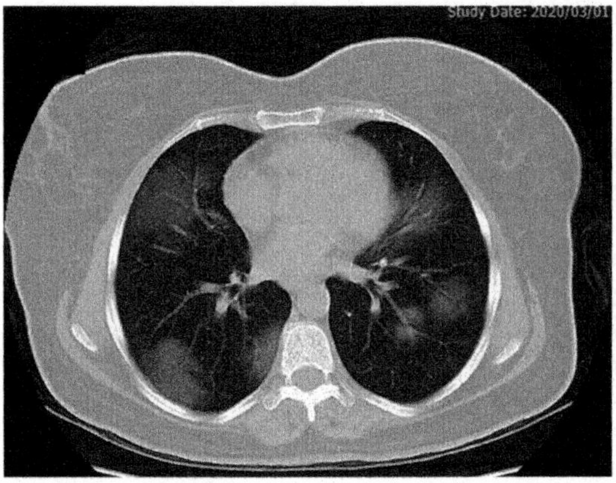

Fig. 24.1 Axial chest CT scan without contrast in a 46-year-old female with COVID-19 pneumonia shows bilateral multifocal patchy ground-glass opacities (GGOs)

24.3.3 Diagnostic Accuracy Test

The ROC-AUC of opacity scores in CT scans was 0.91 (95% CI 0.881–0.937, $p < 0.001$) for distinguishing critical from non-critical COVID-19. When the cutoff value of the opacity scores was set at 7.5, the sensitivity and specificity were 0.87 and 0.93, respectively (Fig. 24.7).

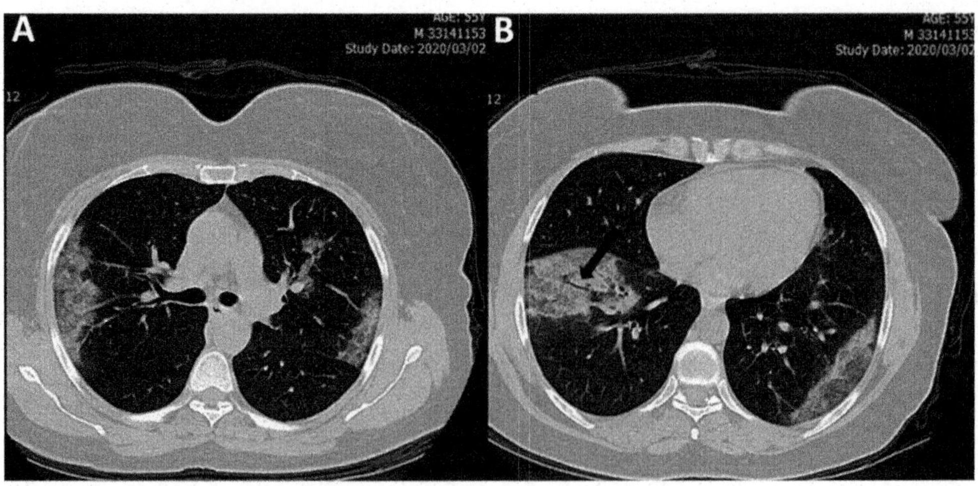

Fig. 24.2 Two axial chest CT scans without contrast in a 55-year-old female with COVID-19 pneumonia show (**a**) a bilateral multifocal and subpleural patchy crazy-paving pattern and (**b**) intralesional bronchiectasis (black arrow) with crazy-paving pattern (**b**)

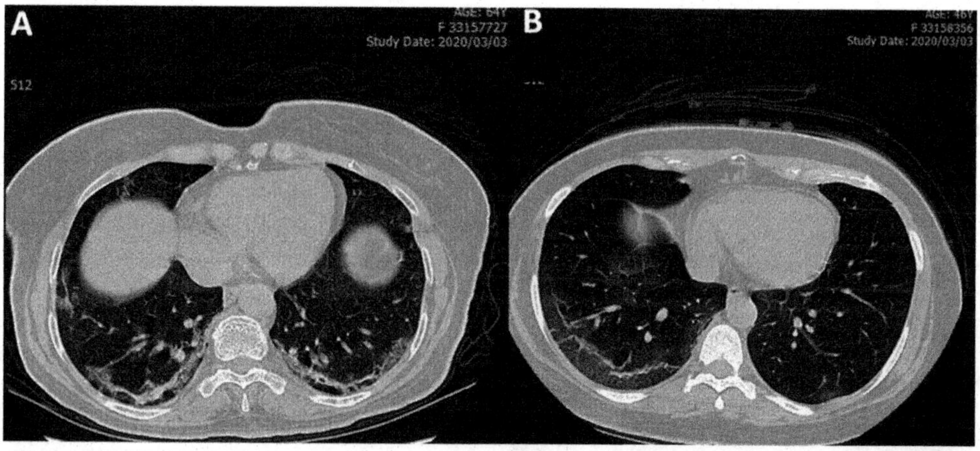

Fig. 24.3 Two axial chest CT scans without contrast in (**a**) a 64-year-old female and (**b**) a 46-year-old male with COVID-19 pneumonia subpleural linear opacities and ground-glass opacities (GGOs)

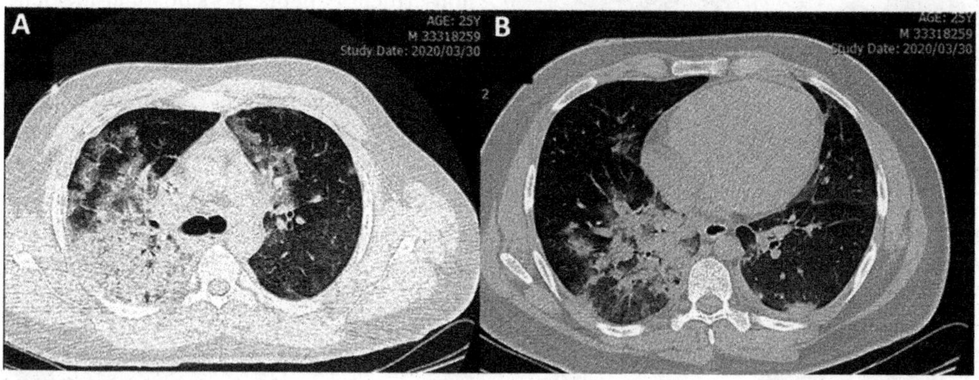

Fig. 24.4 Two axial chest CT scans without contrast in a 25-year-old male with COVID-19 pneumonia show diffuse consolidative opacity in (**a**) right upper lobe and (**b**) right lower lobe with multifocal patchy subpleural consolidative opacities

24.4 Discussion

COVID-19 can develop to severe illnesses including pneumonia, pulmonary edema, acute respiratory distress syndrome, multiple organ failure, or even death [18]. Compared with the non-critical patients, critical patients have poor prognosis and higher mortality [13, 19]. Successful treatment of critical cases is the key to reducing the complications and high mortality rates. This includes basic disease treatment, secondary infection prevention, and timely organ function support. Therefore, it is important to detect the critical cases and related factors of disease in these patients with high accuracy. Studying the characteristics of chest CT scans allows a visual assessment of the type and extent of lung lesions in patients with COVID-19 disease to deepen our understanding of the physiological pathways affected in critical conditions and to promote clinical diagnosis and appropriate tailored treatments [20]. In this large study, the chest CT features of 1078 patients. In general, this showed that chest CT examination was effective in detecting pulmonary parenchymal abnormalities in the natural course of COVID-19 disease.

Bilateral and multifocal involvements of lesions were observed in the majority of individuals in both groups of COVID-19 pneumonia. The most common types of lesions were mixed GGOs with consolidations, pure GGOs, crazy paving, linear opacities, pure consolidations, and mixed GGOs with crazy-paving pattern for both groups of study. Compared with non-critical patients, pure consolidations and mixed GGOs with consolidations were significantly more frequent in critical patients. This indicates that the alveoli are completely filled by inflammatory exudation and usually means that the virus has diffused into the respiratory epithelium, leading to necrotizing bronchitis and diffuse alveolar damage [21, 22]. Critical patients showed more pleural effusion and intralesional traction bronchiectasis compared to the non-critical patients. The extrapulmonary lesions, such as pleural or pericardial effusion, may indicate the occurrence of severe inflammation. Although linear opacities and reversed halo signs were more frequent in non-critical patients, no statistical incidence difference was observed between the two groups based on these criteria. Similar to other chest CT studies, we observed reversed halo signs in both groups [23, 24], and unilateral lung involvement was in found in a small number of patients. In a study by Zhou et al. haloes were observed only in the early stage of COVID-19 and were assumed to have progressed to GGOs within 1 week [25].

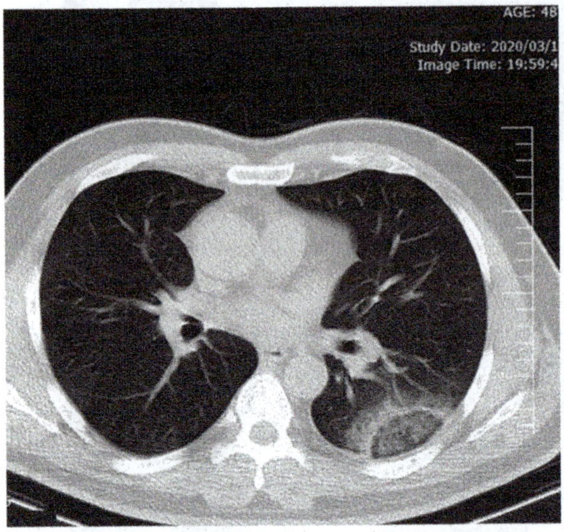

Fig. 24.5 Axial chest CT scan without contrast in a 48-year-old male with COVID-19 pneumonia shows typical reversed halo sign at left lower lobe

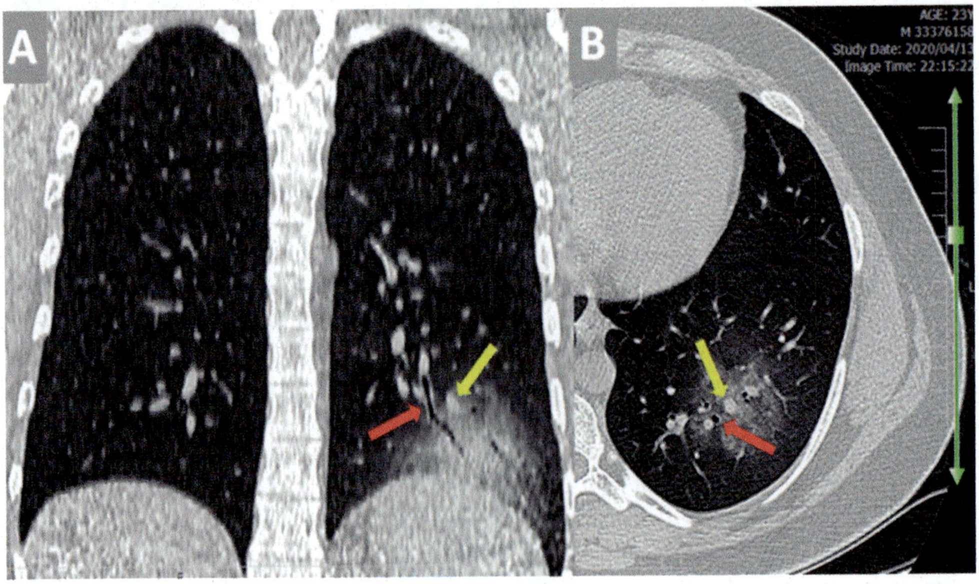

Fig. 24.6 (**a**) Coronal and (**b**) axial chest CT scan without contrast in a 23-year-old female with COVID-19 pneumonia shows unilateral and unifocal ground-glass opacities (GGOs) with intralesional bronchiectasis (red arrow) and vascular enlargement (yellow arrow)

Table 24.2 Univariate and multivariate analysis of characteristics and chest CT scan findings associated with critical and non-critical patients with COVID-19

Variable	Univariate		Multivariate	
	OR (95% CI)	*P*-value	OR (95% CI)	*p*-value
Age	1.052 (1.038–1.056)	<0.001[a]	1.031 (1.015–1.047)	<0.001[a]
Gender				
Male	1			
Female	1.28 (0.89–1.85)	0.180		
Lesion distribution				
Bilateral + multifocal	1		1	
Others	7.70 (2.81–21.1)	<0.001[a]	1.58 (0.527–4.75)	0.413
Lesion type				
Pure GGO #	1		1	
GGO + crazy-paving pattern	0.094 (0.052–0.170)	<0.001[a]	0.193 (0.094–0.394)	<0.001[a]
Pure consolidation	0.485 (0.287–0.822)	0.007[a]	0.379 (0.182–0.792)	0.010[a]
GGO + consolidation	1.165 (0.579–2.345)	0.669	2.15 (0.682–6.775)	0.191
Other findings (%)				
None	1			
Linear opacity	0.430 (0.204–0.904)	0.026[a]	3.651 (1.43–9.29)	0.007[a]
Reversed halo sign	0.544 (0.238–1.243)	0.149	3.024 (1.06–8.55)	0.080
Pleural effusion	0.474 (0.157–1.436)	0.187	2.812 (0.73–10.82)	0.133
Intralesional traction bronchiectasis	0.505 (0.259–1.841)	<0.001[a]	3.552 (1.78–15.87)	<0.001[a]
Lymphadenopathy	1.314 (0.534–3.232)	0.553	4.541 (1.39–14.83)	0.153
Underlying diseases				
None	1			
Pulmonary	0.180 (0.011–2.89)	0.227		
Cardiac	0.167 (0.005–5.45)	0.314		
Kidney	0.4 (0.022–7.201)	0.534		
Presence of diffuse opacity				
Yes	1		1	
No	31.608 (20.81–47.99)	<0.001[a]	5.83 (2.86–11.85)	<0.001[a]
Number of involved lobe with diffuse opacities (%)				
None	1		1	
1	0.044 (0.021–0.094)	<0.001[a]	0.1 (0.01–1.234)	0.073
2	0.147 (0.015–1.406)	0.096	1.22 (0.38–3.52)	0.424
3	1.520 (0.610–3.786)	0.369	1.09 (0.37–3.24)	0.868
4	1.719 (0.687–4.305)	0.247	1.41 (0.47–4.22)	0.536
5	1.700 (0.658–4.391)	0.273	1.34 (0.43–4.18)	0.606
Total opacity score	1.31 (1.26–1.36)	<0.001[a]	1.15 (1.08–1.22)	<0.001[a]

#*GGOs* ground-glass opacities
[a]$p < 0.05$ statistically significant

Potentially the most significant finding of this study was that the total opacity scores of the critical group were significantly higher than those of the non-critical group, and ROC-AUC analysis of CT opacity scores was 0.91 with a sensitivity and specificity of 0.87 and 0.93, respectively, for distinguishing between the critical and non-critical patients with COVID-19 infections. Similarly to our results, previous studies reported that chest CT images gave similar ROC-AUC, sensitivity, and specificity values for detecting the lung lesions in COVID-19 [26–28].

The main strength of the present investigation was the large sample size. However, our study had several limitations. First, because COVID-19 is a novel disease and lacks serial and long-term CT data, we can only analyze the existing information in a retrospective manner. Moreover, the

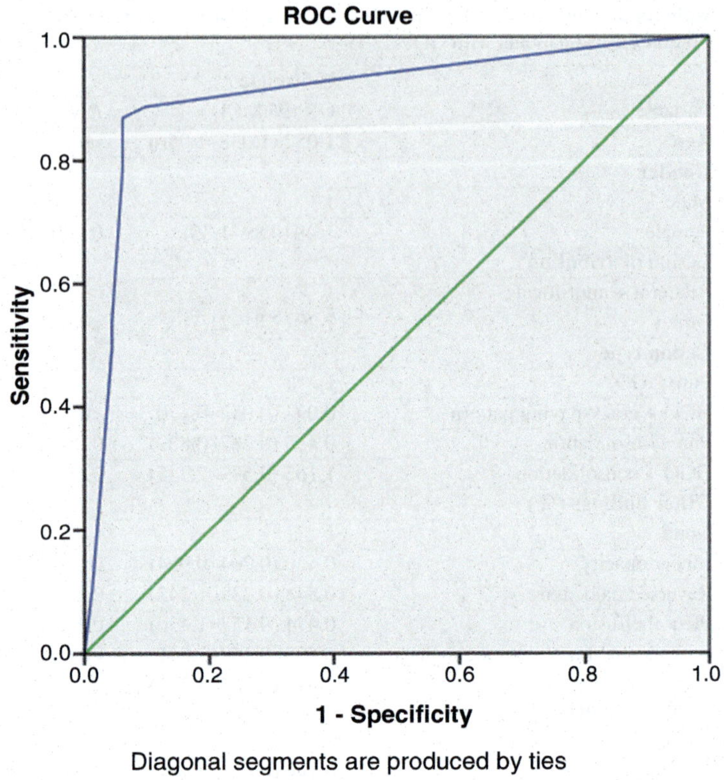

Fig. 24.7 ROC curve for opacity score AUC = 0.91 (sensitivity = 87%, specificity = 93%), cutoff value = 7.5

time of chest CT examination after symptom onset was inconsistent, so it was difficult to summarize the CT appearances that could reflect the whole course of the disease.

In conclusion, our study showed that chest CT examination was very effective in detecting pulmonary parenchymal abnormalities in the natural course of COVID-19. Mixed GGOs with consolidations and pure GGOs in the peripheral lung were the primary CT characteristics of COVID-19. Total opacity scores and the presence of diffuse opacities can accurately distinguish the critical patients from the non-critical patients, which is helpful for the judgment of clinical conditions and has important clinical value for the diagnosis and follow-up of COVID-19 pneumonia.

Acknowledgments Thanks to guidance and advice from the Clinical Research Development Unit of Baqiyatallah Hospital.

References

1. Li H, Liu SM, Yu XH, Tang SL, Tang CK (2020) Coronavirus disease 2019 (COVID-19): current status and future perspectives. Int J Antimicrob Agents 55(5):105951. https://doi.org/10.1016/j.ijantimicag.2020.105951
2. Lai CC, Shih TP, Ko WC, Tang HJ, Hsueh PR (2020) Severe acute respiratory syndrome coronavirus 2 (SARS-CoV-2) and coronavirus disease-2019 (COVID-19): the epidemic and the challenges. Int J Antimicrob Agents 55(3):105924. https://doi.org/10.1016/j.ijantimicag.2020.105924
3. Singhal T (2020) A review of coronavirus Disease-2019 (COVID-19). Indian J Pediatr 87(4):281–286
4. Johns Hopkins University & Medicine. Coronavirus Resource Center. Accessed 24 Apr 2020. https://coronavirus.jhu.edu/map.html
5. Ye Z, Zhang Y, Wang Y, Huang Z, Song B (2020) Chest CT manifestations of new coronavirus disease 2019 (COVID-19): a pictorial review. Eur Radiol 19:1–9. https://doi.org/10.1007/s00330-020-06801-0. Online ahead of print
6. Udugama B, Kadhiresan P, Kozlowski HN, Malekjahani A, Osborne M, Li VYC et al (2020) Diagnosing COVID-19: the disease and tools for detection. ACS Nano 14(4):3822–3835

7. Feng H, Liu Y, Lv M, Zhong J (2020) A case report of COVID-19 with false negative RT-PCR test: necessity of chest CT. Jpn J Radiol 38(5):409–410
8. Xie X, Zhong Z, Zhao W, Zheng C, Wang F, Liu J (2020) Chest CT for typical 2019-nCoV pneumonia: relationship to negative RT-PCR testing. Radiology 12:200343. https://doi.org/10.1148/radiol.2020200343. Online ahead of print
9. Huang P, Liu T, Huang L, Liu H, Lei M, Xu W et al (2020) Use of chest CT in combination with negative RT-PCR assay for the 2019 novel coronavirus but high clinical suspicion. Radiology 295(1):22–23
10. Zhou S, Wang Y, Zhu T, Xia L (2020) CT features of coronavirus disease 2019 (COVID-19) pneumonia in 62 patients in Wuhan, China. AJR Am J Roentgenol 214(6):1287–1294
11. Chung M, Bernheim A, Mei X, Zhang N, Huang M, Zeng X et al (2020) CT imaging features of 2019 novel coronavirus (2019-nCoV). Radiology 295(1):202–207
12. Jacobi A, Chung M, Bernheim A, Eber C (2020) Portable chest X-ray in coronavirus disease-19 (COVID-19): a pictorial review. Clin Imaging 64:35–42
13. Bernheim A, Mei X, Huang M, Yang Y, Fayad Z, Zhang N et al (2020) Chest CT findings in coronavirus Disease-19 (COVID-19): relationship to duration of infection. Radiology 295(3):200463. https://doi.org/10.1148/radiol.2020200463
14. Gorabi AM, Hajighasemi S, Kiaie N, Gheibi Hayat SM, Amialahmadi T, Johnston TP et al (2020) The pivotal role of CD69 in autoimmunity. J Autoimmun 111:102453. https://doi.org/10.1016/j.jaut.2020.102453
15. Hansell DM, Bankier AA, MacMahon H, McLoud TC, Muller NL, Remy J (2008) Fleischner society: glossary of terms for thoracic imaging. Radiology 246(3):697–722
16. Schoen K, Horvat N, Guerreiro NFC, de Castro I, de Giassi KS (2019) Spectrum of clinical and radiographic findings in patients with diagnosis of H1N1 and correlation with clinical severity. BMC Infect Dis 9(1):964. https://doi.org/10.1186/s12879-019-4592-0
17. Chang YC, Yu CJ, Chang SC, Galvin JR, Liu HM, Hsiao CH et al (2005) Pulmonary sequelae in convalescent patients after severe acute respiratory syndrome: evaluation with thin-section CT. Radiology 236(3):1067–1075
18. Chen N, Zhou M, Dong X, Qu J, Gong F, Han Y et al (2020) Epidemiological and clinical characteristics of 99 cases of 2019 novel coronavirus pneumonia in Wuhan, China: a descriptive study. Lancet 395(10223):507–513
19. Wang C, Horby PW, Hayden FG, Gao GF (2020) A novel coronavirus outbreak of global health concern. Lancet 395(10223):470–473
20. Fang Y, Zhang H, Xie J, Lin M, Ying L, Pang P et al (2020) Sensitivity of chest CT for COVID-19: comparison to RT-PCR. Radiology 19:200432. https://doi.org/10.1148/radiol.2020200432. Online ahead of print
21. Koo HJ, Lim S, Choe J, Choi SH, Sung H, Do KH (2018) Radiographic and CT features of viral pneumonia. Radiographics 38(3):719–739
22. Franquet T (2011) Imaging of pulmonary viral pneumonia. Radiology 260(1):18–39
23. Yoon SH, Lee KH, Kim JY, Lee YK, Ko H, Kim KH et al (2020) Chest radiographic and CT findings of the 2019 novel coronavirus disease (COVID-19): analysis of nine patients treated in Korea. Korean J Radiol 21(4):494–500
24. Fang Y, Zhang H, Xu Y, Xie J, Pang P, Ji W (2020) CT manifestations of two cases of 2019 novel coronavirus (2019-nCoV) pneumonia. Radiology 295(1):208–209
25. Zhou Z, Guo D, Li C, Fang Z, Chen L, Yang R et al (2020) Coronavirus disease 2019: initial chest CT findings. Eur Radiol 24:1–9. https://doi.org/10.1007/s00330-020-06816-7. Online ahead of print
26. Li K, Wu J, Wu F, Guo D, Chen L, Fang Z et al (2020) The clinical and chest CT features associated with severe and critical COVID-19 pneumonia. Investig Radiol 55(6):327–331
27. Li L, Qin L, Xu Z, Yin Y, Wang X, Kong B et al (2020) Artificial intelligence distinguishes COVID-19 from community acquired pneumonia on chest CT. Radiology 19:200905. https://doi.org/10.1148/radiol.2020200905. Online ahead of print
28. Liu KC, Xu P, Lv WF, Qiu XH, Yao JL, Gu JF et al (2020) CT manifestations of coronavirus disease-2019: a retrospective analysis of 73 cases by disease severity. Eur J Radiol 126:108941. https://doi.org/10.1016/j.ejrad.2020.108941

The Level of Procalcitonin in Severe COVID-19 Patients: A Systematic Review and Meta-Analysis

25

Farshad Heidari-Beni, Amir Vahedian-Azimi, Sajad Shojaei, Farshid Rahimi-Bashar, Alireza Shahriary, Thomas P. Johnston, and Amirhossein Sahebkar

Abstract

There is data from individual clinical trials suggesting that procalcitonin (PCT) may be a prognostic factor in the severity of COVID-19 disease. Therefore, this systematic review and meta-analysis was performed to investigate PCT levels in severe COVID-19 patients. We searched Embase, ProQuest, MEDLINE/PubMed, Scopus, and ISI/Web of Science for studies that reported the level of PCT of patient with severe COVID-19. We included all studies regardless of design that reported the level of PCT in patients with severe COVID-19. We excluded articles not regarding COVID-19 or not reporting PCT level, studies not in severe patients, review articles,

F. Heidari-Beni
Student of Nursing, Department of Medical Surgical Nursing, School of Nursing and Midwifery, Iran University of Medical Sciences, Tehran, Iran

A. Vahedian-Azimi (✉)
Trauma Research Center, Nursing Faculty, Baqiyatallah University of Medical Sciences, Tehran, Iran
e-mail: amirvahedian63@gmail.com

S. Shojaei
Basic and Molecular Epidemiology of Gastrointestinal Disorders Research Center, Research Institute for Gastroenterology and Liver Diseases, Shahid Beheshti University of Medical Sciences, Tehran, Iran

Gastroenterology and Liver Diseases Research Center, Research Institute for Gastroenterology and Liver Diseases, Shahid Beheshti University of Medical Sciences, Tehran, Iran

F. Rahimi-Bashar
Anesthesia and Critical Care Department, Hamadan University of Medical Sciences, Hamadan, Iran

A. Shahriary
Chemical Injuries Research Center, Systems Biology and Poisonings Institute, Baqiyatallah University of Medical Sciences, Tehran, Iran

T. P. Johnston
Division of Pharmacology and Pharmaceutical Sciences, School of Pharmacy, University of Missouri-Kansas City, Kansas City, MO, USA

A. Sahebkar (✉)
Neurogenic Inflammation Research Center, Mashhad University of Medical Sciences, Mashhad, Iran

Biotechnology Research Center, Pharmaceutical Technology Institute, Mashhad University of Medical Sciences, Mashhad, Iran

Polish Mother's Memorial Hospital Research Institute (PMMHRI), Lodz, Poland

Halal Research Center of IRI, FDA, Tehran, Iran
e-mail: sahebkara@mums.ac.ir; amir_saheb2000@yahoo.com

P. C. Guest (ed.), *Clinical, Biological and Molecular Aspects of COVID-19*, Advances in Experimental Medicine and Biology 1321, https://doi.org/10.1007/978-3-030-59261-5_25

editorials or letters, expert opinions, comments, and animal studies. Nine studies were included in the analysis. The odds of having more severe COVID-19 disease was higher in subjects with elevated PCT (≥0.05 ng/mL) compared with those having low procalcitonin (<0.05 ng/mL) [n = 6, OR(95% CI) = 2.91(1.14, 7.42), p = 0.025). After estimating the mean and standard deviation values from the sample size, median, and interquartile range, a pooled effect analysis indicated higher serum PCT concentrations in patients with severe versus less severe disease [n = 6, SMD(95% CI) = 0.64(0.02, 1.26), p = 0.042]. The results of this study showed that PCT is increased in patients with severe COVID-19 infection.

Keywords

Coronavirus · Procalcitonin · Meta-analysis · SARS-CoV-2 · COVID-19 · Viral infection

25.1 Introduction

Patients with unexplained pneumonia were initially reported in Wuhan, China, in December 2019. A novel coronavirus named severe acute respiratory syndrome coronavirus 2 (SARS-CoV-2) was detected in samples of the lower respiratory tract of the infected patients, and the disease was termed COVID-19 (coronavirus-2019) [1]. This disease is rapidly spreading across the world [2] and has turned into a pandemic [3]. Globally, approximately 5,311,624 confirmed cases of COVID-19 have been reported, including an estimated 342,105 deaths in approximately 209 countries (as of May 23, 2020) [4]. It is presently the greatest health challenge worldwide [5], and the World Health Organization (WHO) has declared the outbreak a global public health emergency [6].

Currently the gold standard for the diagnosis of COVID-19 is reverse transcription polymerase chain reaction (RT-PCR) testing, but, in many countries, testing has been limited due to time and resource issues, and in the early stage of the disease, the positive rate of the test was relatively low [7]. Therefore, there is an urgent need to identify reliable biomarkers that may help predict the risk for illness severity so that appropriate care can be allocated earlier [8].

One of the most studied biomarkers in this field is procalcitonin (PCT), which has both diagnostic and prognostic utility [9]. PCT has become a promising biomarker for early detection of bacterial infections in modern clinical practice [10]. However, the expression of this biomarker may differ in severe COVID-19 infection with some studies reporting that PCT is increased in patients with severe COVID-19 [11, 12] and others reporting that PCT in these patients is normal [13, 14]. These differences could reflect the fact that the clinical characteristics differ in those patients with mild versus severe COVID-19 infections [9].

We have aimed to resolve this issue by carrying out a systematic review of PCT levels in COVID-19 cases.

25.2 Methods

This study was performed according to the Preferred Reporting Items for Systematic Reviews and Meta-Analyses (PRISMA) statement [15]. We aimed to compare the levels of PCT in patients with severe versus non-severe COVID-19 infections, to determine if it can be used as a biomarker to predict disease course. A threshold level of 0.05 ng/mL PCT was taken to discriminate between severe and non-severe infection for each study.

25.2.1 Search Strategy

We searched, Embase, ProQuest, MEDLINE/PubMed, Scopus, and ISI/Web of Science for studies that reported the level of PCT in patients with COVID-19 infections. A date limit was set from December 2019, and the search was performed up to April 5, 2020. The reference list of articles was reviewed using forward and backward citation tracking to identify other eligible documents. No language limits were applied.

25.2.2 Selection Criteria

We included all studies regardless of study design, targeting prospective or retrospective studies that met the following criteria:

1. Reported the level of PCT in COVID-19 patients displaying serious symptoms
2. Reported the level of PCT in COVID-19 patients displaying non-serious symptoms

We excluded:

1. Articles not regarding COVID-19 or not reporting PCT level
2. Studies not including both severe and non-severe COVID-19 patients
3. Review articles, editorials or letters, expert opinions, comments, and animal studies

At least two reviewers independently evaluated titles and abstracts and selected relevant studies for inclusion. If this could not be done reliably using the title and abstract of an article, the full text version was retrieved for detailed analysis. Any disagreement was resolved by a third independent reviewer. The reasons for exclusion of studies were recorded.

At first we searched databases that included Scopus (7), Embase (9), ProQuest (14), PubMed (2), and Web of Science (2) and identified 31 total studies. In the next step we removed the duplicate articles and retained 19 records to include Scopus (7), Embase (2), ProQuest (10), PubMed (1), and Web of Science (2). Subsequently, the records were screened by title and abstract and six further studies were excluded. The remaining 16 records were screened further using the full text, and this left nine articles that met all requirements for inclusion in the study.

25.2.3 Data Extraction

Data were extracted independently by at least two reviewers and included authors, year of publication, clinical setting, sample size, sample type, and levels of PCT. Disagreements on the extracted data were resolved by consensus.

25.2.4 Quality Assessment

The methodological quality assessment of studies was performed independently by two authors using Quality Assessment of Diagnostic Accuracy Studies (QUADAS-2) criteria, as recommended by the Cochrane Collaboration [16]. Disagreements about inclusion criteria, data extraction, and quality assessment were resolved by consensus.

25.2.5 Data Synthesis and Statistical Analysis

We applied random-effects meta-analyses with inverse variance weighting to calculate pooled estimates and 95% confidence intervals (CIs). We estimated the sample mean and standard deviation (SD) from the sample size, median, and interquartile range via the Wan et al. approach [17] to compute a meta-analysis of outcomes and demonstrate effectiveness. In addition, we used odds ratios (ORs) and 95% CIs to compare the rates of disease severity. Heterogeneity (I^2 statistics) was assessed and reported using Cochran's Q-test [18]. We also plotted Galbraith graphs to display heterogeneity. Egger's test and a visual inspection of funnel plots were carried out to evaluate publication bias between studies [19]. In general, a PCT concentration ≥ 0.05 ng/mL was considered as being high in our analysis. Values below 0.05 ng/mL were considered normal or low. Two groups of patients with either low or high disease were also examined. STATA version 14.0 (Stata Corp, College Station, TX) and R version 3.6.3 were used to conduct the analyses.

25.3 Results

A summary of study selection process and characteristics of the included studies are summarized in Fig. 25.1 and Table 25.1, respectively.

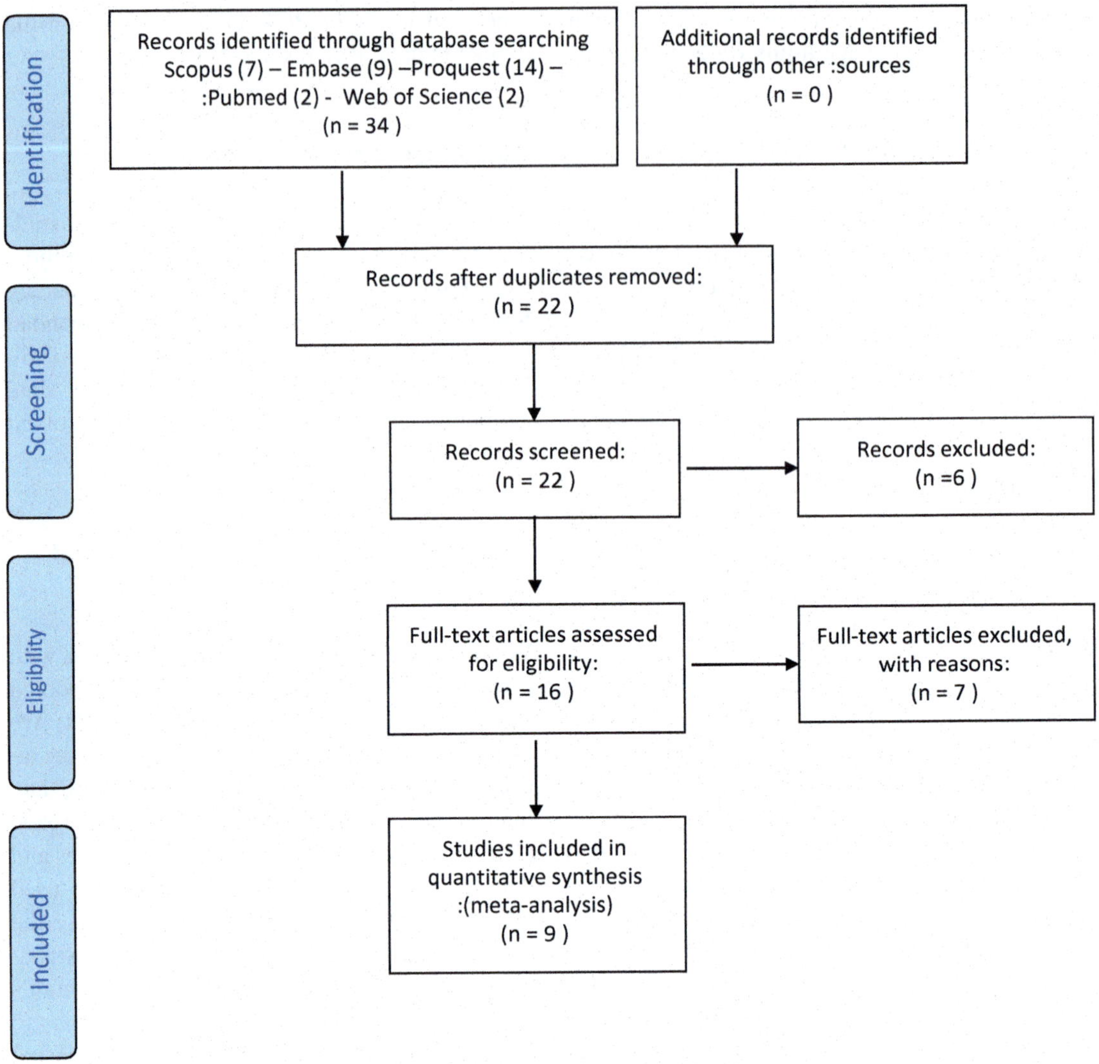

Fig. 25.1 PRISMA flow diagram summarizing the study selection process

25.3.1 Results of PCT Analyses Using Odds Ratios

Results were obtained for PCT with an odds ratio in 6 out of the 9 studies which met the selection criteria. A randomized effect analysis showed significant risk according using a forest plot to visualize the results (Fig. 25.2). The odds of having more severe COVID-19 disease was higher in subjects with elevated PCT compared to those with low PCT levels ($n = 6$, OR(95% CIs) = 2.91 (1.14, 7.42), $p = 0.025$) (Fig. 25.3).

The Q-test showed a significant variations in odds ratios attributable to heterogeneity of the procalcitonin data ($I^2 = 73.5\%$, $p = 0.002$), which can also be seen in the Galbraith plot (Fig. 25.3). There was no publication bias in the PCT studies according to results of Egger's test analyses and by using a funnel plot visualization ($p = 0.434$) (Fig. 25.4).

Table 25.1 Final studies included in meta-analysis

No	Authors	Year	Procalcitonin		Sample size	Setting	Samples	Study design
			Patients with worse condition	All Patients				
1	Qiu et al. [20]	2020	$N = 19$ 0.32 (0.19)	$N = 36$ 0.24 (0.17)	36	Three hospitals in Zhejiang province, China	Children with coronavirus disease 2019	Retrospective
2	Zhang et al. [12]	2020	$N = 56$ 0.1 (0.06–0.3)	$N = 138$ 0.07 (0.04–0.1)	138	No. 7 hospital of Wuhan	Patients diagnosed as COVID-19	Prospective
3	Wang et al. [21]	2020	$N = 36$ 27 (75.0)	$N = 138$ 49 (35.5)	138	Zhongnan Hospital of Wuhan University	Patients with confirmed NCIP	Retrospective
4	Guan et al. [22]	2020	$N = 173$ 16/117 (13.7)	$N = 1099$ 35/633 (5.5)	1099	552 hospitals in 30 provinces, autonomous regions, and municipalities in mainland China	Patients with laboratory-confirmed COVID-19	Retrospective
5	Huang et al. [13]	2020	$N = 13$ 0.1 (0.1–0.4)	$N = 41$ 0.1 (0.1–0.1)	41	Wuhan, China	Patients with laboratory-confirmed 2019 nCoV infection	Prospectively
6	Liu et al. [23]	2020	$N = 69$ 2/69 (2·90)	$N = 80$ 2/80 (2·50)	80	Department of Infectious Diseases, Union Hospital, Tongji Medical College, Huazhong University of Science and Technology	Patients with severe type COVID-19	Retrospective
7	Shi et al. [16]	2020	$N = 82$ 0.06(0.03–0.10)	$N = 416$ 0.07(0.04–0.15)	416	Renmin Hospital of Wuhan University, located in Wuhan, Hubei province, China, was assigned responsibility for the treatment of patients with severe COVID-19 by the Wuhan government	Severe patients admitted to Renmin Hospital of Wuhan University with laboratory-confirmed COVID-19	Retrospective
8	Peng et al. [24]	2020	$N = 16$ 0.20(0.15–0.48)	$N = 96$ 0.11(0.06–0.20)	112	Wuhan Union Hospital West Hospital	Patients with cardiovascular diseases infected with new coronavirus pneumonia	Retrospective
9	Zhou et al. [16]	2020	$N = 11$ 0.21(0.11–0.42)	$N = 9$ 0.05(0.03–0.06)	20	West District of Union Hospital of Tongji Medical College	In-hospital severe patients with COVID-19	Retrospectively

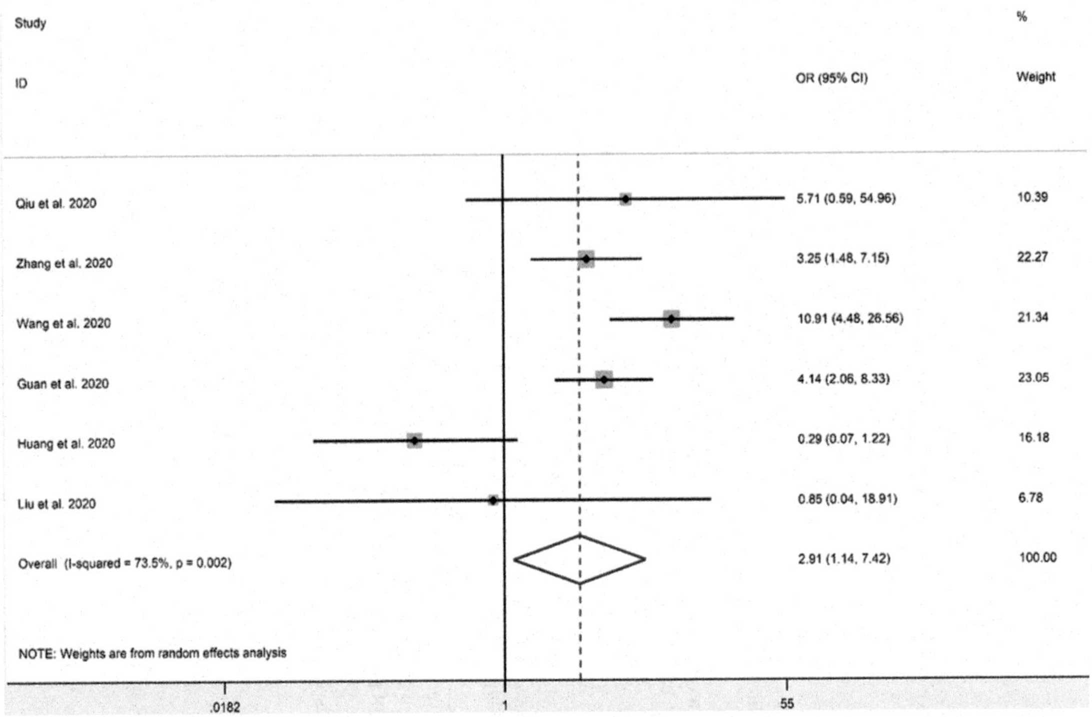

Fig. 25.2 Forest plot results for odds ratio of procalcitonin laboratory data

Fig. 25.3 Galbraith (radial) plot for odds ratio of trials used by procalcitonin laboratory data

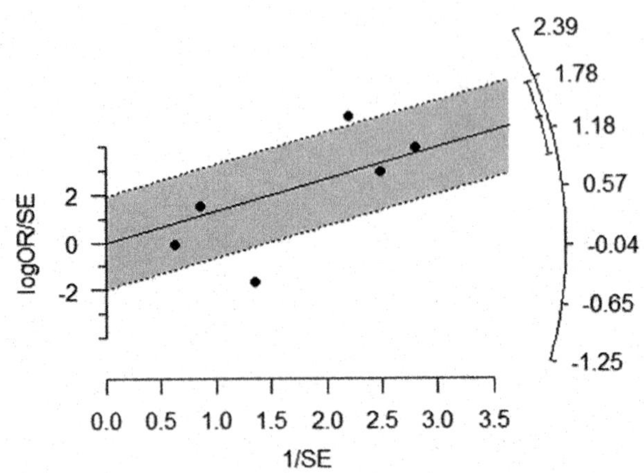

25.3.2 Results of PCT Analyses Using Standardized Mean Difference

All PCT results included in the six studies were analyzed using the standardized mean difference (SMD). For most studies (except the study by Qiu et al. [20]), the median and interquartile range of PCT were reported instead of the mean and standard deviation. After estimating the values of the mean and standard deviation from the sample size, median, and interquartile range, a pooling effect analysis indicated a significantly higher PCT level ($n = 6$, SMD(95% CI) = 0.64(0.02, 1.26), $p = 0.042$) in subjects with severe versus less severe COVID-19 disease (Fig. 25.5).

Fig. 25.4 Funnel plot for verification of publication bias in the meta-analysis of odds ratio of procalcitonin laboratory data

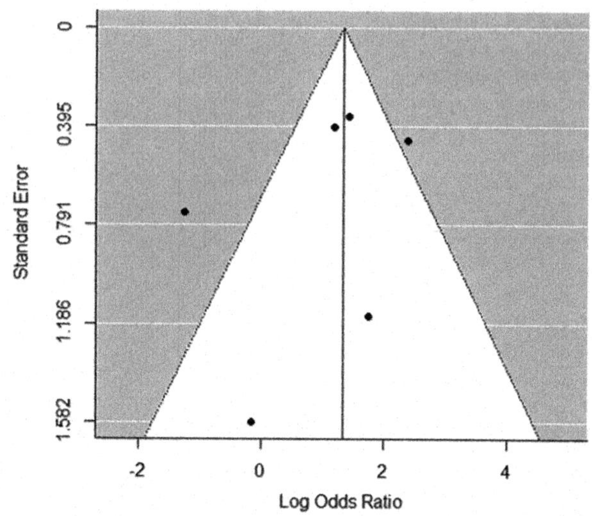

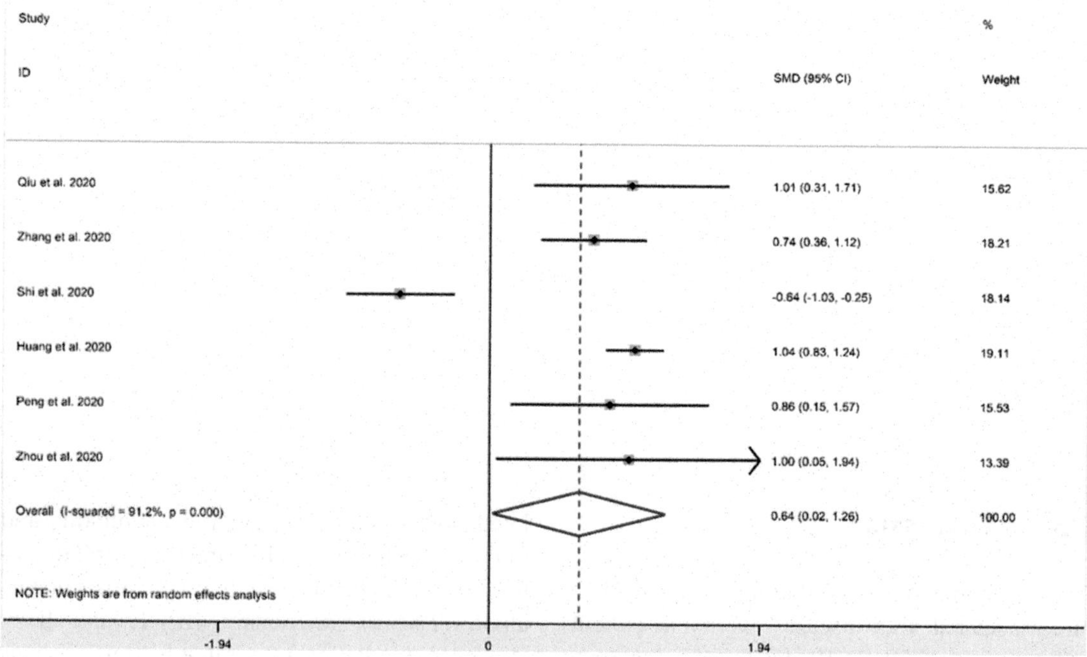

Fig. 25.5 Forest plot results for standardized mean difference of procalcitonin laboratory data

Next, we applied a random effects analysis since there was heterogeneity between the studies for PCT based on the Q-test and radial plot results (I^2 = 91.2%, $p < 0.001$) (Fig. 25.6). Additionally, according to Egger's test and visualization with a funnel plot, there was no publication bias in the results of the meta-analysis ($p = 0.711$) (Fig. 25.7).

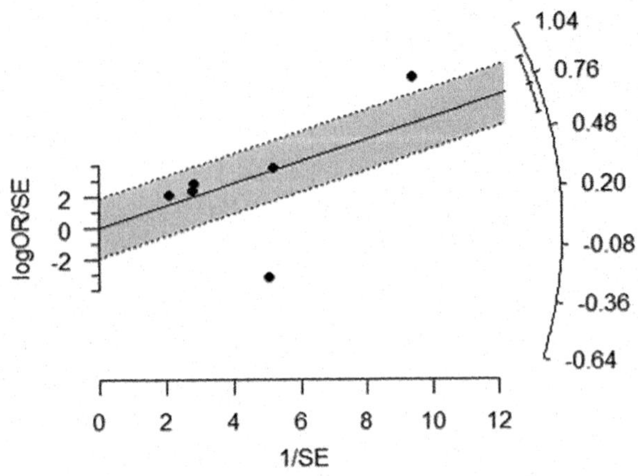

Fig. 25.6 Galbraith (radial) plot for standardized mean difference of trials using procalcitonin laboratory data

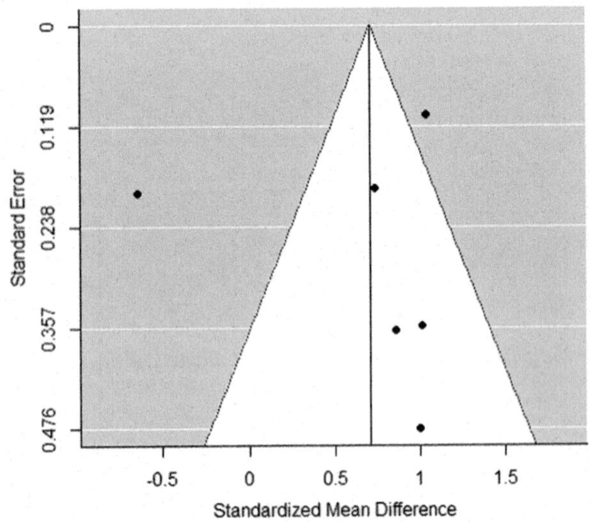

Fig. 25.7 Funnel plot for verification of publication bias in the meta-analysis of standardized mean difference of procalcitonin laboratory data

25.4 Discussion

This study demonstrated that high levels of PCT are associated with disease severity in patients infected with COVID-19. PCT is known to be elevated in bacterial infections and is currently used for diagnosis and decision-making regarding antibiotic treatment duration in respiratory infections [25]. Its synthesis is upregulated in bacterial infections and downregulated in viral infections [26]. PCT is produced by the thyroid C cells in healthy people. In the presence of bacterial infections, PCT production is activated in all parenchymal tissues and its level increases rapidly. PCT production by these tissues is stimulated both directly by bacterial endotoxins and lipopolysaccharides and indirectly by inflammatory mediators that include tumor necrosis factor-alpha (TNF-a), interleukin (IL)-1, and IL-6. However, mediators of viral infection such as interferon-gamma (IFN-γ) decrease the PCT level, which makes it a more specific marker for bacterial infections [26]. Nevertheless, the finding of this study showed that PCT is increased in patients with severe COVID-19. This suggests that in some severe COVID-19 cases, there is a bacterial co-infection that increases their PCT levels. This hypothesis is supported by the work of Zhou et al. in which it was reported that the most of severe COVID-19 patients have viral infection and secondary bacterial infection [16]. Bacterial co-infection is a poor prognostic feature in these patients [27] and may contribute to the death of these patients [28]. Thus, a PCT level determination, in addition to helping with identification of severe patients, may guide physicians in determinations of bacterial co-infection. This would allow them to initiate early antibiotic therapy that may prevent further deterioration of health.

The results of this study are contrary to the results of the study by Lippi and Plebani, which reported that the PCT value would remain within the reference range in severe coronavirus-infected patients [29]. However, the meta-analysis in the present study was of wider scope, including multiple studies that would have accounted for more patient variables. Nevertheless, further studies are required to address this issue.

25.5 Conclusion

The results of this study showed that PCT in patients with severe COVID-19 disease is increased, which suggests that it may play an important role in predicting severity and outcome of infection. Therefore, a PCT level determination may guide physicians in cases of suspected bacterial co-infection to initiate early antibiotic therapies that may prevent further deterioration of health and death.

Acknowledgments Thanks to guidance and advice from the Clinical Research Development Unit of Baqiyatallah Hospital.

Conflict of Interests None.

Funding None.

References

1. Recalcati S (2020) Cutaneous manifestations in COVID-19: a first perspective. J Eur Acad Dermatol Venereol 34(5):e212–e213. https://doi.org/10.1111/jdv.16387
2. Pedersen SF, Ho YC (2020) SARS-CoV-2: a storm is raging. J Clin Invest 130(5):2202–2205
3. Cucinotta D, Vanelli M (2020) WHO declares COVID-19 a pandemic. Acta Biomed 91(1):157–160
4. Johns Hopkins University and Medicine. Coronavirus Resource Center. https://coronavirus.jhu.edu/map.html
5. Ahmadpoor P, Rostaing L (2020) Why the immune system fails to mount an adaptive immune response to a Covid −19 infection. Transpl Int. Apr 1. doi: https://doi.org/10.1111/tri.13611. Online ahead of print. doi: https://doi.org/10.1111/tri.13611
6. Wang LS, Wang YR, Ye DW, Liu QQ (2020) A review of the 2019 Novel Coronavirus (COVID-19) based on current evidence. Int J Antimicrob Agents. Mar 19:105948. doi: https://doi.org/10.1016/j.ijantimicag.2020.105948. Online ahead of print
7. Xu Z, Peng C, Shi Y, Zhu Z, Mu K, Wang X et al (2020) Nelfinavir was predicted to be a potential inhibitor of 2019-nCov main protease by an integrative approach combining homology modelling, molecular docking and binding free energy calculation. bioRxiv. doi:https://doi.org/10.1101/2020.01.27.921627
8. Lippi G, Plebani M, Henry BM (2020) Thrombocytopenia is associated with severe coronavirus disease 2019 (COVID-19) infections: a meta-analysis. Clin Chim Acta 506:145–148. https://doi.org/10.1016/j.cca.2020.03.022
9. Wang X, Fang J, Zhu Y, Chen L, Ding F, Zhou R et al (2020) Clinical characteristics of non-critically ill patients with novel coronavirus infection (COVID-19) in a Fangcang hospital. Clin Microbiol Infect S1198-743X(20):30177–30174. https://doi.org/10.1016/j.cmi.2020.03.032
10. Cleland DA, Eranki AP (2020) Procalcitonin. In StatPearls, StatPearls Publishing Copyright © 2020, StatPearls Publishing LLC; Orlando, FL, USA. https://www.ncbi.nlm.nih.gov/books/NBK539794/
11. Sun D, Li H, Lu XX, Xiao H, Ren J, Zhang FR et al (2020) Clinical features of severe pediatric patients with coronavirus disease 2019 in Wuhan: a single center's observational study. World J Pediatr. Mar 19;1–9. doi: https://doi.org/10.1007/s12519-020-00354-4. Online ahead of print
12. Zhang JJ, Dong X, Cao YY, Yuan YD, Yang YB, Yan YQ et al (2020) Clinical characteristics of 140 patients infected with SARS-CoV-2 in Wuhan, China. Allergy. Feb 19. doi: https://doi.org/10.1111/all.14238.. Online ahead of print
13. Huang C, Wang Y, Li X, Ren L, Zhao J, Hu Y et al (2020) Clinical features of patients infected with 2019 novel coronavirus in Wuhan, China. Lancet 395(10223):497–506
14. Shi S, Qin M, Shen B, Cai Y, Liu T, Yang F et al (2020) Association of Cardiac Injury With Mortality in Hospitalized Patients With COVID-19 in Wuhan, China. JAMA Cardiol. Mar 25;e200950. doi: https://doi.org/10.1001/jamacardio.2020.0950. Online ahead of print
15. Moher D, Liberati A, Tetzlaff J, Altman DG (2009) Preferred reporting items for systematic reviews and meta-analyses: the PRISMA statement. PLoS

Med 6(7):e1000097. https://doi.org/10.1371/journal.pmed.1000097
16. Zhou B, She J, Wang Y, Ma X. (2020) Utility of Ferritin, Procalcitonin, and C-reactive Protein in Severe Patients with 2019 Novel Coronavirus Disease. Preprint from Research Square, 19 Mar 2020. doi: https://doi.org/10.21203/rs.3.rs-18079/v1 PPR: PPR122473
17. Wan X, Wang W, Liu J, Tong T (2014) Estimating the sample mean and standard deviation from the sample size, median, range and/or interquartile range. BMC Med Res Methodol 14:135. https://doi.org/10.1186/1471-2288-14-135
18. Higgins JP, Altman DG, Gøtzsche PC, Jüni P, Moher D, Oxman AD et al (2011) The Cochrane Collaboration's tool for assessing risk of bias in randomised trials. BMJ 343:d5928. https://doi.org/10.1136/bmj.d5928
19. Egger M, Davey Smith G, Schneider M, Minder C (1997) Bias in meta-analysis detected by a simple, graphical test. BMJ 315(7109):629–634
20. Qiu H, Wu J, Hong L, Luo Y, Song Q, Chen D (2020) Clinical and epidemiological features of 36 children with coronavirus disease 2019 (COVID-19) in Zhejiang, China: an observational cohort study. Lancet Infect Dis. Mar 25;S1473–3099(20)30198–5. doi: https://doi.org/10.1016/S1473-3099(20)30198-5. Online ahead of print
21. Wang D, Hu B, Hu C, Zhu F, Liu X, Zhang J et al (2020) Clinical characteristics of 138 hospitalized patients with 2019 novel coronavirus–infected pneumonia in Wuhan, China. JAMA 323(11):1061–1069
22. Guan WJ, Ni ZY, Hu Y, Liang WH, Ou CQ, He JX et al (2020) Clinical characteristics of coronavirus disease 2019 in China. N Engl J Med 382(18):1708–1720
23. Liu T, Zhang J, Yang Y, Ma H, Li Z, Zhang J et al (2020) The potential role of IL-6 in monitoring severe case of coronavirus disease 2019. medRxiv. doi: https://doi.org/10.1101/2020.03.01.20029769
24. Peng YD, Meng K, Guan HQ, Leng L, Zhu RR, Wang BY et al (2020) [Clinical characteristics and outcomes of 112 cardiovascular disease patients infected by 2019-nCoV]. Zhonghua Xin Xue Guan Bing Za Zhi. Mar 2;48(0):E004. doi: https://doi.org/10.3760/cma.j.cn112148-20200220-00105. Online ahead of print
25. de Fonseka D, Maskell NA (2018) The role of procalcitonin in the management of pleural infection. Curr Opin Pulm Med 24(4):380–383
26. Davies J (2015) Procalcitonin. J Clin Pathol 68(9):675–679
27. Khot WY, Nadkar MY (2020) The 2019 novel coronavirus outbreak – a global threat. J Assoc Physicians India 68(3):67–71
28. Li X, Wang L, Yan S, Yang F, Xiang L, Zhu J et al (2020) Clinical characteristics of 25 death cases with COVID-19: a retrospective review of medical records in a single medical center, Wuhan, China. Int J Infect Dis 94:128–132
29. Lippi G, Plebani M (2020) Procalcitonin in patients with severe coronavirus disease 2019 (COVID-19): a meta-analysis. Clin Chim Acta 505:190–191

A Systematic Review of 571 Pregnancies Affected by COVID-19

26

Leila Karimi, Amir Vahedian-Azimi, Somayeh Makvandi, and Amirhossein Sahebkar

Abstract

The outbreak of the novel coronavirus 2019 (COVID-19) disease has been severe and a cause for major concern around the world. Due to immunological and physiological changes during pregnancy, pregnant women have a higher risk of COVID-19 morbidity and mortality. The aim of this study was to collect and integrate the results of previous studies to get an accurate representation and interpretation of the clinical symptoms, laboratory and radiological findings, and characteristics of pregnant women with COVID-19. We conducted a scientific search in main databases with a combination of related MESH terms and keywords. The outcomes included common clinical symptoms at the time of onset of the disease, common laboratory and radiological findings, the rates of vaginal delivery and Cesarean section, Cesarean section indications, maternal complications, and vertical transmission rates. A total of 51 studies comprising 571 pregnant women with COVID-19 pneumonia were included in the study. The most common symptoms were fever, cough, and dyspnea, respectively. Elevated C-reactive protein and ground-glass opacities were the most common laboratory and radiological findings of COVID-19 pneumonia, respectively. A total of 114 Cesarean sections were performed due to COVID-19-related concerns. There were 55 cases of intubation (11.6%) and 13 maternal deaths (2.3%). The vertical transmission rate was 7.9%. We conclude that the characteristics of pneumonia caused by COVID-19 in pregnant women do not appear to be different from those in the general population with COVID-19 infec-

L. Karimi
Behavioral Sciences Research Center, Life Style Institute, Nursing Faculty, Baqiyatallah University of Medical Sciences, Tehran, Iran

A. Vahedian-Azimi
Trauma Research Center, Nursing Faculty, Baqiyatallah University of Medical Sciences, Tehran, Iran

S. Makvandi (✉)
Department of Midwifery, School of Nursing and Midwifery, Islamic Azad university Ahvaz branch, Ahvaz, Iran
e-mail: somayemakvandi@gmail.com

A. Sahebkar (✉)
Neurogenic Inflammation Research Center, Mashhad University of Medical Sciences, Mashhad, Iran

Biotechnology Research Center, Pharmaceutical Technology Institute, Mashhad University of Medical Sciences, Mashhad, Iran

Polish Mother's Memorial Hospital Research Institute (PMMHRI), Lodz, Poland

Halal Research Center of IRI, FDA, Tehran, Iran
e-mail: sahebkara@mums.ac.ir; amir_saheb2000@yahoo.com

P. C. Guest (ed.), *Clinical, Biological and Molecular Aspects of COVID-19*, Advances in Experimental Medicine and Biology 1321, https://doi.org/10.1007/978-3-030-59261-5_26

tions. However, pregnant women with underlying diseases were more likely to develop COVID-19 than others, and, in those infected with the virus, the rate of Cesarean delivery and preterm birth increased.

Keywords

Covid-19 · Pregnancy · Outcomes · Coronavirus

26.1 Introduction

The novel coronavirus 2019 (COVID-19) outbreak is a major concern around the world, and the number of pregnant women with the virus is on the rise. Compared to non-pregnant women, the previous coronavirus outbreaks of severe acute respiratory syndrome (SARS) and Middle East severe respiratory syndrome (MERS) caused higher mortality and morbidity in pregnant women [1, 2]. Due to immunological and physiological changes during pregnancy, pregnant women are at more risk of COVID-19 morbidity and mortality [3]. However, it is still not known at this time if such risks are increased in pregnant women with Covid-19 infections compared to non-pregnant adults.

Because monitoring systems have been developed for COVID-19 cases, it is important to gather and report on pregnancy status as well as the consequences for mothers and fetus, in order to provide initial evidence that can be used to guide the treatment of pregnant women with COVID-19 disease. Chen et al. performed a retrospective study on nine pregnant women with laboratory-confirmed COVID-19 pneumonia, who w ere hospitalized in the Zhongnan Hospital of Wuhan University, Wuhan, China, from Jan 20 to Jan 31, 2020 [4]. In this study, clinical symptoms, laboratory results, and chest computed tomography (CT) scans were retrospectively assessed. The clinical symptoms in seven of the patients included fever, whereas cough was observed in four patients, myalgia in three patients, malaise in two patients, and sore throat in two patients. In addition, laboratory results showed that five patients had lymphopenia, a decrease in blood lymphocytes. None of the patients developed severe COVID-19 pneumonia or died. Findings from this small group of cases showed that the clinical characteristics of COVID-19 pneumonia in pregnant women were similar to non-pregnant patients who developed COVID-19 pneumonia.

In another study of 15 pregnant women in China, Liu et al. found that the most common abnormalities in the laboratory data included lymphopenia and an elevated C-reactive protein (CRP), a biomarker of inflammation [5]. The radiographs of the pregnant mothers with COVID-19 showed that five had multiple bilateral ground-glass opacities in their lungs and three had patchy consolidations. However, no problems were observed in one pregnant woman and her lungs were clear with no infusions.

In another study, Yu et al. showed that out of seven pregnant women with COVID-19 infections, six had fever, one had cough, one had shortness of breath, and one had diarrhea [6]. Different degrees of liver dysfunction, such as an increase in alanine aminotransferase (AAT) or aspartate aminotransferase (AST), were reported in two of the patients. All seven of the patients had Cesarean sections and the outcomes of the women and neonates were good. Three of the neonates were tested for SARS-CoV-2 (the virus that causes COVID-19) and one was infected with SARS-CoV-2 36 h after birth. Examination of the CT scans showed that six of the women had large areas of multiple ground-glass opacities and the rest had less involvement. Biomarker analyses showed increased procalcitonin and erythrocyte sedimentation rate (ESR) in four patients, and all patients had a higher than normal concentration of CRP [6].

The results of the above studies showed that the most common laboratory findings in patients with COVID-19 were a decrease in blood lymphocytes and an increase in CRP. Despite these findings, the number of patients in these studies was too small to draw firm conclusions. The aim of this study was to collect and integrate the results of previous studies to get more accurate interpretation of the clinical symptoms, characteristics, and outcomes of pregnant women with COVID-19 infections and to inform the most appropriate treatment course.

26.2 Methods

The scientific databases of PubMed/MEDLINE, Scopus, ScienceDirect, and CENTRAL (Cochrane Central Register of Controlled Trials) were searched with the following strategy until May 21, 2020 (pregnan* OR birth OR Childbearing OR Prenatal) AND (coronavirus OR "coronavirus disease-19" OR "COVID-19" OR "2019-nCoV disease" OR "2019 novel coronavirus infection" OR "2019-nCoV infection" OR "COVID-19 pandemic" OR "2019 novel coronavirus disease"). Apart from reviews, all studies using any methodology that contained sufficient data about the symptoms and clinical and/or laboratory findings involving pregnant women with COVID-19 infections (laboratory confirmed and/or clinically diagnosed) were included in the study. Articles in languages other than English were excluded. The outcomes included common clinical symptoms at the time of onset of the disease, common laboratory findings, findings related to chest x-ray or CT scan, the rate of vaginal delivery and Cesarean section, Cesarean section indications, maternal complications, and vertical transmission rates.

The output of each database was examined and duplicate articles were identified and deleted. The remaining articles were screened based on titles and abstracts and irrelevant articles were removed. In the next step, the articles were screened based on a full-text evaluation, and finally a number of articles were selected for systematic review. All steps related to data extraction were performed separately by the two researchers, and any disagreement was resolved with discussion.

To extract the data based on the objectives of the study, a form was developed which included information on the following: first author's name, country, sample size, maternal comorbidities, common clinical symptoms, common laboratory and radiological findings, outcomes such as the need for intubation, maternal death, complications, and data about childbirth. The data were extracted independently by two authors, and if there was any disagreement or ambiguity in each case, this was resolved through discussion.

This study involved maximization of research ethics, and stringent attempts were made to avoid any plagiarism or data manipulation.

26.3 Results

A search of scientific databases led to the discovery of 1167 original articles. Duplicate articles were identified using the EndNote software (n = 551), manual screening (n = 67), and by examining titles and abstracts (n = 481), and these were excluded. Of the remaining 68 articles, the full texts were carefully reviewed and a further 17 articles excluded. This left 51 articles which met the criteria for inclusion in this systematic review (Fig. 26.1). A summary of the characteristics of these studies is shown in Table 26.1. These comprised a total of 571 pregnant women in the 51 case reports and case studies. There were 16 studies carried out in China [7–22], 14 in the USA [23–36], six in Italy [37–42], five in Iran [43–47], three in the UK [48–50], two in Portugal [51, 52], one in Thailand [53], one in Turkey [54], one in Peru [55], one in Sweden [56], and one in Australia [57]. Table 26.2 shows the outcomes of these studies in quantitative form. Women ranged in age from 17 to 49 years, and most were in the third trimester of pregnancy at the time the studies were carried out. The most common comorbidities were obesity, gestational diabetes mellitus, chronic hypertension, pulmonary conditions, asthma, diabetes mellitus, and preeclampsia/eclampsia, respectively.

Common symptoms at the onset of disease in the order of prevalence were as follows: fever (65.8%), cough (61.6%), dyspnea (6.4%), fatigue (6.1%), and myalgia (3.8%). Less common symptoms included shortness of breath, anosmia,

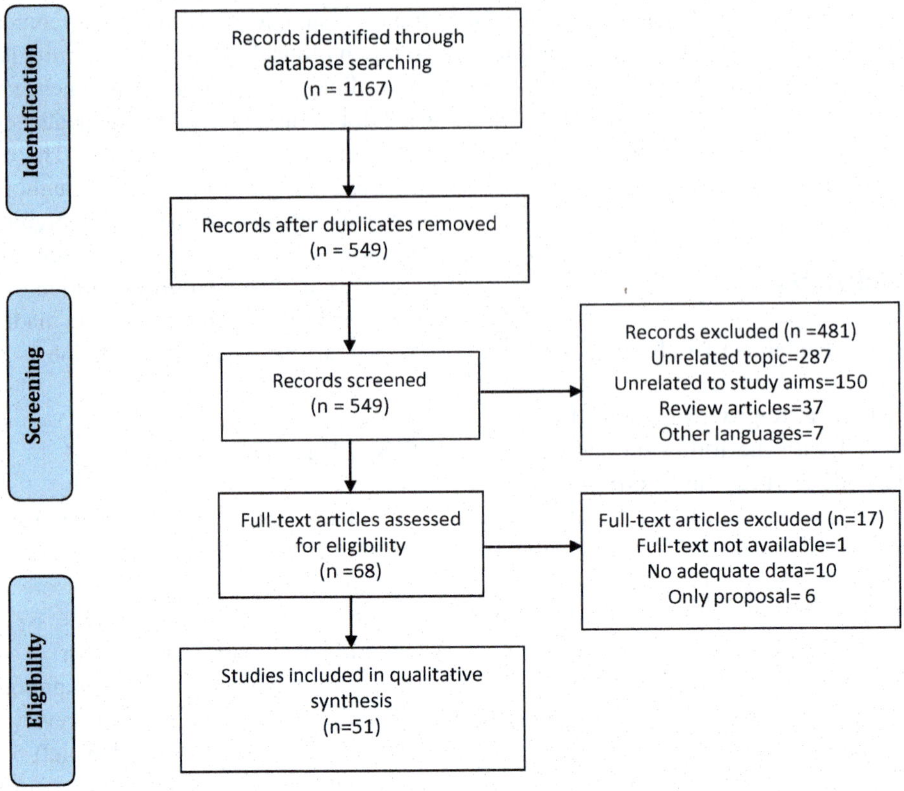

Fig. 26.1 PRISMA flowchart of the review process

malaise, diarrhea, headache, sore throat, emesis, nasal congestion, chills, vomiting, anorexia, rhinitis, and chest discomfort. Common laboratory findings were elevated CRP levels (33.9%), lymphopenia (31.8%), increased serum levels of D-dimer (a sign of significant formation and breakdown of clot; 27.3%), leukocytosis (9.9%), and leukopenia (6.6%). A common radiological finding was patchy shadowing or ground-glass opacities (49.7%).

A total of 321 Cesarean sections, 91 vaginal deliveries, and 20 fetal abortions occurred. In 114 of the Cesarean section cases (35.5%), these were performed due to concerns related to the effects of COVID-19 infections on the mother or fetus. Of the 412 deliveries, 112 were preterm (27.2%).

A number of pregnant women infected with COVID-19 developed complications during treatment. More common complications were acute respiratory distress, septic shock, cardiac dysfunction, multiple organ dysfunction, cardiac arrest, myocardial injury and myocarditis, endocarditis, need for dialysis, pericardial effusion, and pulmonary embolism. There were 55 cases of intubation (11.6%) and 13 maternal deaths (2.3%). The vertical transmission rate was 7.9%.

26.4 Discussion

Due to the immunosuppressive status and physiological changes specific to pregnancy, pregnant women are vulnerable to the effects of respiratory pathogens. The results of present study showed that approximately 15 percent of the pregnant women with COVID-19 were asymp-

Table 26.1 Characteristics of included studies

First author's name	Country	n	Gestational age			Common symptoms	Common abnormal laboratory finding	Common radiological findings	Intubation	Death
			1st trimester	2nd trimester	3rd trimester					
Ahmed [50]	UK	1	–	–	1	Fever	NM	Pulmonary embolism	1	1
Alzamora [55]	Peru	1	–	–	1	General malaise, fatigue	Pancytopenia, elevated CRP	Multiple consolidations	1	–
Anderson [23]	USA	1	–	1	–	Fever, cough	Mild transaminitis	Bilateral consolidation	1	–
Blauvelt [36]	USA	1	–	–	1	Malaise, fever	Lymphopenia, elevated D-dimer	Bilateral opacities	1	–
Breslin [35]	USA	43	–		43	Cough, fever	NM	NM	–	–
Browne [34]	USA	1	–	1		Fever, cough	Leukocytosis	NM	–	–
Buonsens [37]	Italy	4	–	2	2	Fever, cough	Lymphocytopenia, elevated LDH	Bilateral hyperlucency	–	–
Carosso [42]	Italy	1	–	–	1	Fever, cough	NM	NM	–	–
Chen [8]	China	118	22	21	75	Fever, cough	Lymphopenia elevated D-dimer	Bilateral infiltrates	–	–
Chen [9]	China	1	–	–	1	Fever, cough	NM	Patchy ground-glass opacities	–	–
Chen [16]	China	4	–	–	4	Fever, cough	Lymphopenia elevated CRP	Bilateral involvement	1	–
De Socio [39]	Italy	1	–	–	1	Asymptomatic	Elevated fibrinogen and D-dimer	Normal	–	–
De Castro [38]	Italy	1	–	–	1	Fever, chills	Normal	NM	–	–
Dong [18]	China	1	–	–	1	Fever, nasal congestion	Lymphopenia, elevated CRP	Patchy ground-glass opacity	–	–
D'oria [51]	Portugal	12	–	–	12	Headache	NM	NM	–	–
Ferrazzia [41]	Italy	42	NM			Fever, cough	Elevated CRP, leukocytosis	NM	NM	–
Gidlöf [56]	Sweden	1	–	–	1	Headache	Elevated uric acid	Typical signs of pneumonia	–	–
Govind [48]	UK	9	–	9		Cough, anosmia	Lymphopenia	Consolidation	1	1
Hantoushzadeh [43]	Iran	9	–	9		Fever, cough	Lymphopenia, elevated CRP	Patchy ground-glass features	9	7
Hirshberg [33]	USA	5	–	2	3	Fever, dyspnea	Elevated CRP and lactic acid	Multifocal pneumonia	5	–

(continued)

Table 26.1 (continued)

First author's name	Country	n	Gestational age			Common symptoms	Common abnormal laboratory finding	Common radiological findings	Intubation	Death
			1st trimester	2nd trimester	3rd trimester					
Hong [32]	USA	1	–	1		Cough, myalgias	Lymphopenia, elevated AST	Unremarkable	1	–
Huang [19]	China	8	–	–	8	Cough, fever	Elevated CRP	Extensive ground-glass opacities	1	–
Iqbal, S [25]	USA	1	–	–	1	Fever, chills, cough	Lymphopenia	Reticular interstitial opacities	–	–
Juusela, A [28]	USA	2	–	–	2	Emesis, fever	Elevated CRP	Pulmonary edema, ground-glass appearance	1	–
Kalafat, E [54]	Turkey	1	–	–	1	Cough, shortness of breath	NM	Bilateral ground-glass opacities	1	–
Karami, P [44]	Iran	1	–	–	1	Fever, cough	Leukopenia, thrombocytopenia	Faint bilateral patchy opacities	1	1
Kelly, JC [31]	USA	1	–	–	1	Emesis, cough	Lymphopenia, elevation of liver enzymes	Subsegmental atelectasis without consolidation	1	–
Khan, S [20]	China	17	–	–	17	Cough, diarrhea	Increased leukocyte count	NM	–	–
Kuhrt, K [49]	UK	1	–	–	1	Cough, fever,	Lymphopenia, thrombocytopenia	Pleural effusion,	–	–
Lang [13]	China	1	–	–	1	Cough	Elevated CRP	Bilateral lung infiltrates	–	–
Li [14]	China	1	–	–	1	Sore throat, cough	Lymphopenia, elevated CRP	Bilateral ground-glass opacity	1	–
Liu [17]	China	13		2	11	Fever, fatigue	NM	NM	1	–
Lowe [57]	Australia	1	–	–	1	Upper respiratory symptoms (no details)	NM	NM	–	–
Lyra [52]	Portugal	1	–	–	1	Cough	NM	NM	–	–
Martinelli [40]	Italy	1	–	–	1	Fever,dyspnea, rhinitis	Lymphopenia, elevated CRP	Ground-glass opacification	–	–
Mehta [27]	USA	1		1		Nasal congestion, cough	Normal	Patchy ground-glass opacities	1	–
Mulvey [26]	USA	5	–	–	5	Fever (n = 2), sore throat, cough	Decrease of PTT and albumin	NM	–	–

Panichaya [53]	Thailand	1	–	1		Fever, chest discomfort	Normal	Normal	–	–
Peng [21]	China	1	–		1	Fever, fatigue, shortness of breath	NM	Bilateral patchy nodular opacities	–	–
Pierce-Williams [24]	USA	64	–	64		NM	Elevated IL-6 levels, ferritin	NM	19	–
Qiancheng [12]	China	28	3	1	24	Cough, fever	Hypertension, elevated CRP	Patchy ground-glass shadows	NM	–
Schnettler [29]	USA	1	–	–	1	Cough, shortness of breath	Leukopenia, lymphopenia	Ground-glass appearance and consolidations	1	–
Shojaei [45]	Iran	1	–	1	–	Fever, cough	Elevated ESR	Patchy consolidation	1	1
Taghizadieh [46]	Iran	1	–	–	1	Flu-like symptoms (no details)	Increased BUN, Cr	Ground-glass opacity, consolidation	1	–
Vallejo [30]	USA	1	–	–	1	Fever, cough	Increase of D-dimer, LDH	Focal consolidations	1	1
Wu [15]	China	13	5	3	5	Fever, cough	Elevated CRP, leukocytosis,	Patchy ground-glass opacities or consolidation	–	–
Xia [7]	China	1	–	–	1	Fever	Elevated CRP	Consolidation, ground-glass-like shadows, interlobular septal thickening	–	–
Xiong [10]	China	1	–	–	1	Fever, cough	Lymphopenia, elevated CRP	Patchy shadows	–	–
Yan [11]	China	116	4	6	106	Fever, cough	Lymphopenia, elevated CRP	Patchy shadowing or ground-glass opacity	2	–
Yang [22]	China	27	4	–	23	Fever, cough	Lymphopenia, elevated D-dimer	Patchy ground-glass shadows	NM	–
Zamaniyan [47]	Iran	1	–	–	1	Dyspnea, myalgia, anorexia	Elevated CRP, lymphopenia	Unilateral atelectasis, bilateral ground-glass opacities	1	1

Table 26.2 Quantitative form of outcomes of the review study

n		571
Age range (years)		17–49
Gestational age	1st trimester (n)	80
	2nd trimester (n)	48
	3rd trimester (n)	379
	Not mentioned (n)	64
Comorbidities (n)		Obesity (42), gestational diabetes mellitus (26), chronic hypertension (24), pulmonary conditions (16), asthma (15), diabetes mellitus (9), preeclampsia/eclampsia (7), hypothyroidism (7), anemia (5), hepatitis B (4), gestational hypertension (3), coagulopathy (3), polycystic ovary syndrome (2), underweight (1), nephropathy (1), schistosomiasis (1), hypoproteinemia (1), pelvic fracture (1), cholecystitis (1), ulcerative colitis (1), severe scoliosis (1), Behçet syndrome (1), severe myopia (1), renal tubular acidosis (1), vitamin D deficiency (1), migraine headaches (1), autoimmune thyroiditis (1), mitral regurgitation (1), hyperlipidemia (1), thalassemia (1), papillary cell carcinoma (1), myotonic dystrophy (without cardiomyopathy) (1)
Asymptomatic (n)		76
Symptomatic (n)		453
Not mentioned (n)		42
Common symptoms at the onset of disease (n)		Not mentioned (64) Fever (256), cough (240), dyspnea (25), fatigue (24), myalgia (15), anosmia (7), shortness of breath (8), malaise (5), diarrhea (3), headache (3), vomiting (1), emesis (2), sore throat (3), nasal congestion (2), upper respiratory symptoms (no details) (2), chills (2), anorexia (1), rhinitis (1), chest discomfort (1)
Common laboratory findings (n)		Not mentioned (150) Elevated C-reactive protein (143), lymphopenia (134), increased serum levels of D-dimer (115), leukocytosis (42), leukopenia (28), increase of blood urea nitrogen, creatinine (12), impaired liver function (11), thrombocytopenia (6), raised lactic dehydrogenase (4), elevated total creatine phosphokinase (1), elevated ESR (1), elevated uric acid (1), elevated fibrinogen (1), pancytopenia (1)
Common radiological findings (n)		Not mentioned (207) Patchy shadowing or ground-glass opacity (181), bilateral infiltrates (95), unilateral or bilateral pulmonary lesions (30), consolidation (21), multifocal pneumonia (5), not done (4), irregular pleural line on lung ultrasound (4), atelectasis without consolidation (2), pleural effusion (2), pulmonary edema (1), bilateral, basal hyperlucency (1), typical signs of covid-19 pneumonia (no details) (1), interlobular septal thickening (1), pulmonary embolism (1)
Childbirth data (n)	Pregnancy ongoing	139
	Abortion	20
	Vaginal delivery	91
	Cesarean	321
Cesarean indications	Concerns about COVID-19	114/321
	Obstetrical indications	120/321
	Elective	4/321
	Not mentioned	83/321

(continued)

Table 26.2 (continued)

Obstetrical complications	IDFD		9
	Preterm labor		112/412
Maternal complications	Intubation and mechanical ventilation (n)	Yes	55
		No	419
		Not mentioned	97
	Complications during treatment of COVID-19 (n)		Acute respiratory distress syndrome (14), septic shock (4), cardiac dysfunction (2), multiple organ dysfunction syndrome (2), cardiac arrest (2), myocardial injury and myocarditis, endocarditis (2), needs dialysis (2), pericardial effusion (2), pulmonary embolism (2), cerebral emboli (1), *Staphylococcus aureus* bacteremia (1), developing anemia after being admission (1), heart failure and respiratory failure (1), placental abruption (1)
	Maternal death		13/571
Confirmed vertical transmission			25/316

tomatic. In general, the pattern of clinical signs and laboratory findings of COVID-19 infection in pregnant women does not appear to be very different from those reported for COVID-19 non-pregnant female cases. Fever, cough, and dyspnea were the most common symptoms. These findings are consistent with those from a study by Morales et al. of 656 individuals from the general population with COVID-19 infections [58].

Furthermore, there appear to be no major differences between pregnant women and non-pregnant adults with COVID-19 in terms of laboratory findings. Elevated CRP was the most common laboratory finding of SARS-CoV-2 pneumonia. In a study by Han et al., 99% of patients infected by SARS-CoV-2 had elevated CRP levels, indicative of inflammation [59].

Ground-glass opacity in the lungs was the most common radiological finding of COVID-19 infection in pregnant women. The radiological findings from 81 patients with COVID-19 pneumonia in Wuhan, China, showed that in subclinical patients, the common radiological pattern of COVID-19 was unilateral and multifocal ground-glass opacities [60]. Lesions quickly evolved to bilateral, diffuse ground-glass opacities in a week or less from the onset of the disease.

The results of this review showed that in women infected with COVID-19, the rate of Cesarean sections was higher than normal for various reasons. In general, the complications of Cesarean section are much greater than those of vaginal delivery, and there is still insufficient evidence to suggest that vertical transmission in Cesarean section is lower than in vaginal delivery. Therefore, decisions about the type of delivery should be made based on risk of complications as normal [61].

There is conflicting evidence for vertical transmission of the SARS-CoV-2 virus [62]. The vertical transmission rate in our study was 7.9%. In the present study, most pregnant women were in the third trimester of pregnancy, and therefore the time interval from the onset of clinical symptoms of the disease to the time of childbirth was short. Furthermore, approximately one quarter of those in the first trimester chose abortion because of the possible side effects of medications on the fetus. Therefore, the vertical transmission rate obtained from this study cannot be generalized to all pregnant women with COVID-19 infections.

Several studies have now shown that one of the highest predictors of a fatal outcome in COVID-19 cases is the presence of underlying diseases [63]. In our study, a significant percentage of pregnant women suffered from underlying diseases. The most common comorbidity was obesity. In obese individuals, expiratory reserve volume, functional capacity, and compliance of the respiratory system are reduced, and studies of H1N1 influenza cases have shown that caution should be exercised in caring for patients with severe obesity [64].

26.5 Conclusions

Clinical symptoms as well as laboratory and radiological findings of pneumonia caused by SARS-CoV-2 infection in pregnant females do not appear to be different from those found in general COVID-19 cases. However, further studies should be done to investigate the effects of underlying diseases and the causative factors for the higher rate of Cesarean delivery and preterm birth in pregnant women with COVID-19 infections.

Acknowledgments Thanks to guidance and advice from the Clinical Research Development Unit of Baqiyatallah Hospital.

Funding Source None.

References

1. McMurray JJ, Petrie MC, Murdoch DR, Davie AP (1998) Clinical epidemiology of heart failure: public and private health burden. Eur Heart J 19(Suppl P):P9–P16
2. Maiese K, Li F, Chong ZZ (2005) New avenues of exploration for erythropoietin. JAMA 293(1):90–95
3. De Boer RA, Pinto YM, Van Veldhuisen DJ (2003) The imbalance between oxygen demand and supply as a potential mechanism in the pathophysiology of heart failure: the role of microvascular growth and abnormalities. Microcirculation 10(2):113–126
4. Chen H, Guo J, Wang C, Luo F, Yu X, Zhang W et al (2020) Clinical characteristics and intrauterine vertical transmission potential of COVID-19 infection in nine pregnant women: a retrospective review of medical records. Lancet 395(10226):809–815
5. Liu D, Li L, Wu X, Zheng D, Wang J, Yang L et al (2020) Pregnancy and perinatal outcomes of women with coronavirus disease (COVID-19) pneumonia: a preliminary analysis. AJR Am J Roentgenol 18:1–6. https://doi.org/10.2214/AJR.20.23072. Online ahead of print
6. Yu N, Li W, Kang Q, Xiong Z, Wang S, Lin X et al (2020) Clinical features and obstetric and neonatal outcomes of pregnant patients with COVID-19 in Wuhan, China: a retrospective, single-centre, descriptive study. Lancet Infect Dis 20(5):559–564
7. Xia H, Zhao S, Wu Z, Luo H, Zhou C, Chen X (2020) Emergency Caesarean delivery in a patient with confirmed COVID-19 under spinal anaesthesia. Br J Anaesth 124(5):e216–e218
8. Chen L, Li Q, Zheng D, Jiang H, Wei Y, Zou L et al (2020) Clinical characteristics of pregnant women with Covid-19 in Wuhan, China. N Engl J Med:NEJMc2009226. https://doi.org/10.1056/NEJMc2009226. Online ahead of print
9. Chen R, Chen J, Meng QT (2020) Chest computed tomography images of early coronavirus disease (COVID-19). Can J Anaesth 67(6):754–755
10. Xiong X, Wei H, Zhang Z, Chang J, Ma X, Gao X et al (2020) Vaginal delivery report of a healthy neonate born to a convalescent mother with COVID-19. J Med Virol. https://doi.org/10.1002/jmv.25857. Online ahead of print
11. Yan J, Guo J, Fan C, Juan J, Yu X, Li J et al (2020) Coronavirus disease 2019 (COVID-19) in pregnant women: a report based on 116 cases. Am J Obstet Gynecol 23(20):30462–30462
12. Qiancheng X, Jian S, Lingling P, Lei H, Xiaogan J, Weihua L et al (2020) Coronavirus disease 2019 in pregnancy. Int J Infect Dis 95:376–383
13. Lang GJ, Zhao H (2020) Can SARS-CoV-2-infected women breastfeed after viral clearance? J Zhejiang Univ Sci B 21(5):405–407
14. Li J, Wang Y, Zeng Y, Song T, Pan X, Jia M et al (2020) Critically ill pregnant patient with COVID-19 and neonatal death within two hours of birth. Int J Gynaecol Obstet. https://doi.org/10.1002/ijgo.13189. Online ahead of print
15. Wu Y, Liu C, Dong L, Zhang C, Chen Y, Liu J et al (2020) Coronavirus disease 2019 among pregnant Chinese women: Case series data on the safety of vaginal birth and breastfeeding. BJOG. https://doi.org/10.1111/1471-0528.16276. Online ahead of print
16. Chen Y, Peng H, Wang L, Zhao Y, Zeng L, Gao H et al (2020) Infants born to mothers with a new coronavirus (COVID-19). Front Pediatr 16(8):104. https://doi.org/10.3389/fped.2020.00104
17. Liu Y, Chen H, Tang K, Guo Y (2020) Clinical manifestations and outcome of SARS-CoV-2 infection during pregnancy. J Infect. https://doi.org/10.1016/j.jinf.2020.02.028. Online ahead of print
18. Dong L, Tian J, He S, Zhu C, Wang J, Liu C et al (2020) Possible vertical transmission of SARS-CoV-2 from an infected mother to her newborn. JAMA 323(18):1846–1848
19. Huang W, Zhao Z, He Z, Liu S, Wu Q, Zhang X et al (2020) Unfavorable outcomes in pregnant patients with COVID-19 outside Wuhan. China J Infect. https://doi.org/10.1016/j.jinf.2020.05.014. [Epub ahead of print]
20. Khan S, Jun L, Nawsherwan SR, Li Y, Han G et al (2020) Association of COVID-19 with pregnancy outcomes in health-care workers and general women. Clin Microbiol Infect 26(6):788–790
21. Peng Z, Wang J, Mo Y, Duan W, Xiang G, Yi M et al (2020) Unlikely SARS-CoV-2 vertical transmission from mother to child: a case report. J Infect Public Health 13(5):818–820
22. Yang H, Hu B, Zhan S, Yang LY, Xiong G (2020) Effects of SARS-CoV-2 infection on pregnant women and their infants: a retrospective study in Wuhan, China. Arch Pathol Lab Med 18(10):2020–0232

23. Anderson J, Schauer J, Bryant S, Graves CR (2020) The use of convalescent plasma therapy and remdesivir in the successful management of a critically ill obstetric patient with novel coronavirus 2019 infection: A case report. Case Rep Womens Health. e00221. https://doi.org/10.1016/j.crwh.2020.e00221. Online ahead of print
24. Pierce-Williams RAM, Burd J, Felder L, Khoury R, Bernstein PS, Avila K et al (2020) Clinical course of severe and critical COVID-19 in hospitalized pregnancies: a US cohort study. Am J Obstet Gynecol MFM. 100134. https://doi.org/10.1016/j.ajogmf.2020.100134. Online ahead of print
25. Iqbal SN, Overcash R, Mokhtari N, Saeed H, Gold S, Auguste T et al (2020) An uncomplicated delivery in a patient with Covid-19 in the United States. N Engl J Med 382(16):e34. https://doi.org/10.1056/NEJMc2007605
26. Mulvey JJ, Magro CM, Ma LX, Nuovo GJ, Baergen RN (2020) A mechanistic analysis placental intravascular thrombus formation in COVID-19 patients. Ann Diagn Pathol 46:151529. https://doi.org/10.1016/j.anndiagpath.2020.151529
27. Mehta H, Ivanovic S, Cronin A, VanBrunt L, Mistry N, Miller R et al (2020) Novel coronavirus-related acute respiratory distress syndrome in a patient with twin pregnancy: A case report. Case Rep Womens Health. e00220. https://doi.org/10.1016/j.crwh.2020.e00220. Online ahead of print
28. Juusela A, Nazir M, Gimovsky M (2020) Two cases of coronavirus 2019-related cardiomyopathy in pregnancy. Am J Obstet Gynecol MFM. 100113. https://doi.org/10.1016/j.ajogmf.2020.100113. Online ahead of print
29. Schnettler WT, Al Ahwel Y, Suhag A (2020) Severe ARDS in COVID-19-infected pregnancy: obstetric and intensive care considerations. Am J Obstet Gynecol MFM. 100120. https://doi.org/10.1016/j.ajogmf.2020.100120. Online ahead of print
30. Vallejo V, Ilagan JG (2020) A Postpartum Death Due to Coronavirus Disease 2019 (COVID-19) in the United States. Obstet Gynecol. https://doi.org/10.1097/AOG.0000000000003950. Online ahead of print
31. Kelly JC, Dombrowksi M, O'Neil-Callahan M, Kernberg AS, Frolova AI, Stout MJ (2020) False-Negative COVID-19 Testing: Considerations in Obstetrical Care. Am J Obstet Gynecol MFM. 100130. https://doi.org/10.1016/j.ajogmf.2020.100130. Online ahead of print
32. Hong L, Smith N, Keerthy M, Lee-Griffith M, Garcia R, Shaman M et al (2020) Severe COVID-19 infection in pregnancy requiring intubation without preterm delivery: a case report. Case Rep Womens Health 27:e00217. https://doi.org/10.1016/j.crwh.2020.e00217
33. Hirshberg A, Kern-Goldberger AR, Levine LD, Pierce-Williams R, Short WR, Parry S et al (2020) Care of critically ill pregnant patients with COVID-19: a case series. Am J Obstet Gynecol. S0002–9378(20)30515–9. https://doi.org/10.1016/j.ajog.2020.04.029. Online ahead of print
34. Browne PC, Linfert JB, Perez-Jorge E (2020) Successful treatment of preterm labor in association with acute COVID-19 infection. Am J Perinatol. https://doi.org/10.1055/s-0040-1709993. Online ahead of print
35. Breslin N, Baptiste C, Gyamfi-Bannerman C, Miller R, Martinez R, Bernstein K et al (2020) COVID-19 infection among asymptomatic and symptomatic pregnant women: Two weeks of confirmed presentations to an affiliated pair of New York City hospitals. Am J Obstet Gynecol MFM. 100118. https://doi.org/10.1016/j.ajogmf.2020.100118. Online ahead of print
36. Blauvelt CA, Chiu C, Donovan AL, Prahl M, Shimotake TK, George RB et al (2020) Acute Respiratory Distress Syndrome in a Preterm Pregnant Patient With Coronavirus Disease 2019 (COVID-19). Obstet Gynecol. https://doi.org/10.1097/AOG.0000000000003949. Online ahead of print
37. Buonsenso D, Raffaelli F, Tamburrini E, Biasucci DG, Salvi S, Smargiassi A et al (2020) Clinical role of lung ultrasound for the diagnosis and monitoring of COVID-19 pneumonia in pregnant women. Ultrasound Obstet Gynecol. https://doi.org/10.1002/uog.22055. Online ahead of print
38. De Castro A, Abu-Hishmeh M, El Husseini I, Paul L (2019) Haemophilus parainfluenzae endocarditis with multiple cerebral emboli in a pregnant woman with coronavirus. IDCases. 18:e00593. https://doi.org/10.1016/j.idcr.2019.e00593
39. De Socio GV, Malincarne L, Arena S, Troiani S, Benedetti S, Camilloni B et al (2020) Delivery in asymptomatic Italian woman with SARS-CoV-2 infection. Mediterr J Hematol Infect Dis 12(1):e2020033. https://doi.org/10.4084/MJHID.2020.033
40. Martinelli I, Ferrazzi E, Ciavarella A, Erra R, Iurlaro E, Ossola M et al (2020) Pulmonary embolism in a young pregnant woman with COVID-19. Thromb Res 191:36–37
41. Ferrazzi E, Frigerio L, Savasi V, Vergani P, Prefumo F, Barresi S et al (2020) Vaginal delivery in SARS-CoV-2 infected pregnant women in Northern Italy: a retrospective analysis. BJOG. https://doi.org/10.1111/1471-0528.16278. Online ahead of print
42. Carosso A, Cosma S, Borella F, Marozio L, Coscia A, Ghisetti V et al (2020) Pre-labor anorectal swab for SARS-CoV-2 in COVID-19 pregnant patients: is it time to think about it? Eur J Obstet Gynecol Reprod Biol 249:98–99
43. Hantoushzadeh S, Shamshirsaz AA, Aleyasin A, Seferovic MD, Aski SK, Arian SE et al (2020) Maternal death due to COVID-19 disease. Am J Obstet Gynecol 28(20):30516–30510
44. Karami P, Naghavi M, Feyzi A, Aghamohammadi M, Novin MS, Mobaien A et al (2020) Mortality of a pregnant patient diagnosed with COVID-19: A case report with clinical, radiological, and histopathological findings. Trav Med Infect Dis. 101665. https://doi.

org/10.1016/j.tmaid.2020.101665. Online ahead of print
45. Shojaei S, Kouchek M, Miri MM, Salarian S, Sistanizad M, Ariana S et al (2020) Twin pregnant woman with COVID-19: a case report. Journal of Cellular & Molecular Anesthesia 5(1):43–46
46. Taghizadieh A, Mikaeili H, Ahmadi M, Valizadeh H (2020) Acute kidney injury in pregnant women following SARS-CoV-2 infection: A case report from Iran. Respir Med Case Rep. 30:101090. https://doi.org/10.1016/j.rmcr.2020.101090. Online ahead of print
47. Zamaniyan M, Ebadi A, Aghajanpoor Mir S, Rahmani Z, Haghshenas M, Azizi S (2020) Preterm delivery in pregnant woman with critical COVID-19 pneumonia and vertical transmission. Prenat Diagn. https://doi.org/10.1002/pd.5713. Online ahead of print
48. Govind A, Essien S, Kartikeyan A, Fakokunde A, Janga D, Yoong W et al (2020) Re: Novel Coronavirus COVID-19 in late pregnancy: Outcomes of first nine cases in an inner city London hospital. Eur J Obstet Gynecol Reprod Biol. S0301–2115(20)30257–8. https://doi.org/10.1016/j.ejogrb.2020.05.004. Online ahead of print
49. Kuhrt K, McMicking J, Nanda S, Nelson-Piercy C, Shennan A (2020) Placental abruption in a twin pregnancy at 32 weeks' gestation complicated by COVID-19, without vertical transmission to the babies. Am J Obstet Gynecol MFM. 100135. https://doi.org/10.1016/j.ajogmf.2020.100135. Online ahead of print
50. Ahmed I, Azhar A, Eltaweel N, Tan BK (2020) First Covid-19 maternal mortality in the UK associated with thrombotic complications. Br J Haematol. https://doi.org/10.1111/bjh.16849. Online ahead of print
51. Dória M, Peixinho C, Laranjo M, Varejão AM, Silva PT (2020) Covid-19 during pregnancy: a case series from an universally tested population from the north of Portugal. Eur J Obstet Gynecol Reprod Biol. https://doi.org/10.1016/j.ejogrb.2020.05.029. Online ahead of print
52. Lyra J, Valente R, Rosário M, Guimarães M (2020) Cesarean Section in a Pregnant Woman with COVID-19: First Case in Portugal. Acta Med Port. https://doi.org/10.20344/amp.13883. Online ahead of print
53. Panichaya P, Thaweerat W, Uthaisan J (2020) Prolonged viral persistence in COVID-19 second trimester pregnant patient. Eur J Obstet Gynecol Reprod Biol. https://doi.org/10.1016/j.ejogrb.2020.05.030. Online ahead of print
54. Kalafat E, Yaprak E, Cinar G, Varli B, Ozisik S, Uzun C et al (2020) Lung ultrasound and computed tomographic findings in pregnant woman with COVID-19. Ultrasound Obstet Gynecol 55(6):835–837
55. Alzamora MC, Paredes T, Caceres D, Webb CM, Valdez LM, La Rosa M (2020) Severe COVID-19 during pregnancy and possible vertical transmission. Am J Perinatol https://doi.org/10.1055/s-0040-1710050. Online ahead of print
56. Gidlöf S, Savchenko J, Brune T, Josefsson H (2020) COVID-19 in pregnancy with comorbidities: more liberal testing strategy is needed. Acta Obstet Gynecol Scand https://doi.org/10.1111/aogs.13862. Online ahead of print
57. Lowe B, Bopp B (2020) COVID-19 vaginal delivery – a case report. Aust N Z J Obstet Gynaecol. https://doi.org/10.1111/ajo.13173. Online ahead of print
58. Rodriguez-Morales AJ, Cardona-Ospina JA, Gutiérrez-Ocampo E, Villamizar-Peña R, Holguin-Rivera Y, Escalera-Antezana JP et al (2020) Clinical, laboratory and imaging features of COVID-19: a systematic review and meta-analysis. Travel Med Infect Dis 34:101623. https://doi.org/10.1016/j.tmaid.2020.101623
59. Han R, Huang L, Jiang H, Dong J, Peng H, Zhang D (2020) Early clinical and CT manifestations of coronavirus disease 2019 (COVID-19) pneumonia. AJR Am J Roentgenol 17:1–6
60. Shi H, Han X, Jiang N, Cao Y, Alwalid O, Gu J et al (2020) Radiological findings from 81 patients with COVID-19 pneumonia in Wuhan, China: a descriptive study. Lancet Infect Dis 20(4):425–434
61. Chen D, Yang H, Cao Y, Cheng W, Duan T, Fan C et al (2020) Expert consensus for managing pregnant women and neonates born to mothers with suspected or confirmed novel coronavirus (COVID-19) infection. Int J Gynaecol Obstet 149(2):130–136
62. Amouroux A, Attie-Bitach T, Martinovic J, Leruez-Ville M, Ville Y (2020) Evidence for and against vertical transmission for SARS-CoV-2 (COVID-19). Am J Obstet Gynecol. S0002–9378(20)30524-X. https://doi.org/10.1016/j.ajog.2020.04.039. Online ahead of print
63. Ruan Q, Yang K, Wang W, Jiang L, Song J (2020) Clinical predictors of mortality due to COVID-19 based on an analysis of data of 150 patients from Wuhan, China. Intensive Care Med 46(5):846–848
64. Dietz W, Santos-Burgoa C (2020) Obesity and its implications for COVID-19 mortality. Obesity (Silver Spring) 28(6):1005. https://doi.org/10.1002/oby.22818

The 2019 Novel Coronavirus Disease in Pregnancy: A Systematic Review

27

Somayeh Makvandi, Mitra Mahdavian, Goli Kazemi-Nia, Amir Vahedian-Azimi, Paul C. Guest, Leila Karimi, and Amirhossein Sahebkar

Abstract

In December 2019, a respiratory disease caused by a coronavirus called SARS-CoV-2 (COVID-19) began in Wuhan, China, and quickly became a pandemic. In such situations, pregnant women are suspected of being among the vulnerable groups. The aim of this study was to report clinical symptoms, laboratory findings, and obstetrical complications, maternal, fetal, and neonatal complications of COVID-19 infection in pregnant women. We searched the Cochrane library, MEDLINE/PubMed, and Web of Sciences from their inception to April 5, 2020. Any study involving pregnant women with COVID-19 which evaluated the effect of the disease on pregnancy outcomes and fetal and neonatal complications was included in the study. The outcomes were the symptoms and laboratory findings, obstetrical complications, mode of delivery, and maternal, fetal, and neonatal complications. The search resulted in 69 titles and abstracts, which were narrowed down to

S. Makvandi
Department of Midwifery, School of Nursing and Midwifery, Islamic Azad university Ahvaz branch, Ahvaz, Iran

M. Mahdavian
Department of Midwifery, School of Nursing and Midwifery, Islamic Azad university Bojnourd branch, Bojnourd, Iran

G. Kazemi-Nia
Sina Hospital, Ahvaz Jundishapur University of Medical Sciences, Ahvaz, Iran

A. Vahedian-Azimi
Trauma Research Center, Nursing Faculty, Baqiyatallah University of Medical Sciences, Tehran, Iran

P. C. Guest
Laboratory of Neuroproteomics, Department of Biochemistry and Tissue Biology, Institute of Biology, University of Campinas (UNICAMP), Campinas, Brazil

L. Karimi (✉)
Behavioral Sciences Research Center, Life style institute, Nursing Faculty, Baqiyatallah University of Medical Sciences, Tehran, Iran
e-mail: leilakarimi1487@gmail.com

A. Sahebkar (✉)
Neurogenic Inflammation Research Center, Mashhad University of Medical Sciences, Mashhad, Iran

Biotechnology Research Center, Pharmaceutical Technology Institute, Mashhad University of Medical Sciences, Mashhad, Iran

Polish Mother's Memorial Hospital Research Institute (PMMHRI), Lodz, Poland

Halal Research Center of IRI, FDA, Tehran, Iran
e-mail: sahebkara@mums.ac.ir; amir_saheb2000@yahoo.com

P. C. Guest (ed.), *Clinical, Biological and Molecular Aspects of COVID-19*, Advances in Experimental Medicine and Biology 1321, https://doi.org/10.1007/978-3-030-59261-5_27

12 studies involving 68 women. The three most common symptoms of patients were fever, cough, and fatigue. The most common laboratory findings were an increase in C-reactive protein (CRP) and lymphopenia. The most common obstetrical complication was preterm labor (33.3%). No maternal deaths were reported. The Cesarean section rate was 83.3% and the vertical transition rate was 2.23%. The findings showed that the clinical symptoms and laboratory measures of pregnant women affected by COVID-19 did not differ from the general population. In general, the prognosis of mothers who suffered from COVID-19 and their newborns was satisfactory. However, there is a need for further rigorous studies to confirm these findings as the pandemic progresses.

Keywords

COVID-19 · Novel coronavirus infection · Pregnancy outcome

27.1 Introduction

Coronaviruses are one of the largest viral families that can cause common cold to more severe diseases such as severe acute respiratory syndrome (SARS) and Middle East respiratory syndrome (MERS) [1]. Coronaviruses have been prevalent in some birds and mammals, but there have been several examples of their occurrence in humans in recent years. The new coronavirus disease, which began in December 2019 with a widespread human outbreak in Wuhan, China, is known as COVID-19 [2]. The disease has quickly crossed borders between countries and continents, forcing the World Health Organization to declare a state of emergency. This has now become a pandemic and has attracted the attention of many countries [3].

In cases of disease epidemics, pregnant women should be classified as a vulnerable group due to lower body immunity. Changes in the cardiovascular system during pregnancy, including an increase in heart rate, increased oxygen consumption, and decreased lung capacity, increase the risk of developing severe respiratory diseases in pregnant women [4, 5].

In the context of the 2019 coronavirus pandemic, there are several unanswered questions about the effects of the disease on a pregnant woman and her fetus and neonate:

1. Are the clinical symptoms and laboratory findings of pregnant women suffering from COVID-19 infection similar to those in general patients?
2. Do pregnant women with COVID-19 infection have a higher risk for pregnancy complications?
3. What is the rate of vertical transmission of the disease to the fetus or neonate?

To answer these questions, a systematic review of the available related evidence was conducted.

27.2 Methods

27.2.1 Data Sources and Search Strategy

Two investigators (SM and LK) conducted searches of the Cochrane library, Cochrane Central Register of Controlled Trials (CENTRAL), MEDLINE/PubMed, and Web of Sciences up to April 5, 2020. Additionally, Google Scholar was searched with the aim of exploring the citations of final studies included in the systematic review. The search strategy used was:

"COVID-19" OR "COVID19" OR "2019 novel coronavirus infection" OR "2019-nCoV infection" OR "COVID-19 pandemic" OR "coronavirus disease-19" OR "2019-nCoV disease" OR "2019 novel coronavirus disease" OR "coronavirus disease 2019."

AND

"Pregnancy" OR "Pregnancy Outcomes" OR "Pregnancy Outcome" OR "Outcome, Pregnancy" OR "Outcomes, Pregnancy."

The reference lists of articles were checked for further relevant publications.

27.2.2 Inclusion Criteria

Any type of study involving pregnant women with COVID-19 and evaluating the effect of the disease on pregnancy outcomes and fetal and neonatal complications was included in the study. Only studies published in English were reviewed.

27.2.3 Outcomes

Our outcomes were the symptoms and laboratory characteristics of COVID-19, obstetrical complications, mode of delivery, maternal complications, detection of coronavirus in products of conception, and the infants and fetal and neonatal complications.

27.2.4 Study Selection and Data Extraction

Two authors independently reviewed the search output. We screened titles and abstracts of search results to exclude irrelevant studies. We then retrieved full text articles of seemingly relevant studies and examined these to see whether or not they met the inclusion criteria. The reviewers (SM and LK) resolved any disagreement through discussion and consensus agreement. We designed a data extraction form. Data were extracted by the two reviewers using the agreed form. Any disagreement was resolved by consensus.

27.3 Results

27.3.1 Results of the Search

Figure 27.1 shows the PRISMA flowchart for study inclusion and exclusion. A total of 69 studies corresponding to our search strategy were identified. Of these, 14 irrelevant records were excluded based on the title and abstract review. The full-text articles for the remaining 55 articles were retrieved. After review of the full-text articles, 43 articles were excluded and 12 studies met our inclusion criteria for systematic review. The characteristics of these studies are presented in Table 27.1.

27.3.2 Participants and Settings

A total of 68 participants were included across all 12 studies, ranging from one to 15 persons per study. Mothers ranged in age from 22 to 41 years old. One patient was in the first trimester, four women were in the second trimester of pregnancy, and another 63 were in the third trimester. Five women had a history of past medical conditions. Eleven studies were conducted in China [6–16], and one was performed in Honduras [17].

27.3.3 Outcomes

27.3.3.1 Symptoms of COVID-19 at Admission

The three most common symptoms of patients at the time of hospitalization were fever (76.5%), cough (39.7%), and fatigue (13.2%), respectively. Other symptoms included dyspnea (11.8%), myalgia (10.3%), sore throat (8.8%), diarrhea (5.9%), nasal congestion (5.9%), and skin rash (1.5%; one patient) (Table 27.1).

27.3.3.2 Laboratory Characteristics

Examination of lung computerized tomography (CT) scans or chest X-rays showed some evidence in favor of coronavirus disease in 45 of the 51 patients tested (88.2%). The most common laboratory finding was an increase in C-reactive protein (CRP), which was detected in 31 of the 44 cases tested (74.3%). The next most common biomarker tested was for the presence of lymphopenia, which occurred in 30 out of 43 cases (70.5%). Other laboratory findings included decreased platelet counts (23.1%) and increased levels of liver enzymes (11.62%) (Table 27.1).

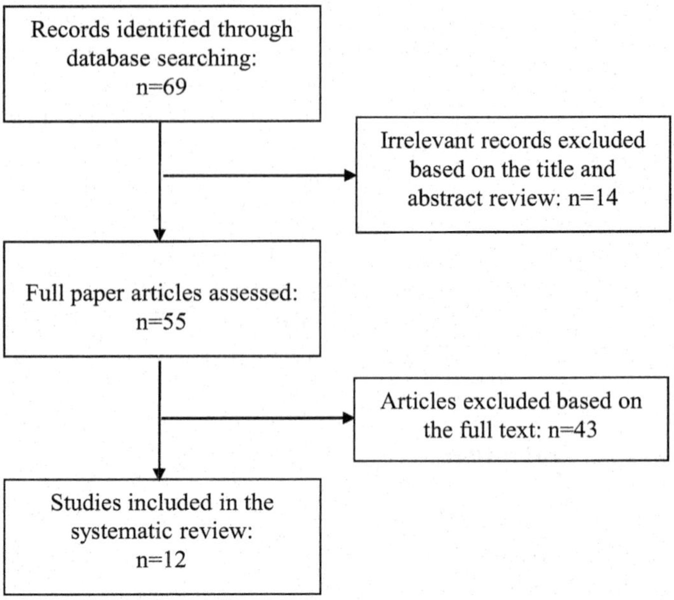

Fig. 27.1 Flowchart for systematic review

27.3.3.3 Mode of Delivery

Fifty women had a Cesarean section (83.33%). Ten had a vaginal delivery (16.66%) and eight were still pregnant by the end of the study.

27.3.3.4 Obstetrical Complications

Of the 60 cases that gave birth, 20 were preterm (33.3%). Other complications included 11 cases of fetal distress (18.3%), six cases of premature rupture of membranes (PROM) (10%), three cases of gestational diabetes (4.5%), two cases of preeclampsia (3%), and two cases of gestational hypertension (3%).

27.3.3.5 Maternal Complications

There were no reports of maternal deaths in the studied cases. The need for mechanical ventilation was found in only one case (1.5%).

27.3.3.6 Fetal and Neonatal Complications

From 45 tests for detection of coronavirus in products of conception and the infants, there were 44 negative tests (97.7%) and one test was positive (2.2%). One case of stillbirth [10], one case of neonatal death [11], and seven cases of lung infection were reported [11, 12, 14]. No severe asphyxia was reported.

27.4 Discussion

In the present review, the most common symptoms of the disease in pregnant women were fever, cough, and fatigue. These findings suggest that the symptoms at the onset of the disease in pregnant women do not differ from those in the normal population. A report of 72,314 records in China showed that in patients with coronavirus, typical symptoms were fever, cough, and fatigue [18]. In another study, Huang et al. showed that common symptoms at onset of illness were fever (98%), cough (76%), and myalgia or fatigue (44%) and less common symptoms were sputum production, headache, hemoptysis, and diarrhea [2]. These findings are consistent with our study.

In a study of 149 patients with COVID-19 infection, Yang et al. found that among the laboratory biomarker measures, elevated CRP and lymphopenia were the most common findings [2]. These results are also consistent with present study. Therefore, in terms of laboratory measures, it seems that there is no difference between pregnant women and non-pregnant patients.

The findings showed that the rate of Cesarean section for termination of pregnancy was much higher than for vaginal delivery. An expert

Table 27.1 Characteristics of included studies

	First author	Cuifang Fan	Dehan Liu	Huijun Chen	Nan Yu	Siyu Chen	Xiaotong Wang	Yangli Liu	Lysien Ivania Zambrano	Huaping Zhu	Yang Li	Lan Dong	Suliman Khan
Characteristics	Number of cases	2	15	9	7	5	1	13	1	10	1	1	3
	Age of mothers (years)	29–34	23–40	26–40	29–34	25–31	28	22–36	41	25–35	30	29	27–33
	Gestational age (weeks)	36–37	12–38	36–39	37–41	38–41	30	25–38	31	31–39	35	34	34–38
Symptoms of COVID-19	Fever	2	13	7	6	–	+	10	+	9	–	+	2
	Cough	–	9	4	1	1	–	2	+	5	+	–	3
	Fatigue	–	4	–	–	–	–	4	+	–	–	–	–
	Dyspnea	–	1	1	1	–	–	3	–	–	–	+	1
	Myalgia	–	3	3	–	–	–	–	+	–	–	–	–
	Sore throat	1	1	2	–	–	–	1	–	1	–	–	–
	Diarrhea	–	1	1	1	–	–	–	–	1	–	–	–
	Nasal congestion	2	–	–	–	1	–	–	–	–	–	+	–
	Skin rash	1	–	–	–	–	–	–	–	–	–	–	–
Laboratory characteristics	Evidence on lung CT or CXR in favor of coronavirus disease	2	15	8	7		+	NM	NM	10	+	+	NM
	Lymphopenia	1	12	5	5	4	+	NM	NM	NM	–	+	2
	Low platelet	NM	NM	NM	2	–	NM	NM	NM	NM	+	NM	NM
	Elevated CRP	NM	10	6	7	–	+	NM	NM	NM	NM	+	2
	Elevated liver enzymes	NM	NM	3	2	–	–	NM	NM	NM	NM	–	–

(continued)

Table 27.1 (continued)

	First author	Cuifang Fan	Dehan Liu	Huijun Chen	Nan Yu	Siyu Chen	Xiaotong Wang	Yangli Liu	Lysien Ivania Zambrano	Huaping Zhu	Yang Li	Lan Dong	Suliman Khan
Obstetrical complications	Gestational hypertension	–	–	1	–	–	–	–	+	–	–	NM	–
	Gestational diabetes	–	1	–	–	2	–	–	–	–	–	NM	–
	Preeclampsia	–	–	1	–	1	–	–	–	–	–	NM	–
	Fetal distress	–	–	2	–	–	–	3	–	6	–	NM	–
	PROM	–	–	2	–	–	–	1	–	3	–	NM	–
	Premature delivery	–	–	4	–	–	+	6	–	6	+	+	1
Mode of delivery	Vaginal delivery	–	1	–	–	3	–	–	–	3	–	–	3
	Cesarean	2	10	9	7	2	+	10	–	7	+	+	–
	Pregnant at end of study	–	4	–	–	–	–	3	+	–	–	–	–
Maternal complications	Death	–	–	–	–	–	–	–	–	–	–	–	–
	Mechanical ventilation	–	–	–	–	–	–	1	–	–	–	–	–
Fetal and neonatal complications	Detection of coronavirus in products of conception and the infants	Negative	NM	Only six were checked and were negative	Of the threecases checked, one was positive	Negative	Negative	Negative	NA	Negative	Negative	Negative	Negative
	Stillbirth	–	–	–	–	–	–	1	NA	–	–	–	–
	Low birth weight	–	–	2	–	–	+	–	NA	7	–	–	–
	Lung infection	2	NM	–	1	–	–	–	NA	4	–	–	–
	Severe asphyxia	–	–	–	–	–	–	–	NA	–	–	–	–
	Neonatal death	–	–	–	–	–	–	–	NA	1	–	–	–
	In hospital	–	–	–	–	–	–	–	–	4	–	–	–

NM not mentioned, *NA* not applicable, *CRP* C-reactive protein

consensus for managing pregnant women and neonates born to mothers with suspected or confirmed novel coronavirus infection stated that at present, there is no conclusive evidence of the best delivery method to reduce the risk of vertical transmission [19]. In other words, whether or not Cesarean section can reduce the risk of vertical transmission in COVID-19 has not yet been confirmed. According to the evidence, the decision on the time and type of delivery in pregnant women suffering from COVID-19 infections is a multidisciplinary effort influenced by several factors such as the patient's clinical condition and obstetrical factors [20].

Examination of obstetric complications in women with COVID-19 infection showed that preterm delivery was more common than other complications. The findings of a study by Mascio et al. showed that in 41% of pregnant women with the disease, preterm delivery occurred as the most common obstetrical complication [20]. This could be due to premature termination of pregnancy due to the potential risks of COVID-19 infection to the mother and fetus.

There were no reports of maternal deaths in the studied cases. In line with this finding, a review of 41 pregnant women with COVID-19 infection did not report any maternal deaths [5]. This suggests that the mortality rate of pregnant mothers due to COVID-19 is lower than that which occurred with SARS and MERS. A review by Dashraath et al. concluded that the mortality rate in pregnant women due to SARS and MERS was 18 to 25 percent [5].

In the present study, the vertical transition rate of COVID-19 was 2.2%. However, it was not possible to judge whether or not this finding is conclusive. This is because the majority of the women studied were in the last trimester of pregnancy, and it is not clear what the rate of transmission to the fetus would have been if the disease had occurred earlier in the pregnancy.

Finally, it is important to remember that there is little data on the effect of 2019 novel coronavirus on pregnancy and its maternal, fetal, and neonatal complications. The articles reviewed above were all from small studies and quality was not assessed. Therefore, the interpretation of the findings of this study should be considered with caution. Another limitation of the present study was that only English-language studies were included. Due to the fact that the outbreak of this disease has been Wuhan City, China, articles in Chinese may also have been published in this regard.

Another potential limitation of this study relates to the fact that analysis of biomarkers was limited to measurement of circulating CRP, lymphocyte levels, and markers of liver damage. This may be a critical factor to consider in future studies as reports indicate different manifestations of COVID-19 infection due to different patient susceptibilities and disease severities [21]. The *Handbook of COVID-19 Prevention and Treatment* has recommended the use of serum antibody testing for monitoring immune response of the host to infection [22]. It has also recommended testing for circulating CRP levels as most cases of COVID-19 infection with significantly elevated CRP have been linked to more severe disease, consistent with the Centers for Disease Control and Prevention (CDC) guidelines in the USA [23]. In line with this, an elevation in CRP levels was the most consistent finding in the studies analyzed in this review. Lymphopenia is considered a cardinal laboratory biomarker with prognostic potential in systemic infections such as in COVID-19 cases [24]. In addition, lactate dehydrogenase (LDH) may serve as a biomarker of liver damage, and CRP and IL-6 may be used to identify cases with severe inflammation and poor prognosis. Another finding in COVID-19 patients has been reports of hypercoagulation [25]. Thus biomarkers associated with activation of the clotting cascade should be assessed. This could be performed rapidly using test strips which measure clotting time in combination with a metered readout and monitoring using a Smartphone app, as described by Vegt and Guest in 2018 [26].

27.5 Conclusions

The findings of this review study showed that the clinical symptoms and laboratory findings of pregnant women affected by COVID-19 infection did not differ from the general population. However, preterm labor and Cesarean section appeared more likely to occur in pregnant women suffering from COVID-19 infection than their counterparts. In general, the prognosis of newborns of infected mothers was satisfactory. Due to the lack of data, the authors strongly recommend that more quality studies be performed on pregnant women affected by COVID-19 infection in all three trimesters of pregnancy to achieve more accurate and definitive results. We also recommend the incorporation of additional biomarker readouts in these studies to detect and monitor COVID-19-related disease complications.

Acknowledgments Thanks to guidance and advice from the Clinical Research Development Unit of Baqiyatallah Hospital.

Funding Source None

References

1. Guarner J (2020) Three emerging coronaviruses in two decades: the story of SARS, MERS, and now COVID-19. Am J Clin Pathol 153(4):420–421
2. Huang C, Wang Y, Li X, Ren L, Zhao J, Hu Y et al (2020) Clinical features of patients infected with 2019 novel coronavirus in Wuhan, China. Lancet 395(10223):497–506
3. Xu Z, Shi L, Wang Y, Zhang J, Huang L, Zhang C et al (2020) Pathological findings of COVID-19 associated with acute respiratory distress syndrome. Lancet Respir Med 8(4):420–422
4. Wong SF, Chow KM, Leung TN, Ng WF, Ng TK, Shek CC et al (2004) Pregnancy and perinatal outcomes of women with severe acute respiratory syndrome. Am J Obstet Gynecol 191(1):292–297
5. Dashraath P, Jeslyn WJL, Karen LMX, Min LL, Sarah L, Biswas A et al (2020) Coronavirus Disease 2019 (COVID-19) Pandemic and Pregnancy. Am J Obstet Gynecol. pii: S0002–9378(20)30343–4. https://doi.org/10.1016/j.ajog.2020.03.021. [Epub ahead of print]
6. Chen H, Guo J, Wang C, Luo F, Yu X, Zhang W et al (2020) Clinical characteristics and intrauterine vertical transmission potential of COVID-19 infection in nine pregnant women: a retrospective review of medical records. Lancet 395(10226):809–815
7. Chen S, Liao E, Shao Y (2020) Clinical analysis of pregnant women with 2019 novel coronavirus pneumonia. J Med Virol. https://doi.org/10.1002/jmv.25789. [Epub ahead of print]
8. Dong L, Tian J, He S, Zhu C, Wang J, Liu C et al (2020) Possible vertical transmission of SARS-CoV-2 from an infected mother to her newborn. JAMA. https://doi.org/10.1001/jama.2020.4621. [Epub ahead of print]
9. Li Y, Zhao R, Zheng S, Chen X, Wang J, Sheng X et al (2020) Lack of vertical transmission of severe acute respiratory syndrome coronavirus 2, China. Emerg Infect Dis 26(6). https://doi.org/10.3201/eid2606.200287. [Epub ahead of print]
10. Liu Y, Chen H, Tang K, Guo Y (2020) Clinical manifestations and outcome of SARS-CoV-2 infection during pregnancy. J Infect. pii: S0163–4453(20)30109–2. https://doi.org/10.1016/j.jinf.2020.02.028. [Epub ahead of print]
11. Zhu H, Wang L, Fang C, Peng S, Zhang L, Chang G et al (2020) Clinical analysis of 10 neonates born to mothers with 2019-nCoV pneumonia. Translat Pediatr 9(1):51–60
12. Fan C, Lei D, Fang C, Li C, Wang M, Liu Y et al (2020) Perinatal transmission of COVID-19 associated SARS-CoV-2: should we worry? Clin Infect Dis. Pii: ciaa226. https://doi.org/10.1093/cid/ciaa226. [Epub ahead of print]
13. Liu D, Li L, Wu X, Zheng D, Wang J, Yang L et al (2020) Pregnancy and perinatal outcomes of women with coronavirus disease (COVID-19) pneumonia: a preliminary analysis. AJR Am J Roentgenol 18:1–6
14. Yu N, Li W, Kang Q, Xiong Z, Wang S, Lin X et al (2020) Clinical features and obstetric and neonatal outcomes of pregnant patients with COVID-19 in Wuhan, China: a retrospective, single-Centre, descriptive study. Lancet Infect Dis 20(5):559–564
15. Wang X, Zhou Z, Zhang J, Zhu F, Tang Y, Shen X (2020) A case of 2019 novel coronavirus in a pregnant woman with preterm delivery. Clin Infect Dis. Pii: ciaa200. https://doi.org/10.1093/cid/ciaa200. [Epub ahead of print]
16. Khan S, Peng L, Siddique R, Nabi G, Xue M, Liu J et al (2020) Impact of COVID-19 infection on pregnancy outcomes and the risk of maternal-to-neonatal intrapartum transmission of COVID-19 during natural birth. Infect Control Hosp Epidemiol 19:1–3
17. Zambrano LI, Fuentes-Barahona IC, Bejarano-Torres DA, Bustillo C, Gonzales G, Vallecillo-Chinchilla G et al (2020) A pregnant woman with COVID-19 in Central America. Travel Med Infect Dis. 101639. https://doi.org/10.1016/j.tmaid.2020.101639. [Epub ahead of print]

18. Tabatabaeizadeh SA, Avan A, Bahrami A, Khodashenas E, Esmaeili H, Ferns GA et al (2017) High-dose supplementation of vitamin D affects measures of systemic inflammation: reductions in high-sensitivity C-reactive protein level and neutrophil to lymphocyte ratio (NLR) distribution. J Cell Biochem 118(12):4317–4322
19. Chen D, Yang H, Cao Y, Cheng W, Duan T, Fan C et al (2020) Expert consensus for managing pregnant women and neonates born to mothers with suspected or confirmed novel coronavirus (COVID-19) infection. Int J Gynaecol Obstet 149(2):130–136
20. Di Mascio D, Khalil A, Saccone G, Rizzo G, Buca D, Liberati M et al (2020) Outcome of coronavirus spectrum infections (SARS, MERS, COVID 1-19) during pregnancy: a systematic review and meta-analysis. Am J Obstet Gynecol MFM 25:100107. https://doi.org/10.1016/j.ajogmf.2020.100107
21. Singhal T (2020) A review of coronavirus Disease-2019 (COVID-19). Indian J Pediatr 87(4):281–286
22. Handbook of COVID-19 Prevention and Treatment. Chinese guidelines (2020). alnap.org/help-library/handbook-of-covid-19-prevention-and-treatment
23. Interim Clinical Guidance for Management of Patients, CDC guidelines (2020). www.cdc.gov/coronavirus/2019-ncov/hcp/clinical-guidance-management-patients.html
24. Terpos E, Ntanasis-Stathopoulos I, Elalamy I, Kastritis E, Sergentanis TN, Politou M et al (2020) Hematological findings and complications of COVID-19. Am J Hematol. https://doi.org/10.1002/ajh.25829. [Epub ahead of print]
25. Violi F, Pastori D, Cangemi R, Pignatelli P, Loffredo L (2020) Hypercoagulation and Antithrombotic Treatment in Coronavirus 2019: A New Challenge Thromb Haemost. https://doi.org/10.1055/s--0040-1710317. [Epub ahead of print]
26. Vegt J, Guest PC (2018) A user-friendly app for blood coagulation disorders. Methods Mol Biol 1735:499–504

Acute Kidney Injury and Covid-19: A Scoping Review and Meta-Analysis

28

Mehdi Jafari-Oori, Marco Fiorentino, Giuseppe Castellano, Abbas Ebadi, Farshid Rahimi-Bashar, Paul C. Guest, Amir Vahedian-Azimi, and Amirhossein Sahebkar

Abstract

Acute kidney injury (AKI) is a fatal complication of the new severe acute respiratory syndrome coronavirus (SARS-CoV-2) which causes COVID-19 disease. Here, we performed a scoping review and meta-analysis including clinical studies on patients with SARS-CoV-2 infection with data on AKI assessment and characteristics, and the overall prevalence of AKI was estimated using a random-effects model. We identified 21 articles which passed the search criteria. All were quantitative observational studies which used a cross-sectional, retrospective, case report, or cohort methodology. This showed that aging, diabetes, cardiovascular disease, previous chronic disease, and other comorbidities were risk factors of AKI. Although the prevalence of proteinuria, hematuria, and increased serum creatinine was reported for up to 60% of the patients with COVID-19, the overall prevalence of AKI was estimated to be 8%. We conclude that although approximately two-thirds of patients with COVID-19 had symptoms of kidney damage, most of these did not meet the diagnostic criteria for AKI. Further studies should be performed to validate biomarkers

M. Jafari-Oori · A. Vahedian-Azimi (✉)
Trauma Research Center, Nursing Faculty, Baqiyatallah University of Medical Sciences, Tehran, Iran
e-mail: amirvahedian63@gmail.com

M. Fiorentino
Department of Emergency and Organ Transplantation, Nephrology Dialysis and Transplantation Unit, "Aldo Moro" University, Bari, Italy

G. Castellano
Nephrology, Dialysis and Transplantation Unit, Department of Medical and Surgical Science, University of Foggia, Foggia, Italy

A. Ebadi
Behavioral Sciences Research Center, Life style institute, Nursing Faculty, Baqiyatallah University of Medical Sciences, Tehran, Iran

F. Rahimi-Bashar
Anesthesia and Critical Care Department, Hamadan University of Medical Sciences, Hamadan, Iran

P. C. Guest
Laboratory of Neuroproteomics, Department of Biochemistry and Tissue Biology, Institute of Biology, University of Campinas (UNICAMP), Campinas, Brazil

A. Sahebkar (✉)
Biotechnology Research Center, Pharmaceutical Technology Institute, Mashhad University of Medical Sciences, Mashhad, Iran

Neurogenic Inflammation Research Center, Mashhad University of Medical Sciences, Mashhad, Iran

Polish Mother's Memorial Hospital Research Institute (PMMHRI), Lodz, Poland

Halal Research Center of IRI, FDA, Tehran, Iran
e-mail: sahebkara@mums.ac.ir; amir_saheb2000@yahoo.com

P. C. Guest (ed.), *Clinical, Biological and Molecular Aspects of COVID-19*, Advances in Experimental Medicine and Biology 1321, https://doi.org/10.1007/978-3-030-59261-5_28

for improved AKI diagnosis in COVID-19 patients and new treatment options are required to reduce the rate of mortality.

Keywords

Acute kidney injury · COVID-19 · Meta-analysis · Scoping review

28.1 Introduction

In December 2019, a serious respiratory infectious disease caused by a novel coronavirus (SARS-CoV-2) occurred in China, which has now been formally named by the World Health Organization (WHO) as "coronavirus 2019" (COVID-19) [1]. The disease has spread quickly from Wuhan to other countries around the world, and the WHO announced the outbreak of this new coronavirus as an international health emergency concern [2]. Lung involvement represents the main complication of the disease, as coronavirus predominantly affects lung epithelial alveolar cells inducing an interstitial pneumonia and, consequently, acute respiratory distress syndrome (ARDS) [3, 4]. However, the occurrence of acute kidney injury (AKI) following COVID-19 disease has been widely reported, as up to 25% of patients who died developed AKI [5–7].

The AKI definition is based on standard criteria according to the Kidney Disease: Improving Global Outcomes (KDIGO) recommendations: (1) serum creatinine (SCR) ≥26 μmol/L (0.3 mg/dL) within 48 h or SCR >1.5 times of baseline value in the past 7 days or (2) a urine volume <0.5 mL/kg/h for 6 h [8, 9]. Interestingly, patients who developed AKI during the course of the disease exhibited a higher mortality rate than other patients [7]. Because most of the medications given to treat patients with COVID-19 are excreted by the kidneys, injuries to this organ can interfere with the metabolism, excretion, dosage, and expected concentration of the drugs, which increases their toxicity [6].

AKI was observed in 5% to 15% of patients with severe acute respiratory syndrome (SARS) and Middle East respiratory syndrome (MERS) coronavirus infections, with 60–90% mortality [7, 10]. However, different findings regarding COVID-19 and risk for AKI have been reported by clinical studies [11]. The present study aims to review the main available data on this issue through a systematic research of the current literature, with a special focus on the incidence of AKI, and clinical characteristics and outcomes of patients developing AKI during the course of the disease.

28.2 Methods

28.2.1 Design of Study

A scoping literature review is a repetitive method of available literature in the field to point out the width and depth of an issue [12]. This study was performed according to the Preferred Reporting Items for Systematic Reviews and Meta-Analyses (PRISMA) guidelines [13–15]. A flowchart of study selection procedure is provided in Fig. 28.1, and the methodological quality of primary studies was estimated.

28.2.2 Search Strategy

In this review, clinical studies conducted worldwide and published in the English or Persian languages in internal and external databases were assessed up to April 2020. To search for studies related to AKI and COVID-19, the national databases including Magiran, IranMedex, Iranian Archive for Scientific Documents Center (IASD), and Iranian National Library (INL) and international databases such as MEDLINE (PubMed, Ovid), Scopus, Web of Science, Embase, ProQuest, and Google and Google Scholar as search engines were used. Gray literature and reference lists of the extracted primary articles were also reviewed to find related studies. The keywords and subject headings used to search these

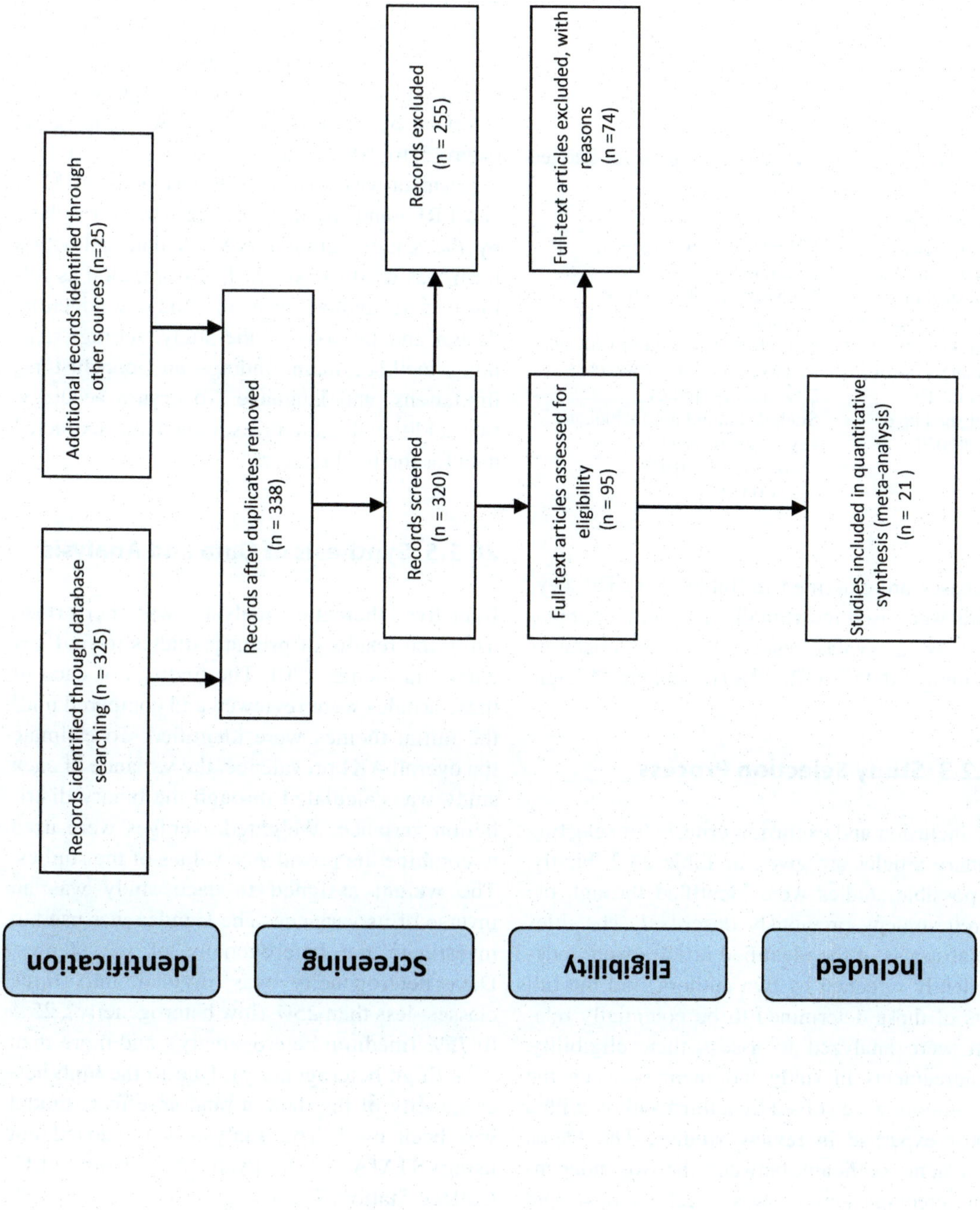

Fig. 28.1 The process of identification and selection of studies

Table 28.1 Keywords and subject headings used during the search

Search terms
"Novel coronavirus" OR "novel coronavirus 2019" OR "2019 novel coronavirus" OR "2019 nCoV" OR "Wuhan coronavirus" OR "Wuhan pneumonia" OR COVID-19 OR "2019-nCoV" OR "SARS-CoV-2" OR "coronavirus 2019" OR "2019-nCoV" and "acute kidney injury" OR "acute renal injury" OR "acute renal insufficiencies" OR "acute kidney insufficiencies" OR "acute kidney failure" OR "acute kidney tubular necrosis"

Table 28.2 Inclusion and exclusion criteria for selected primary articles

Inclusion criteria	Exclusion criteria
Peer-reviewed, primary studies	Not peer-reviewed, primary research
Published in English or Persian	Not written in English or Persian
Pointing to the AKI on patients with COVID-19	Not related to the empirical data (letters, editorials, news, etc.)
Published up to April 2020	Low-quality studies
	Studies that did not include the keywords of "KAI," "coronavirus 2019," "COVID-19"

databases are classified in Table 28.1. The keywords were used individually and with combinations as a syntax using Boolean operators including "AND," "OR," "NOT," and the "*" sign.

28.2.3 Study Selection Process

The inclusion and exclusion criteria for selecting primary articles are given in Table 28.2. Firstly, all possible studies were identified through the search strategy previously described. The titles and abstracts of the identified articles were independently screened by two authors, and the full texts of those determined to be potentially relevant were analyzed to assess their eligibility. Disagreements in study judgment between the two authors were solved by a third author, with a robust expertise in review studies. The kappa agreement coefficient between the two independent screening authors was excellent ($r = 0.98$ and P-value <0.0001).

28.2.4 Quality Assessment and Data Extraction

To evaluate the quality of studies, a five-item tool was applied as described in previous studies [16–19]. The five items were related to the research design, sampling method, sample size, comparison group, and psychometric properties. Each item scaled from 0 to 3 with an overall score from 0 to 15 [17]. Based on this approach, the studies were divided into three quality categories, characterized by weak (score 4 or lower), moderate (score 5 to 10), or strong (score over 10). The assessment was performed by two authors (MJO and FRB) and the disagreements were resolved by the senior author (AVA). A data extraction form was used to extract the information as follows: first author, year of study conduction, design and purpose of the study, setting, sampling method, main findings and conclusions, limitations, and language. To ensure accuracy, two additional authors examined the extracted data for the final review.

28.2.5 Synthesis of Data and Analysis

Inductive thematic analysis was performed using the results of primary studies to find key emerging themes [20]. The findings of each of these articles were reviewed and compared until the initial themes were identified. To estimate the overall AKI prevalence, the variance of each study was calculated through the binary distribution variance. Weighted averages were used to combine the prevalence values of the studies. The weight assigned to each study was an inverse of its variance. The I^2 index was used to investigate the heterogeneity of the studies. Data heterogeneity was divided into three classes: less than 25% (low heterogeneity), 25% to 75% (medium heterogeneity), and more than 75% (high heterogeneity). Due to the high heterogeneity of the data, a random-effects model was been used. The analysis was carried out using STATA 12 software (StataCorp LLC; College Station, TX, USA).

28.3 Results

28.3.1 Literature Search

The flow diagram of the selection process is illustrated in Fig. 28.1. A total of 350 potentially relevant references were obtained after reviewing databases and search engines and the reference list of relevant studies. After a first screening for title and abstract, 95 studies were selected for full-text evaluation, while duplicate and irrelevant articles were excluded. Finally, 21 primary studies which met the criteria were selected for review and analysis [3, 7, 11, 21–38].

28.3.2 Study Characteristics

The characteristics of the 21 final articles are shown in Table 28.3. Study design was different between the selected articles which were either cross-sectional, retrospective cohort, or case reports. All studies were written in English language, and most were conducted in China during the early phase of the COVID-19 epidemic.

28.3.3 Methodological Quality Appraisal

The quality of all articles was examined by the five-item tool, as reported in Table 28.3. Based on this scale, a total of 17 studies had a strong quality, three studies were moderate [22, 37, 38], and only one was classified as weak [24].

28.3.4 Narrative Summary of Themes

28.3.4.1 AKI and COVID-19

Data from the primary articles were classified and discussed as follows: (1) pathology of COVID-19, (2) clinical and laboratory features, (3) diagnosis, (4) risk factors, (5) prevalence and mortality, and (6) treatments.

28.3.4.2 Pathology of the COVID-19 Virus

In all primary studies, except in one [11], a relationship between AKI and the COVID-19 virus was described. AKI occurred in both patients with and without preexisting chronic kidney disease [27, 39]. Various possible hypotheses have been reported regarding the pathogenesis of acute renal impairment. The most frequent mechanism described is the presence of angiotensin-converting enzyme II (ACE-2) receptors in kidney cells. The SARS-CoV-2 virus uses ACE-2 enzymes as receptors to facilitate its entry into renal target cells. ACE-2 is located at the renal epithelial tubular cells and may cause an inflammatory response, which leads to AKI [35]. In addition, COVID-19 infection induces a systemic inflammatory response, especially in critically ill patients, which could increase the risk of AKI episodes [35]. However, a direct pathogenic role of SARS-CoV-2 in the human kidney should be considered, as a postmortem kidney biopsy study reported the presence of particles related to COVID-19 in the cytoplasm of epithelial cells in the tubules [38]. This mechanism could explain the presence of the virus in the kidneys and urine of these patients. In addition to ACE-2, transmembrane serine protease (TMPRSSs) is another receptor that facilitates SARS-CoV-2 entry into kidney cells. The kidney epithelial cluster is divided into seven subgroups, including proximal convoluted tubule cells, loop of Henle, distal tubules, podocyte, collecting duct, proximal straight tubule cells, and proximal tubule cells. Genetic analysis of the ACE-2 and TMPRSS genes showed a relatively high expression in glomerular epithelial cells (podocytes) and proximal straight tubule cells, which are host to the entrance of COVID-19 and other coronaviruses [26, 35, 38, 40]. In addition to epithelial kidney tubules, ACE-2 is also present in bladder epithelial cells, thus suggesting that the urinary tract is a potential route for SARS-CoV-2 virus infection along with the respiratory and gastrointestinal tract [40].

Table 28.3 Characteristics of the primary articles

Number	First author	Year	Aim of the study	Study design	Sample size	Country	Type of data gathering	Main findings	Study limitations	Quality score
1	Wang [11]	2020	Investigate effects of SARS-CoV-2 infection on renal function by analyzing the clinical data of 116 patients with COVID-19 hospitalization	Cross-sectional	116 COVID-19-confirmed patients (111 COVID-19-confirmed patients without chronic kidney disease (CKD) and five with CKD)	China	Demographic data, medical history, contact history, potential comorbidities, symptoms, signs, laboratory test results, chest computed tomography (CT) scans, and treatment measures	Of the 111 patients without CKD, 12 patients (10.8%) showed a slight increase in blood urea or creatinine nitrogen, and eight (7.2%) showed trace or + 1 albuminuria, and zero patients get AKI. Kidney function indicators showed stable status in five patients with CKD, without exacerbation of CKD, and were treated and survived	There were no pre-dialysis patients with CKD which barred monitoring of the patients because they were at high threat of AKI	Strong
2	Cheng [32]	2020	Demonstration of the prevalence and in-hospital outcome of AKI in COVID-19 patients	Consecutive cohort	710 COVID-19 patients	China	Hematuria, proteinuria, serum creatinine concentration, and other clinical parameters	Of the 710 COVID-19 patients: 89 (12.3%) died, 44% had proteinuria and hematuria, and 26.9% had hematuria. The prevalence of serum creatinine and blood urea nitrogen was 15.5% and 14.1%. AKI resulted in 3.2% of cases	Unmeasured or unknown confounders, lack of long-term evaluation effect of the virus on the kidneys due to lack of access to the clinical information of patients	Strong

3	Cheng [7]	2020	Prevalence and in-hospital outcome of AKI in COVID-19 patients	Consecutive cohort	701 COVID-19 patients	China	Hematuria, proteinuria, serum creatinine concentration, and other clinical parameters	Mortality rate: 16.1% (113 pts) On admission: 43.9% of patients had proteinuria and 26.7% hematuria. Elevated serum creatinine: 14.4% Elevated blood urea nitrogen: 13.1% AKI: 5.1% stage 1: 2%, stage 2: 1%, stage 3: 2%, CRRT information NA	Unmeasured or unknown confounders, lack of long-term evaluation effect of the virus on the kidneys due to lack of access to clinical information of patients	Strong
4	Xu [35]	2020	Investigation of potential mechanism of SARS-CoV-2 in AKI at single-celled level		15 normal human kidney samples	China	Single-cell RNA sequencing (scRNA-seq)	NA	NA	Strong
5	Chu [3]	2005	Describing the clinical, pathologic, and laboratory features of AKI	Retrospective	536 SARS patients with normal plasma creatinine at first clinical presentation and kidney tissues from seven other patients postmortem	China	Laboratory test, light microscopy, and electron microscopy	36 (6.7%) developed AKI from 5 to 48 days after the beginning of viral disease regardless of a normal plasma CR level at first clinical day. 33 SARS cases (91.7%) with AKI passed away. The mortality rate of patients with SARS and AKI compared with those with SARS and no renal impairment: 91.7% vs 8.8%. Renal tissue mainly showed acute tubular necrosis and there was no evidence of glomerular pathology ARDS and age: risk factors of AKI	NA	Strong

(continued)

Table 28.3 (continued)

Number	First author	Year	Aim of the study	Study design	Sample size	Country	Type of data gathering	Main findings	Study limitations	Quality score
6	CHEN [22]	2003	To assess relationship of rhabdomyolysis with AKI with severe acute respiratory syndrome	Cross-sectional	Thirty possible SARS patients	Taiwan	Hematological and biochemistry data	Three (10%) patients had rhabdomyolysis with associated AKI	NA	Moderate
7	Li [23]	2020	Evaluation of kidney damage in patients with COVID-19	Retrospective	59 patients infected by SARS-CoV-2 (including 28 diagnosed as severe cases and three deaths)	China	Laboratory test results, chest CT scans	63% (32/51) of the patients: proteinuria, 19% (11/59) of the patients: elevated level of plasma creatinine and urea nitrogen, respectively, and 27% (16/59) of patients: elevated level of urea nitrogen abnormalities of the kidneys according to CT scan: (27/27 patients)100%	Only 27 of the 51 patients underwent CT scans	Strong
8	AlGhamdi [24]	2015	Reporting of clinical presentations and outcomes of two renal transplant recipients with MERS-CoV infection	A case report	Two cases of MERS-CoV infections in two renal transplant recipients	China	Laboratory test results	NA	The results of two patients with MERS-CoV kidney transplant recipients were as follows: one with poor results and the other with optimal results	Weak
9	Deng [25]	2020	Analyzing basic characteristics of patients who succumbed to and who recovered from COVID-19	Retrospective	109 COVID-19 patients	China	Laboratory test results	The incidence of AKI was higher in the non-recovery group compared to patients who recovered (18.3% vs 0, $p < 0.001$)	NA	Strong

10	Yang [34]	2020	Describing clinical course and consequences of extremely ill patients with SARS-CoV-2 pneumonia	Retrospective observational	52 critically ill adult patients with SARS-CoV-2 pneumonia	China	Demographic data, symptoms, laboratory values, comorbidities, treatments, and clinical outcomes	32 (61·5%) patients died at 28 days. 15 (29%) had acute kidney injury	Sample size, lack of control group	Strong
11	Lin [26]	2020	Assessing angiotensin-converting enzyme II (ACE-2) in kidney/bladder		Gene expression matrices of scRNA-Seq data from normal kidneys of three healthy donors	China	NA	ACE-2 expression distributed across multiple cell types. Particularly, ACE-2 was mainly boosted in proximal tubule cells, including both convoluted tubule and straight tubule	NA	Strong
12	Chen [33]	2020	To describe clinical properties of patients with COVID-19 who passed away	Retrospective, observational	From a cohort of 799 cases, 113 who passed away and 161 who revived with a diagnosis of COVID-19 were analyzed	China	Clinical characteristics and laboratory findings	AKI in deceased patients (n = 28; 25%)	Some laboratory tests (e.g., cardiac troponin I, N-terminal pro-brain natriuretic peptide, and arterial blood gas tests) were not done in all the patients, and missing data or important tests might lead to bias of clinical characteristics	Strong

(continued)

Table 28.3 (continued)

Number	First author	Year	Aim of the study	Study design	Sample size	Country	Type of data gathering	Main findings	Study limitations	Quality score
13	Wang [27]	2020	Describe epidemiological and clinical characteristics of COVID-19 pneumonia	Retrospective, single-center case series	138 consecutive hospitalized patients with confirmed COVID-19 pneumonia	China	Epidemiological, demographic, clinical, laboratory, radiological, and treatment data	AKI in five (3.6%), 3/36 in ICU (8%), 2/102 outside ICU (2%). CRRT rate: 1.5%	NA	Strong
14	Choi [21]	2003	Assess characteristics and consequences of patients with SARS	Retrospective cohort	267 patients (227 cases of confirmed SARS and 40 cases of probable SARS)	China	Clinical, laboratory, and radiographic measures and 3-month mortality rate	The 3-month mortality rate was 12%. Factors contributing to mortality were respiratory failure, acute renal failure (6%), and nosocomial sepsis	Retrospective and relied on data collected from case records	Strong
15	Diao [36]	2020	Assess effect of COVID-19 on human kidney	Retrospective	85 patients	China	Immunohistochemistry	27.06% (23/85) of patients showed acute renal failure (ARF). Elderly patients and patients with underlying diseases such as high blood pressure and heart failure were more likely to show ARF (65.22% vs 24.19%, $p < 0.001$; 69.57% vs 11.29%, $p < 0.001$, respectively). Virus particles were seen in the kidneys	NA	Strong
16	Guan [28]	2020	Identifying clinical characteristics of COVID-19 in China	Cross-sectional	1099 patients	China	NA	AKI: six patients	COVID-19 caused AKI	Strong
17	Pacciarini [38]	2008	Sensitivity of kidney glomerulus tubes to SARS-CoV infection	Interventional	NA	NA	Laboratory devices	These findings suggest that proximal tubules of human kidney nephrons may be a site of SARS-CoV reproduction and persistent reproduction	NA	Moderate

18	Yeung [37]	2016	To show mechanism of kidney cell damage in a laboratory setting	Interventional	NA	NA	NA	Cellular genes that are significantly disturbed by MERS-CoV have been implicated in kidney disease. Furthermore, MERS-CoV induced apoptosis through upregulation of Smad7 (a type of protein that regulates cell growth) and fibroblast growth factor 2 (FGF2) expression in both kidney and lung cells	NA	Moderate
19	Drosten [29]	2013	Clinical features and virological analysis of a case with COVID-19	Case report	1	Germany	Clinical, laboratory, and radiographic measures	Maximum virus concentration in urine samples, and stool samples obtained on days 12 and 16 contained the virus. AKI was observed	NA	Strong
20	Chen [30]	2020	Explaining epidemiological and clinical features of COVID-19 pneumonia	Retrospective, single-center	99 patients with COVID-19	China	RT-PCR, clinical, and radiological features and laboratory data	Three patients (3%) with AKI	Low sample size, only confirmed case	Strong
21	Guery [31]	2013	Reporting of detailed clinical and virological data for two related cases of MERS-CoV disease	Case report	Two patients with MERS-CoV with underlying immunosuppressive disorders	France and Dubai	Whole blood, plasma, and serum specimens for MERS-CoV by real-time RT-PCR	Both patients developed acute renal failure	NA	Strong

28.3.4.3 Clinical Characteristics and Diagnosis of AKI

In patients with AKI due to COVID-19 disease, an increase of serum creatinine and low urine output have been described according to the KDIGO criteria [7, 11, 23, 32]. In addition, a significant proportion of patients exhibited proteinuria and hematuria. In addition, some histological data is available that typically describes the presence of acute tubular necrosis, and kidney abnormalities have also been described through computed tomography (CT) scans [23]. The classic diagnosis of AKI is based on serum creatinine and urine output and is divided into three stages based on these criteria [8, 9].

28.3.4.4 Risk Factors of AKI

In some of the primary articles, the risk factor of AKI was discussed. Old age [3], ARDS [3], underlying diseases such as high blood pressure and heart failure [36], taking immunosuppressive drugs, male gender, and high serum creatinine levels [7] predisposed patients to acute kidney damage. Increased leukocytes and decreased lymphocytes and platelets, coagulation disorders such as prolonged activation of partial thromboplastin, high D-dimer, and increased procalcitonin, aspartate aminotransferase, and lactate dehydrogenase are also predictive of AKI [7].

28.3.4.5 The Prevalence of AKI and Associated Mortality

The prevalence of AKI was reported in 13 initial studies [3, 7, 11, 22, 23, 27–29, 31, 33, 34, 36]. Apart from two case report studies [29, 31] and another study reporting of 0% of patients with AKI [11], the overall prevalence was calculated for ten studies at 8% (Fig. 28.2). Although the overall prevalence of AKI was low, a high mortality rate was reported for those patients who developed AKI [25]. In a study of the 2003 SARS coronavirus, although the prevalence of AKI was reported to be 6.7%, 91.7% of patients with AKI died [3]. The mortality rate in patients without AKI was only 8.8% [3]. A recent study of the SARS-CoV-2 coronavirus showed that the mortality was significantly higher in patients with high baseline serum creatinine (33.7%) than in patients with normal baseline serum creatinine (13.2%) levels [7].

28.3.4.6 Treatment of AKI

Current treatment of AKI in patients with COVID-19 and other coronaviruses includes general management and supportive and continuous renal replacement therapy (CRRT). There is currently no effective antiviral therapy. Supportive care such as complete bed rest, giving enough nutrients and fluid, maintaining good blood pressure and oxygenation, prevention, and treatment of potential complications such as secondary infections is important [3, 7, 32, 33]. Early hospitalization in an intensive care unit is recommended for critically ill patients. Preventive measures of AKI in patients with COVID-19 include the use of diuretics, serum electrolyte adjustment, corticosteroids, plasma infusions, and administration of monoclonal antibody therapeutics.

28.4 Discussion

In this study, we conducted a comprehensive review of renal involvement in patients with COVID-19 and other coronavirus infections, reporting the main findings on the pathogenesis of the virus in the kidney, clinical manifestation of AKI, the risk or underlying factors, prevalence and mortality data, and treatments.

Regarding the pathogenesis of the COVID-19 infection, there are two hypothesized mechanisms: (1) a systemic inflammatory response with a cytokine storm and (2) direct cell damage by the virus. The ACE-2 and dipeptidyl peptidase 4 (DPP4) enzymes expressed in renal tubular cells, were identified as the binding partners for SARS-CoV and MERS-CoV, respectively [41, 42]. The RNA of SARS-CoV-2 has been found in kidney and urinary tissues, and it has previously been isolated in urine [43, 44]. A study of the SARS-CoV virus suggested the importance of the systemic inflammatory syndrome in inducing multi-organ failure and death due to the cytokine storm [45]. In this study, Huang et al. reported that the cytokine storm was related to interferon-

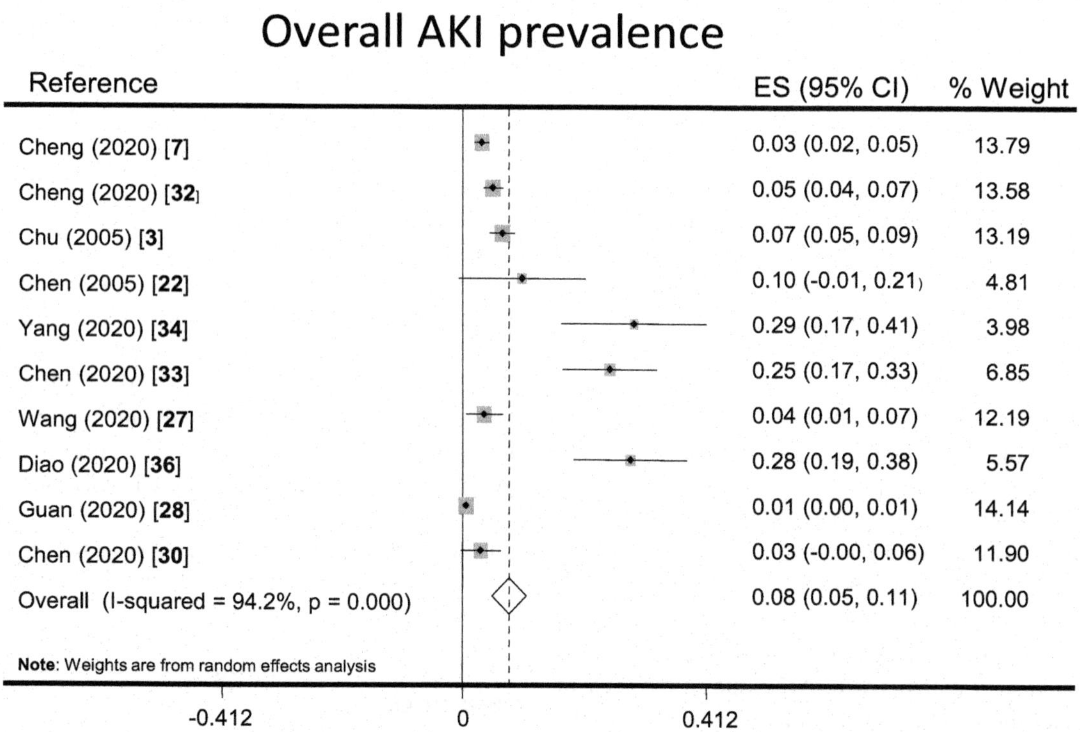

Fig. 28.2 Forest plot diagram for estimating the overall prevalence of AKI using the random model

gamma, which causes severe organ damage like AKI in SARS patients [45]. Moreover, other studies have shown that cytokine-induced inflammatory AKI has been reported in several clinical conditions [46]. ARDS-associated AKI may be caused by a variety of reasons, including gas exchange disorders, hemodynamic changes such as fluid overload with right heart failure and systemic congestion, harmful mechanical ventilation strategies, and the development of secondary infections. Age, severity of the disease, diabetes, and previous renal failure have been described as the main risk factors of AKI [47].

All the studies analyzed used the classical AKI definition, which is based on serum creatinine level and urine output [8]. In addition, some studies reported prevalence data on the exact stage of AKI. Considering the limitations of an AKI diagnosis based on only these parameters indicative of kidney damage, the adoption of new molecular biomarkers for AKI detection [48]. As examples, a study identified and validated measurement of urinary tissue inhibitor of metalloproteinases 2 (TIMP-2) and insulin-like growth factor binding protein 7 (IGFBP7) as a way of predicting moderate to severe AKI within 12 h of hospitalization in intensive care [49]. This is currently the only test approved by the Food and Drug Administration (FDA) in this setting, suggesting that a wider application of the test may help physician in preventing progression to AKI. A later study showed that urinary [TIMP-2]·[IGFBP7] greater than 0.3 $ng/mL^2/1{,}000$ identified patients at highest risk for imminent AKI [50].

Several studies have shown that, although a large number of patients with COVID-19 infection experienced increased serum creatinine levels, only a limited proportion of patients met the diagnostic criteria of AKI. It is important to note that studies including only critically ill COVID-19 patients are characterized by higher rate of AKI episodes. A study carried in New York of 5,449 patients admitted into hospital with COVID-19 found that 36.6% of these developed AKI based on KDIGO criteria [51].

Conversely, our meta-analysis suggested that the incidence of AKI in COVID-19 patients was lower at around 8%. In addition to the standard criteria for AKI, several studies have reported laboratory findings suggesting renal damage. In particular, hematuria and proteinuria have been reported in several clinical studies. In one of the early studies carried out in China, 10.8% of patients with COVID-19 had a slight increase in serum urea and creatinine levels, and 7.2% of these patients had albuminuria, while none had AKI [11]. In addition, Cheng et al. examined 710 patients with COVID-19 and reported that 44% of the patients had proteinuria and 26.9% had hematuria [32]. In previous reports related to the SARS-CoV and MERS-CoV infections, AKI was reported in 5% to 15% of patients, with a high mortality rate (60–90%) [10].

Our study has some limitations. We were only able to provide an estimate of the overall mortality rate in patients with AKI, due to the lack of relevant data in primary studies. Also, due to the novelty of the disease and the lack of related knowledge, we were unable to identify the main mechanism of renal impairment. Therefore, we have only provided a summary of the results of the preliminary studies in this area. Also, due to these limitations, we did not perform statistical methods of meta-analysis such as subgroup analysis and publication bias assessment.

In conclusion, this review updates the current evidences of AKI in patients with COVID-19 infection. Although no definitive data are available to date about the pathogenesis of AKI in these patients, a direct effect on kidney cells and the systemic inflammatory syndrome induced by viral infection are the two-leading cause of AKI in this setting. The incidence and prevalence of AKI in patients with COVID-19 are low and may be as low as 8%. However, most patients also presented other urinary abnormalities such as hematuria and albuminuria. Finally, the onset of AKI is predictive of worse short-term outcomes, as mortality rate in these patients is significantly higher than in patients without AKI. Thus special attention should be made to identify patients at risk of AKI using appropriate biomarker testing so these patients can be placed on appropriate treatment programs for the best possible outcome.

It is possible that recent global efforts to identify potential treatments of COVID-19 using repurposed drugs will be successful in curbing the incidence and effects of AKI [52]. Of these candidates, the Recovery (Randomised Evaluation of COVID-19 Therapy) trial, conducted by the University of Oxford, UK, has just shown that the anti-inflammatory steroid dexamethasone improved survival by approximately 30% in critically ill patients [53]. Although further research is required, this is the most promising drug identified to date for improving outcomes of patients with COVID-19 disease.

Acknowledgments Thanks to guidance and advice from the Clinical Research Development Unit of Baqiyatallah Hospital.

References

1. Lu H, Stratton CW, Tang YW (2020) Outbreak of pneumonia of unknown etiology in Wuhan, China: the mystery and the miracle. J Med Virol 92(4):401–402
2. World Health Organization (2020) Novel coronavirus (2019-nCoV) situation report – 1, 21 Jan 2020. https://www.who.int/docs/default-source/coronaviruse/situation-reports/20200121-sitrep-1-2019-ncov.pdf?sfvrsn=20a99c10_4
3. Chu KH, Tsang WK, Tang CS, Lam MF, Lai FM, To KF et al (2005) Acute renal impairment in coronavirus-associated severe acute respiratory syndrome. Kidney Int 67(2):698–705
4. ARDS Definition Task Force, Ranieri VM, Rubenfeld GD, Thompson BT, Ferguson ND, Caldwell E et al (2012) Acute respiratory distress syndrome: the Berlin Definition. JAMA 307(23):2526–2533
5. Perico L, Benigni A, Remuzzi G (2020) Should COVID-19 concern nephrologists? Why and to what extent? The emerging impasse of angiotensin blockade. Nephron 144(5):213–221
6. Rismanbaf A, Zarei S (2020) Liver and kidney injuries in COVID-19 and their effects on drug therapy; a letter to editor. Arch Acad Emerg Med 8:e17
7. Cheng Y, Luo R, Wang K, Zhang M, Wang Z, Dong L et al (2020) Kidney disease is associated with in-hospital death of patients with COVID-19. Kidney Int 97(5):829–838
8. Kidney Disease Improving Global Outcomes (2012) KDIGO clinical practice guideline for acute kidney injury. Kidney Int Suppl 2:1–138. https://kdigo.org/

wp-content/uploads/2016/10/KDIGO-2012-AKI-Guideline-English.pdf
9. Fiorentino M, Castellano G, Kellum JA (2017) Differences in acute kidney injury ascertainment for clinical and preclinical studies. Nephrol Dial Transplant 32:1789–1805. https://doi.org/10.1093/ndt/gfx002
10. Naicker S, Yang CW, Hwang SJ, Liu BC, Chen JH, Jha V (2020) The novel coronavirus 2019 epidemic and kidneys. Kidney Int 97(5):824–828
11. Wang L, Li X, Chen H, Yan S, Li D, Li Y et al (2020) Coronavirus disease 19 infection does not result in acute kidney injury: an analysis of 116 hospitalized patients from Wuhan, China. Am J Nephrol 51(5):343–348
12. Levac D, Colquhoun H, O'Brien KK (2010) Scoping studies: advancing the methodology. Implement Sci 5:69. https://doi.org/10.1186/1748-5908-5-69
13. Moher D, Shamseer L, Clarke M, Ghersi D, Liberati A, Petticrew M et al (2015) Preferred reporting items for systematic review and meta-analysis protocols (PRISMA-P) 2015 statement. Syst Rev 4(1):1. https://doi.org/10.1186/2046-4053-4-1
14. Chung C, McKenna L, Cooper SJ (2020) Patients' experiences of acute deterioration: a scoping review. Int J Nurs Stud 101:103404. https://doi.org/10.1016/j.ijnurstu.2019.103404
15. Liberati A, Altman DG, Tetzlaff J, Mulrow C, Gøtzsche PC, Ioannidis JP et al (2009) The PRISMA statement for reporting systematic reviews and meta-analyses of studies that evaluate health care interventions: explanation and elaboration. PLoS Med 6(7):e1000100. https://doi.org/10.1371/journal.pmed.1000100
16. Hoodin F, Weber S (2003) A systematic review of psychosocial factors affecting survival after bone marrow transplantation. Psychosomatics 44(3):181–195. https://doi.org/10.1176/appi.psy.44.3.181
17. Silverman MK, Kopf AW, Grin CM, Bart RS, Levenstein MJ (1991) Recurrence rates of treated basal cell carcinomas. Part 2: curettage-electrodesiccation. J Dermatol Surg Oncol 17(9):720–726
18. Jafari Oori M, Mohammadi F, Norouzi-Tabrizi K, Fallahi-Khoshknab M, Ebadi A (2019) Prevalence of medication adherence in patients with hypertension in Iran: a systematic review and meta-analysis of studies published in 2000-2018. ARYA Atheroscler 15(2):82–92
19. Oori MJ, Mohammadi F, Norozi K, Fallahi-Khoshknab M, Ebadi A, Gheshlagh RG (2019) Prevalence of HTN in Iran: meta-analysis of published studies in 2004–2018. Curr Hypertens Rev 15(2):113–122
20. Braun V, Clarke V (2012) Thematic analysis. In: Cooper H, Camic PM, Long DL, Panter AT, Rindskopf D, Sher KJ (eds) APA handbook of research methods in psychology, Vol 2: research designs: quantitative, qualitative, neuropsychological, and biological. American Psychological Association, Washington, DC, pp 57–71
21. Princess Margaret Hospital SARS Study Group, Choi KW, Chau TN, Tsang O, Tso E, Chiu MC et al (2003) Outcomes and prognostic factors in 267 patients with severe acute respiratory syndrome in Hong Kong. Ann Intern Med 139(9):715–723
22. Chen LL, Hsu CW, Tian YC, Fang JT (2005) Rhabdomyolysis associated with acute renal failure in patients with severe acute respiratory syndrome. Int J Clin Pract 59(10):1162–1166
23. Volunteers A-2019-nCoV; Li Z, Wu M, Yao J, Guo J, Liao X et al (2020) Caution on Kidney Dysfunctions of 2019-nCoV Patients. medRxiv. https://doi.org/10.1101/2020.02.08.20021212
24. Al Ghamdi M, Mushtaq F, Awn N, Shalhoub S (2015) MERS CoV infection in two renal transplant recipients: case report. Am J Transplant Off J Am Soc Transplant Am Soc Transplant Surg 15:1101–1104. https://doi.org/10.1111/ajt.13085
25. Deng Y, Liu W, Liu K et al (2020) Clinical characteristics of fatal and recovered cases of coronavirus disease 2019 (COVID-19) in Wuhan, China: a retrospective study. Chin Med J (Engl). https://doi.org/10.1097/CM9.0000000000000824
26. Lin W, Hu L, Zhang Y, Ooi JD, Meng T, Jin P et al (2020) Single-cell analysis of ACE2 expression in human kidneys and bladders reveals a potential route of 2019-nCoV infection. bioRxiv. https://doi.org/10.1101/2020.02.08.939892
27. Wang D, Hu B, Hu C, Zhu F, Liu X, Zhang J et al (2020) Clinical characteristics of 138 hospitalized patients with 2019 novel coronavirus-infected pneumonia in Wuhan, China. JAMA 323(11):1061–1069
28. Guan WJ, Ni ZY, Hu Y, Liang WH, Ou CQ, He JX et al (2020) Clinical characteristics of coronavirus disease 2019 in China. N Engl J Med 382(18):1708–1720
29. Drosten C, Seilmaier M, Corman VM, Hartmann W, Scheible G, Sack S et al (2013) Clinical features and virological analysis of a case of Middle East respiratory syndrome coronavirus infection. Lancet Infect Dis 13(9):745–751
30. Chen N, Zhou M, Dong X, Qu J, Gong F, Han Y et al (2020) Epidemiological and clinical characteristics of 99 cases of 2019 novel coronavirus pneumonia in Wuhan, China: a descriptive study. Lancet (London, England) 395(10223):507–513
31. Guery B, Poissy J, el Mansouf L, Séjourné C, Ettahar N, Lemaire X et al (2013) Clinical features and viral diagnosis of two cases of infection with Middle East respiratory syndrome coronavirus: a report of nosocomial transmission. Lancet 381(9885):2265–2272
32. Cheng Y, Luo R, Wang K, Zhang M, Wang Z, Dong L et al (2020) Kidney impairment is associated with in-hospital death of COVID-19 patients. medRxiv. https://doi.org/10.1101/2020.02.18.20023242
33. Chen T, Wu D, Chen H, Yan W, Yang D, Chen G et al (2020) Clinical characteristics of 113 deceased patients with coronavirus disease 2019: retrospective study. BMJ 368:m1091. https://doi.org/10.1136/bmj.m1091

34. Yang X, Yu Y, Xu J, Shu H, Xia J, Liu H et al (2020) Clinical course and outcomes of critically ill patients with SARS-CoV-2 pneumonia in Wuhan, China: a single-centered, retrospective, observational study. Lancet Respir Med 8(5):475–481
35. Pan X-W, Xu D, Zhang H, Zhou W, Wang LH, Cui XG (2020) Identification of a potential mechanism of acute kidney injury during the COVID-19 outbreak: a study based on single-cell transcriptome analysis. Intensive Care Med 46(6):1114–1116
36. Diao B, Wang C, Wang R, Feng Z, Tan Y, Wang H et al (2020) Human kidney is a target for novel severe acute respiratory syndrome coronavirus 2 (SARS-CoV-2) infection. medRxiv. https://doi.org/10.1101/2020.03.04.20031120
37. Yeung ML, Yao Y, Jia L, Chan JF, Chan KH, Cheung KF et al (2016) MERS coronavirus induces apoptosis in kidney and lung by upregulating Smad7 and FGF2. Nat Microbiol 1(3):16004. https://doi.org/10.1038/nmicrobiol.2016.4
38. Pacciarini F, Ghezzi S, Canducci F, Sims A, Sampaolo M, Ferioli E et al (2008) Persistent replication of severe acute respiratory syndrome coronavirus in human tubular kidney cells selects for adaptive mutations in the membrane protein. J Virol 82(11):5137–5144
39. Huang C, Wang Y, Li X, Ren L, Zhao J, Hu Y et al (2020) Clinical features of patients infected with 2019 novel coronavirus in Wuhan, China. Lancet 395(10223):497–506
40. Zou X, Chen K, Zou J, Han P, Hao J, Han Z et al (2020) Single-cell RNA-seq data analysis on the receptor ACE2 expression reveals the potential risk of different human organs vulnerable to 2019-nCoV infection. Front Med 14(2):185–192
41. Li W, Moore MJ, Vasilieva N, Sui J, Wong SK, Berne MA et al (2003) Angiotensin-converting enzyme 2 is a functional receptor for the SARS coronavirus. Nature 426(6965):450–454
42. Raj VS, Mou H, Smits SL, Dekkers DH, Müller MA, Dijkman R et al (2013) Dipeptidyl peptidase 4 is a functional receptor for the emerging human coronavirus-EMC. Nature 495(7440):251–254
43. Peiris JSM, Chu CM, Cheng VCC, Chan KS, Hung IF, Poon LL et al (2003) Clinical progression and viral load in a community outbreak of coronavirus-associated SARS pneumonia: a prospective study. Lancet 361(9371):1767–1772
44. Ding Y, He L, Zhang Q, Huang Z, Che X, Hou J et al (2004) Organ distribution of severe acute respiratory syndrome (SARS) associated coronavirus (SARS-CoV) in SARS patients: implications for pathogenesis and virus transmission pathways. J Pathol 203(2):622–630
45. Huang KJ, Su IJ, Theron M, Wu YC, Lai SK, Liu CC et al (2005) An interferon-gamma-related cytokine storm in SARS patients. J Med Virol 75(2):185–194
46. Perazella MA, Shirali AC (2018) Nephrotoxicity of Cancer immunotherapies: past, present and future. J Am Soc Nephrol 29(8):2039–2052
47. Fanelli V, Fiorentino M, Cantaluppi V, Gesualdo L, Stallone G, Ronco C et al (2020) Acute kidney injury in SARS-CoV-2 infected patients. Crit Care 24(1):155. https://doi.org/10.1186/s13054-020-02872-z
48. Pike F, Murugan R, Keener C, Palevsky PM, Vijayan A, Unruh M et al (2015) Biomarker enhanced risk prediction for adverse outcomes in critically ill patients receiving RRT. Clin J Am Soc Nephrol 10(8):1332–1339
49. Kashani K, Al-Khafaji A, Ardiles T, Artigas A, Bagshaw SM, Bell M et al (2013) Discovery and validation of cell cycle arrest biomarkers in human acute kidney injury. Crit Care 17(1):R25. https://doi.org/10.1186/cc12503
50. Bihorac A, Chawla LS, Shaw AD, Al-Khafaji A, Davison DL, Demuth GE et al (2014) Validation of cell-cycle arrest biomarkers for acute kidney injury using clinical adjudication. Am J Respir Crit Care Med 189(8):932–939
51. Hirsch JS, Ng JH, Ross DW, Sharma P, Shah HH; Northwell Nephrology COVID-19 Research Consortium (2020) Acute kidney injury in patients hospitalized with COVID-19. Kidney Int. May 16:S0085-2538(20)30532-9. https://doi.org/10.1016/j.kint.2020.05.006. Online ahead of print
52. No authors listed (News in Brief) (2020) Biggest COVID-19 trial tests repurposed drugs first. Nat Biotech 38:510. https://doi.org/10.1038/s41587-020-0528-x
53. BBC News (2020) Coronavirus: what is dexamethasone and how does it work? https://www.bbc.co.uk/news/health-53077879

Cardiac Injury in COVID-19: A Systematic Review

29

Malihe Sadat Moayed, Farshid Rahimi-Bashar, Amir Vahedian-Azimi, Thozhukat Sathyapalan, Paul C. Guest, Tannaz Jamialahmadi, and Amirhossein Sahebkar

Abstract

Coronavirus 2019 (COVID-19) is responsible for the current pandemic which has already resulted in considerable mortality worldwide. This systematic review was conducted to summarize the results of the published articles assessing the incidence of heart diseases in patients infected with COVID-19. The electronic databases Scopus, Web of Science, Pubmed, Science Direct, and ProQuest were used to search for potentially relevant articles. Articles published from Dec 2019 to April 2020 were included. All cross-sectional, retrospective or prospective observational cohort and case-control studies were selected which reported the incidence or prevalence of myocardial injury, myocardial infarction, or cardiovascular disease in patients with confirmed COVID-19 infection. Based on the inclusion criteria, 12 articles were selected. The incidence of cardiac injury was reported in 8 articles and 8 articles focused on the cardiovascular outcomes of COVID-19 infection. The incidence of new cardiac injury was reported to be 7.2–77% in live and dead patients, respectively. The results showed that patients with cardiac injury had worse outcomes

M. S. Moayed · A. Vahedian-Azimi (✉)
Trauma Research Center, Nursing Faculty, Baqiyatallah University of Medical Sciences, Tehran, Iran
e-mail: amirvahedian63@gmail.com

F. Rahimi-Bashar
Anesthesia and Critical Care Department, Hamadan University of Medical Sciences, Hamadan, Iran

T. Sathyapalan
Academic Diabetes, Endocrinology and Metabolism, Hull York Medical School, University of Hull, Hull, UK

P. C. Guest
Laboratory of Neuroproteomics, Department of Biochemistry and Tissue Biology, Institute of Biology, University of Campinas (UNICAMP), Campinas, Brazil

T. Jamialahmadi
Department of Food Science and Technology, Quchan Branch, Islamic Azad University, Quchan, Iran

Department of Nutrition, Faculty of Medicine, Mashhad University of Medical Sciences, Mashhad, Iran

A. Sahebkar (✉)
Biotechnology Research Center, Pharmaceutical Technology Institute, Mashhad University of Medical Sciences, Mashhad, Iran

Neurogenic Inflammation Research Center, Mashhad University of Medical Sciences, Mashhad, Iran

Halal Research Center of IRI, FDA, Tehran, Iran

Polish Mother's Memorial Hospital Research Institute (PMMHRI), Lodz, Poland
e-mail: sahebkara@mums.ac.ir

P. C. Guest (ed.), *Clinical, Biological and Molecular Aspects of COVID-19*, Advances in Experimental Medicine and Biology 1321, https://doi.org/10.1007/978-3-030-59261-5_29

including higher mortality than those without cardiac injury. The most common cardiac injury outcomes were shock and malignant arrhythmias. The most common radiographic findings in patients with cardiac injury were multiple mottling and ground-glass opacities in the lungs (64.6%). A significant number of patients with cardiac injury required noninvasive mechanical ventilation (46.3%) or invasive mechanical ventilation (22.0%). Acute respiratory distress syndrome was seen in 58.5%, acute kidney injury in 8.5%, electrolyte disturbances in 15.9%, hypoproteinemia in 13.4%, and coagulation disorders in 7.3% of patients with cardiac injuries. In addition, survival days were negatively correlated with cardiac troponin I levels ($r = -0.42$, 95%, $p = 0.005$). The results of this review showed that myocardial injury in patients with COVID 19 has a poor prognosis. Hence, cardiac investigation and management in these patients are crucial.

Keywords

COVID-19 · Cardiac Injury · Mortality · Cardiovascular · Prognosis

29.1 Introduction

SARS-CoV-2 is the pathogen responsible for the current pandemic causing severe pneumonia worldwide [1, 2]. The disease is known as coronavirus 2019 (COVID-19) as it first erupted in Wuhan China at the end of 2019. Cardiovascular complications can occur in patients with COVID-19 infection, and one of the major causes of death among patients hospitalized with COVID-19 is viral myocarditis or myocardial injury [3].

The studies done during the previous historical coronavirus outbreaks of Severe Acute Respiratory Syndrome (SARS) and Middle East Respiratory Syndrome (MERS) found that one of the important cause of mortality was the development of cardiovascular complications including acute myocardial infarction (AMI) and myocarditis [4, 5]. Various mechanisms including systemic inflammation, relative ischemia, and pathogen-mediated damage have been attributed to cardiovascular complications in individuals with coronavirus infections [6]. Studies have shown that arrhythmia, bradycardia, tachycardia, hypotension, and sudden cardiac death are common in SARS pneumonia [7–10]. Various investigations including electrocardiographic changes, troponin elevation, and echocardiography or invasive diagnostics methods were used to demonstrate myocarditis, cardiac injury, myocardial infarction, or other subclinical left ventricular diastolic impairment in these studies.

The circulating levels of cardiac troponin I (cTnI) have been used as an independent determinant of clinical disease [3, 6] and the presence of cardiovascular diseases is a risk factor for mortality in patients with Covid-19 pneumonia [11]. These findings underscore the importance of biomarkers not only in the detection of the disease but also in the treatment of COVID-19 illness [12]. One study showed that 16.7% of confirmed patients with COVID-19 disease had myocardial damages [13]. In a clinical cohort study of COVID-19 infections, 7.2% of patients hadacute myocardial injury (AMI), 8.7% showed signs of shock, and 16.7% had arrhythmia [14]. Another cohort study with a follow-up found that 82 patients (19.7%) had a cardiac injury. In one study, 27.8% of patients with COVID-19 disease had elevated cardiac troponin T (cTnT) levels [15].

There is still a large knowledge gap regarding cardiac complications due to COVID-19 infection [6]. This is critical given the large proportion of cases with such complications and the impact this can have on current treatment strategies and disease outcomes. Therefore the current systematic review was conducted to investigate the prevalence and outcome of cardiac injury in COVID-19 patients.

29.2 Methods

In the current systematic review, we selected the studies that reported prevalence of myocardial injury or which assessed the association between cardiac injury and outcome in patients with COVID- 19 disease. This systematic review included all cross-sectional, retrospective, or prospective observational cohorts, and case-controlled studies that assessed the prevalence and complication of cardiac injury in COVID-19 patients.

29.2.1 Information Source

An extensive search was conducted on the electronic database PubMed, Scopus, Web of Science, Science Direct, and ProQuest for articles fulfilling the inclusion criteria (see Appendix 1). Publications from 2019 to 20 April 2020 were included. The Endnote software (Thomson Reuters, X9, Bld 9325) was used to screen the studies. Related studies were identified and screened by searching references of the included studies.

29.2.2 Data Collection Process

After excluding duplicated references of the 308 initial search publications, 206 articles remained based on Preferred Reporting Items for Systematic Reviews and Meta-Analyses (PRISMA) guidelines (Fig. 29.1). Two authors screened all of the abstracts and titles. Full texts for all potentially relevant articles were reviewed. Twelve articles had full text available and were included in this systematic review. A meta-analysis of the findings was not conducted due to high heterogeneity of these studies and most of the papers lacked a common set of attributes that could be combined. Therefore, a systematic review was conducted.

29.2.3 Ethical Considerations

The Ethics Committee of Baqiyatallah University of Medical Sciences approved this study with code IR.BMSU.REC.1399.116.

29.3 Results

Elevation in cTnI or cTnT levels was used as a biomarker of myocardial injury in many studies. Also in some studies, the cardiac injury was confirmed by evaluating circulating levels of α-hydroxybutyrate dehydrogenase (HBDB), N-terminal pro-B-type natriuretic peptide (NT-proBNP), creatine kinase (CK), or lactate dehydrogenase (LDH).

The 12 articles which met the search criteria were categorized into two separate groups.

29.3.1 First Group

The prevalence of cardiac injury in COVID-19 infected patients was reported in the first group. The occurrence rate of new cardiac damage was reported to be between 7.2% [14] and 77% [16] in living and dead patients, respectively. In addition, 44.4% of COVID-19 patients developed a new cardiac injury during the length of hospital stay [17].

29.3.2 Second Group

The relationship between the cardiac injury and outcome of patients was analyzed in the second group. The results of the second group of publications showed that the patients with cardiac injury had worse outcomes than other infected patients. The mortality rate was reported to be 60.9 vs. 25.8% ($p = 0.013$) in the study by He et al. [17]. There were also significant differences in mortality rate between patients with cardiac injury compared to patients who did not have any

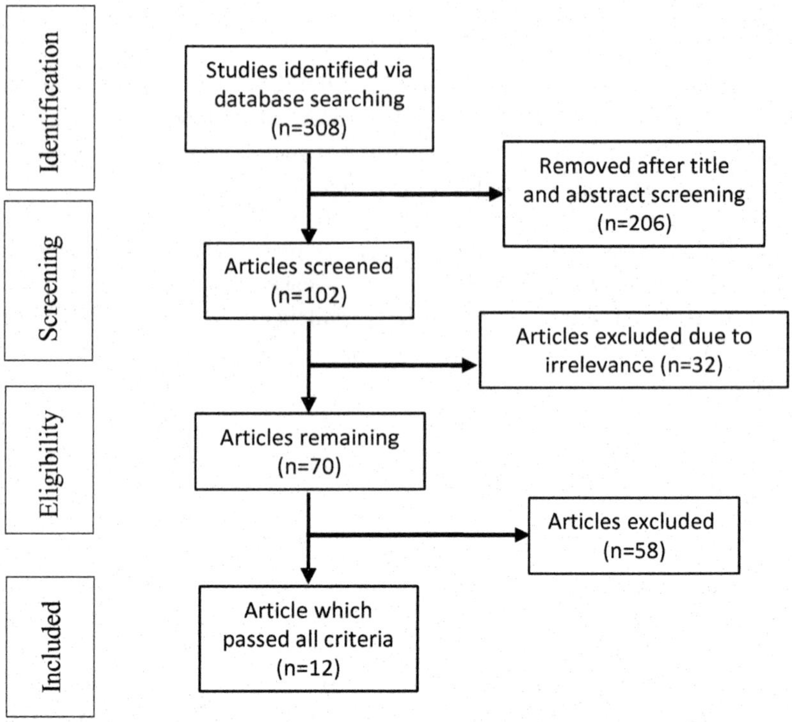

Fig. 29.1 Screening and selection of articles based on PRISMA guidelines

cardiac injury in both the Wu et al. (59.6 vs. 0.8%, $p < 0.001$) [18] and Shi et al. (51.2 vs. 4.5%) ($p < 0.001$) [19] studies.

Wu et al. showed that days of survival were negatively correlated with cTnI ($r = -0.42$, $p = 0.005$) and LDH ($r = -0.35$, $p = 0.022$) levels on admission [20]. He et al. demonstrated that the level of C-reactive protein (CRP) [153.6 (80.3, 240.7) vs. 49.8 (15.9, 101.9) ng/L] and NT-proBNP [852.0 (400.0, 2315.3) vs. 197.0 (115.3, 631.0) ng/L] were significantly higher in COVID-19 infected patients with myocardial damage than patients without myocardial damage [17]. Another study demonstrated that a higher level of cTnI of ≥6.126 pg/mL on admission was associated with a higher mortality compared to individuals with lower levels of cTnI [20]. Also, Guo et al. showed that in patients who died, cTnT and NT-proBNP plasma levels had increased significantly prior to death compared with the admission values [15].

Several studies examined the side effects and comorbidities of cardiac injury and the level of the cardiac biomarkers in COVID-19 patients [15, 21–23]. High levels of cTnT were more frequently associated with the need for mechanical ventilation (59.6 vs. 10.4%) and the occurrence of malignant arrhythmias (71.2 vs. 51.1%) [15]. Increased levels of cTnI (46.8 vs. 4.8 ng/L), HBDB (453 vs. 245 U/L), and CK (199 vs. 88 U/L) were also found to be common in severe or critically ill patients [24] and Hui et al. found that cTnI was 40-fold higher in critical COVID-19 patients [21].

Results of a meta-analysis by Lippi et al. showed that the severity of disease in COVID-19 patients was correlated with higher levels of cTnT [23]. A meta-analysis found that a significantly higher incidence of acute myocardial injury occurred in patients who were hospitalized in the intensive care unit (ICU) in comparison to non-ICU patients [22].

Shi et al. demonstrated an increase in reports of ground-glass opacities and multiple mottling (64.6% vs. 4.5%) in chest radiographs of patients with cardiac involvement. More patients (46.3%) required noninvasive mechanical ventilation or invasive mechanical ventilation [15, 19] who had

Table 29.1 First group: the prevalence cardiac injury in COVID-19 infected patients

	Author	Number	Prevalence of cardiac injury
1	Shi et al. [19]	416 hospitalized patients	82 patients with cardiac injury (19.7%)
2	Guo et al. [15]	187 patients	Elevated cTnT level in 52 infected patients (27.8%) with myocardial injury
3	Chen et al. [16]	274 patients [113 died and 161 recovered]	Acute myocardial injury more common in dead patients (77%)
4	Deng et al. [18]	9 dead patients and 116 recovered	Dead patients had more complications such as acute myocardial injury 59.6% versus recovered group 0.8% ($p < 0.001$).
5	Wang et al. [14]	138 patients	Most patients required intensive care. In this study acute cardiac injury was present in 7.2% patients, shock in 8.7% patients, and arrhythmia in 16.7% patients
6	Li al [22]	1527 patients	Acute cardiac injury became manifest in at least 8% of patients with COVID-19 infection
7	Liu et al. [13]	30 patients 26 common and 4 severe	Myocardial damage detected in 5 patients (16.7%).
8	Huang et al. [25]	41 confirmed patients	Acute cardiac injury in 12% of patients.

cardiac involvement compared to COVID-19 patients without cardiac involvement. In addition, acute respiratory distress syndrome (ARDS) [14, 19], hypo-proteinemia, coagulation disorders, electrolyte disturbances [19], tachycardia [21], shock and arrhythmia [14], and disseminated intravascular coagulation were common in patients with cardiac involvement compared to patients without cardiac involvement [18]. These results are shown in Tables 29.1 and 29.2.

29.4 Discussion

The current systematic review was conducted to investigate the prevalence and outcomes of cardiac injury in COVID-19 patients. The results of this study showed that there is a currently a high incidence of cardiac injury in patients with COVID-19 infection. Some of the studies included in this review reported an increase in markers of cardiac injury associated with poor outcomes such as increased chances of death in COVID-19 patients. We found that cardiac injury was found in 77% of patients who died compared with only 7% in those who survived. One study found that high levels of cardiac injury markers were found in 8–12% of COVID-19 cases [26]. Another study reported that 11.8% of patients with no history of cardiovascular disease had a significant myocardial injury, with raised markers of cardiac injury [27]. The studies demonstrated that cardiac injury with a higher level of cTnI was associated with an increase in mortality and other comorbidities.

There are now numerous biomarkers which can be used to assess or predict COVID-19 disease severity, such as in those cases that progress to cardiac injury. These include an increase in lung ground glass opacities on computed tomography scans, the requirement of invasive ventilation, acute kidney injury, electrolyte imbalance, hypo-proteinemia, increased levels of cardiac and inflammatory biomarkers, disorders of blood coagulation, and increased hospital stay.

The response of the inflammatory system and the presence of a disorder in the immune system due to progression of COVID-19 can lead to the development of cardiac damage in these patients [20]. The results of these studies confirmed that COVID-19 patients with myocardial injury often had poor prognosis because of multi-organ failure. It is likely that the increase in the incidence of shock and hemodynamic disorder resulting from cardiac injury due to myocardial ischemia or necrosis is linked to the higher mortality rate. In one study, 80% of patients with myocardial injury were admitted to the ICU [27]..

Table 29.2 Second group: the association between cardiac injury and outcome in COVID-19 infected patients

	Author	Number	Comparison	Heart injury biomarker	Outcome
1	Lippi et al. [23]	341	Severe vs. non-severe disease	cTnI or cTnT values	Levels increased in COVID-19 patients with severe disease versus non-severe disease (SMD, 25.6 ng/L, 6.8–44.5 ng/L; $p < 0.001$).
2	Shi et al. [19]	416	Cardiac injury vs. without cardiac injury	CKMB, cTnI, NT-proBNP[a]	Need for invasive mechanical ventilation in high level cardiac biomarkers was 22.0%; $p < 0.001$ compared to group with lower levels. Higher report of ground-glass opacity and multiple mottling (64.6%), acute respiratory distress syndrome (58.5%), acute kidney injury (8.5%), electrolyte disturbances (15.9%), hypo-proteinemia (13.4%), DIC (7.3%), and higher mortality (51.2%) than those without
3	Guo et al. [15]	187	Myocardial injury	TnTs plasma cTnT levels	High levels of heart enzyme had more repeated persistent dysrhythmias and mechanical ventilation was 59.6% vs. 10.4% in patients with ordinary enzyme levels
4	Zeng et al. [17]	54	Myocardial injury	Serum cTnI level	Mortality was higher in patients with high level of enzymes (14 vs. 8)
5	Zhou et al. [24]	34	Very severe and severe COVID-19 group	cTnI, CK[b], HBDB[c], LDH	Increased levels of cTnI (46.8 ng/L vs. 4.8 ng/L), LDH (513 U/L vs. 287 U/L), HBDB (453 U/L vs. 245 U/L), and CK (199 U/L vs. 88 U/L) in the very severe group compared to the severe group
6	Wu et al. [20]	188 infected patients	Mortality and inpatients days	cTnI[d]; CK; CK-MB; LDH[e]; α-HBDH).	The higher mortality (50.0%) in high levels of cTNI on reception (10.0% or 9.1%). Also, the cTNI level was correlated with survival days negatively ($r = -0.42$, $p = 0.005$)
7	Hui et al. [21]	41 consecutive corona	Comparison across degrees of disease severity	HR[f], cTnI and EAT[g], CT[h]	Peak level of cTnI 40-fold higher in severe cases compared to normal levels. 4 (9.75%) of patients had high levels of cTnI. AMI of COVID-2019 was common in severe and critical patients
8	Deng et al. [18]	Enrolled 109 COVID-19 patients	Who died during hospitalization and 116 recovered	cTnI	More dead patients had acute cardiac injury (59.6% vs. 0.8%, $p < 0.001$), DIC (6.4% vs. 0, $p = 0.006$), and shock (11.9% vs. 0, $p < 0.001$) compared to those that recovered

[a]N-terminal pro-B-type natriuretic peptide
[b]Creatine kinase
[c]α-Hydroxybutyrate dehydrogenase
[d]High-sensitivity cardiac troponin I
[e]Lactate dehydrogenase
[f]HR: Heart Rate,
[g]EAT:Epicardial Adipose Tissue
[h]CT: Chest Computed Tomography

An earlier study showed that MERS-CoV infection resulted in increased incidence of cardiac dysfunction and myocarditis in patients [10]. Patients with an increased level of cTnI had increased morbidity and mortality [15, 17, 20, 23, 25]. Some studies reported a similar rate of complications between MERS and COVID-19 [28, 29]. However, another study reported that COVID-19 infection was associated with more cardiac complications, hypotension, bradycardia, and cardiac arrest, suggesting that the etiology of these complications is complex [30].

Therefore, monitoring of cardiac biomarkers as soon as COVID-19 patients are diagnosed and also continuous assessment during the hospitalization will help to identify which patients are at increased risk of suffering myocardial damage [20, 23]. This monitoring is more critical for patients who have high risk factors including demographic characteristics such as advanced age (older patients), presence of cardiovascular comorbidities such as hypertensive or obese patients, male gender, and smoking [20, 31]. The collaboration of several medical specialists, such as cardiologists, diabeticians, epidemiologists, and emergency room doctors, is therefore needed [32]. Health professionals should focus not only on respiratory parameters but also on cardiac and other metabolic parameters, which are also predictors of mortality [33].

29.5 Conclusion

Acute cardiac injury damage to the myocardium was found to be associated with worse outcomes and increased chances of death in patients with COVID-19 infections. However, the specific mechanisms underlying this are still not certain. Therefore further studies are required to investigate which patients are more likely to experience cardiac injury so that these individuals can be targeted with specific intervention strategies. For example, recent studies in the UK have shown that treatment with dexamethazone increased the survival rate of patients with the most critical stage of the disease [34]. In addition, anticoagulation therapy has been applied with some success [35]. We suggest that testing patients for biomarkers associated with early signs of cardiac damage should be implemented upon COVID-19 patient admissions so that appropriate treatments can be tempered in a precision medicine manner. Until a vaccine is available, such an approach should improve COVID-19 patient outcomes and thereby increase chances of survival.

Acknowledgments Thanks to the guidance, and advice from the Clinical Research Development Unit of Baqiyatallah Hospital.

Appendix 1

Scopus: 44 (TITLE-ABS-KEY ("Novel coronavirus" OR "Novel coronavirus 2019" OR "2019 novel coronavirus" OR "2019 nCoV" OR "Wuhan coronavirus" OR "Wuhan pneumonia" OR covid-19 OR "2019-nCoV" OR "SARS-CoV-2" OR "coronavirus 2019" OR "2019-nCoV") AND TITLE-ABSKEY (cardiovascular OR myocardial OR myocarditis OR myocardium OR heart OR cardiac OR hypotension OR tachycardia OR bradycardia OR fibrillation OR ventricular OR arrhythmia OR hypertension OR "blood pressure")

EMBASE: 65 ('novel coronavirus':ti,ab,kw OR 'novel coronavirus 2019':ti,ab,kw OR '2019 novel coronavirus':ti,ab,kw OR 'wuhan coronavirus':ti,ab,kw OR 'wuhan pneumonia':ti,ab,kw OR 'covid 19':ti,ab,kw OR 'sars cov 2':ti,ab,kw OR 'coronavirus 2019':ti,ab,kw OR '2019 ncov':ti,ab,kw) AND (cardiovascular:ti,ab,kw OR myocardial:ti,ab,kw OR myocarditis:ti,ab,kw OR myocardium:ti,ab,kw OR heart:ti,ab,kw OR cardiac:ti,ab,kw OR hypotension:ti,ab,kw OR tachycardia:ti,ab,kw OR bradycardia:ti,ab,kw OR fibrillation:ti,ab,kw OR ventricular:ti,ab,kw OR arrhythmia:ti,ab,kw OR hypertension:ti,ab,kw OR 'blood pressure':ti,ab,kw)

ProQuest: 94 ab ("Novel coronavirus" OR "Novel coronavirus 2019" OR "2019 novel coronavirus" OR "2019 nCoV" OR "Wuhan coronavirus" OR "Wuhan pneumonia" OR covid-19 OR "2019-nCoV" OR "SARS-CoV-2" OR "coronavirus 2019" OR "2019-nCoV") AND ab (cardiovascular OR myocardial OR myocarditis OR myocardium OR heart OR cardiac OR hypotension OR tachycardia OR bradycardia OR fibrillation OR ventricular OR arrhythmia OR hypertension OR "blood pressure")

Web of science: 12 ("Novel coronavirus" OR "Novel coronavirus 2019" OR "2019 novel coronavirus" OR "2019 nCoV" OR "Wuhan coronavirus" OR "Wuhan pneumonia" OR covid-19 OR "2019-nCoV" OR "SARS-CoV-2" OR "coronavirus 2019" OR "2019-nCoV")

AND **TOPIC:** (cardiovascular OR myocardial OR myocarditis OR myocardium OR heart OR cardiac OR hypotension OR tachycardia OR bradycardia OR fibrillation OR ventricular OR arrhythmia OR hypertension OR "blood pressure")

PubMed: 105 ("Novel coronavirus" [Title/Abstract] OR "Novel coronavirus 2019" [Title/Abstract] OR "2019 novel coronavirus" [Title/Abstract] OR "2019 nCoV" [Title/Abstract] OR "Wuhan coronavirus" [Title/Abstract] OR "Wuhan pneumonia" [Title/Abstract] OR covid-19 [Title/Abstract] OR "2019-nCoV" [Title/Abstract] OR "SARS-CoV-2" [Title/Abstract] OR "coronavirus 2019" [Title/Abstract] OR "2019-nCoV"[Title/Abstract])) AND (cardiovascular [Title/Abstract] OR myocardial [Title/Abstract] OR myocarditis [Title/Abstract] OR myocardium [Title/Abstract] OR heart [Title/Abstract] OR cardiac [Title/Abstract] OR hypotension [Title/Abstract] OR tachycardia [Title/Abstract] OR bradycardia [Title/Abstract] OR fibrillation [Title/Abstract] OR ventricular [Title/Abstract] OR arrhythmia [Title/Abstract] OR hypertension [Title/Abstract] OR "blood pressure"[Title/Abstract])

References

1. Hui DS, Azhar EI, Madani TA, Ntoumi F, Kock R, Dar O et al (2020) The continuing 2019-nCoV epidemic threat of novel coronaviruses to global health—the latest 2019 novel coronavirus outbreak in Wuhan, China. Int J Infect Dis 91:264–266
2. Riou J, Althaus CL (2020) Pattern of early human-to-human transmission of Wuhan 2019 novel coronavirus (2019-nCoV), December 2019 to January 2020. Euro Surveill 25(4):2000058. https://doi.org/10.2807/1560-7917.ES.2020.25.4.2000058
3. Deng Q, Hu B, Zhang Y, Wang H, Zhou X, Hu W et al (2020) Suspected myocardial injury in patients with COVID-19: evidence from front-line clinical observation in Wuhan, China. Int J Cardiol 311:116–121
4. Nguyen JL, Yang W, Ito K, Matte TD, Shaman J, Kinney PL (2016) Seasonal influenza infections and cardiovascular disease mortality. JAMA Cardiol 1(3):274–281
5. Oudit G, Kassiri Z, Jiang C, Liu P, Poutanen S, Penninger J et al (2009) SARS-coronavirus modulation of myocardial ACE2 expression and inflammation in patients with SARS. Eur J Clin Investig 39(7):618–625
6. Xiong TY, Redwood S, Prendergast B, Chen M (2020) Coronaviruses and the cardiovascular system: acute and long-term implications. Eur Heart J 41(19):1798–1800
7. Yu C, Wong RS, Wu E, Kong S, Wong J, Yip GW et al (2006) Cardiovascular complications of severe acute respiratory syndrome. Postgrad Med J 82(964):140–144
8. Pan S, Zhang H, Li C, Wang C (2003) Cardiac arrest in severe acute respiratory syndrome: analysis of 15 cases. Zhonghua Jie He He Hu Xi Za Zhi 26(10):602–605
9. Li SSL, Cheng CW, Cl F, Chan YH, Lee MP, Chan JWM et al (2003) Left ventricular performance in patients with severe acute respiratory syndrome: a 30-day echocardiographic follow-up study. Circulation 108(15):1798–1803
10. Alhogbani T (2016) Acute myocarditis associated with novel Middle East respiratory syndrome coronavirus. Ann Saudi Med 36(1):78–80
11. Corrales-Medina VF, Suh KN, Rose G, Chirinos JA, Doucette S, Cameron DW et al (2011) Cardiac complications in patients with community-acquired pneumonia: a systematic review and meta-analysis of observational studies. PLoS Med 8(6):e1001048. https://doi.org/10.1371/journal.pmed.1001048
12. World Health Organization (2020) Report of the WHO-China Joint Mission on Coronavirus Disease 2019 (COVID-19). https://www.who.int/docs/default-source/coronaviruse/who-china-joint-mission-on-covid-19-final-report.pdf
13. Liu M, He P, Liu H, Wang X, Li F, Chen S et al (2020) Clinical characteristics of 30 medical workers infected with new coronavirus pneumonia. Zhonghua Jie He He Hu Xi Za Zhi 43(3):209–214
14. Wang D, Hu B, Hu C, Zhu F, Liu X, Zhang J et al (2020) Clinical characteristics of 138 hospitalized patients with 2019 novel coronavirus–infected pneumonia in Wuhan, China. JAMA 323(11):1061–1069
15. Guo T, Fan Y, Chen M, Wu X, Zhang L, He T et al (2020) Cardiovascular implications of fatal outcomes of patients with coronavirus disease 2019 (COVID-19). JAMA Cardiol. Mar 27: e201017. https://doi.org/10.1001/jamacardio.2020.1017. Online ahead of print
16. Chen T, Wu D, Chen H, Yan W, Yang D, Chen G et al (2020) Clinical characteristics of 113 deceased patients with coronavirus disease 2019: retrospective study. BMJ368: m 1091. https://doi.org/10.1136/bmj.m1091
17. He X, Lai J, Cheng J, Wang M, Liu Y, Xiao Z et al (2020) Impact of complicated myocardial injury on the clinical outcome of severe or critically ill COVID-19 patients. Zhonghua Xin Xue Guan Bing Za Zhi Mar 15; 48(0): E011. https://doi.org/10.3760/cma.j.cn112148-20200228-00137. Online ahead of print

18. Deng Y, Liu W, Liu K, Fang YY, Shang J, Wang K et al (2020) Clinical characteristics of fatal and recovered cases of coronavirus disease 2019 (COVID-19) in Wuhan, China: a retrospective study. Chin Med J 133(11):1261–1267
19. Shi S, Qin M, Shen B, Cai Y, Liu T, Yang F et al (2020) Association of cardiac injury with mortality in hospitalized patients with COVID-19 in Wuhan, China. JAMA Cardiol Mar 25:e200950. doi: https://doi.org/10.1001/jamacardio.2020.0950. Online ahead of print
20. Wu C, Hu X, Song J, Du C, Xu J, Yang D et al (2020) Heart injury signs are associated with higher and earlier mortality in coronavirus disease 2019 (COVID-19). MedRxiv. https://doi.org/10.1101/2020.02.26.20028589
21. Hui H, Zhang Y, Yang X, Wang X, He B, Li L et al (2020) Clinical and radiographic features of cardiac injury in patients with 2019 novel coronavirus pneumonia. medRxiv. https://doi.org/10.1101/2020.02.24.20027052
22. Li B, Yang J, Zhao F, Zhi L, Wang X, Liu L et al (2020) Prevalence and impact of cardiovascular metabolic diseases on COVID-19 in China. Clin Res Cardiol 109(5):531–538
23. Lippi G, Lavie CJ, Sanchis-Gomar F (2020) Cardiac troponin I in patients with coronavirus disease 2019 (COVID-19): evidence from a meta-analysis. Prog Cardiovasc Dis. Mar 10. doi:https://doi.org/10.1016/j.pcad.2020.03.001. Online ahead of print
24. Zhou B, She J, Wang Y, Ma X (2020) The clinical characteristics of myocardial injury in severe and very severe patients with 2019 novel coronavirus disease. J Infect 81(1):147–178
25. Huang C, Wang Y, Li X, Ren L, Zhao J, Hu Y et al (2020) Clinical features of patients infected with 2019 novel coronavirus in Wuhan, China. Lancet 395(10223):497–506
26. Lippi G, Plebani M (2020) Laboratory abnormalities in patients with COVID-2019 infection. Clin Chem Lab Med 58(7):1131–1134
27. Zheng YY, Ma YT, Zhang JY, Xie X (2020) COVID-19 and the cardiovascular system. Nat Rev Cardiol 17(5):259–260
28. Saad M, Omrani AS, Baig K, Bahloul A, Elzein F, Matin MA et al (2014) Clinical aspects and outcomes of 70 patients with Middle East respiratory syndrome coronavirus infection: a single-center experience in Saudi Arabia. Int J Infect Dis 29:301–306
29. Al-Abdallat MM, Payne DC, Alqasrawi S, Rha B, Tohme RA, Abedi GR et al (2014) Hospital-associated outbreak of Middle East respiratory syndrome coronavirus: a serologic, epidemiologic, and clinical description. Clin Infect Dis 59(9):1225–1233
30. Kochi AN, Tagliari AP, Forleo GB, Fassini GM, Tondo C (2020) Cardiac and arrhythmic complications in patients with COVID-19. J Cardiovasc Electrophysiol 31(5):1003–1008
31. Ruan Q, Yang K, Wang W, Jiang L, Song J (2020) Clinical predictors of mortality due to COVID-19 based on an analysis of data of 150 patients from Wuhan, China. Intensive Care Med 46(5):846–848
32. Jing ZC, Zhu HD, Yan XW, Chai WZ, Zhang S (2020) Recommendations from the Peking Union Medical College Hospital for the management of acute myocardial infarction during the COVID-19 outbreak. Eur Heart J 41(19):1791–1794
33. Sun X, Wang J, Liu Z, Zhou X, Yan XW, Li T et al (2020) Characteristics of patients with COVID-19 pneumonia admitted to the intensive care unit and predictors of mortality in Wuhan. China: A Single-Centered Retrospective Cohort Study SSRN Electron J. https://doi.org/10.2139/ssrn.3559629
34. https://www.newscientist.com/article/2244804-covid-19-news-archive-uk-using-dexamethasone-to-treat-patients/
35. Atallah B, Mallah SI, AlMahmeed W (2020) Anticoagulation in COVID-19. Eur Heart J Cardiovasc Pharmacother. Apr 30;pvaa036. https://doi.org/10.1093/ehjcvp/pvaa036

A Systematic Review of the Assessment of the Presence of SARS-CoV-2 in Human Semen

30

Amir Vahedian-Azimi, Leila Karimi, Somayeh Makvandi, Tannaz Jamialahmadi, and Amirhossein Sahebkar

Abstract

Theoretically, human testes are highly expressive organs for angiotensin converting enzyme 2 (ACE2), the SARS-CoV-2 receptor. This study aimed to investigate whether the causative agent of COVID-19 is found in semen. The databases of PubMed/MEDLINE, Scopus, Web of Science, and Google Scholar were searched using a combination of related keywords. All studies with original data, involving detection of SARS-CoV-2 in semen of male patients with COVID-19 or in those who have recovered from it, were included in the study. Six articles, including 136 samples, entered the systematic review. Most of the studies were performed in the recovery phase of COVID-19. In four articles, SARS-CoV-2 was not detected in semen, while in the other two articles semen testing showed the presence of the virus in some samples. Testicular discomfort, testicular cell damage, and spermogram disruption were also reported in some studies. We conclude that the study question cannot be answered with this number of studies. Since most of the samples were mild to moderate forms of COVID-19, it is not yet clear what the presence of the virus in semen will be in severe cases. The long-term effects are also vague. More original articles with bet-

A. Vahedian-Azimi
Trauma Research Center, Nursing Faculty, Baqiyatallah University of Medical Sciences, Tehran, Iran

L. Karimi
Behavioral Sciences Research Center, Life style institute, Nursing Faculty, Baqiyatallah University of Medical Sciences, Tehran, Iran

S. Makvandi (✉)
Department of Midwifery, School of Nursing and Midwifery, Islamic Azad university Ahvaz branch, Ahvaz, Iran
e-mail: somayemakvandi@gmail.com

T. Jamialahmadi
Biotechnology Research Center, Pharmaceutical Technology Institute, Mashhad University of Medical Sciences, Mashhad, Iran

Department of Food Science and Technology, Quchan Branch, Islamic Azad University, Quchan, Iran

Department of Nutrition, Faculty of Medicine, Mashhad University of Medical Sciences, Mashhad, Iran

A. Sahebkar (✉)
Biotechnology Research Center, Pharmaceutical Technology Institute, Mashhad University of Medical Sciences, Mashhad, Iran

Neurogenic Inflammation Research Center, Mashhad University of Medical Sciences, Mashhad, Iran

Polish Mother's Memorial Hospital Research Institute (PMMHRI), Lodz, Poland

Halal Research Center of IRI, FDA, Tehran, Iran
e-mail: sahebkara@mums.ac.ir

P. C. Guest (ed.), *Clinical, Biological and Molecular Aspects of COVID-19*, Advances in Experimental Medicine and Biology 1321, https://doi.org/10.1007/978-3-030-59261-5_30

ter design and in different phases of the disease are needed to draw robust conclusions.

Keywords

SARS-CoV-2 · COVID-19 · Semen · Angiotensin-converting enzyme 2

30.1 Introduction

In December 2019, new-origin pneumonia in Wuhan, Hubei Province, China, was identified and given the name of coronavirus disease 2019 (COVID-19) from the World Health Organization. The cause of the disease is a new coronavirus called SARS-CoV-2, which causes acute respiratory syndrome and carries the risk of death [1, 2]. Although the predominant manifestations of the SARS-CoV-2 are respiratory manifestations, the growing body of scientific evidence suggests that other systems in the body, such as the gastrointestinal, cardiovascular, urinary, and reproductive systems, may also be affected [3–6]. Guan stated that most people with COVID-19 are male [7]. Chen believed that most of these men are of childbearing age [8]. Therefore, a significant question is whether SARS-CoV-2 can attack the testes.

According to the evidence, this virus has many similarities with the SARS-CoV, the corona virus that caused Severe Acute Respiratory Syndrome (SARS) in 2002–2004, and most likely uses the same receptor to enter host cells. Recent evidence suggests that seminiferous ducts of the testis, adult Leydig cells, and the prostate are highly expressive organs for the coronavirus receptor, angiotensin converting enzyme 2 (ACE2) (Fig. 30.1) [9]. A study of a testicular autopsy obtained from patients who died of SARS-CoV showed that the virus could cause orchitis [10]. So it is not unreasonable to think that one of the ways to transmit COVID-19 is via semen. In addition, if the virus causes damage to the testes, then the world may face a fertility crisis in the future. Therefore special attention should be paid to the evaluation and appropriate interventions in the fertility of young patients, including sperm analysis and sperm freezing in the early stages of the disease.

This study aimed to collect and integrate the results of previous studies on the presence of SARS-CoV-2 in human semen.

30.2 Methods

The databases of PubMed/MEDLINE, Scopus, and Web of science were searched up to June 14, 2020. The search strategy used the terms: *"COVID-19" OR "COVID19" OR "2019 novel coronavirus infection" OR "2019-nCoV infection" OR "COVID-19 pandemic" OR "coronavirus disease-19" OR "2019-nCoV disease" OR "2019 novel coronavirus disease" OR "coronavirus disease 2019" AND (Sperm* OR semen.* To further identify potentially related studies, Google Scholar was searched using a combination of related keywords, and the references of primary articles were also reviewed. All relevant articles were evaluated using predefined selection criteria. All studies with original data, involving the detection of the SARS-CoV-2 in the semen of men with COVID-19 or men who recovered from it were included in the study. Articles in non-English languages were not included.

The primary output of the Search Strategy in all databases was entered into the EndNote software X5 and, in the first step, the duplicate articles were deleted. Then, all the remaining articles were evaluated in terms of title and abstract. The unrelated articles were removed and the full texts of the rest of the articles were carefully evaluated to see if they met the entry criteria. The data retrieval tool was a researcher-made form that included the data: name of the first author, country, type of the study, population, age, disease severity, the period between disease onset or disease recovery and semen collection, and the findings. All of the above steps were performed by two independent authors, and any disagreement between them was resolved through discussion.

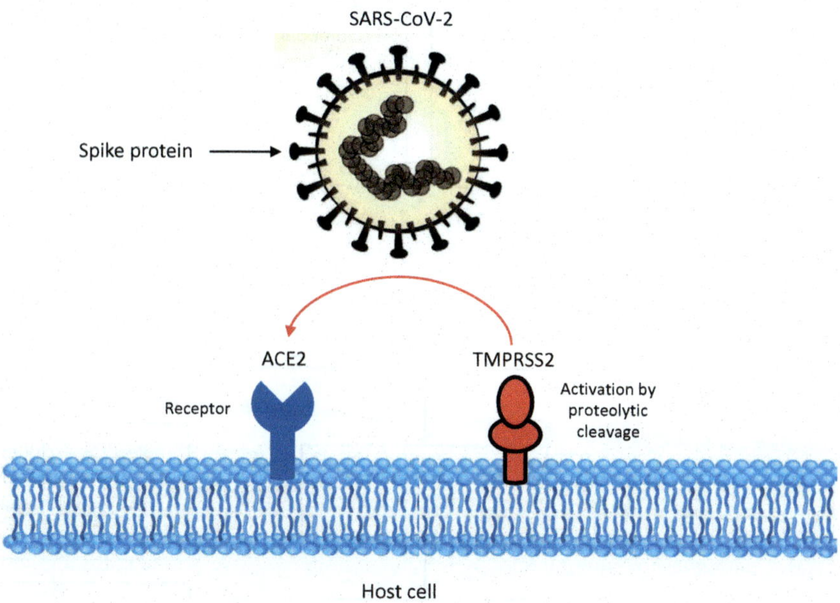

Fig. 30.1 Receptor-mediated infection by SARS-CoV-2

30.3 Results

The PRISMA flow chart of inclusion and exclusion process is shown in Fig. 30.2. Of the 62 primary articles after the elimination of duplicates, 35 articles entered the first phase and 23 articles entered the second phase of screening. Finally, six papers entered qualitative synthesis. Table 30.1 shows a summary of the final articles. There were four papers from China [6, 11–13], one from Germany [14], and one from Italy [15]. The studies included two prospective cohort articles [12, 14], a retrospective cohort [6], a case report [15], a case series [13], and a cross-sectional study [11]. The sample size included a total of 136 cases, ranging from one sample [15] to 38 per study [12]. Five studies were performed on the semen of living men who had recovered or suffered from COVID-19 disease [11–15], but one study was performed on the testes of COVID-19 patients who had died [6]. In four studies, the severity of COVID-19 disease ranged from asymptomatic to moderate [11, 13–15] and, in one study, it resulted in death [6]. Li et al. did not report the severity of the disease [12].

30.3.1 Outcome

Contradictory data were been obtained in studies on the presence of SARS-CoV-2 in semen samples. Three studies did not find the presence of SARS-CoV-2 RNA in semen samples of male patients recovering from COVID-19 disease [11, 13, 14]. Pan et al. believed that although their findings did not identify SARS-CoV-2 in semen, they could not definitively rule out the presence of the virus in the seminal fluid during acute infection with more severe COVID-19 symptoms [11]. Paoli et al. reported that in a 31-year-old man who was found positive for SARS-CoV-2, in addition to semen, a urine sample was negative for the viral RNA when symptoms were partially improved [15]. On the contrary, Li et al. reported that 6 of 38 patients had results positive for SARS-CoV-2 in semen samples, including 4 of 15 patients who were at the acute stage of infection and 2 of 23 patients who were recovering [12]. In a unique assessment, a postmortem examination of the testes from 12 COVID-19 male patients showed that in one case the SARS-CoV-2 virus was detected in the semen; however, due to fibro-

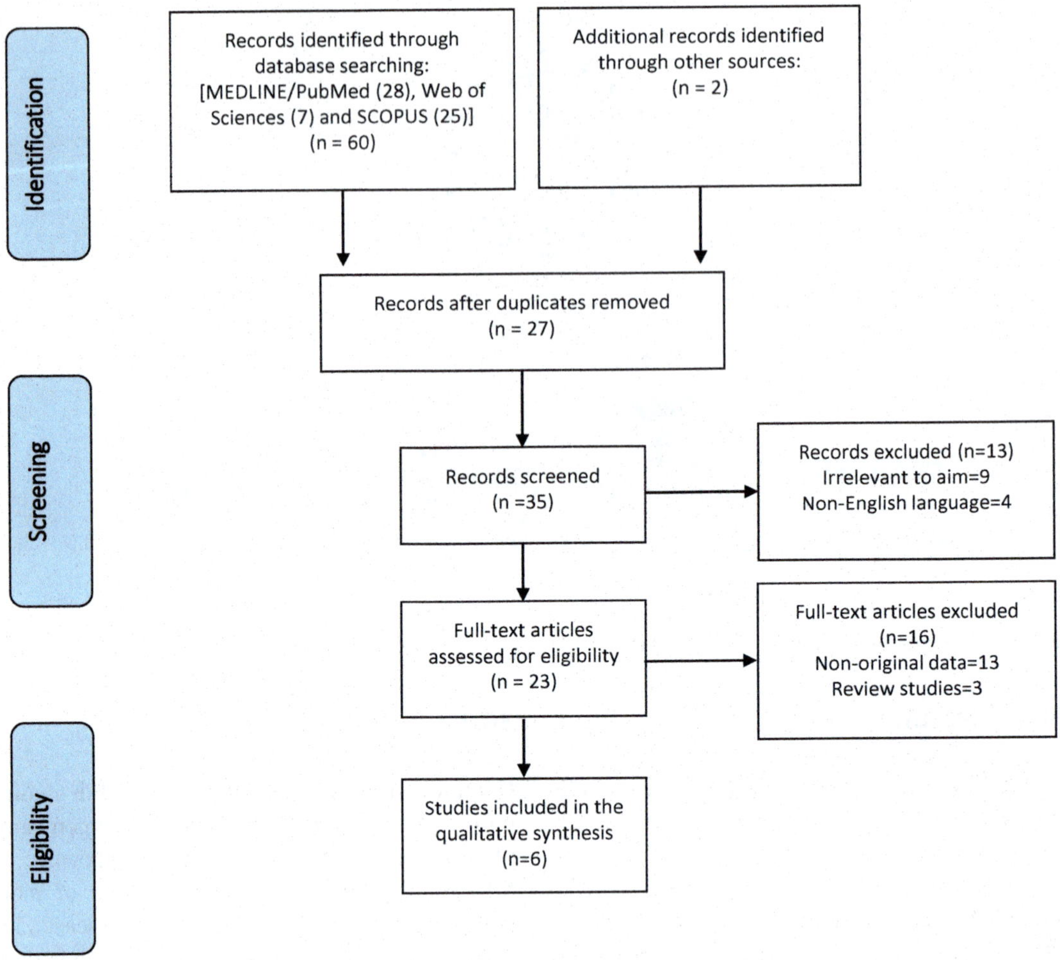

Fig. 30.2 PRISMA flow chart of the study selection process

vascular tissue and few seminiferous tubules, the authors were not sure whether the virus was detected in the blood or in the testes [6].

30.3.2 Other Findings

In the studies of Holtmann et al. [14] and Pan et al. [11], one and six patients, respectively, had testicular discomfort at the time of COVID-19 diagnosis. In a microscopic evaluation, the Sertoli cells and interstitium of the testes of men who died of COVID-19 disease were also affected by the disease, and the number of Leydig cells was significantly lower than in the control group [6]. Holtmann et al. stated that in cases of moderate infection with SARS-CoV-2, there was significant impairment of sperm quality (concentration, count, total progressive motility, and complete motility) in comparison with cases of mild infections and the control group. In their study, the definition of mild disease was the cases where it was possible to take care of the patient at home, and moderate disease included cases in which the patient was hospitalized and received up to 6 liters of oxygen. Also, febrile patients had less volume and complete motility in the semen analysis than afebrile patients ($p < 0.05$) [14].

Table 30.1 Characteristics of included studies

First author's name	Country	Type of study	Population	Age (years; mean or range)	Disease severity	Period between disease onset or recovery and semen collection	Findings
Holtmann [14]	Germany	Prospective cohort	Study group: 18 men recovered from infection with SARS-CoV-2 and 2 current patients Control group: 14 healthy men with no reported andrological pathology	42.2 for recovered 33.4 for controls	1 without symptoms, 15 mild, and 4 moderate	8–54 days after recovery (32.7 days)	SARS-CoV-2 RNA was not detected in semen samples from recovered or acutely infected subjects
Li [12]	China	Prospective cohort	23 participants clinically recovered and 15 at acute stage of infection	Not reported	Not reported	Not reported	6 patients were positive for SARS-CoV-2, including 4 of 15 patients at acute stage of infection and 2 of 23 patients who were recovering
Pan [11]	China	Cross-sectional	34 adult males diagnosed with SARS-CoV-2	18–55	All patients had mild-moderate symptoms (no details)	One month after diagnosis of COVID-19	SARS–CoV-2 was not detected in semen of patients recovering from COVID-19
Paoli [15]	Italy	Case report	A man found positive for SARS-CoV-2	31	Mild (isolation at home)	8 days after diagnosis of COVID-19 (when symptoms were partially improved)	Semen and urine samples negative for SARS-CoV-2 RNA
Song [13]	China	Case series	12 patients in recovery	23–38	11 mild and 1 without symptoms	Not reported	All semen samples negative for SARS-CoV-2 detection

(continued)

Table 30.1 (continued)

First author's name	Country	Type of study	Population	Age (years; mean or range)	Disease severity	Period between disease onset or recovery and semen collection	Findings
Yang [6]	China	Retrospective cohort	Study group: Testes of 12 patients who died of COVID-19 Control group: 5 testes resulting from orchiectomy due to penile cancer, prostate cancer, or perineal trauma	65 for the study group (range 42–87)	Resulted in death	An hour after death	RT-PCR detection of virus in 1/12 cases. Sertoli cells showed swelling, vacuolation, and cytoplasmic rarefaction, detachment from tubular basement membranes, and loss and sloughing into lumens of the intratubular cell mass. Mean number of Leydig cells in COVID-19 testes significantly lower than controls. Edema in interstitium and mild inflammatory infiltrates composed of T lymphocytes and histiocytes

30.4 Discussion

To the best of our knowledge, the present study is the first systematic review to assess the presence of SARS-CoV-2 in human semen. The population we studied mostly consisted of males who had improved from mild to moderate forms of COVID-19 disease. The findings of some studies indicated the presence of SARS-CoV-2 RNA and some of the findings indicated that the virus was absent in semen. Some studies also showed abnormalities in sperm analysis and microscopic damage to testicular cells due to COVID-19 infection. Therefore, we cannot draw a definite conclusion regarding the primary aim of the study. There is still a need for more quality research in this area, given the potential implications of viral damage to this tissue.

The virus that causes COVID-19 disease is SARS-CoV-2, which has more than 70% similarity in amino acid sequence with SARS-CoV, the cause of the earlier SARS epidemic. In addition, both SARS-CoV and SARS-CoV-2 appear to gain entry into cells via the ACE2 and through the action of the transmembrane protease serine 2, which primes the viral spike protein (Fig. 30.1) [9]. This suggests that the organs or tissues most vulnerable to COVID-19 infection are those with high expression levels of ACE2 and TMPRSS. The evidence to date shows that the respiratory, cardiovascular, gastrointestinal, and urinary systems are currently reported as possible targets for SARS-CoV-2 [16]. Some scientific evidence has shown that there is a high expression of ACE2 in the testis [17, 18]. There is evidence that SARS-CoV causes orchitis in humans [10]. In the present review, in some cases, orchitis was reported at the time of diagnosis of COVID-19. Because testicular damage and germ cell destruction are seen in the SARS infection [10], SARS-CoV-2 may be transmitted by semen in addition to the conventional route. This transmission may occur even in asymptomatic carriers or during the incubation period.

Another concern is about individuals who have been infected by SARS-CoV-2 who are planning to initiate a pregnancy. The Society for Assisted Reproductive Technologies (SART) and the American Society for Reproductive Medicine (ASRM) have recommended that ART candidate patients, gamete donors, and gestational carriers who meet SARS-CoV-2 diagnostic criteria avoid getting pregnant or do not participate in reproductive programs until they are disease-free [19].

Dimensions related to fertility in the COVID-19 are ambiguous and original data in this area are urgently needed. Here, the reviewed studies had small sample sizes. In addition, semen was collected at the time of recovery in most of the studies. We also excluded studies in languages other than English. In the future, it is suggested to design studies with higher sample sizes and perform semen testing at all stages of the infection.

30.5 Conclusions

The findings of some studies identified the SARS-CoV-2 virus in semen and others did not find it. However, damage to testicular tissue was confirmed by microscopic examination and spermogram disruption was reported in another study. Also, since most of the samples were obtained from individuals with mild to moderate forms of COVID-19 infection, it is not yet clear regarding the viral presence in semen from more severe cases. The long-term effects are also not possible to determine as only 6 months have passed since the beginning of this pandemic and no end is in sight (time or writing June 29, 2000). The number of cases around the world still appears to be increasing and some countries, such as the USA, already appear to be experiencing a second wave [20, 21]. More original articles with better design and in different phases of the disease are needed for definitive conclusions and many of these studies will have to be conducted in the years to come.

Acknowledgements Thanks to guidance and advice from the Clinical Research Development Unit of Baqiyatallah Hospital.

References

1. Huang C, Wang Y, Li X, Ren L, Zhao J, Hu Y et al (2020) Clinical features of patients infected with 2019 novel coronavirus in Wuhan, China. Lancet 395(10223):497–506
2. Xu Z, Shi L, Wang Y, Zhang J, Huang L, Zhang C et al (2020) Pathological findings of COVID-19 associated with acute respiratory distress syndrome. Lancet Respir Med 8(4):420–422
3. Wong SH, Lui RN, Sung JJ (2020) Covid-19 and the digestive system. J Gastroenterol Hepatol 35(5):744–748
4. Pei G, Zhang Z, Peng J, Liu L, Zhang C, Yu C et al (2020) Renal involvement and early prognosis in patients with COVID-19 pneumonia. J Am Soc Nephrol 31(6):1157–1165
5. Hendren NS, Drazner MH, Bozkurt B, Cooper J, Leslie T (2020) Description and proposed management of the acute COVID-19 cardiovascular syndrome. Circulation 141(23):1903–1914
6. Yang M, Chen S, Huang B, Zhong JM, Su H, Chen YJ et al (2020) Pathological findings in the testes of COVID-19 patients: clinical implications. Eur Urol Focus. May 31;S2405–4569(20)30144–9. https://doi.org/10.1016/j.euf.2020.05.009
7. Guan WJ, Ni ZY, Hu Y, Liang WH, Ou CQ, He JX et al (2020) Clinical characteristics of coronavirus disease 2019 in China. N Engl J Med 382(18):1708–1720
8. Chen F, Lou D (2020) Rising concern on damaged testis of COVID-19 patients. Urology. Apr 25; S0090–4295(20)30459–3. https://doi.org/10.1016/j.urology.2020.04.069. Online ahead of print
9. Lukassen S, Chua RL, Trefzer T, Kahn NC, Schneider MA, Muley T et al (2020) SARS-CoV-2 receptor ACE 2 and TMPRSS 2 are primarily expressed in bronchial transient secretory cells. EMBO J 39(10):e105114. https://doi.org/10.15252/embj.20105114
10. Xu J, Qi L, Chi X, Yang J, Wei X, Gong E et al (2006) Orchitis: a complication of severe acute respiratory syndrome (SARS). Biol Reprod 74(2):410–416
11. Pan F, Xiao X, Guo J, Song Y, Li H, Patel DP et al (2020) No evidence of severe acute respiratory syndrome–coronavirus 2 in semen of males recovering from coronavirus disease 2019. Fertil Steril 113(6):1135–1139
12. Li D, Jin M, Bao P, Zhao W, Zhang S (2020) Clinical characteristics and results of semen tests among men with coronavirus disease 2019. JAMA Netw Open 3(5):e208292. https://doi.org/10.1001/jamanetworkopen.2020.8292
13. Song C, Wang Y, Li W, Hu B, Chen G, Xia P et al (2020) Absence of 2019 novel coronavirus in semen and testes of COVID-19 patients. Biol Reprod Apr 16: ioaa050. Published online 2020 Apr 16. doi: https://doi.org/10.1093/biolre/ioaa050
14. Holtmann N, Edimiris P, Andree M, Doehmen C, Baston-Buest D, Adams O et al (2020) Assessment of SARS-CoV-2 in human semen-a cohort study. Fertil Steril May 29. https://doi.org/10.1016/j.fertnstert.2020.05.028 [Epub ahead of print]
15. Paoli D, Pallotti F, Colangelo S, Basilico F, Mazzuti L, Turriziani O et al (2020) Study of SARS-CoV-2 in semen and urine samples of a volunteer with positive naso-pharyngeal swab. J Endocrinol Invest Apr 23;1–4. https://doi.org/10.1007/s40618-020-01261-1. Online ahead of print
16. Zou X, Chen K, Zou J, Han P, Hao J, Han Z (2020) Single-cell RNA-seq data analysis on the receptor ACE2 expression reveals the potential risk of different human organs vulnerable to 2019-nCoV infection. Front Med 14(2):185–192
17. Wang Z, Xu X (2020) scRNA-seq profiling of human testes reveals the presence of the ACE2 receptor, a target for SARS-CoV-2 infection in spermatogonia, Leydig and Sertoli cells. Cell 9(4):920. https://doi.org/10.3390/cells9040920
18. Fan C, Li K, Ding Y, Lu WL, Wang J (2020) ACE2 expression in kidney and testis may cause kidney and testis damage after 2019-nCoV infection. MedRxiv. https://doi.org/10.1101/2020.02.12.20022418
19. American Society for Reproductive Medicine (2020) SART and ASRM issue advice for infertility patients concerning the novel coronavirus (COVID-19). https://www.asrm.org/news-and-publications/news-and-research/press-releases-and-bulletins/sart-and-asrm-issue-advice-for-infertility-patients-concerning-the-novel-coronavirus-covid-19/
20. https://www.worldometers.info/coronavirus/#countries
21. https://coronavirus.jhu.edu/map.html

Index

P. C. Guest (ed.), *Clinical, Biological and Molecular Aspects of COVID-19*, Advances in Experimental Medicine and Biology 1321, https://doi.org/10.1007/978-3-030-59261-5

D

T

Printed by Printforce, United Kingdom